Basics of Solid and Hazardous Waste Management Technology

Kanti L. Shah
Ohio Northern University

Prentice Hall

Upper Saddle River, New Jersey   Columbus, Ohio

Library of Congress Cataloging-in-Publication Data

Shah, Kanti L.
    Basics of solid and hazardous waste management technology / Kanti L. Shah.
        p. cm.
    Includes bibliographical references and index.
    ISBN 0-13-960378-6
    1. Refuse and refuse disposal. 2. Hazardous wastes–Management. I. Title.

TD791.S53 2000
628.4–dc21   99–051543

Editor: Stephen Helba
Assistant Editor: Michelle Churma
Production Editor: Louise N. Sette
Production Supervision: York Production Services
Design Coordinator: Karrie Converse-Jones
Cover Designer: Mark Schumaker
Cover Photo: © H. Armstrong Roberts
Production Manager: Matthew Ottenweller
Marketing Manager: Chris Bracken

This book was set in Times Roman by York Graphic Services, Inc., and was printed and bound by R.R. Donnelley & Sons Company. The cover was printed by Phoenix Color Corp.

© 2000 by Prentice-Hall, Inc.
Pearson Education
Upper Saddle River, New Jersey 07458

Printed in the United States of America

10 9 8 7 6 5 4 3 2 1

ISBN: 0-13-960378-6

Prentice-Hall International (UK) Limited, *London*
Prentice-Hall of Australia Pty. Limited, *Sydney*
Prentice-Hall of Canada, Inc., *Toronto*
Prentice-Hall Hispanoamericana, S. A., *Mexico*
Prentice-Hall of India Private Limited, *New Delhi*
Prentice-Hall of Japan, Inc., *Tokyo*
Prentice-Hall (Singapore) Asia Pte. Ltd., *Singapore*
Editora Prentice-Hall do Brasil, Ltda., *Rio de Janeiro*

Dedicated to my father, the late Dr. Swarupchand M. Shah—
scholar, teacher, and university administrator.

Preface

This book on solid and hazardous waste management technology offers a pragmatic introduction to solid and hazardous wastes and the issues associated with managing these wastes. The book is designed primarily for use by students in civil engineering, environmental technology, environmental studies, and similar disciplines at engineering colleges, community colleges, and technical institutes. In addition, the engineers, scientists, and others in different disciplines who may become involved in any aspect of solid and hazardous waste management for the first time will find this book of value as an initial reference. It has been designed to encourage self-teaching by providing completely worked-through examples and detailed case studies.

The qualities that distinguish this book are its clear, easy-to-read style and its logical and systematic treatment of the subject based on technical and scientific fundamentals. The goal is to present an integrated approach to teaching and learning about solid and hazardous waste management technology. To accomplish this, the book encompasses the entire spectrum of basic concepts and tools needed for making decisions. These include current scientific concepts and technology and the complex social, political, legal, and ethical issues associated with management of the solid and hazardous wastes.

Because the field of solid and hazardous waste management technology is multidisciplinary, a review of primer material such as chemistry, biology, geology, hydraulics, and engineering calculations is included so that readers can easily comprehend the subject matter. Numerous diagrams and photographs are included for clarification.

To help readers learn and understand the material in this book, solved examples, problems, and case studies (both real and contrived) centering on technical and social problems are provided as well. This is because the cognitive theory of learning holds that maturation of knowledge takes time and repetition of the same ideas in different guises.

This book is divided into three parts. The first covers the basic science and engineering principles so that readers with little or no previous knowledge of these principles can comprehend and use the book. More advanced students and readers may also wish to refer to this material as a refresher.

The second and third parts present a simple style, clear approach to the understanding of the solid and hazardous waste management technology. The second part is devoted to solid waste, with chapters on environmental legislation and regulations; sources, composition, and characteristics; physical, chemical, and biological properties; storage, collection, and transportation; processing technologies; source reduction and reuse; disposal; and management and control of landfill leachate and gas. The third part details all aspects of hazardous waste, with chapters on characteristics and quantities; transportation of waste; waste minimization; treatment, storage, and disposal; site remediation; and sources and types of hazardous waste in municipal solid waste.

Acknowledgements

During the preparation of this book, I was assisted by many professional engineers, government agencies, former students, and recycling and recovery facilities. I acknowledge the assistance of Hull and Associates, Consulting Engineers, Toledo, Ohio; Environmental Consultants, Inc., Maumee, Ohio; the U.S. Environmental Protection Agency; the Ohio Environmental Protection Agency, Bowling Green, Ohio; and Scott Strahley, Engineer, Hancock County Landfill. Acknowledgement is also due to the editor of *Waste Age Recycling,* The Marion County Recycling Center, San Rafel, California; The National Recovery Technologies, Inc., Nashville, Tennessee; and the Miami County Sanitary Engineering Department, Troy, Ohio.

In addition, I thank Ms. Michelle Churma, Assistant Editor, Career and Technology, Prentice-Hall, for valuable editorial suggestions.

I would also like to acknowledge the reviewers of this text: Jeff Bates, Columbus State Community College (OH); Rhonda L. Howard, Coconino Community College (AZ); Michael A. McKenna, Mission College (CA); Robert R. Treloar, Paradis Valley Community College (AZ); and Phillip S. Waldrop, Georgia Southern University (GA).

On a more personal note, I thank my dear wife, Pushpa, who tolerated long hours of writing as well as papers, journals, and books everywhere in the family room. Finally, the seeds for this project were no doubt planted by my late father, who taught me the importance of engineering in improving the quality of life.

Kanti L. Shah

Contents

I FUNDAMENTALS 1

1 Basic Concepts 1

Overview   1
Ecology   2
Public Health   8
Geology and Soil   9
Historical Perspective   12
Review Questions   13
References   13

**2 Factors Affecting Municipal
Waste Management
Decision-Making 14**

Basis for Decisions   14
Methods of Financing   16
Cost Components   18
Types of Analysis   20
Environmental Impact
   Statements   35
Environmental Ethics   39
Review Questions   41
References   42

3 Engineering Calculations 43

Dimensions and Units
   of Measure   43
Detention Time   48
Measurements and
   Approximations   48
Technical Information Analysis   50
Review Questions   57

4 Hydraulics 58

Pressure   58
Flow   64
Pumps   78
Review Questions   81

5 Hydrology 82

The Hydrologic Cycle   82
The Hydrologic Cycle for
   a Landfill   84
Review Questions   88

**6 Basic Scientific Concepts of Solid
and Hazardous Waste
Management 89**

Chemistry and Physics   89
Microbiology   101
Review Questions   108

**II SOLID WASTE MANAGEMENT
TECHNOLOGY 109**

7 Overview 109

Definition of Solid Waste
   Management   110
Development of Solid Waste
   Management   110
Functional Elements of a Solid
   Waste Management System   112
Review Questions   114
References   114

8 Environmental Laws and Regulations   115

Development of Environmental
    Regulations   115
National Environmental Policy
    Act   119
Brownfields   124
Case Studies   125
Review Questions   126
References   126

9 Environmental Effects and Public Health Aspects   127

Sources of Diseases   127
Physical and Chemical
    Hazards   127
Disease and Other Hazard
    Prevention   128
Transfer, Processing, Recovery, and
    Disposal Facilities   128
Composting   129
Air Pollution and Water Pollution
    from Solid Waste Management
    129
Environmental Impact Statements
    130
Review Questions   131
References   131

10 Sources, Composition, and Characteristics of Solid Waste   132

Sources   132
Composition   135
Hazardous Waste in Municipal Solid
    Waste   148
Review Questions   151
References   152

11 Physical, Chemical, and Biological Properties of Municipal Solid Waste   153

Physical Properties   153
Chemical Properties   159
Biological Properties   167
Review Questions   169
References   170

12 Storage, Collection, and Transportation of Solid Waste   171

Storage at the Source   171
Collection   174
Transfer Stations and
    Transportation   192
Case Study   200
Review Questions   203
References   204

13 Solid Waste Basic Processing Technologies   205

Preprocessing   205
Physical Processing   214
Chemical Transformation
    (Combustion)   230
Biological Transformation   250
Life Cycle Assessment   257
Review Questions   260
References   260

14 Source Reduction, Reuse, Recycling, and Recovery of Municipal Solid Waste   262

Reducing Quantity and Toxicity   263
Recycling   264
Economic Analysis of MRFs   276
Case Study   281
Review Questions   286
References   287

15 Disposal of Solid Waste and Residual Matter in Landfills   288

Landfill Classification   288
Siting Considerations   289
Site Suitability   293
Landfilling Techniques   298
Landfill Cover Design   304
Landfill Liners   308
Moisture in Landfills   310
Estimation of Leachate Generation
    Rates   315
Landfill Operation   316
Site Operations   322
On-Site Operation Facilities   323
Review Problems   327
References   328

16 Management and Control of Landfill Leachate and Gases 329

Leachate Control   329
Leachate Removal System   329
Leachate Characteristics   332
Leachate Treatment   336
Landfill Closure and Postclosure
  Considerations   361
Subtitle D Regulations   362
Case Study   365
Landfill Gas Generation   378
Landfill Gas Migration   380
Gas Production Potential   381
Gas Recovery and Migration
  Control   383
Landfill Gas Composition and
  Characteristics   387
Landfill Gas Utilization   387
Review Questions   389
References   390

III HAZARDOUS WASTE MANAGEMENT TECHNOLOGY 391

17 Characteristics of Hazardous Waste 391

Definition of Hazardous Waste   392
Generation of Hazardous Waste   392
Identification and Characterization
  of Hazardous Waste   395
Priority Toxic Pollutants and
  Hazardous Waste   400
Review Questions   403
References   404

18 Transportation of Hazardous Waste 405

Regulations   405
Emergency Response   407
Safety   412
Review Questions   413
Reference   413

19 Hazardous Waste Minimization and Environmental Audits 414

Pollution Prevention   415
Environmental Audits   415

Case Study   422
Review Questions   423
References   423

20 Hazardous Waste Treatment, Storage, and Disposal 424

Treatment Technologies   424
Waste Stabilization and
  Solidification Techniques   425
Storage Techniques   427
Land Disposal Techniques   430
Review Questions   434
References   434

21 Site Remediation 435

The Superfund Program   435
Site Contamination   436
Remedial Investigation/Feasibility
  Study   440
Gas Collection and Recovery
  Systems   451
Treatment Technologies for
  Remediation of Contaminated
  Sites   460
Case Studies   480
Review Questions   486
References   486

Appendix A   Interest Tables 487
**Appendix B   Useful Conversion
            Factors 489**
**Appendix C   Saturation Values of
            Dissolved Oxygen in
            Water at Various
            Temperatures 492**
**Appendix D   Soil Moisture Retention
            for Various Amounts
            of Potential
            Evapotranspiration for a
            Root Zone Water-
            Holding Capacity of
            4 Inches 493**
Appendix E   Glossary 496
**Appendix F   For Further Information
            510**

Index 513

I FUNDAMENTALS

Chapter 1   Basic Concepts

This chapter introduces the problems associated with solid waste generation, the relevant questions discussed in this book, and the basic and pertinent topics in the fields of public health, ecology, and geology. It also provides an historical perspective.

OVERVIEW

A community generates both municipal and industrial waste materials. Once considered not to be much of a problem, the disposal of these wastes recently has become an issue of immense proportions. More and more, these problems are in the news, with political and social concerns overshadowing technical and economic issues.

Management of the vast quantities of solid waste generated by urban communities is a very complex process. Ordinarily, the collection and disposal of solid waste is the responsibility of the local municipality. If these wastes are not handled and disposed of properly, however, serious public health and environmental problems can occur. Hazardous waste (e.g., poisonous or ignitable materials from industrial processes) requires special methods for storage, collection, transport, treatment, and final disposal. Improper disposal of hazardous materials carries a great potential for contamination, and Part III of this book is devoted to hazardous wastes.

To appreciate the scale of the problem faced by federal, state, and local governments as well as by private enterprise, a thorough knowledge of the problem and its individual elements is necessary. What are solid and hazardous wastes? What is the difference between the two? Who produces what? Why is so much waste generated? What are the effects of waste on the environment and public health? Why are residents opposed to the location of disposal facilities in their area? Why is the management of solid and hazardous wastes so costly? What is the role of regulating agencies? What are the opportunities for reduction, reuse, and recycling, and how do they affect the management costs? What technologies are available to manage such wastes? What ethical, political, social, and financial issues are involved? In this book, all these questions are discussed. It is very difficult—if not impossible—to provide complete answers, but readers should find sufficient information with which to make sounder decisions regarding the complexities of solid and/or hazardous waste management. Readers are encouraged to supplement this material with information available on the Internet.

ECOLOGY

Ecology is the science that seeks to answer questions about how nature works. It is the branch of biological sciences concerned with the relationships or interactions between organisms and their living (i.e., biotic) and nonliving (i.e., abiotic) environment. Components of the biotic environment include air, water, minerals, and soil. An ecosystem includes living organisms and the environment with which they exchange materials and energy. An essential component of most natural ecosystems is energy, usually sunlight. Examples of land-based ecosystems include jungles, forests, and meadows; water-based or aquatic ecosystems include streams, rivers, lakes, ponds, estuaries, and marshes, which are a part of the hydrologic cycle (see Chapter 5). An ecosystem can be small or large (e.g., a pond or an ocean). Even the entire surface of the earth can be viewed as an ecosystem, and the term *biosphere* is often used in this context. The layer of air surrounding the earth and the land and water of the earth itself provide the abiotic components that support life in the biosphere. Energy from the sun sustains the life cycles within the biosphere.

Industrialization and population have affected the natural balances in the biosphere. Global problems related to environmental pollution (e.g., acid rain, smog, the greenhouse effect) are examples. These problems have the potential to seriously damage the environment.

In addition to natural ecosystems such as lakes or forests, there are engineered (i.e., man-made) systems to dispose of and treat wastes generated by human activities. These systems are based on the processes found in natural systems. For example, common methods of wastewater treatment based on biological systems include the trickling filter process and the activated sludge process; for solid waste disposal, they include the sanitary landfill and composting. These are engineered ecosystems with microbes as the biotic component and the tank (i.e, wastewater container), air (for input to wastewater), and liners for the landfill as the abiotic components.

Energy, Food Chains, and Metabolism in Ecosystems

Energy is the capacity to do work. It cannot be created or destroyed; it can only be transformed. Both materials and energy flow inside ecosystems. The difference is that energy flows only in one direction and materials flow in a cyclical manner. Figure 1–1 shows the one-way flow of energy and the circulation of nutrients. In addition to energy, living organisms need nutrients. All organisms need water, and most need gaseous oxygen. In addition, plants and animals require carbon, hydrogen, phosphorus, potassium, iodine, nitrogen, sulfur, calcium, iron, and magnesium. Certain other elements are required in smaller amounts as well. (See Chapter 6 for a brief review of the chemistry involved.)

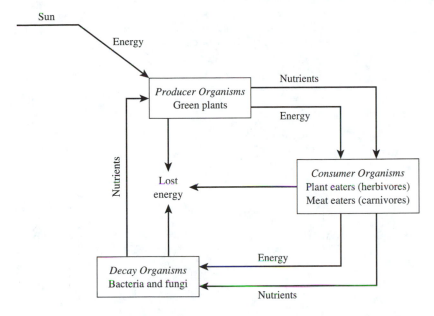

Figure 1–1   Simplified version of the food chain.

Figure 1–2 represents the nutrient flow through an ecosystem. Initially, dead organics (e.g., dead vegetation, dead animals, organic waste) is decomposed by microorganisms into products such as ammonia (NH_3), carbon dioxide (CO_2), and hydrogen sulfide (H_2S) for nitrogenous, carbonaceous, and sulfur matter. These products are decomposed further as well, and the process continues until they are stabilized as nitrate (NO_3), carbon dioxide (CO_2), sulfate (SO_4^{-2}), and phosphate (PO_4^{-3}). Carbon dioxide is a carbon source for the plants, and nitrate, phosphate, and sulfate are used as nutrients for building new tissues by the plants.

Biogeochemical Cycles

Any element or compound needed by an organism for its survival, growth, and reproduction is called a nutrient. The nutrients on which life depends can be recycled indefinitely within an ecosystem, which needs a constant source of energy. This recycling of nutrients to and from the environment occurs in biogeochemical cycles (*bio* meaning "living"; *geo* for water, rocks, and soils; and *chemical* for the changing of matter from one form to another). These nutrient cycles, which are driven directly or indirectly by incoming energy from the sun, include the carbon, sulfur, oxygen, nitrogen, phosphorus, and hydrologic (i.e., water) cycles. (The hydrologic cycle is discussed in Chapter 5.)

Nutrient materials are constantly recycled or circulated through the ecosystem. All energy on earth originates from the sun as light energy. Plants get this energy through photosynthesis, and using both nutrients (in the soil) and carbon dioxide, plants con-

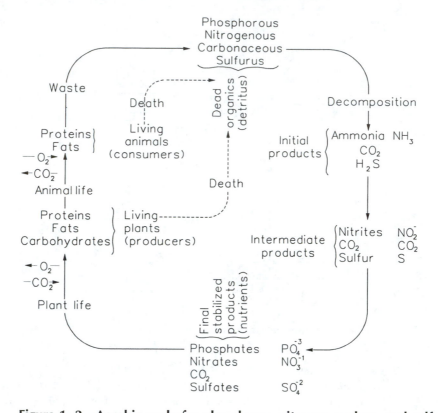

Figure 1–2   Aerobic cycle for phosphorus, nitrogen, carbon, and sulfur.
(Adapted from McGaughy, P.H., Engineering Management and Water Quality. *New York: McGraw-Hill, 1968; with permission.)*

vert this light energy to chemical energy and store it in the form of starch, sugar, protein, and fat.

Carbon Cycle.   Carbon dioxide, both in the air and dissolved in water, is the primary source of carbon. The carbon is removed from the carbon dioxide and incorporated with other chemical elements in complex organic molecules through photosynthesis. Carbon dioxide escapes into the atmosphere when the organics are later broken down during decomposition.

The combustion of fossil fuels for energy increases the concentration of carbon dioxide in the atmosphere. Carbon dioxide absorbs radiation at wavelengths almost the same as that of heat radiation going back into space. Because this radiation is adsorbed by the gases in the atmosphere, the temperature of the atmosphere increases, thus heating the earth. This system is similar to a greenhouse: the energy from the sunlight passes through the greenhouse, but the heat radiation is prevented from escaping. The atmospheric gases that adsorb heat energy are termed *greenhouse gases,* because they cause the earth to heat up just like a greenhouse.

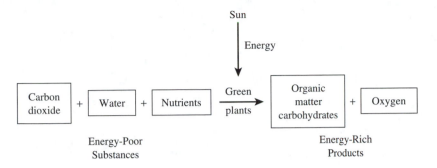

Figure 1–3 Photosynthesis.

Nitrogen Cycle. Nitrogen gas (N_2) forms approximately 78 percent of the atmosphere. To be used by plants during photosynthesis, however, nitrogen gas must be changed to nitrate (NO_3). Plant tissues are consumed by animals, and decomposition of dead animals and wastes results in the formation of ammonia (NH_3). Then, through nitrification, bacteria converts the ammonia to nitrate (NO_3). This latter process is significant in the control of water pollution.

Phosphorus Cycle. Phosphorus, mainly in the form of phosphate ions, is an essential nutrient for both plants and animals. Various forms of phosphorus cycle through the lower atmosphere, water, soil, and organisms. Unlike carbon and nitrogen, however, which come primarily from the atmosphere, little phosphorus comes from the atmosphere. Instead, phosphorus largely occurs as a mineral (in phosphate rocks) and enters the cycle through mining and erosion.

Animal wastes and the decay products of dead animals and plants return much of the phosphorus to the soil and surface waters and, eventually, to the ocean bottom. In addition, phosphate ions are added to aquatic ecosystems by the runoff of animal wastes (e.g., wastes from livestock feedlots), runoff of commercial phosphate fertilizers from farms, and the discharge of both untreated and treated municipal sewage and leachate from landfills.

An excessive supply of this nutrient causes heavy growth of algae and other aquatic plants, which disrupts life in aquatic ecosystems. Excessive growth of plants and algae can also cause taste and odor problems in sources of drinking water, especially lakes and reservoirs.

Photosynthesis, Respiration, and Metabolism

The biological and chemical processes by which an organism sustains its life are called *metabolism*. The two fundamental metabolic processes of living organisms are photosynthesis (Figure 1–3) and respiration (Figure 1–4).

Plants manufacture high-energy carbohydrate molecules and are called producers. These green plants are autotrophic; an autotrophic organism uses solar energy (green plants) or chemical energy (some bacteria) to manufacture its own organic nutrients from inorganic nutrients. The simplified equation is:

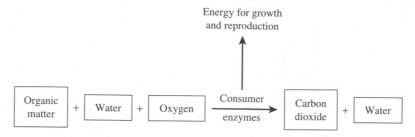

Figure 1–4   Respiration.

nutrients + water + CO_2 + sunlight energy + chlorophyll in green plant

$\qquad \rightarrow$ organic matter (carbohydrate) + gaseous oxygen (O_2)   **(1-1)**

All other organisms must use this organic matter for nourishment and growth. Animals using these energy-giving molecules are called consumers (or heterotrophes). They cannot produce their own food, and they must get it by eating or by decomposing organic matter. The process can be shown by a simplified equation:

organic matter (mostly high-energy molecules) + O_2

$\qquad \rightarrow$ consumer's energy for growth and reproduction + CO_2 + nutrients   **(1-2)**

Both plants and animals produce waste and eventually die, thus forming waste organic matter known as detritus, which still contains considerable energy. This is one of the reasons we need to treat wastewater to remove or destroy organic wastes. Organisms that use detritus (e.g., bacteria, fungi) are known as decomposers. They get nutrients by breaking down organics in the wastes and dead bodies of other organisms and animals into simpler chemicals. Most of these end products are returned to the soil and water for use by producers.

The energy from the sun received by plants is used by consumers and decomposers to make cellular material for maintenance of the organism. Figure 1–5 illustrates this one-way flow. The slope of the line indicates the rate at which energy is extracted. As the energy level decreases, the rate of energy extraction slows, which is an important concept in waste treatment (e.g., leachate).

Aerobic and Anaerobic Decomposition

The simplified food chain shown in Figure 1–1 is closed by the decomposers or decay organisms, as explained earlier. Decomposition of waste products and dead organisms by microorganisms produces energy and fully oxidized forms of nutrients. These are CO_2, NO_3, SO_4^{-2}, and PO_4^{-3}. Carbon dioxide is used by plants as a source of carbon, whereas the later three compounds are used as nutrients or as the building blocks of new plant tissue.

Decomposers are further subdivided into three groups—aerobic, anaerobic, and facultative microorganisms—on the basis of whether they require molecular oxygen in their metabolic activity. Decomposition that occurs in the presence of free oxygen is called aerobic decomposition, and the microorganisms that thrive in oxygen are called

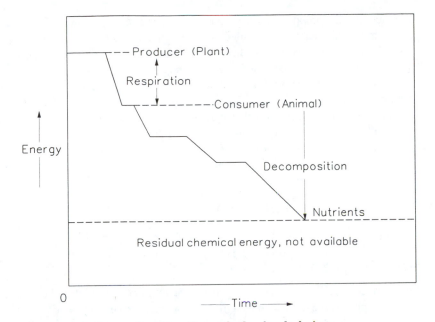

Figure 1–5 Loss of energy through the food chain.
(Adapted from McGaughy, P.H., Engineering Management and Water Quality. *New York: McGraw-Hill, 1968; with permission.)*

aerobes. Aerobic decomposition results in oxidation of the carbon, hydrogen, sulfur, nitrogen, and phosphorus in complex organic molecules. These elements then combine with oxygen, forming carbon dioxide, water sulfate, nitrate, and other simple substances that can be taken up by green plants for photosynthesis. The microbes use the energy released from the organic molecules in this process for growth and reproduction. Aerobic decomposition is a "clean" biochemical process that does not produce the offensive odors often associated with decay. The equation of aerobic decomposition is:

$$\text{organics} + O_2 \rightarrow CO_2 + H_2O + \text{energy} \tag{1-3}$$

Most wastewater treatment plants (including landfill leachate treatment) use aerobic process. This process is also used for composting.

Certain species of microorganisms can decompose organic material without free available oxygen. These organisms are called anaerobes, and the process is called anaerobic decomposition. As shown in the equations (1-4) and (1-5), the end products of anaerobic decomposition include methane (CH_4), ammonia, hydrogen sulfide (H_2S), and volatile organic acids. Anaerobic decomposition is an inefficient biochemical process. Many of these compounds have unpleasant odors associated with putrefaction; for example, hydrogen sulfide causes the rotten-egg odor. The equations of anaerobic decomposition are:

$$\text{organics} \rightarrow \text{organic acids} + CO_2 + H_2O + \text{energy} \tag{1-4}$$

$$\text{organic acids} \rightarrow CH_4 + CO_2 + \text{energy} \tag{1-5}$$

Organics first decompose to form organic acids and other products. The anaerobes responsible for this reaction are called acid formers. The organic acid formed in this reaction is then converted into methane and other products. This reaction is performed by anaerobic bacteria called methane formers.

Anaerobic decay is used in some wastewater treatment processes. Methane is one of the few odorless products, and this gas has high energy value, thus making it useful as a fuel. It is collected for that purpose at some sewage treatment plants and sanitary landfills used for garbage disposal. Facultative bacteria are active in both anaerobic and aerobic treatment units. When no dissolved oxygen is available, these organisms can use oxygen from other sources, such as nitrates and sulfates, as indicated in equations (1-6) and (1-7):

$$\text{organics} + NO_3 \rightarrow CO_2 + N_2 + \text{energy} \qquad \textbf{(1-6)}$$

$$\text{organics} + SO_4 \rightarrow CO_2 + H_2S + \text{energy} \qquad \textbf{(1-7)}$$

Stable Ecosystem

The more complex the ecosystem regarding the number of and interrelationships among different species, the more stable it will be. Diversity of species provides a buffer against ecological disruptions by increasing the likelihood of adaption to changing environmental conditions. In general, the diversity of species is indicative of a healthier ecosystem.

Aquatic ecosystems (e.g., streams, lakes) generally are sensitive to pollution from human activity. Among the effects of water pollution are reduced levels of dissolved oxygen in water and altered pH (an indicator of alkalinity or acidity). This type of pollution changes the ecological balance. In heavily polluted water, only hardy organisms such as maggots and sludge worms are found. In a clean stream or lake, many different species of organisms and game fish (e.g., trout) may be found, but in a polluted stream, only a few species of more tolerant organisms (e.g., catfish) may survive.

PUBLIC HEALTH

The public health and aesthetic problems associated with improper disposal of solid and hazardous waste are the driving motivation for their removal from the human environment. The United States and other developed countries can afford to construct and operate relatively well-designed treatment and disposal facilities, which in general have allowed their citizens to enjoy good health and a clean environment. Inadequate sanitary facilities in underdeveloped countries has been a major factor in the spread of disease.

Garbage is a major portion of domestic solid waste, and it results from food marketing, preparation, and consumption in residential housing units. Garbage contains putrescible organic material that needs careful storage, collection, and disposal because of the attraction to rodents and flies. Rodent-borne diseases (e.g., typhus, bubonic plague) are more readily controlled through a good garbage-disposal technique. Broken glass, rusty metals, household pesticides, paints, and so on are sources of the chemical and physical hazards associated with solid waste.

Hazardous waste is generated by industrial and commercial establishments and by the consumption of their products. For example, improper disposal of motor oil, batteries, paint, and so on can harm public health and the environment. Environmental pollution and the resulting incidence of disease are part of the problem referred to as solid and hazardous waste management. Examples of diseases include—but not limited to—cancer, asthma, bronchitis, emphysema, and nervous system related problems. Because it is impossible to shut down domestic industries or sacrifice the lifestyles of citizens, waste should be managed safely. Along these lines, the U.S. Congress passed the Pollution Prevention Act in 1990 to prevent or reduce pollution at the "source" whenever and wherever possible (see Chapters 8 and 19).

GEOLOGY AND SOIL

In most cases, both liquid and solid wastes are disposed either on top of or below the ground. Such wastes are likely to cause serious contamination of naturally occurring bodies of water; therefore, protecting such bodies of water is of great concern. Soil types and characteristics play an important role in this area.

Most soils begin as rock. The three major types of rocks are igneous, sedimentary, and metamorphic. A common type of igneous rock is granite, which is composed mainly of the mineral quartz and feldspar. Exposure to the elements gradually breaks the rock down into small bits and pieces because of physical disintegration and chemical decay. There are gradual but constant changes in the rock's composition and structure, and this process is called weathering. Physical weathering is the breaking down of rock mostly by temperature changes and by the physical action of moving ice, water, and wind. Chemical weathering is the dissolution of rock by reactions resulting from exposure to various chemicals. For example, carbon dioxide and oxygen can cause oxidation of rock, and acidic rainwater and acidic bacterial secretions can cause chemical weathering.

Thus, the solid rock, which is made of consolidated minerals, is broken down into small unconsolidated pieces, or soil. These soil particles are then transported by wind or water, deposited elsewhere, and form sediments. These sediments may be covered under additional deposits of material. Eventually, the sediments are consolidated and compacted under the load of overlying layers, and with time, the mineral may recrystallize and the rock particles cement together, thus forming a sedimentary rock. The chief kinds of sedimentary rock are conglomerate, which is formed from gravel; sandstone, which is formed from sand; shale, which is formed from clay and silt; and limestone, which is formed from masses of calcium carbonate. Some wind-deposited sand is consolidated into sandstone as well.

Excessive heat and pressure can cause the mineral structure of igneous and sedimentary rock to change. This newly formed rock is called metamorphic rock. Both sedimentary and metamorphic rocks may, again, undergo the weathering and deposition process.

In general, sedimentary rocks are porous and relatively permeable, whereas most igneous and metamorphic rocks are impermeable. Sedimentary rocks such as limestone are soluble, and solution cavities increase permeability.

Rock formations can gradually bend and shatter because of environmental changes and movements of the earth. When stresses cause rocks to shatter and crack, fracture lines and fissures are formed. These are called joints, and joints in which two sides of a rock mass move relative to each other are called faults. Joints and faults allow the flow of water through them.

Types of Soil

Soil, which is the unconsolidated rock fragments formed from weathering, is generally classified by the percentage of clay, silt, and sand. Even so, there are several different soil classification schemes. The one in Figure 1–6 is from the U.S. Department of Agriculture (USDA) Soil Conservation Service (1). Texture is given in the standard terms used by the USDA, which are defined according to the percentage of sand, silt, and clay in the fraction of the soil that is 7 to 27 percent clay, 28 to 50 percent silt, and less than 52 percent sand. Soil types are characterized by their particle size as follows:

clay <0.005 mm
silt 0.005–0.050 mm
sand 0.05–2.00 mm
gravel >2.0 mm

Clay differs from gravel, sand, and silt. Clay is composed of very fine particles, which are colloidal in nature. In part, these sizes are responsible for the soil characteristics. Gravel, sand, and silt consist of relatively coarse, grained, bulky particles. The very finely divided clay particles have a strong affinity for water. Most clays are plastic and cohesive, and the presence of organic compounds tends to increase their compressibility. The surface properties of soil are different because of the presence of ions, which can undergo exchange and alter the behavior and properties of the soil.

Soil Properties

The physical properties of soil relevant to a sanitary landfill relate to the activities associated with construction and operation of the landfill. The following properties are important:

- *Permeability:* The quality of the soil or rock that enables water to flow downward through the pore spaces.
- *Porosity:* The percentage of total soil or rock volume occupied by voids or pore spaces.

The permeability of soil decreases with the particle size. Silt is considerably less permeable because of its small particle and void size. Clay is porous but fairly impervious to the flow of water.

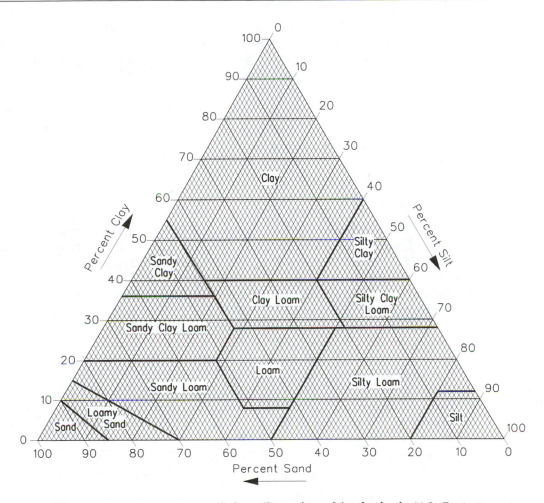

Figure 1–6   Percentages of clay, silt, and sand in the basic U.S. Department of Agriculture soil textural classes.

The impermeability of clay soil can be an advantage when building a sanitary landfill, because it can prevent the passage of water into the fill and the loss of leachate from it. A tight clay is very effective in this case, because the soil particles are well packed. A silty clay is less effective but is usually considered to be satisfactory. For landfill gas control, soil with a large particle size that is relatively uniform and porous is desirable, because the gas flows through the path of least resistance.

Swelling and Cracking

Swelling is the potential for soil to change in volume and gain in moisture. Volume change mainly occurs because of the interaction of clay minerals with water, and the change varies with the amount and type of clay minerals in the soil. When these soils dry, they crack, thus breaking the integrity of soil layer; clay and silty clays are especially susceptible to this problem. The size of the load on the soil and the magnitude of the change in soil moisture content influence the amount of swelling of soils. When wet, clay and silty clay soil do not support heavy vehicles.

Support of Vegetation

Soils such as silt, sandy silt, and clay–sandy silt tend to absorb and retain water as well as nutrients essential for the growth of grass. When a landfill is full, a final cover of soil that can support vegetation to protect against erosion and infiltration of water is very important.

Soil Survey Maps

The Soil Conservation Service (SCS) of the USDA has prepared countrywide soil survey maps. Such maps are available from county SCS offices and describe the soil types for the various areas. These maps are superimposed on aerial photographs of the area. In addition to the soil series distribution, they show bodies of surface water, roads, and other physical features. Such maps are valuable in planning the use and management of areas of interest.

Soil survey reports contain useful information regarding soil characteristics. SCS data include soil index properties, permeability, gradation, and suitability of land for a certain type of development. Depths to bedrock and water table are given as well. (Bedrock is the unweathered rock formation underlying the surface soils, and the water table represents the depth at which the soil is saturated with water. These are discussed in Chapter 5.) Both depth to bedrock and depth to water table are important in locating, designing, and building engineering works, especially solid or liquid waste disposal facilities. For example, the P_m (Pewamo) Series is described as a soil type having poor draining and slow permeability characteristics. This makes it unsuitable as ground for landfills or for absorption of septic tank effluent.

HISTORICAL PERSPECTIVE

Broad ecological awareness started during the 1960s, when many people expressed their interest in environmental quality. They desired clean air and water, streams and lakes that could be used for swimming, fishing, and for safe drinking water. The quest for environmental quality now includes solving problems of solid waste disposal, noise, radioactive material, and chemical wastes as well.

With the publication of Rachel Carson's *Silent Spring* in 1962, a dramatic turnabout occurred in American thought and attitude. *Silent Spring* essentially awoke the United States to the fact that chemicals being sprayed one day to rid crops of insects and other pests were being eaten the next day at the breakfast table with bowls of cornflakes and milk. The book focused attention on the environmental damage caused by improper use of pesticides. This text is widely regarded as having launched the environmental movement of the 1970s, and this movement focused the attention of politicians, lawmakers, and governmental agencies on appropriate legal and regulatory measures for environmental quality control.

In 1970, President Nixon signed the National Environmental Policy Act (NEPA) into law. NEPA established a national policy to "maintain conditions under which humans and nature can exist in productive harmony, and fulfill the social, economic, and other requirements of present and future generations of Americans"(2). This act also resulted in the birth of a new regulatory agency: the Environmental Protection Agency (EPA). In addition to establishing and enforcing pollution controls and environmental quality standards, the EPA is a research organization that studies the causes, effects, and methods to control environmental pollution. It also provides technical and financial assistance to the state governments.

In the past, solid waste management of garbage, trash, and other refuse consisted of dumping on the ground or burial in landfill. Until the mid-1970s, the federal government and the general public were unaware there was a hazardous waste problem. This world was shattered forever by environmental disasters such as Love Canal, where in 1976 puddles of improperly disposed of chemicals began surfacing in the backyards of many residents.

In the years since 1970, the U.S. Congress has passed many laws to establish and implement standards of environmental quality for water, land, and air, including the Resource Conservation and Recovery Act (RCRA), which deals with solid and hazardous waste management from cradle to grave. (The historical development of solid and hazardous waste management and of environmental legislation and regulations are discussed in Chapters 7 and 8, respectively.)

REVIEW QUESTIONS

1. List the natural land-based and water-based ecosystems.
2. What are the man-made ecosystems?
3. Discuss the difference between the biotic and abiotic components of an ecosystem.
4. Name the two basic processes of metabolism, and state the difference between them.
5. What are the autotrophic and heterotrophic organisms?
6. What is the relationship between the rate of energy extraction and the energy level?
7. What is the difference between aerobic and anaerobic decomposition?

    a. List the end products of each.
    b. Which type of decomposition is preferred?
8. What type of processes (aerobic or anaerobic) occur in a sanitary landfill? List the end products in a landfill decomposition.
9. With the help of equations, explain the production of CO_2 and CH_4.
10. What is the primary motivation for solid waste removal from the human environment?
11. When and how did the environmental movement start?
12. List the characteristics of a good soil for a septic tank leachate system.

REFERENCES

1. U.S. Department of Agriculture. *Soil Survey of Hardin County of Ohio*. Washington, D.C. Soil Conservation Service, 1994.

2. Council on Environmental Quality. *The Fifth Annual Report of Council on Environmental Quality*. Washington, D.C., U.S. Govt. Printing Office, 1974.

Chapter 2 Factors Affecting Municipal Waste Management Decision-Making

This chapter presents various aspects of solid waste management that affect the decision-making process and briefly describes how engineers, professionals, and community leaders make decisions. It begins with local, multijurisdictional, state, and federal factors, and it continues with the technical, economic, risk, and environmental impact analyses associated with decision-making. Finally, ethical analysis, which is a relatively new decision-making tool is introduced. Factors such as political and social considerations—and the ethical analysis—are subjective.

BASIS FOR DECISIONS

Local Factors

One of the basic steps decision makers should take is to base their planning process according to the political, institutional, and economic realities of their own and of neighboring communities (1). The political setting in which the decision makers must operate can be very complex. Groups such as elected officials, the news media, and citizen organizations (e.g., business, political, environmental, and neighborhood groups) may have different points of view. Within the solid and hazardous waste industry itself, waste haulers, recyclers, vendors and others will have relevant points of view as well. Decision makers should consider using the input from all these groups and identify groups whose assistance with planning an integrated program would be desirable.

The fiscal impact of any municipal waste program must be carefully considered. Planning for municipal solid waste management requires an analysis of budgetary constraints and the potential impact of alternatives. Financing options, including state and federal funding, as well as technical alternatives should be assessed. Decision makers must review the current outlays for waste management, including the sources of current funding. For example, the extent to which user fees, special assessments, and the general fund are used should be analyzed. Later in the planning process, the level and stability of these funding programs must be matched with the future needs. Funding requirements should be developed for various combinations of program approaches and technical options. Decision makers must review the feasibility of user charges, tipping fees, and various options to finance the proposed project or programs.

When developing a waste management plan, the existing institutional infrastructure should be examined to determine what personnel resources (i.e., technical and other staff) are available and needed for the new programs or for changes in existing programs. The decision makers must be very familiar with state and federal regulations. It is not unusual for the administrative tasks associated with waste management (e.g., permitting, enforcement, collection, processing, disposal, recycling, contracting) to be divided among the Division of Public Works and the Department of Public Health. Local solid waste programs may be administered by more than one level of government (e.g., cities, counties, states, regional authorities) across several jurisdictions.

Multijurisdictional Factors

Increasingly, many communities are turning to regional approaches for solid waste management. A regional approach allows communities to achieve economies of scale through better use of capital and more efficient management. This approach enables member communities to provide large-scale services that otherwise would not be financially possible and to centralize waste processing and disposal.

Because of rising costs, the goal of environmentally acceptable landfills may be better met by regional facilities that can tap the resources of several communities. Multijurisdictional approaches also may benefit composting and recycling programs, which may require a large waste generation and collection base to produce a marketable product.

There are some problems in the regional approach, however. Member communities lose autonomy in locating waste disposal sites. In the past, local jurisdictions have often had difficulty cooperating on joint waste management. One way to shield the integrated waste management program from the political arena is to establish quasigovernmental authorities or similar agencies with independent bonding authority. In general, such an authority is a corporate body with a charter authorized by the state legislature. It can be established by municipalities and counties to operate outside the regular structure of government; can finance, construct, and operate revenue-producing public enterprises; and may have regulatory powers as well.

State and Federal Factors

Local decision makers must understand and anticipate both state and federal requirements and what these requirements mean to the local solid waste program. State policy or law often limits local initiative and suggests the appropriate programs for communities. One common mechanism used by states to ensure these issues receive attention at the local level is to require all levels of government beneath the state level to develop, adopt, and implement a solid waste management plan. State plans emphasize integrated solid waste management as an "hierarchy of approaches" yielding an integrated solid waste program through some combination of reduction, reuse, recycling, composting, energy recovery, incineration, and landfilling.

A state will undertake several regulatory activities and require actions by communities through legislation. Most state laws require local governments to create recycling programs that will achieve certain levels of recycling. Legislation may contain provisions for grants or funds to support waste reduction activities as well.

Several federal regulations have also been enacted to improve the performance of sanitary landfills. The principal federal requirements for municipal solid waste landfills are contained in Subtitle D of the Resource Conservation and Recovery Act (RCRA) and in EPA regulations. (Requirements under Subtitle D are discussed in Chapters 8 and 15.)

METHODS OF FINANCING

Operating Revenues

The cost of solid waste collection and disposal is increasing rapidly. Difficulty in finding a suitable site for a new landfill within a reasonable distance from a community, transportation costs, and strict regulatory requirements, including added costs of closure, postclosure, and remediation, are the principal reasons for the high cost of waste disposal.

Tax Financing.   Faced with the rising cost of solid waste disposal, many communities have found new sources of funding through tax revenues or have allowed the private waste disposal agencies to handle the job. Many communities have used a portion of the property tax to support the solid waste management program. Some state statutes also give communities or counties the authority to levy special taxes to fund projects. A municipal utility tax is another way to raise funds; utilities commonly subject to such a tax are telephone, gas, electricity, water and cable franchises, and solid waste systems themselves.

As mentioned, many communities have private companies collect their solid waste. The operation of these companies is regulated by local and state governments. Citizens contract the services with the waste haulers, who bill them for the service. This arrangement is becoming very popular as communities find it increasingly difficult to deal with the complex problems associated with the funding of and the regulatory requirements for solid waste systems.

User Fees.   User fees can be an equitable means of funding solid waste management services if such fees are properly administered. Fees can be based on the actual cost of collection and disposal, or they can be assessed at a uniform or a variable rate (depending on the amount and kind of services provided). A straight user charge allocates an equal share of the cost to all users within a service-level group. A variable rate based on a progressive user charge represents an attempt to correlate cost and service by charging residents according to the amount of waste generated. Decision makers should ensure that tipping fees should reflect the full cost of the facilities, including compliance with more stringent environmental controls than in the past. Also, some jurisdictions (e.g., New Jersey) have added a surcharge to landfills as a way of discouraging disposal and thus encouraging alternatives (e.g., reduce, reuse, recover, and recycle).

Capital Financing

The funds needed to finance construction of publicly or privately owned systems are borrowed from the commercial lending market. These funds are obtained through bonds that are sold by bonding institutions at an interest rate based on current conditions and the security used to back the bonds. A brief discussion on the possible methods of capital financing is needed to understand the bonding methods and their impact on a solid waste management project.

General Obligation Bonds.   Among the various public borrowing mechanisms, general obligation (GO) bonds are usually the most flexible and least costly alternative. The issuing municipal corporation or agency guarantees a GO bond with its credit, which is based on its property tax base. A tax based on the assessed valuation of the property is levied annually. Because general obligation bonds are considered to be the safest of all municipal securities, they carry lower interest rates than other forms of municipal debt having a similar maturity. Most states require voter approval to issue state or local GO bonds.

The property tax is collected by the county in which the municipal agency is located and is transferred to this agency for making the annual payment on the bonds. In several cases, this has led to abuse of the funding method. Because the revenue for retiring the bonds was guaranteed, technically inefficient and unsound systems were built, and little or no effort was made to improve them. Issuing a GO bond may not require any direct technical or economic analysis of the particular projects to be funded, and small projects may be grouped together to obtain capital, thus making GO bonds an ideal funding mechanism for solid waste facilities in small- and medium-sized communities.

Revenue Bonds.   One mechanism to circumvent the constraints associated with GO bonds is the municipal revenue bond. Revenue bonds do not require voter approval, and they do not affect a city's legal debt limit. A revenue bond is used to finance a single project with defined, revenue-producing services. Therefore, it is becoming the preferred method of financing solid waste management systems. Revenue bonds do not have the full faith and credit of the community, but they do pledge the net revenue generated by the project as a guarantee of payment. Revenue bonds are riskier than GO bonds, because they rely on the income from the operation of the facility to retire the bonds. Thus, these bonds cannot be sold unless an analysis shows that revenue income is adequate to cover both bond payments and system expenses. Here, the flow of incoming waste takes on significance. Reliable projections of processed solid waste are essential, because income projections are based on it. For example, if the nominal capacity is 100 tons per day but only 75 tons per day are received for processing, the income generated from the service is only 75 percent of the projected income.

Revenue bonds can also be used to finance private projects, but the capital required for private companies is higher than for public agencies. There are several reasons for this difference, including tax status. Municipal corporations do not pay federal tax or property tax. Also, because of the tax-exempt nature of the income from the municipal bonds, the interest rates are lower. The income from bonds used to finance private projects is not exempt from federal income tax; therefore, the interest rates are higher.

Other Debt Instruments.   The broad categories of GO and revenue bonds include a wide variety of individual debt instruments, such as tax increment and tax allocation bonds. These bonds are secured by the "additional" or "incremental" tax revenues generated by the new capital projects financed by those bonds. They are often used by local governments to finance redevelopment projects.

Other Financing Methods

Bank Loans.   A municipal bank loan is not a viable alternative to long-term bond financing. Relatively small-scale capital requirements, however, can be met in the short term (≤ 5 years) at low cost by securing a bank loan (1). In the solid waste field, bank loans have typically been used to supply short-term funding for vehicles, trailers, and so on.

Leasing.   In lease agreements, the leasing company usually purchases and holds title to the asset, and the municipality pays rent for its use during the term of the lease. In the solid waste field, lease agreements are usually arranged by an equipment company, which places the financing with a bank or leasing company. Often, the contract agreement stipulates that the company purchase the equipment at a "fair market value" at the end of the lease. Use of leasing agreements by private solid waste companies is quite prevalent and is often worthwhile, because the cost of leasing can be deducted as a business expense.

Industrial Revenue Bonds and Pollution Control Revenue Bonds.   Several methods have been developed to provide a financial incentive to encourage development and attract new business to a community. Industrial revenue bonds (IRBs) and pollution control revenue bonds (PCRBs) can be issued by a municipality either for or on behalf of a private enterprise. The municipality technically owns the facility and equipment, which it leases to the private firm. The lease payments are specified, and the municipality thus acts as a vehicle through which a corporation may obtain low-cost financing. Both IRBs and PCRBs are similar to other municipal bonds in that the interest paid is exempt from federal tax. Therefore, these interest rates are lower than the market rate. In the solid waste field, PCRBs have seldom been used.

COST COMPONENTS

To consider and compare alternative systems for solid waste management, some common economic basis is needed for cost comparisons. Costs include capital costs as well as operation and maintenance costs, which include closure and postclosure costs.

Capital Costs

Predevelopment Costs.   Predevelopment costs are usually associated with data collection, site selection, investigation, and permitting. First three items are professional services that involve land surveys, land prices, geotechnical investigations, future needs, environmental impact statements, and feasibility reports. Landfills in remote areas generally have lower land costs but higher transportation costs. The cost of obtaining a permit or license (and the cost of the engineering or legal support associated with permitting) depend on the individual state requirements.

Construction Costs.   Typical construction items include:

- Site development
- General excavation
- Liner construction
- Leachate collection and treatment
- Landfill gas management
- Groundwater monitoring
- Surface water drainage controls
- Equipment (e.g., vehicles, scales)
- Other facilities (e.g., maintenance building, access roads, utilities, fencing)

The impact of these factors can vary considerably from site to site. For example, in many cases, the cost of hauling material and supplies is expected to be high because the landfill is a remote area. The cost of road construction for borrowed materials (e.g., clay) must be included in the unit cost. The liner and leachate collection and treatment systems are expensive components of a landfill.

Operation and Maintenance Costs

Operation and maintenance are variable costs and include personnel, equipment, maintenance, utilities, administration costs, and fuel. These costs depend on how the facility operates, the quantity of material processed, and the site. For example, the energy required depends on how many hours the facility operates and the quantity of solid waste it processes. Maintenance costs also depend on the hours operated and the quantity processed. These costs are site specific as well. Both operating and maintenance costs will be higher in areas of high precipitation because of leachate production and erosion.

Labor.   The requirements for operating and maintenance personnel are specific for each facility. The base rate represents the hourly wage, which is based on job skill requirements and experience. Certain fringe benefits (e.g., health insurance, clothing allowance) are applied to all employees as well.

Administrative Personnel.   Administrative personnel include the general or facilities manager, business manager, and professional and nonprofessional support staff. These employees also receive base pay plus fringe benefits.

Equipment and Supplies.   This category includes the operation and maintenance of equipment and the supplies required to operate the facility. Equipment includes transport vehicles, wheeled equipment used on site, pumps, and so on. Maintenance supplies are significant for certain processing systems (e.g., fuel and oil for vehicles).

Maintenance of the transport and collection vehicles, loaders, tractors, and other equipment is an important part of solid waste management. Equipment failure can result in significant economic loss and bad public relations. Failure of a truck used to transport waste, for example, will result in a financial loss, because the investment (i.e., the truck) is not producing a return.

Utilities.   Utilities such as electricity, gas, and water are needed at any solid waste management facility for operating machinery (e.g., a shredder or hammer mill at a waste processing facility), for heating buildings, for running pumps and sprayers at a landfill, and for the personal needs of the employees. Utility costs depend on location; for example, costs are likely to be high in certain metropolitan areas and states. These costs should be evaluated to determine the economic feasibility of a given solid waste management system.

Administration Costs.   Administrative costs include the salary and fringe benefits of the administrative personnel (discussed earlier), insurance, and taxes (imposed on privately owned facilities). The amount of salary and fringe benefits paid to administrative employees depends on the size and location of the facility. In the waste management field, insurance is very important and can be expensive. In addition to workers' compensation, many agencies are self-insured against potential liability. There is always the possibility of an accident at a processing plant or landfill (e.g., an explosion or injury while operating machinery). There is also a significant potential for accidents during the collection and transportation of solid waste. In a hazardous waste disposal facility, accidental exposure or contact with the waste can cause injury or even death.

Legal and Fiscal Services.   Lawyers and accountants are needed in the operation of solid waste facilities, and the costs associated with these services depend on the size, location, and type of facility. Lawyers are needed because of legal challenges from the public. Accountants perform the audits required to check the financial records of the agency. The facility may be owned by a private or a semiprivate agency, and the community may have a profit-sharing agreement with this agency. If so, such audits are of interest to the community having the facility with a profit sharing agreement with the agency. Costs associated with both the legal and fiscal services can be significant if such services are needed for an extended period.

Taxes are owed by privately owned but not by municipal facilities. Private agencies must pay both the state and federal taxes.

TYPES OF ANALYSIS

Environmental engineers, other professionals, and community leaders base their solid waste management decisions on technical analysis, economical analysis, risk analysis, and environmental analysis. In recent years, ethical considerations have received attention as well, so in some cases, an ethical analysis is also performed.

Technical Analysis

In engineering practice, a variety of technical decisions may be right, because a problem may have more than one equally correct solution. It is the task of the engineer, however, to make every effort to find the best solution. For example, a landfill may have a clay liner or one made of some synthetic material (e.g., plastic). With correct design and construction, a liner will prevent the leachate from moving into the groundwater; thus, a liner made of either clay or synthetic material can be technically correct. Technical decisions made by an engineer are evaluated, checked, and modified by another competent engineer who has more experience than the former.

Example 2–1

A solid waste transfer station/recycling facility receives 80,000 lbs of waste per day. Fifty percent of this waste is processed for recycling, and the other 50% is transported to a landfill after compaction to 400 lbs/yd^3. The facility needs a semi-trailer to haul the waste to a landfill. The manager of this facility can purchase one of the following semitrailers:

1. 105-yd^3 capacity.
2. 100-yd^3 capacity.
3. 96-yd^3 capacity.

Which semitrailer is suitable for the transport of this facility's waste?

Solution

Total waste received = 80,000 lbs/day
Waste recycled = 40,000 lbs/day
Waste to landfill = 40,000 lbs/day

$$\text{Volume of waste after compaction} = \frac{40,000 \text{ lbs}}{\text{day}} \times \frac{\text{yd}^3}{400 \text{ lbs}} = 100 \text{ yd}^3/\text{day}$$

Obviously, the semitrailer with the 100-yd^3 capacity should serve the purpose. However, a semitrailer with a 105-yd^3 capacity will allow a little more capacity, if and when it is needed. Engineers usually provide a factor of safety; therefore, both capacities are correct. If the cost differential is insignificant, a 105-yd^3-capacity truck obviously is preferred. Consideration of the costs of engineering works now leads to economic analysis.

Economic Analysis

Whether working for a public or a private entity, engineers analyze and recommend various alternatives for engineering works to the decision makers. The public expects engineers to come up with the best engineering design and the most economical system. For example, a city sanitary engineer plans to purchase a waste collection vehicle and is considering two types of such vehicles, each with different features and price. The vehicle with a mechanized loading system is more expensive than the one without this system. Which vehicle should the city engineer buy for the municipality to meet the desired objectives? Obviously, in the absence of any other factors, the lowest-total-cost alternative would be the sensible decision.

Analysis of cost-effectiveness is difficult, because money changes in value with time. For example, if we invest $100 in an interest-bearing account with an annual interest rate of 6%, then a year later that $100.00 investment will become $106.00. A dollar today has a different value than it will a year from now. In addition, the operating cost of the facility will vary each year because of this change in the value of money. Therefore, the operational cost of a facility or equipment should not be added over the life of that facility or equipment.

To estimate the costs of a facility or equipment and compare the costs of different alternatives, it is necessary first to resolve such costs to a common base (i.e., into a form that can be compared). This means we must use the equivalence concept to convert from a cash flow representing the alternative into an equivalent sum or cash flow.

Present Worth and Annual Cost Analysis.   Two techniques commonly used for this purpose are annual cost and present worth analysis. The conversion of capital to annual cost as well as the calculation of the present worth of the operating cost is done using tables or preprogrammed calculators. (Appendix A provides sample interest tables.)

In present worth calculations, capital costs are the funds needed to construct a facility or buy equipment. Operating costs are calculated as if the money to pay such costs is available and is deposited in a bank or some other institution to pay the operation costs over the expected life of the facility. Higher operating costs require a larger initial sum to pay for the cost of operation.

The mathematical relations used in engineering economy employ the following symbols:

P = value or sum of money at a time denoted as the present (dollars)
A = a series of consecutive, equal, end-of-period amounts of money
     (dollars per month, dollars per year, and so on)
n = number of interest periods (months, years, and so on)
i = interest rate per interest period (percent per month, percent per year, and so on)

P represents single-time-occurrence values, and A occurs at each interest period for a specified number of periods with the same value. For a series to be represented by the symbol A, it must be uniform (i.e., the dollar value must be the same for each period), and the uniform dollar amounts must extend through consecutive interest periods.

Without a formal proof, the two relevant equations are as follows:

$$P = A\left[\frac{(1 + i)^n - 1}{i(1 + i)^n}\right] \qquad i \neq 0 \tag{2-1}$$

The term in brackets is called the *uniform-series present worth factor (PWF)*. By re-arranging, we can express A in terms of P:

$$A = P\left[\frac{i(1 + i)^n}{(1 + i)^n - 1}\right] \tag{2-2}$$

The term in the brackets is called the *capital-recovery factor (CRF)*.

Annual costs are converted to present worth by calculating what the money spent as the annual cost would be worth at the present time. From Appendix A, at 6% annual interest, $1.00 a year from now would be worth only $0.9434 today; conversely, $0.9434 invested today at 6% interest would yield $1.00 in a year. If another $1.00 is invested at the beginning of the second year, the total investment today would be worth $1.8333, as shown in the present worth column of the table. Thus, if a constant sum A is invested for n time periods at i rate of annual compound interest, the present worth P equals

$$A\left[\frac{(1 + i)^n - 1}{i(1 + i)^n}\right] \tag{2-1}$$

Example 2–2

A municipal engineer wants to buy a truck to haul solid waste to the disposal site. The engineer can buy a truck with a mechanized loading system, which only needs one driver to operate the vehicle and the loading system, or a truck without a mechanized loading system, which needs a second person to load and unload the vehicle. Each truck is expected to have a useful life of 10 years. Which truck should the engineer purchase based only on the costs listed here? Calculate the costs on both an annual and a present worth basis assuming an interest rate of 7% (see Appendix A).

	Nonmechanized	Mechanized
Purchase price of the vehicle	$103,530	$137,760
Fuel and oil per year	$7,000	$9,000
Maintenance cost per year	$5,000	$6,000
Driver's pay and fringe benefits per year	$50,000	$50,000
Semiskilled laborer's pay and fringe benefits per year	$40,000	—
Insurance per year	$2,000	$2,500

Solution

Calculation of Annual Cost for the Nonmechanized Truck

Interest on capital per year ($P \times CRF$)	$=$	$14,743
$= 103,530 \times 0.1424$		
Fuel and oil	$=$	$7,000
Maintenance cost	$=$	$5,000
Driver's pay and fringe benefits	$=$	$50,000
Semiskilled laborer's pay and fringe benefits	$=$	$40,000
Insurance	$=$	$2,000
Total	$=$	$118,743/year

Calculation of Annual Cost for the Mechanized Truck

Interest on capital per year ($P \times CRF$)	$=$	$19,617
$= 137,760 \times 0.1424$		
Fuel and oil	$=$	$9,000
Maintenance cost	$=$	$6,000
Driver's pay and fringe benefits	$=$	$50,000
Semiskilled laborer's pay and fringe benefits	$=$	—
Insurance	$=$	$2,500
Total	$=$	$87,117/year

Thus, on an annual cost basis, the mechanized truck is the choice, because its annual cost is lower than that of the nonmechanized truck.

Calculation of Present Worth for the Nonmechanized Truck

Capital cost	$=$	$103,530
Present worth of maintenance + fuel and oil costs + driver's and semiskilled laborer's pay and fringe benefits + insurance for 10 years ($PWF \times A$) = 7.024(7,000 + 5,000 + 50,000 + 40,000 + 2,000)	$=$	$730,496
Total	$=$	$834,026

Calculation of Present Worth for the Mechanized Truck

Capital cost	$=$	$137,760
Present worth of maintenance + fuel and oil costs + driver's pay and fringe benefits + insurance for 10 years ($PWF \times A$) = 7.024(9,000 + 6,000 + 50,000 + 2,500)	$=$	$474,120
Total	$=$	$611,880

Thus, on a present worth basis, the mechanized truck is also the choice, because the community will spend less money overall for this truck. It is interesting to note that the cost of operation for the nonmechanized loading truck is higher than that for the mechanized truck (because the need for a laborer for loading). In other words, in certain cases, mechanization can be economical.

Cost Indexing

Because costs change both nationally and locally, cost data must be referenced to some index. In general, these costs are referenced to a specific year and have a cost index value associated with them. There are two indices: one that covers heavy construction (i.e., concrete and earth moving), and one that covers building construction. These indices are published weekly in *Engineering–News Record*. The U.S. EPA has developed a cost index for small and large wastewater treatment plants as well, and *Chemical Engineering* magazine publishes an index that shows how much chemical plant construction costs have increased with time. Some typical indices are reported in Table 2–1. No index listed in this table is totally satisfactory for use with the solid waste management equipment, but the *Engineering–News Record* construction cost (ENRCC) index is most commonly used to prepare general estimates. Cost data reported in the literature can be adjusted to a common basis for comparison with the indices reported in Table 2–1 through the following relationship:

$$\text{current cost} = \frac{\text{current value of index}}{\text{value of index at the time it was noted}} \times (\text{cost cited in document}) \qquad \textbf{(2-3)}$$

If the month in which the equipment is purchased or the building completed is not given, it is common practice to use the June end-of-the-month index value.

Table 2–1   Indices for Adjusting Cost Data

Index	Base Year (Index = 100)
Engineering–News Record	
Building	1913
Construction	1913
EPA	
Sewage Treatment Plant Construction	1957–1959
Chemical Engineering	
Plant	1957–1959
Equipment, machinery, and supports	1957–1959

Example 2–3

A landfill was constructed at a cost of $5,700,000 and completed in December 1986 in Cincinnati, Ohio. Estimate the construction cost of the same landfill in March 1997 dollars.

Solution

Applying the equation:

$$\text{Cost in March 1997 dollars} = \frac{\text{March 1997 value of index}}{\text{value of index at the time it was noted}} \times \text{cost cited in document}$$

$$= \frac{5482}{4567} \times 5,700,000$$

$$= \$6,841,997$$

Benefit/Cost Analysis

Governmental economic analyses use the benefit/cost ratio method, largely because of federal legislation from the 1930s. The U.S. Army Corps of Engineers and Bureau of Reclamation frequently use this method to justify projects. If the total benefits/total cost ratio is more than 1.0, the project is considered to be worthwhile. Obviously, the projects with the highest ratios should receive priority.

As with cost-effectiveness analysis, benefit/cost analysis is in terms of dollars. Each benefit or cost is expressed in monetary terms. For example, the benefits of a sanitary landfill may include revenues from tipping fees and from selling landfill gas, and the costs may include construction, operation, and maintenance costs. These benefits and costs can be clearly quantified, and a dollar value can be assigned.

Costs such as groundwater pollution and damage to vegetation and wildlife because of construction and operation, however, cannot be easily expressed in monetary terms. Even so, some of these costs and benefits are real and should be included in the benefit/cost analysis. In most cases, monetary values are forced on these benefits and costs. This approach has several problems, of course. If community leaders or politicians have an interest in a project, the benefits of items that are difficult to quantify or express in monetary terms are bloated and the costs deflated. This approach results in a benefit/cost ratio greater than 1.0, and the project is thus justified. In addition, the monetary value varies from person to person and group to group, because some benefit much more from a public project than others even though all may have to share the costs. For example, people who like watersports would benefit more from a lake project than people who do not enjoy watersports. Many benefits are highly subjective as well. In the case of a new landfill project, citizens who generate large quantities of waste would benefit more than citizens who generate small quantities of waste if the assessment of the costs of construction and operation are levied equally.

Example 2–4

A community plans to build a landfill to dispose of wastes for the next 20 years. The cost estimates and expected revenue are:

Landfill construction cost	=	$7,600,000
Start-up costs	=	$500,000
Operation and maintenance costs per year	=	$1,000,000
Tipping fees received per year	=	$4,500,000

Assume that citizens will only support the new landfill facility if the benefit/cost ratio is greater than 1.5. At a 7% interest rate over the 20 years life, determine if the landfill should be built.

Solution

Present worth of capital costs =		
7,600,000 + 500,000	=	$8,100,000
Present worth of operation and maintenance costs		
$(PWF \times A) = 10.594 \times 1,000,000$	=	$10,594,000
Total cost	=	$18,694,000
Present worth of tipping fees received		
$(PWF \times A) = 10.594 \times 4,500,000$	=	$47,673,000
Total benefits (revenues)	=	$47,673,000

$$\frac{\text{total benefits}}{\text{total costs}} = \frac{\$47.673 \times 10^6}{\$18.694 \times 10^6} = 2.55$$

Because the benefit/cost ratio is greater than the desired 1.5, the facility should be approved.

Risk Analysis

Benefit/cost analysis becomes very complex when life and health issues enter into the calculations. These issues have then considered by performing a risk analysis, which is divided into two phases: risk assessment, and risk management. Risk assessment involves the data collection study and the analysis of certain hazards to human health and the environment. Using the gathered data (and after statistical analysis), a risk assessment is made, which then is used in the decision-making process. Risk management is the process of deciding what to do (using the risk analysis as a tool). A decision is then made about how to allocate limited resources to protect public health and the environment (i.e., about reducing risks).

The usual explanation of risk is to point out that some exists in everything we do. After all, our lifetime risk of death is 100 percent. Reliable data on the various causes of death, however, gives us some idea about what our cause of death may be and when it might occur. There are many ways of calculating the risk of death associated with some cause. For example, the risk from an environmental hazard can be defined as the ratio of the number of deaths in a given population exposed to a pollutant or hazard to the number of deaths in a population not exposed to that pollutant or hazard. That is,

$$\text{risk} = \frac{P_1}{P_0} \tag{2-4}$$

where P_1 is the number of deaths in a given population exposed to a specific pollutant per unit of time and P_0 is the number of deaths in a similar-size population not exposed to the pollutant per unit of time.

A second method of calculating risk is to determine the number of deaths from a certain cause and then divide this number by the total number of deaths:

$$\text{risk of dying from cause } x = \frac{P_x}{P_{\text{total}}} \tag{2-5}$$

where P_x is the number of deaths from cause x per unit of time and P_{total} is the total number of deaths in the population per unit of time.

Example 2–5

In 1992, there were 2.177×10^6 deaths in the United States of which approximately 521×10^3 were from cancer. What was the risk of dying from cancer relative to that of dying from other causes?

Solution

$$P_x = 521 \times 10^3$$
$$P_{\text{total}} = 2.177 \times 10^6$$

Using equation (2-5),

$$\frac{P_x}{P_{\text{total}}} = \frac{521 \times 10^3}{2.177 \times 10^6} = 0.239$$

That is, neglecting age and so forth, the risk or probability of a U.S. resident dying of cancer was, on average, approximately 24 percent.

Example 2–6

Considering the following causes of death for 1996 in Allen County, Ohio:

Heat disease	=	309
Cancer	=	246
Stroke	=	50
Lung disease	=	46
Accidents	=	21
Suicides	=	6
Homicides	=	6
Total	=	684

What was the risk of dying from cancer in Allen County relative to dying from other causes?

Solution

The total number of deaths in Allen County in 1996 was 684. The number of deaths from cancer in that same year was 246. Thus,

$$P_x = 246$$
$$P_{total} = 684$$

Using equation (2-5),

$$\frac{P_x}{P_{total}} = \frac{246}{684} = 0.359$$

That is, neglecting age and so forth, the risk or probability of an Allen County resident dying from cancer was, on average, approximately 36 percent.

The comparison between the probabilities of dying from cancer in Allen County, Ohio, and in the United States as a whole indicates a higher probability in Allen County. Obviously, such conclusions should be viewed with caution. More data are needed, such as the age range of people living in the county, their habits (e.g., smoking), and environmental factors (e.g., water supply, air).

Example 2–7

Allen County had a 1996 population of 49,500 people. The county has a refinery and a chemical plant. It is suspected that the emissions from these plants have resulted in an increased incidence of cancer, which has killed 246 people in the area close to these plants. Nearby, a community of 28,000 inhabitants without a refinery or chemical plant has lost 13 people to cancer. What was the risk of dying of cancer in Allen County?

Solution

In this case, risk is defined as in equation (2-4):

$$P_1 = \frac{\text{cancer deaths in Allen County}}{\text{Allen County population}} = \frac{246}{49,500} = 0.00497$$

$$P_0 = \frac{\text{cancer deaths in nearby community}}{\text{population of community}} = \frac{13}{28,000} = 0.000464$$

$$\text{risk} = \frac{P_1}{P_0} = \frac{0.00497}{0.000464} = 10.71$$

That is, a person was more than ten times more likely to die of cancer in Allen County than in the community without the emissions from the chemical and refinery plants.

A third method of calculating risk is to determine the number of deaths from various causes based on population group. That is,

$$\text{relative risk of dying from cause} = \frac{P_c}{P} \tag{2-6}$$

where P_c is the number of deaths due to a cause, c, and P is the population.

Example 2–8

There were 246 cancer related deaths in 1996 in Allen County, Ohio, and the population was 49,500. Determine the risk of dying from cancer in Allen County per population.

Solution

$$\text{relative risk of dying from cancer} = \frac{P_c}{P} = \frac{246}{49,500} = 0.00496$$

That is, approximately 0.5 percent.

Risk Assessment.   Risk assessment is the probability that exposure to some mix or variety of contaminants will cause some disease (i.e., response), such as birth defects, reproductive problems, tumors, cancer of various kinds, and so on. The National Academy of Sciences (2) recommends that risk assessment be divided into four steps: hazard identification, dose–response assessment, exposure assessment, and risk characterization. After a risk assessment has been completed, risk management is performed.

Hazard identification means determining and defining the source of a pollutant and whether it is likely to have adverse health effects on the population exposed. The hazard identification process includes examining the pathways a pollutant may take. A pollutant can enter the body through any of three pathways: ingestion with food or drinks, inhalation, or contact with skin. For example, a solvent stored in a pond may produce fumes that may be inhaled, or the liquid may seep into the groundwater and thus contaminate the drinking water.

Dose–response assessment means determining the potential health impact of a pollutant on the receptor. The goal of a dose–response assessment is to find a relationship between the amount of pollutant a human is exposed to and the associated risk of an adverse health response. Most dose–response relationships are based on high doses given to animals; assessment includes a method of extrapolating these data to humans. This necessity to extrapolate data presents a problem when interpreting the data for humans. All humans are not of one standard size and health. To simplify this analysis, however, the EPA (3) suggests using the assumption that all adult humans weigh 70 kg, live for 70 years, and drink 2 L of water and breathe 20 m^3 of air each day. These values are used for comparing different risks. The problem is that the average life expectancy is now greater than 75 years, and many people do not drink 2 L of water per day.

Another problem is that the shape of the dose–response curve for a cancer-causing chemical is not known. Some experts suggest a linear curve, starting with zero effect at zero concentration, in which the harmful effect increases linearly. Others suggest that small doses of carcinogens do not have a harmful effect and that each chemical has a threshold below which there is no effect.

To protect public health, the EPA chooses to err on the safe side. It has developed what is called the *potency factor (PF) for carcinogens:*

$$PF = \frac{\text{incremental lifetime cancer risk}}{\text{chronic daily intake (mg/kg/day)}} \qquad (2\text{-}7)$$

The dose response relationship therefore is:

$$\text{lifetime risk} = \text{average daily dose} \times PF$$

The units of average daily dose are (mg pollutant/kg body weight/day), and the units for the potency factor are (mg pollutant/kg body weight/day)$^{-1}$. This is because lifetime risk is unitless. The dose is an exposure averaged over an entire lifetime. For humans, the EPA assumes a lifetime of 70 years, as mentioned. The EPA has calculated the *PF* for many common chemicals (Table 2–2).

Table 2–2   Toxicity Data for Selected Potential Carcinogens

Chemical	Category	Potency Factor Oral Route $(mg/kg/day)^{-1}$	Potency Factor Inhalation Route $(mg/kg/day)^{-1}$
Arsenic	A	1.75	50
Benzene	A	2.9×10^{-2}	2.9×10^{-2}
Benzol (a) pyrene	B2	11.5	6.11
Cadmium	B1	—	6.1
Carbon tetrachloride	B2	0.013	—
Chloroform	B2	6.1×10^{-3}	8.1×10^{-2}
Chromium VI	A	—	41
DDT	B2	0.34	—
1,1-Dichloroethylene	C	0.58	1.16
Dieldrin	B2	30	—
Heptachlor	B2	3.4	—
Hexachloroethane	C	1.4×10^{-2}	—
Methylene chloride	B2	7.5×10^{-3}	1.4×10^{-2}
Nickel and compounds	A	—	1.19
Polychlorinated biphenyls (PCBs)	B2	7.7	—
2,3,7,8-TCDD (dioxin)	B2	1.56×10^{5}	—
Tetrachloroethylene	B2	5.1×10^{-2}	$1.0 - 3.3 \times 10^{-3}$
1,1,1-Trichloroethane (1,1,1-TCA)	D	—	—
Trichlorethylene (TCE)	B2	1.1×10^{-2}	1.3×10^{-2}
Vinyl chloride	A	2.3	0.295

Source: U.S. Environmental Protection Agency, "Guidelines for Carcinogen Risk Assessment," *Federal Register*, Vol. 51, No. 185, 1986.

Example 2–9

A landfill leachate with benzene as one of its components infiltrates into the groundwater. This groundwater is a water supply for 60,000 adults with a daily intake of 0.12 mg of benzene each. What are the maximal number of excess lifetime cancer cases?

Solution

From Table 2–2, the potency factor for benzene is 0.029 $(mg/kg/day)^{-1}$. Assuming an adult weight of 70 kg, the lifetime cancer risk from benzene intake for an individual is:

$$0.12 \text{ mg/day} \times 0.029 \text{ mg/kg/day} \div 70 \text{ kg} = 0.0000497\%$$

Maximal cases i.e., risk for the entire adult population is

$$\text{risk} \times \text{exposed population} = 0.0000497 \times 60,000$$
$$= 2.98$$

That is, three excess lifetime cancer cases.

The assessment of human exposure—one of the elementary concepts of risk assessment—is often overlooked in public debate. The risk has two components: the toxicity of the substance or its hazardous nature, and the amount and length of the exposure to the substance. It should be obvious that unless individuals are exposed to the toxic or hazardous substances, there is no risk.

Figure 2–1 illustrates some of the transport mechanisms that form the pathways for pollutants (4). Pollutants exposed to the atmosphere may volatilize and be transported by the wind. Pollutants exposed to soil may leach into groundwater and end up in drinking water wells. During their transport, pollutants may change their characteristics as well.

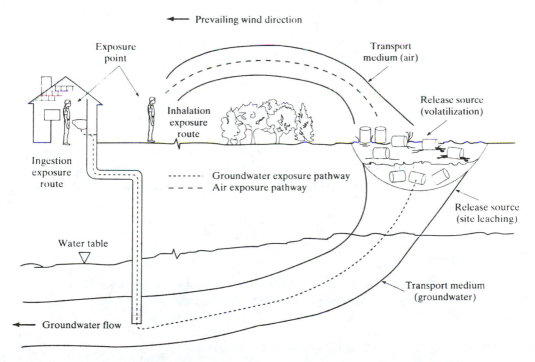

Figure 2–1   Exposure pathways.

(From U.S. Environmental Protection Agency. Superfund Public Health Evaluation Manual. Washington, D.C.: U.S. Govt. Printing Office, 1986.)

Risk characterization is the final step in risk assessment. Various observations and studies are put together to achieve the overall risk characterization (e.g., an estimation of overall risk to some population, as shown previously). This measure of risk is a single number; it is not very descriptive.

Deciding what is an acceptable risk is difficult and controversial. Is a risk of one in a million acceptable? Who decides if it is? One can be sure that the person or a group who would be harmed from such a risk would answer with an emphatic "No."

The whole concept of *acceptable risk* is based on the value system and studies used by engineers and scientists in the field of environmental management. This value system is often different than that held by the public, especially the people who are affected.

Any analysis where risk and the associated costs are unequally assigned, for example, one person in a million will suffer adverse health may be considered unfair. Many people would argue that it is unethical to place a value on human life. Read discussion on ethics in this chapter.

Risk Based on Epidemiological Studies.   A very important source of human risk information is epidemiological studies. Epidemiology is the study of the incidence of diseases in real populations. Epidemiologists find the correlations between disease rates and various environmental factors, and they attempt to show quantitatively the relationship between exposure and risk.

Data analysis involves setting up a 2 × 2 matrix. For example:

	With Disease	**Without Disease**
Exposed	x	y
Not Exposed	s	z

The rows divide the populations according to those who have and those who have not been exposed to the factor in question. The columns divide the populations according to those who have and those who do not have the disease. To check for an association between the exposure and the disease, the relative risk and attributable risk are calculated.

$$\text{relative risk} = \frac{\dfrac{x}{x + y}}{\dfrac{s}{s + z}} \qquad (2\text{-}8)$$

If the ratios in the numerator and the denominator are the same, the relative risk is 1.0. In other words, the odds of having the disease would not depend on the exposure being studied.

The attributable risk is the difference between the odds of having the disease with exposure and of having the disease without exposure.

$$\text{attributable risk} = \frac{x}{x + y} - \frac{s}{s + z} \qquad (2\text{-}9)$$

If the difference is 0.0, there is no relationship between exposure and risk.

Example 2–10

An evaluation of the health records of 80 workers at a hazardous waste disposal site finds that 14 workers developed lung cancer. In another group with backgrounds similar to the exposed workers but who never worked at a hazardous waste site, 20 individuals developed lung cancer and 380 did not. Find the relative risk and the attributable risk.

Solution

Put the data into a matrix:

	With Disease	Without Disease
Exposed	14	66
Not Exposed	20	380

Using equations (2-8) and (2-9),

$$\text{relative risk} = \frac{\dfrac{14}{14 + 66}}{\dfrac{20}{20 + 380}} = 3.5$$

and

$$\text{attributable risk} = \frac{14}{80} - \frac{20}{400} = 0.125$$

The relative risk is greater than 1; therefore, a relationship between exposure and risk is indicated. For the workers who were exposed, the risk of cancer has increased by 0.125, which is the attributable risk, over that of the people who were not exposed.

Such studies should be interpreted with caution. Many variables and data collection methods may lead to invalid conclusions.

ENVIRONMENTAL IMPACT STATEMENTS

The first comprehensive environmental legislation in the United States, the National Environmental Policy Act (NEPA), took effect on January 1, 1970. Consequently, the environmental impact of a project was considered in the planning process in addition to the traditional economic and technical problems.

The NEPA regulations focused on major federal projects that could affect the environment. Soon after NEPA took effect, many states adopted similar laws, and local municipalities began incorporating these laws through environmental planning requirements regarding land use and zoning ordinances. Today, when a new project or development is planned that might affect environmental quality, an environmental impact assessment is performed.

The first step in the process is to determine if a project falls within the relevant act or regulations and if it is likely to cause environmental problems. If so, an assessment is undertaken, leading to preparation of an environmental impact statement (EIS). An EIS for a project is a written report that summarizes the findings of a detailed environmental review process. It is a public document and is written in a format specified by authorized national, state, and/or local agencies. Eventually, a decision is made at the political level to accept the proposed project, to accept an amended form of the project, to reject it, or to accept an alternate proposal. A typical EIS consists of three parts: an inventory, an assessment, and an evaluation.

Inventory

An inventory is the first investigation that is conducted for the site and vicinity of the proposed project. It is necessary to have details on existing (i.e., predevelopment) environmental conditions. A basic objective of an environmental study is to anticipate potential impacts of a proposed construction project. If an environmental inventory report is not already available for the area of interest, the interdisciplinary team preparing the EIS should conduct a detailed environmental survey.

The inventory of existing natural resources and various facilities in the area of the proposed project site typically includes the following categories:

- Soils, topography, and geology
- Vegetation
- Wildlife
- Public utilities
- Transportation
- Air quality
- Surface and groundwater quality
- Population and land use
- Economic conditions
- Any natural or man-made features of interest

In addition to a detailed environmental inventory, a preliminary plan should be made available by the project authority. Though preliminary, this plan must be comprehensive enough to provide a realistic and easy-to-comprehend assessment of the environmental impact. For example, a preliminary plan for a proposed recycling facility would show its distance from the surrounding communities, elevation of the area, road to the facility, utilities (e.g., water, gas, electricity), and so on.

Assessment

The second step of an EIS is assessment. Certain environmental impacts can be evaluated objectively, because they generally are not subjective or based on conflicting personal opinions. For example, the expected stormwater runoff from an area with a completed sanitary landfill can be compared to existing runoff rates. Because the landfill is capped, the runoff will increase, and as discussed in Chapter 15, these effects might include soil erosion and flooding.

Impacts on humans, vegetation, and wildlife are difficult to evaluate objectively. Their extent, value, or importance is difficult to agree on because of the diverse interests among various groups of people. In addition, there is a difference between short-term and long-term impacts.

Several methodologies for conducting an environmental assessment are available. All have the goal of providing a systematic environmental evaluation of the project. Procedures range from simple checklists to complex matrices. Two such procedures are described here: the checklist, and the numerical matrix.

In the checklist method, all potential environmental impacts for the various project alternatives are listed. This involves first listing those areas of the environment likely to be affected by the proposed project and then listing the importance and magnitude of the impact. Usually, the importance of the negative impacts (0 to 5) are indicated by a minus sign ($-$). That is, small to moderate importance can be shown with a lower minus number, whereas a greater importance (i.e., causing a severe impact) can be indicated with a higher minus number. Beneficial or positive impacts are indicated by a plus sign ($+$). If the environmental impact is not applicable for a project alternative, a zero is indicated.

Weighting factors are also used in the matrix method to indicate the relative importance of a specific impact. Weighting factors are established by the assessment team, and they are based on both the project and the site. For example, in the case of a landfill in a windy area, the impact on an area's litter problem may be considered more important than the impact on that area's noise level. In this case, the litter problem may be assigned a weighting factor of, say, 0.4, compared with 0.2 for noise.

Weighting factors can be multiplied by the respective impact magnitudes. For example, the impact on the litter problem has a magnitude of, say, 3, and the impact on the noise level has a magnitude of, say, 5. By weighting the impacts (i.e., multiplying by weighting factors), however, it is evident that the significance of the impact on the litter problem ($0.4 \times 3 = 1.2$) is more important than the impact on the noise level ($0.2 \times 5 = 1.0$). If the weighted impacts for all the listed areas of concern are added together, an environmental impact index can be obtained for each project alternative. The alternative with the lowest index is considered to be least harmful in its environmental impact.

Example 2–11

A plant manager at a small landfill has two alternatives: transport leachate by a tanker truck to a wastewater treatment plant 11 miles away, or build a wastewater treatment facility and treat leachate at the site. Analyze these two alternatives using the checklist method.

Solution

First the areas of environmental impact are listed. (Only a few areas are considered; a detailed and thorough assessment would include many others.) Values indicating importance and magnitude are assigned (0–5), and the two columns are then multiplied.

Alternative I: Transport Leachate

Areas of Impact	Importance (I)	Magnitude (M)	I × M
Air pollution (truck)	−4	2	−8
Traffic problems	−3	3	−9
Water quality degradation	−4	2	−8
Land pollution	−2	2	−4
Odor	−2	2	−4
Environmental impact			= −33

Notes

- Exhaust from the truck will cause air pollution.
- Traffic problems include congestion and accident risk.
- Water quality degradation depends on the treatment applied before discharge.
- Land pollution may occur because of spills and sludge disposal.
- Odors are possible, during transport and treatment.

Alternative II: Treatment at Site

Areas of Impact	Importance (I)	Magnitude (M)	I × M
Air pollution	−4	1	−4
Traffic problems	−3	0	0
Water quality degradation	−4	3	−12
Land pollution	−2	1	−2
Odor	−2	1	−2
Environmental impact			= −20

Notes

- Air pollution, if any, may occur during treatment and disposal.
- Water quality degradation depends on treatment and disposal. Treatment may not be as good as that applied at the conventional, large plant in alternative I.
- Land pollution may occur through sludge disposal or leaks from treatment units, but the magnitude should be lesser than that in alternative I. This is because no truck transport is involved.
- Odors may be a problem during treatment. The magnitude is small, however, because of the open landfill site.

On the basis of this analysis, treatment at the site would have the lower adverse impact. This evaluation is based only on the environmental impact; economic aspects are not considered.

Evaluation

Comparison of the environmental impact assessment results and conclusions are included in the evaluation part of the EIS. The evaluation should be based on an understanding of the assumptions and calculations used in the assessment step.

Obviously, different people can draw different conclusions from an EIS. People who do not live close to a proposed site are always interested in a facility that will permit economical and convenient ways of waste disposal. People living close to a proposed site are likely to oppose the project because of perceived as well as real environmental problems associated with a disposal facility and the possible decline of their real estate values. Conflicting values are difficult to resolve, and the decision-making process is complex. How can the technical decisions, which are based on engineering and science, be made in light of conflicting values? At this point, a relatively new tool—the ethical analysis—is helpful.

ENVIRONMENTAL ETHICS

The decision-making tools discussed so far have been economic analysis, for making decisions based on money; technical analysis, for good engineering design; risk analysis, for calculating potential damage to the health and well-being of humans; and the environmental impact analysis, for considering the long-term effects on resources and the environment. Ethical analysis is a similar framework, but for making decisions based on values (5).

Ethics is the philosophical study of morality. Moral philosophy is the attempt to achieve a systematic understanding of, in Socrates' words, "how we ought to live" and why. Morals are the values that people choose to guide how they treat each other (e.g., telling the truth).

Environmental ethics concerns itself with the attitude of people toward other living things and the natural environment. Humans, nonhuman animals, and plants share a common environment, a common ecosphere. Concern for the environment has increased rapidly, becoming what is known as environmental ethics. It is a framework that allows us to make decisions about our environment, cutting across national boundaries, and that overlaps with engineering ethics.

Environmental ethics poses questions about the morality of the relationships between humans and nature. Do humans have obligations, duties, and responsibilities to the natural environment? Many philosophers and ethicists have expressed their views.

American philosopher Paul Taylor (6) holds that all living things have an intrinsic good in and of themselves; therefore, they are candidates for inclusion in the moral community. He suggests that once we can admit that we humans are not superior, we will recognize that all life has a right to moral protection. Taylor thinks humans are no more—and no less—important than other organisms. This line of reasoning is not totally convincing, however. There are viruses that cause diseases such as polio, AIDS, and so on. Should these microorganisms be included in the moral community? Should the lives of all creatures be equal, including a human life? Is there an hierarchy of value for each living creature? Who has the right to establish such values? Again, these questions and more need to be debated and discussed in environmental ethics.

According to Dr. Albert Schweitzer (7), a person is ethical "only when life, as such, is sacred to him, that of plants and animals as that of his fellow men" (p. 16). He believed an ethical person would not maliciously harm anything that grows but would exist in harmony with nature. He recognized, of course, that humans must kill other organisms to eat but felt this should be done with a sense of compassion and sacredness toward all life. Schweitzer believed that human beings are simply one part of the natural system.

Forester Aldo Leopold (8) suggests that ecosystems should be preserved; otherwise, nothing can survive. In what is termed *ecocentric environmental ethics*, a "circle of moral concern" is used, within which various life forms exist. This is a principle of "non-destruction, non-interference, and generally non-meddling" (p. 16). Each person must change his or her character or ethic, and the social system and social attitude must change to become compatible with the global ecology. The acceptable system is one in which we learn to conserve and share exhaustible resources. This requires that we reduce our needs and that the materials we use be replenishable. We must treat all the earth as a trust, to be used so its content is neither diminished nor permanently changed. We must release no substances that cannot be incorporated by the ecosystem without damage to the environment.

Many scientists, engineers, and conservationists warn that we are misusing our scarce resources and damaging the environment. Technology and its practitioners have engaged us in ongoing global development, an experiment that needs to be carefully monitored, modified, and guided so that unacceptable changes in the environment do not occur.

From these philosophical discussions and observations of professionals in the environmental area, a pragmatic view of environmental ethics emerges:

All living beings, humans and nonhumans, have the same moral worth. Every living thing has a right to exist independent of its potential use to us. Humans should not cause the extinction of any species or the elimination of their habitat, because everything is connected to and intermingled with everything else.

We need to protect ourselves from harmful organisms and kill animals for food, but we should not harm or kill for frivolous wants or by polluting the environment. If we need to alter nature to meet our basic needs, we should cause the least possible harm to the environment. And we should use nature and animals only to meet the basic needs.

All present and future inhabitants of this earth are entitled to their fair share of resources and have a right to live in a clear and healthy environment. Need should not become greed. To prevent further damage to the earth's environment and depletion of its resources, population control and changes in lifestyle are needed.

Civil rights leaders and environmental activists have long claimed that poor minorities suffer most from pollution and benefit least from clean-up programs. In such communities, it took longer to clean up the Superfund toxic waste sites, and the job was done with less stringent standards than in predominantly well-to-do communities. In 1983, the U.S. General Accounting Office reported that three out of every five commercial hazardous waste disposal sites in several states were located in predominantly poor communities. Every person has a right to live in a clean and healthy environment. Unfortunately, greed and lack of ethical consideration by certain groups make that impossible.

Environmental professionals (i.e., engineers, technologists, scientists) are engaged in a truly challenging and worthwhile mission. Along with technical, economical, and scientific analyses, environmental ethical analysis can guide us in the professional decision-making process. In a very general sense, these professionals are watchdogs of the environment. The objective of such professionals should be to protect the planet, to preserve nature and its resources for the sake of our progeny. These professionals are expected to work to the best of their ability, with zeal, and in good faith to protect the earth.

REVIEW QUESTIONS

1. Assume you are a city council member and chairperson of a committee to plan a landfill.
   a. What preliminary steps would you take to start the process toward planning such a facility?
   b. What role would you expect the city engineer to play in the process?
2. Assume you are the city council member from question 1.
   a. Describe a regional approach to solid waste management.
   b. List the advantages and disadvantages of such an approach.
   c. Would you encourage a regional approach to the solid waste management problems? State the reasons for your answer.
3. Assume you are the city council member from question 1.
   a. What role do federal and state regulations play in solid waste management?
   b. What federal or state agency would you most likely have to deal with?
4. What is the present worth of the landfill with annual operating costs of $200,000 for the next 20 years? Assume an interest rate of 8%. (Obtain the factor from any economics book.)

5. Make an economic analysis of the following. Assume an interest rate of 7% and useful life of 7 years. Which alternative is cheaper to operate in the long run?
   a. A $160,000 truck requiring maintenance and fuel costs of $10,000 per year.
   b. A $210,000 truck requiring maintenance and fuel costs of $7,000 per year.
6. What is cost indexing?
7. A landfill waste compactor cost was $180,000 in 1986 dollars (ENR cost index). What would it cost in 1997 dollars (ENR cost index)?
8. a. Should a project with $\dfrac{\Sigma \text{ benefit}}{\Sigma \text{ cost}} = 0.85$ be undertaken? Explain your answer.

   b. What are the drawbacks of relying on the benefit/cost ratio for justifying a project?
9. a. The average number of deaths due to respiratory problems in a town during a given time period is 50. The average number of total deaths during the same period is 725. What is the risk of dying from respiratory problems?
   b. Determine the relative risk of dying from respiratory failure if the town's population is 50,100.

10. Define the potency factor.

A hazardous waste disposal site leaks carbon tetrachloride into the surface water, which happens to be a town's water supply for 50,000 adults. The daily intake of carbon tetrachloride is estimated to be 0.14 mg. What are maximal number of excess lifetime deaths due to the toxicity of carbon tetrachloride?

11. a. What are the main components of an environmental impact statement?
    b. What is the significance of such a report?

12. a. What is the significance of environmental ethics?
    b. Describe a situation in which environmental ethics may guide the decision-making process.

REFERENCES

1. U.S. Environmental Protection Agency. *Decision Makers Guide to Solid Waste Management.* Washington, D.C.: U.S. Govt. Printing Office, 1989.

2. National Academy of Sciences. *Risk Assessment in the Federal Government.* Washington, D.C.: National Academy Press, 1983.

3. U.S. Environmental Protection Agency. "Guidelines for Carcinogen Risk Assessment," *Federal Register,* Vol. 51, No. 185, 1986.

4. U.S. Environmental Protection Agency. *Superfund Public Health Evaluation Manual.* Washington, D.C.: U.S. Govt. Printing Office, 1986.

5. Vesiland P.A. *Introduction to Environmental Engineering.* Boston: PWS Publishing, 1996.

6. Taylor, P.W. *Respect for Nature: A Theory of Environmental Ethics.* Princeton, N.J.: Princeton University Press, 1986.

7. Schweitzer, A. *Out of My Life and Thought: An Autobiography.* New York: Henry Holt, 1949.

8. Leopold, A. *A Sand County Almanac.* New York: Oxford University Press, 1966.

Chapter 3   Engineering Calculations

This chapter reviews the units of measures, dimensions, and principles of basic calculation and analysis used in technical and scientific work to obtain and process the information needed to make decisions.

DIMENSIONS AND UNITS OF MEASURE

Fundamental dimensions describe basic characteristics such as length (L), time (T), mass (M), and force (F). Such dimensions are used to describe physical quantities. Dimensions are independent of units, and they are described in units of measure such as feet, meters, or fathoms. When given values, these units describe something quantitatively (e.g., 6.56 ft, 2 m). Derived dimensions are obtained from one or more fundamental dimension; for example, area has the dimensions of $L \times L$ (L^2), volume the dimensions of $L \times L \times L$ (L^3), and velocity $\left(\dfrac{L}{T}\right)$.

The systems of units commonly used are the American or English engineering system and the Systeme International (SI). The American system is based on feet (length), pounds (mass or force), seconds (time), and temperature in degrees Fahrenheit (°F). The SI system is based on meters (length), kilograms (mass), seconds (time), and temperature in degrees kelvin (°K).

Both the American and SI systems are used in the United States, and both systems are used in this book. For selected units and conversions, see Appendix B.

Pressure

Pressure is defined as force per unit area, or

$$P = \frac{F}{A} \tag{3-1}$$

where P represents pressure, F represents force, and A represents the area over which the force is distributed.

In the American system, pressure is usually expressed in terms of pounds per square inch (lb/in^2 or psi). In SI units, pressure is expressed in terms of newtons per square meter (N/m^2) or pascals. To convert,

$$1 \text{ kPa} = 1000 \text{ Pa} = 0.147 \text{ psi}$$

Power

Power is defined as the rate at which work is done. In the American system, it may be expressed as foot-pounds per second (ft-lb/sec), and it is often expressed as horsepower (hp). To convert,

$$1 \text{ hp} = 550 \text{ ft-lb/sec}$$

In the SI system, power may be expressed as newton-meters per second (N-m/sec) or watts. To convert,

$$1000 \text{ N-m/sec} = 1 \text{ kilowatt (kw)}.$$

Density

The density of a substance is defined as its mass divided by unit volume, or

$$\rho = \frac{M}{V} \tag{3-2}$$

where ρ represents density, M represents mass, and V represents volume.

In the American system, density is commonly expressed as lb/ft³. Thus, water has a density of 62.4 lb/ft³. In the SI system, the basic unit for density is kg/m³. Thus, in this system, water has a density of 1×10^3 kg/m³, which is equal to 1 g/cm³.

Concentrations

Concentrations are expressed in terms of mass per unit volume, parts per million or billion, or percent.

Mass per Unit Volume.   One commonly used term for concentration is milligrams per liter (mg/L). For example, if 0.4 g of salt is dissolved in 2500 mL of water, then the concentration is expressed as

$$\frac{0.4 \text{ g}}{2500 \text{ mL}} = \frac{400 \text{ mg}}{2.5 \text{ L}} = 160 \text{ mg/L}$$

The concentration of material m is expressed as the mass of the material in a unit volume consisting of the material m and of some other material n. The concentration of m in a mixture of m and n therefore is

$$C_m = \frac{M_m}{V_m + V_n} \tag{3-3}$$

where C_m represents concentration of material m, M_m represents mass of material m, V_m represents volume of material m, and V_n represents volume of material n.

Example 3–1

Granular charcoal with a mass of 0.65 kg and a volume of 0.05 m^3 is put in a container, and 100 L of water are poured into it. Determine the concentration of charcoal granules.

Solution

Using the equation (3-3),

$$C_m = \frac{M_m}{V_m + V_n}$$

where m represents the charcoal granules and n represents the water. Thus,

$$C_m = \frac{0.65}{0.05 + 100 \times 10^{-3}} = 4.34 \; kg/m^3 = 4.34 \; \frac{10^6 \; mg/kg}{10^3 \; l/m^3} = 4340 \; mg/L$$

In this example, the volume of water (100 *L*) is added to the volume of granules already in the container. Thus, the total volume of water and granules is more than 100 *L*. If, with the granules (volume, 0.05 m^3) in the container, the water is poured so that the container is filled to the 100-*L* mark, the total volume is

$$V_m + V_n = 100 \; L$$

and the concentration of the granules is

$$C_m = \frac{0.65}{100 \times 10^{-3}} = 6.5 \; kg/m^3 = 6500 \; mg/L$$

That is, the concentration of granules is higher, because the total volume is lower.

Parts per Million.   Another measure of concentration is parts per million (ppm). This expression conveys just how small the concentrations encountered in the environmental field really are. If the small variation in the density of water because of substances dissolved in it is neglected (i.e., $\rho = 1.0$ g/cm^3), then ppm is numerically equivalent to milligrams per liter by the following conversion:

$$1.0 \; mg/L = \frac{0.001 \; g}{1000 \; mL} = \frac{0.001 \; g}{1000 \; cm^3} = \frac{0.001 \; g}{1000 \; g} = \frac{1 \; g}{1,000,000 \; g}$$

or one gram in one million grams (i.e., 1 ppm).

Percentage Concentration.   For convenience, some concentrations, usually those in excess of 10,000 mg/L, are generally expressed in terms of percentages. A concentration may be determined as a percentage through the following equation:

$$P_m = \frac{M_n}{M_m + M_n} \times 100 \qquad (3\text{-}4)$$

where P_m represents the percentage of material m, M_m represents the mass of material m, and M_n represents the mass of material n.

Example 3–2

A 1000-mL sample of wastewater has 200 mg of solids in it. Express the concentration of this solution.

Solution

Note that 1000 mL of water has a mass of 1000 g. Thus, applying equation (3-4),

$$P_m = \frac{0.2 \ g}{1000 \ g} \times 100 = 0.02\%$$

Flow Rate

In many contexts, it is important to know "how much" fluid is flowing in a stream or a pipe. The parameter that tells us "how much" is known as the *flow rate*, which is either the mass flow rate or the volume flow rate. The former is expressed in lb/sec or kg/sec, whereas the latter is expressed as m^3/sec or ft^3/sec. The mass (M) of liquid flowing at a point in a unit time relates to the volume (V) of that liquid as

$$\text{mass} = \text{density} \times \text{volume}$$

Therefore,

$$Q_M = \rho \times Q_v \qquad (3\text{-}5)$$

where Q_M represents mass flow rate, Q_v represents volume flow rate, and ρ represents density of fluid.

Weight flow rate (Q_w) may be calculated through equation (3-6):

$$Q_w = \gamma \times Q_v \qquad (3\text{-}6)$$

where Q_w represents weight flow rate and γ represents the specific weight of the fluid ($\gamma = \rho g$).

The relationship between the mass flow of some component, say, substance S_1, and its concentration and total volume flow of (S_1 + water) is

$$Q_{Ms_1} = C_{S_1} \times Q_v (S_1 + water) \qquad (3\text{-}7)$$

where, Q_{Ms_1} is the mass flow rate of S_1, C_{S_1} is the concentration of S_1, and Q_v is the volume flow rate of S_1 and water.

It is important to note that equation (3-5) or (3-6) is applicable to only one substance in a flow, that is, water (or any other fluid). Equation (3-7), however, considers two substances; fluid, and the substance the flowing fluid carries with it.

Example 3–3

A wastewater treatment plant effluent has a concentration of 0.0025% solids (i.e., 25 mg/L, or 25 mg of solids per 1L of flow [solids plus water]) at a flow of 0.6 million gallons per day (mgd). How many pounds per day of solids are discharged in the plant effluent?

Solution

Using equation (3-7),

$$\text{mass flow} = \text{concentration} \times \text{volume flow}$$

or

$$Q_{Ms_1} = Cs_1 \times Q_V \ (S_1 + water)$$
$$= 25 \ mg/L \times [2.2 \times 10^{-6} \ lb/mg] \times 3.78 \ L/gal \times (0.6 \times 10^6 \ gal/day)$$
$$= 124.74 \ lb/day$$

Example 3–4

A wastewater treatment plant effluent has an organics concentration of 30 mg/L at a flow rate of 0.027 m^3/sec (organics plus water). How much organic material is in the plant effluent?

Solution

Using equation (3-7),

$$\text{mass flow} = \text{concentration} \times \text{volume flow}$$

or

$$Q_{Ms_1} = Cs_1 \times Q_V \ (S_1 + water)$$
$$= 30 \ mg/L \times [1 \times 10^{-6} \ kg/mg] \times (0.027 \ m^3/sec \times 10^3 \ L/m^3 \times 86,400 \ sec/day)$$
$$= 70 \ kg/day$$

DETENTION TIME

Detention time, which is also called retention time or residence time, is defined as the theoretical amount of time that an average particle of fluid resides in a container through which a fluid flows. The detention time can be computed as follows:

$$T_D = \frac{V}{Q}$$

(3-8)

where T_D represents detention time, V represents the volume of water in tank, and Q represents the average flow rate (volume per unit of time).

Detention time is usually expressed in terms of hours. The units for V and Q should be consistent so that equation (3-8) will be dimensionally correct.

Example 3–5

A holding pond for a landfill leachate has a volume of 600 m³, and the flow into the pond is 2.3 m³/hour. Calculate the detention time in the pond.

Solution

$$T_D = \frac{600 \text{ m}^3}{2.3 \text{ m}^3/\text{hour}} = 261 \; hours \; 10.87 \; days$$

MEASUREMENTS AND APPROXIMATIONS

Measurements

Engineers and other technical personnel measure various physical quantities that control the design. Care and skill in both making and interpreting these measurements is essential.

Any physical measurement cannot be assumed to be exact; errors are likely to be present. Quantities determined by analytical techniques or observations are also likely to have errors. These errors have many causes, such as human error or bias, changes in the environment during the procedure, (e.g., temperature, wind direction), a defect in the instrument, and so on.

A method of expressing results and measurements is needed that will indicate how reliable these numbers are, and the use of significant figures provides some capability. Significant figures are those that transfer information based on the value of the digit. Zeros that are used only for location of the decimal point or that have no nonzero digit to their left are not significant; they can be eliminated without loss of information. For example, in the number 0.0018, only the digits *1* and *8* are significant, because the zeros have no nonzero digit to their left. The zeros are holding a place and can be eliminated by writing $0.0018 = 1.8 \times 10^{-3}$.

Zeros at the end of a number can be a problem and should be used with care. Suppose the manager of a waste collection company states that the daily estimate of solid waste generated by a town is 50,670 lbs. The last digit (i.e., zero) may be significant if in fact, the quantity of waste is exactly 50,670 lbs. Because the number is an estimate, however, it is always desirable to state the number as approximately 51,000 lbs. The three zeros here are holding their place and are not significant, because this number can also be stated as 5.1×10^4 lbs.

Approximations

In many instances, an engineer or other technical person is expected to provide approximate information (i.e., estimate the result with reasonable accuracy). For example, he or she may be asked by a client, such as a town's mayor, what it might cost to build a landfill in which to dispose of the town's solid waste for a 20-year period. The engineer is being asked for a ballpark estimate, not for an exact figure. In other words, the mayor wants an approximate figure—and right away. Engineers are expected to be able to make rough estimates in the area of their competency, to provide figures that can be used for making tentative decisions. To do this, engineers rely on their basic understanding of the problem and their own experience.

Example 3–6

The city's sanitary engineer is asked to meet with the mayor and city manager to discuss the disposal of solid waste. The city administration wants to provide collection and disposal service to the city's 20,000 residents for the first time. The mayor's immediate concern is how much land will be needed, so she asks the sanitary engineer how many acres will be required for the next 20 years.

Solution

The national average of solid waste production is 6 lbs per person per day, and because of the state EPA mandate, 25% of this waste is to be recycled. Thus, the waste to be landfilled is 4.5 lbs/person/day. Each citizen will therefore produce

$$4.5 \text{ lbs/day} \times 365 \text{ days/year} = 1642.5 \text{ lbs/year}$$

Based on the current practice, the engineer estimates the waste will probably be compacted to a density of approximately 1000 lbs/yd^3. On this basis, the per capita landfill volume is

$$\frac{1{,}642.5 \text{ lbs}}{\text{year}} \times \frac{\text{yd}^3}{1000 \text{ lbs}} = 1.64 \text{ yd}^3/\text{year}$$

Therefore, the per capita requirement for a 20-year period is

$$1.64 \text{ yd}^3/\text{year} \times 20 \text{ years} = 32.8 \text{ yd}^3/\text{person}$$

and the total requirement for 20,000 persons is

$$32.8 \text{ yd}^3 \times 20,000 \text{ persons} = 656,000 \text{ yd}^3$$

The engineer's knowledge of the area's geology indicates that bedrock occurs at approximately 12 ft below the surface. To allow for possible variations in the bedrock depth, the engineer conservatively estimates its depth as 10 ft. Thus, the required area for the landfill itself is

$$\frac{656,000 \text{ yd}^3}{10 \text{ ft}} \times 27 \text{ ft}^3/\text{yd}^3 = 1,771,200 \text{ ft}^2$$

or

$$\frac{1,771,200 \text{ ft}^2 \times \text{acre}}{43,560 \text{ ft}^2} = 40.66 \text{ acres} = 41 \text{ acres}$$

The patterns of the recent past indicate some growth in the city's population and, consequently, an increase in solid waste generation. The engineer thus allows for a 10% increase in waste generation during the life of the landfill. Allowance is also made for utility buildings, an access road, and a fence around the landfill. Standard practice is to increase the required area by a factor of approximately 1.25. Thus, an approximate total area is

$$(41 \text{ acres} + [0.10 \times 41]) \times 1.25 = 56.4 \text{ acres} = 57 \text{ acres}$$

TECHNICAL INFORMATION ANALYSIS

Engineers and scientists base their work on numerical data derived from surveys and experiments, which constitute the raw material on which interpretations, analyses, and decisions are based. Engineers and scientists take chances, and armed with the probabilities for events of interest, they can decide on prudent action. For example, an engineer may design a landfill for a town based on 5-year data. If data for more than 5 years are available, however, would different design parameters be established? That is the risk the engineer must take.

This section provides basic ideas of data and graphical analysis as well as some tools that are available. Probability and statistics are not developed here, however a course in these subjects is recommended. The material in this section is to aid in the understanding and solution of problems involving solid and hazardous waste management.

The solution of many engineering problems involving large amounts of data can be made easy by determining single numbers that describe the unique characteristics of the data. Common measures are the average or arithmetic mean, the median, the mode, and the standard deviation (SD).

The mean or average is a popular measure and lends itself to further statistical manipulation. A mean is the sum of the observations divided by the number of observations. A gross error or even a few abnormal data points can have a pronounced effect on this value. An average is a single value that typifies the whole group, but it does not generally give adequate information about the distribution of observations within that group.

To avoid these problems, it is possible to describe the "center" of a set of data with other kinds of statistical descriptions. For example, the median is the value at the middle of data arranged in increasing or decreasing order. The term *middle observation* refers to the distance from extremes. If the number of observations is odd, the median is the middle value; for an even number of observations, the median is the average of two middle values. Both the mean and the median are rarely the same. These terms both describe the center of a set of data, but they describe it in different ways. The mean can be considered the "center of gravity" for the data, whereas the median divides the data so that half the items are equal to or greater than the median.

There is yet another measure of a data set as well: the mode. The mode is the item that occurs with the highest frequency.

Example 3–7

The refuse generated by a family of four, based on a 5-day sampling, is 20, 24, 19, 18, and 19 lbs/day. Calculate the mean, median, and mode.

Solution

The mean is

$$\overline{X} = \frac{\Sigma x}{n} \tag{3-9}$$

where $\overline{X}$ is the mean of observations, Σx is the sum of the observations, and n is the number of observations. Therefore,

$$\overline{X}_1 = \frac{100}{5} = 20 \; lbs$$

The median is the middle number when the observations are arranged in increasing or decreasing order. In this example, that order is 24, 20, 19, 19, and 18. Thus, 19 is the median.

The observation with the highest frequency in a data set is the mode. Thus, the mode in this example is 19.

The central tendency of observation sets and average calculations has been considered, but the average does not give sufficient information about the distribution of observations. For example, consider the following:

Sample 1 (weight in pounds): 20, 24, 19, 18, 19 (mean, 20 lbs)

Sample 2 (weight in pounds): 20, 28, 20, 14, 18 (mean, 20 lbs)

Despite having the same mean, the two sets of observations clearly differ appreciably in the scatter of values about their mean. Scatter may be important, because this knowledge may indicate the proportion of samples whose value may be critical.

Scatter

There are several measures of scatter or dispersion. One is the SD, which can be estimated from a sample as follows:

$$S = \sqrt{\frac{\sum_{i=1}^{n}(x_i - \bar{x})^2}{n-1}} \tag{3-10}$$

where x_i is the value of observation in which $i = 1, 2, 3, \ldots n$, $\bar{X}$ is the observed sample mean, n is the sample size, and S is the standard deviation.

To compare the scatter of the two samples, calculate the SD of each sample. Again, the samples are:

Sample 1			Sample 2		
Observation (x_i)	Mean $(\bar{X})$	$(x_i - \bar{X})^2$	Observation (x_i)	Mean $(\bar{X})$	$(x_i - \bar{X})^2$
20	20	$(20 - 20)^2$	20	20	$(20 - 20)^2$
24	20	$(24 - 20)^2$	28	20	$(28 - 20)^2$
19	20	$(19 - 20)^2$	20	20	$(20 - 20)^2$
18	20	$(18 - 20)^2$	14	20	$(14 - 20)^2$
19	20	$(19 - 20)^2$	18	20	$(18 - 20)^2$

Thus, using equation (3-10),

$$S = \sqrt{\frac{22}{4}} = 2.35 \; lbs$$

and

$$S = \sqrt{\frac{104}{4}} = 5.1 \; lbs$$

The SD of sample 2 is little more than twice the SD of sample 1, even though the means of both samples are equal. This is because the scatter of the observations in Sample 2 is greater than that in sample 1. This means that sample 2 has much more variable generation of waste than sample 1.

This example shows that the SD is a very important statistic, because it gives a measure of the dispersion of numbers from the mean value in a data set. In other words, the SD measures variation in terms of how much each number deviates from the mean value of the sample.

As early as the eighteenth century, scientists observed that plots of measurements showed a bell shape. Much of the statistical analysis in engineering is based on the bell-shaped curve, which is known as the normal curve. The assumption is that many types of data sets (e.g., mean annual discharge of a stream, annual rainfall) can be described by a normal curve, and that the location of the curve can be defined by the average (i.e., mean) and a measure of spread of the curve (i.e., how widely the data is distributed) by the SD. The bell-shaped curve is described by a mathematical function called the Gauss function. The distribution described by this function is commonly known as the normal distribution;

$$p(x) = \frac{1}{\sqrt{2\pi}\sigma} \exp\left[-\frac{1}{2}\left(\frac{x - \mu}{\sigma}\right)\right]^2 \tag{3-11}$$

where $p(x)$ is the probability density and represents the rate of change of probability with x; μ is the mean, estimated as $\bar{X}$; $\bar{X}$ is the observed sample mean $\left(\frac{\Sigma x}{n}\right)$; σ is the standard deviation, estimated as S; S is the observed standard deviation, or $\sqrt{\frac{\Sigma(x_i - \bar{x})^2}{n - 1}}$; and n is the sample size.

The SD is a measure of the curve's spread, and it is the value of x at which 68.26% of all values of x fall within plus and minus the mean μ (estimated by $\bar{X}$). As shown, a small SD means there is little variability in the data. In contrast, a large SD means the data are widely spread.

Another useful statistic in engineering and scientific work is the coefficient of variation, which is expressed by

$$V = \frac{\sigma}{\mu} \times 100 \tag{3-12}$$

and is estimated as

$$V = \frac{S}{\bar{x}} \tag{3-13}$$

In engineering works, data described by the normal curve are often plotted as a cumulative function, in which the vertical axis is the cumulative fraction of observations. A straight line on a probability paper implies that data are normally distributed and that statistics such as the mean and SD can be read off plot. These curves are used in hydrology, resource management, and several other technical data analyses. Example 3–8 shows the statistical calculations and graphical analysis.

Example 3–8

A waste management company plans to add an apartment complex to its solid waste collection route. Therefore, the company has collected data on the weekly amount of solid waste generated by the residents.

Week	Solid Waste (lbs)
1	2625
2	2125
3	2375
4	2875
5	2625
6	2875
7	3125
8	3375
9	3125
10	3125
11	3375

Answer the following questions:

a. Calculate the mean, median, and mode.

b. Plot a histogram.

c. Calculate the SD.

d. Plot the fractional cumulative frequency on a normal probability paper.

e. Estimate the mean and SD from the plot.

f. Based on an analysis of these limited data, what preliminary conclusions can be drawn by the waste management company?

Solution

First, arrange the data in increasing order of magnitude:

2125

2375

2625

2625

2875

2875

3125

3125

3125

3375

3375

a. Using equation (3-9), the mean is

$$\bar{x} = \frac{\Sigma x_i}{n} = \frac{31,625}{11} = 2,875 \; lbs$$

By definition, the median is the middle observation (distance from the extremes). Thus, the median is 2875 lbs.

The mode is defined as the observation that occurs most frequently. Thus, the mode is 3125 lbs.

b. A frequency distribution is represented by a histogram. The data in this example can be plotted as shown in Figure 3–1. From the plot of these limited data, an approximately normal distribution is noted.

c. Using equation (3-10), the SD is

$$\sqrt{\frac{\Sigma(x_i - \bar{x})^2}{n - 1}} = \sqrt{\frac{1,625,000}{10}} = 403 \; lbs$$

d. The plot of a fractional cumulative frequency on a probability paper is shown in Figure 3–2. The plot tends to be a straight line.

e. From the probability plot (Figure 3–2), the mean and SD can be read as approximately 2825 lbs and 465, respectively.

Note that the calculated mean in (a) and the SD in (c) are 2875 lbs and 403 lbs, respectively. The difference between these values and those from the graph results from scatter and error in plotting a line based on very few observations.

The SD can also be roughly approximated as follows:

$$S = \frac{2}{5} (x_{90} - x_{10})$$

$$= \frac{2}{5} (3125 - 2160)$$

$$= 386$$

This value is in the ballpark of the value calculated through equation (3-10).

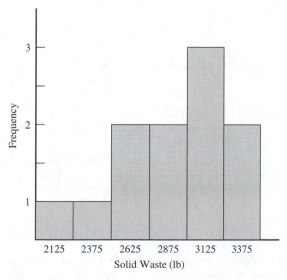

Figure 3–1   Histogram of a frequency distribution.

Figure 3–2 Probability paper plotting a fractional cumulative frequency.

f. Based on this analysis of limited data and information, the company can draw the following preliminary conclusions about the weight of solid waste generated by the residents of this complex:

The median weight of the waste is likely to be 2875 lbs. That is, over an extended period, half the time the weight may equal or exceed this value, and the other half it may not. In the long run, the residents may generate approximately 2875 ± 403 lbs per week (i.e., in the range of 2472 to 3278 lbs per week).

The waste management company can use this information to plan its collection services (e.g., containers, labor, collection vehicles) with 68-percent reliability over the long run. In other words, if the waste collection is done over, say, 100 weeks, it is likely that for 68 weeks the quantity of waste will be within the predicted range; for the remaining weeks, it probably will not be.

Return Period

The reciprocal of the probability is termed the *return period* or, sometimes, the *recurrence interval.* This period is how often a precipitation event of a certain magnitude is likely to recur. If the annual probability of an event occurring is, say, 10 percent, then that event can be expected to recur once in 10 years. In other words, the return period is 10 years. This can be stated mathematically as

$$T_R = \frac{1}{P} = \frac{1}{10} = 0.1$$

where T_R is the return period and p is the fractional probability.

Another way of stating this is that an event of this magnitude will, on average, occur 1 year in 10 years and has a probability of $0.1 \times 100 = 10\%$. It most certainly does not mean that an event of this magnitude will occur every 10 years.

REVIEW QUESTIONS

1. Write the units of measure for the following in both American and SI units:
   a. Force.
   b. Area.
   c. Pressure.
   d. Power.
   e. Density.

2. a. Define *concentration*.
   b. Determine the concentration of a chemical with a mass of 0.55 kg and a volume of 0.06 m³ that is put in a container with 1000 L of water.

3. Show by calculations that parts per million is equal to milligrams per liter.

4. An industrial landfill leachate has a concentration of 0.0038% solids at a flow rate of 1 million gallons per day. How many pounds of solids per day are discharged from the landfill?

5. a. What is the significance of the detention time of leachate in a pond?
   b. Calculate the retention time in a pond that has a volume of 1000 m³ and a flow rate of 3 m³/hour.

6. Assume you are a manager of a waste management company that wants to submit a proposal for its services to the city manager of Toledo. Discuss the data collection and analysis needed before submitting your proposal.

7. What is the significance of the mean and the SD in planning a waste collection route?

8. Find out from a town's public works or sanitation department the weekly amount of solid waste generated for the last year. Perform the statistical analysis, and interpret the results for use in designing a solid waste collection system.

Chapter 4   Hydraulics

Hydraulics is the study of the behavior of liquids, both at rest and in motion. This chapter presents the fundamental concepts of hydraulics, which are necessary to understand the wastewater flow. It can also serve as a primer for students who have not yet been exposed to this subject and as a review for students who have.

For practical purposes, water is considered to be an incompressible liquid with a unit weight of 62.4 lb/ft^3. In the SI system, the term *weight* (i.e., force due to gravity) is termed *newton* (N), and the unit weight of water is 9800 newtons per cubic meter (9800 N/m^3) or 9.8 kilonewtons per cubic meter (9.8 kN/m^3). The prefix *kilo-* stands for 1000.

PRESSURE

Whether standing still in a tank or flowing in a pipe, water exerts force against the walls of its container. Pressure is defined as a force per unit area (see Chapter 3). In equation form, this is expressed as

$$P = \frac{F}{A} \tag{4-1}$$

where P is pressure, F is force, and A is the area over which the force is exerted (i.e., distributed). In American units, pressure is expressed in terms of pounds per square inch (lb/in^2 or psi). In SI metric units, pressure is expressed as newtons per square meter N/m^2. The unit N/m^2 is called a pascal and is abbreviated as P_a. One kilopacal (kP$_a$) is frequently used in practical applications. To convert,

$$1 \text{ kP}_a = 1000 \text{ P}_a = 0.147 \text{ psi}$$

Hydrostatic Pressure

The pressure of water at rest is called hydraulic pressure. The important principles of hydraulic pressure are:

1. The pressure increases in direct proportion to the depth.
2. The pressure depends only on the depth of the water above the element or point in question, not on the water surface area.
3. The pressure in a continuous volume of liquid is the same among points at the same depth.
4. The magnitude of pressure at any point is the same in all directions.

58

Figure 4–1 shows two connected tanks. The water surface in both tanks is at the same elevation and at atmospheric pressure (i.e., barometric pressure). The pressure at point p_1 is the same as that at point p_2, because both points are at the same depth in the water. Similarly, the hydrostatic pressures at points p_1^1, p^{11}, and p_2^1 are equivalent, because they are at the same depth as well.

The pressure at point p_1^1 is greater than the pressure at point p_1. The pressure at point p_1 is greater than zero as well. Because the pressure at any point in a container is directly proportional to the depth, then the gauge pressure at point p_1^1 would be twice the pressure at point p_1 if point p_1^1 were twice as deep as point p_1. Pressure varies in direct proportion to the depth of the point, regardless of the volume or surface area of the water.

Atmospheric pressure at sea level is approximately 101 kPa, or 14.7 psi. The use of "conversion factors", whose physical significance is often lost, may be avoided by remembering that standard atmospheric pressure at sea level is 14.7 psi, 101.3 kPa absolute, 29.92 in. or 760 mm of mercury (32°F or 0°C), and 33.9 ft or 10.3 m of water (60°F or 15.6°C). This standard atmospheric pressure also allows pressures to be quoted in atmospheres; for example, a pressure of 147 psi or 1013 kPa could be stated as 10 atmospheres. In practical hydraulic problems, however, atmospheric pressure usually is neglected. Atmospheric pressure is considered to be a zero reference point, and the pressure is termed a *gauge pressure*. If the atmospheric pressure is taken into account, then the pressure is termed an *absolute pressure*.

A complete vacuum would have a pressure of absolute zero. Pressures greater than absolute zero but less than atmospheric pressure are termed *partial vacuums*. A partial vacuum has a negative sign when expressed in terms of gauge pressure; absolute pressures are always positive. For example, an absolute pressure of 10 psi (68.7 kPa) is equivalent to a gauge pressure of -4.7 psi (32.3 kPa). The quantitative relationship between gauge pressure and absolute pressure is given by equation (4-2) and is illustrated in Figure 4–2.

$$p_{absolute} = p_{atmospheric} + p_{gauge} \tag{4-2}$$

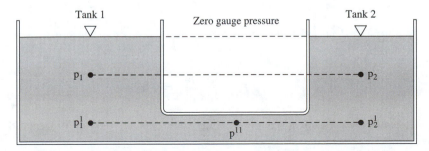

Figure 4–1   Hydrostatic pressure at various points in a tank.

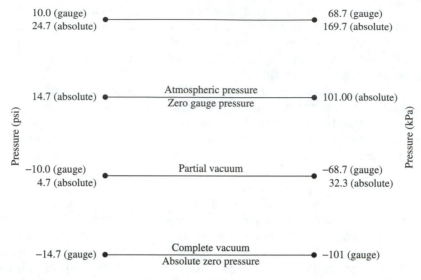

Figure 4–2   Relationship between gauge pressure and atmospheric pressure.

Computation of Pressure

The force that causes the pressure in a container is the weight of the water. Consider the container in Figure 4–3(a), which is in the shape of a cube and has an open top. If this container is filled with water and then placed on a level surface, pressure will be exerted against the bottom of the container. This pressure is called *hydrostatic pressure* and results from the weight of water above the bottom. The pressure at any point in the water depends on the depth of that point. In American units, the weight of 1 ft^3 of water is approximately 62.4 lbs. The area of the container bottom shown in Figure 4–3(a) is 1 ft^2, and from equation (4-1), the pressure at the bottom is:

$$P = \frac{F}{A} = \frac{62.4 \; lbs}{1.00 \; ft^2} \; or \; 62.4 \; lbs/ft^2$$

With a 1-ft depth of water on an area of 1 /ft^2 (144 in^2), the pressure at the bottom would be

$$P = \frac{62.4 \; \text{lbs}}{144 \; \text{in}^2} = 0.43 \; \text{psi}$$

Because the pressure at this point in the water depends only on its depth, the pressure can be represented as:

$$p = 0.43h \tag{4-3}$$

where p is the hydrostatic pressure and h is the depth in feet.

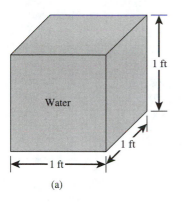
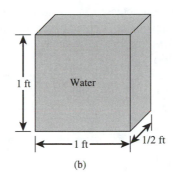
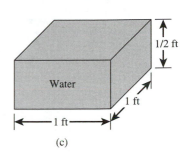

(a)   (b)   (c)

Figure 4–3   Example container dimensions.

In SI units, if the container is filled with water to a depth of 1 m, the volume of water would be 1 m^3, and its weight 9.8 kN. From equation (4-1), the pressure at the bottom of the tank is

$$P = \frac{F}{A} = \frac{9.8 \text{ kN}}{1 \text{ m}^2} = 9.8 \text{ kPa}$$

Following the reasoning given earlier,

$$p = 9.8h \qquad\qquad (4\text{-}4)$$

where p is hydrostatic pressure (kPa) and h is depth (m).

Now suppose the container in Figure 4–3(b), which has a bottom area of $\frac{1}{2}$ ft^2 (1 ft $\times$ $\frac{1}{2}$ ft), is filled with water. The weight of the water will then be half of 62.4 lbs, or 31.2 lb. From equation (4-1), the pressure at the bottom is

$$P = \frac{F}{A} = \frac{31.2 \text{ lbs}}{0.50 \text{ ft}^2} \text{ or } 62.4 \text{ lbs/ft}^2$$

which is the same as in the previous case.

Now suppose the original container is cut in half horizontally by moving the bottom halfway toward the top (Figure 4–3[c]). In this case, the weight of the water in the container will be 31.2 lbs and the area of the bottom 1.00 ft^2. Hence, the pressure at the bottom will be $\frac{31.2 \text{ lbs}}{1.00 \text{ ft}^2}$, or 31.2 lbs/ft^2.

From these examples, when the depth of the water in tank is 0.5 ft, the pressure clearly is 31.21 lbs/ft^2, and when the depth is 1.0 ft, the pressure is 62.4 lbs/ft^2. It can be stated that pressure varies linearly with depth, and that the pressure at any depth is equal to the product of the specific (or unit) weight of the water (or other liquid) and the depth. In equation form,

$$p = \gamma h \qquad\qquad (4\text{-}5)$$

where p is pressure, γ is the specific (or unit) weight of the liquid, and h is depth.

If the specific weight is expressed in pounds per cubic foot and the depth in feet, then pressure will be expressed in pounds per square foot (lb/ft^2). If the specific (or unit) weight is in newtons per cubic meter and the depth in meters, then pressure will be expressed in newtons per square meter (N/m^2) or pascals.

The relationship expressed by equation (4-5) is true regardless of the shape of the container or the specific liquid. The pressure determined by the equation results only from the weight of the liquid. Additional pressure may be present from other sources, however, such as air pressure applied to the surface of the liquid in an open container.

Example 4–1

A large tank is 25 ft in depth and holds leachate (i.e., wastewater) from a landfill. Calculate the pressure at the bottom of the tank and at a point 10 ft below the water surface.

Solution

From equation (4-5),

$$p = \gamma h$$

The specific weight of water is 62.4 lbs/ft^3, the depth of the liquid in the tank is 25 ft. Thus,

$$p = 62.4 \; lbs/ft^3 \times 25 \; ft = 1560 \; lbs/ft^2$$

or

$$1560 \; lbs/ft^2 \times \frac{1 \; ft^2}{144 \; in^2} = 10.8 \; psi$$

In SI units, 1 lb/ft^2 = 47.88 Pa (N/m^2), so p = 74,693 N/m^2 (Pa)
At a point 10 ft below the water surface,

$$p = 62.4 \; lbs/ft^3 \times 10 \; ft = 624 \; lbs/ft^2 = 4.3 \; psi$$

In SI units, p = 29,877 N/m^2 (Pa)

Pressure Head

It is also possible to express pressure as the height of a column of water (feet or meters) instead of as psi or kPa. This pressure head is the actual or equivalent height of water above a selected point.

Example 4–2

A closed tank containing leachate (Figure 4–4) is 3.5-ft deep. The leachate gives off gas, which accumulates above its surface. A pressure gauge at the bottom of the tanks reads 18 psi. Determine:

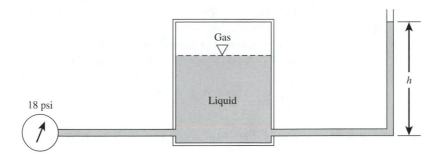

Figure 4–4 Leachate storage tank.

    a. The pressure head of leachate at the tank bottom.

    b. The height to which the leachate will rise in the vertical tube.

    c. The pressure of the trapped gas.

Solution

From equation (4-5),

$$h = \frac{p}{\gamma}$$

where p is hydrostatic pressure (psi), h is depth from the surface of the water (ft), and γ is the specific weight of water (lb/ft^3). Thus,

$$h = \frac{p\,\dfrac{lb}{in^2}}{62.4\,\dfrac{lb}{ft^3}} = \frac{p\,\dfrac{lb}{in^2}}{62.4\,\dfrac{lb}{ft^3} \times \dfrac{ft^2}{144\ in^2}} = \frac{p\,\dfrac{lb}{in^2}}{0.43\,\dfrac{lb}{ft \times in^2}} = \frac{p}{0.43}$$

    a. $h = \dfrac{p}{0.43} = \dfrac{18}{0.43} = 41.86\,ft.$

    b. The leachate liquid would rise 41.86 ft in the vertical tube, which is a height equal to pressure head at the tank bottom.

    c. If the tank were open to the atmosphere, then 3.5 ft of water depth would cause a pressure of 0.43 × 3.5 ft, or 1.5 psi, to register on the gauge. The difference, or 18 − 1.5 = 16.5 psi, must be exerted by the gas in the closed tank. Therefore, the gas pressure in the tank is 16.5 psi. (The pressure in a small volume of gas does not depend on the height or depth of the gas.)

Pressure measurements are made and recorded automatically with electromechanical instruments at pollution control facilities. A mercury manometer or a piezometer is used to measure pressures in the field. A mercury manometer is more practical, however, because mercury is 13.6 times as heavy as water. Therefore, the equivalent pressure head of water in the system is 13.6 times the measured height of the column of mercury. If the column of mercury is 12-in high (Figure 4–5), the pressure in the system is

$$p = 13.6 \times 0.43 \times \frac{12}{12} = 5.85 \text{ psi}$$

Application of Hydraulic Pressure for Mechanical Advantage

A hydraulic cylinder converts fluid pressure into mechanical advantage—or power—to perform useful work. The principle of this operation is based on liquids, for practical purposes, being incompressible. This incompressibility of liquids permits a piston to transfer power within a hydraulic system (e.g., a hydraulic actuator).

A simple type of linear actuator converts fluid energy into a mechanical force or motion (Figure 4–6). As fluid enters the port, it forces the piston to extend. In a dump-truck hoist, for example, the cylinder raises the loaded waste container. The piston later retracts under the weight of the box, some cylinders have a mechanical spring to retract the piston. Hydraulic fluid enters and returns through the same port.

FLOW

Water may flow in closed pipes (i.e., conduits) under pressure or in open channels under the force of gravity. A volume of water flowing past any given point in a pipe or channel per unit of time is called the *flow rate* or *discharge*.

In American units, the flow rate may be expressed as cubic feet per second (ft³/s or cfs) gallons per minute (gpm), or million gallons per day (mgd). To convert, 1 mgd = 1.55 cfs = 700 gpm.

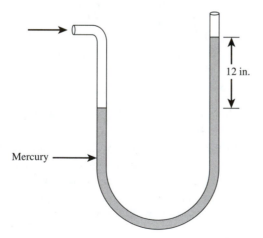

12 in.

Mercury

Figure 4–5   Mercury manometer for measuring pressure.

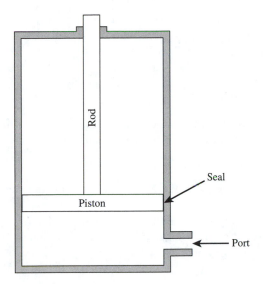

Figure 4–6   Linear actuator.

In SI units, the flow rate unit is cubic meters per second (m³/s). The term for a relatively small flow rate is liters per second (L/s). To convert,

$$1 \text{ m}^3 = 1000 \text{ L} = 35.315 \text{ ft}^3$$

Example 4–3

Convert a flow rate of 40 ft³/s to gallons per minute and million gallons per day.

Solution

Using a conversion of 1 ft³ = 7.48 gal, See Appendix B

$$40 \, ft^3/s \times 7.48 \, gal/ft^3 \times 60 \, s/min = 17,952 \, gal/min$$

Thus,

$$17,952 \, gal/min \times 60 \, min/hour \times 24 \, hours/day$$
$$= 25.85 \times 10^6 \, gal/day = 25.85 \, mgd$$

Alternate Solution

Because 1.55 cfs = 700 gpm = 1 mgd,

$$40 \, cfs = \frac{700 \, gpm \times 40 \, cfs}{1.55 \, cfs} = 18,065 \, gpm$$

and

$$40 \; cfs = \frac{1 \; mgd \times 40 \; cfs}{1.55 \; cfs} = 25.81 \; mgd$$

Flow Rate and Velocity

Flow rate represents the volume per unit of time, and velocity represents the distance per unit of time. The relationship is expressed by the formula

$$Q = A \times V \tag{4-6}$$

where Q is the flow rate or discharge, A is the cross-sectional flow area, and V is the velocity of flow.

In American units, V is usually expressed in terms of ft/s and A in terms of ft^2; thus, the units for Q would be ft^3/s. In SI units, V is expressed in terms of m/s and A in terms of m^2; thus, the units for Q would be m^3/s.

Example 4–4

Wastewater is flowing with a velocity of 3 ft/s in a 15-in-diameter pipe. The pipe is flowing full. Compute the flow rate in terms of cfs.

Solution

Because the pipe is flowing full, the flow area is equal to the cross-sectional area. The cross-sectional area of a round pipe is $A = \dfrac{\pi}{4} D^2$, where D is the diameter and π is approximately 3.14. Thus,

$$D = 15 \; in. \times \frac{1 \; ft}{12 \; in.} = 1.25 \; ft$$

$$A = \frac{\pi}{4} \times (1.25)^2 = 1.23 \; ft^2$$

Therefore, from equation (4-6),

$$Q = A \times V = 1.22 \; ft^2 \times 3 \; ft/s = 3.66 \; ft^3/s$$

Example 4–5

Calculate the diameter of a pipe that can carry a discharge of 0.6 m^3/s of wastewater at a velocity of 2 m/s.

Solution

From equation (4-6),

$$A = \frac{Q}{V}$$

$$A = \frac{0.6 \, m^3/s}{2 \, m/s} = 0.3 \, m^2$$

but $A = \frac{\pi}{4} D^2$, where D is the diameter. Therefore, rearranging the terms,

$$D = \left(\frac{4 \times A}{\pi}\right) = \left(\frac{4 \times 0.3}{\pi}\right)^{\frac{1}{2}} = 0.618 \, m$$

Continuity of Flow

The method for calculating the velocity of flow in a closed-pipe system is based on the principle of continuity. For practical purposes, water is considered to be an incompressible fluid; in other words, its volume changes little. The quantity of fluid flowing past any section in a given amount of time is constant, regardless of changes in flow or velocity. This is termed *steady flow*.

In Figure 4–7, flow rate Q_1 at section 1 must equal to flow rate Q_2 at section 2 if no water is added, stored, or removed between section 1 and section 2. Obviously, the flow velocity is higher at section 2 than at section 1. Because Q is constant and equal to AV, V must get larger when A gets smaller, such that $A \times V$ must always equal Q. Also, if the area of flow increases, the velocity of flow must decrease. The principal of flow continuity can be expressed as:

$$Q = A_1V_1 = A_2V_2 \tag{4-7}$$

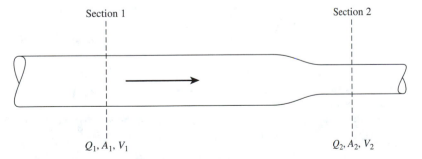

Section 1 Section 2

Q_1, A_1, V_1 Q_2, A_2, V_2

Figure 4–7   Constricted pipeline.

Example 4–6

In the pipeline shown in Figure 4–7, assume the area at section 1 is 0.6 m^2 and the area at section 2 is 0.4 m^2. For a flow rate of 1 m^3/s, compute the velocities at sections 1 and 2.

Solution

The following values are known:

Section 1	Section 2
$Q_1 = 1$ m^3/s	$Q_2 = 1$ m^3/s
$A_1 = 0.6$ m^2	$A_2 = 0.4$ m^2

Using equation (4–7),

$$Q = A_1 V_1 = A_2 V_2$$

$$1 \ m^3/s = 0.6 V_1 = 0.4 V_2$$

Therefore,

$$V_1 = \frac{1 \ m^3/s}{0.6 \ m^2} = 1.67 \ m/s \quad V_2 \frac{1 \ m^3/s}{0.4 \ m^2} = 2.5 \ m/s$$

Note that the velocity is inversely proportional to the area $\left(V \propto \dfrac{1}{A} \right)$ and, consequently, is also inversely proportional to the square of the diameter $\left(V \propto \dfrac{1}{d^2} \right)$.

Conservation of Energy

Energy can be neither created nor destroyed, but it can be converted from one form to another. This is called the law of conservation of energy. This law is very useful when solving problems involving flow of water.

Hydraulic systems involve three forms of mechanical energy: potential energy due to pressure, potential energy due to elevation, and kinetic energy due to velocity. Hydraulic energy is expressed in terms of energy head, in feet or meters of water. This is equivalent to foot-pounds per pound of water (ft-lb/lb = ft) or newton-meters per newton of water (N-m/N = m).

The total energy head in a hydraulic system is the sum of the energy heads (Figure 4–8) which is expressed mathematically as:

total head = elevation head + pressure head + velocity head

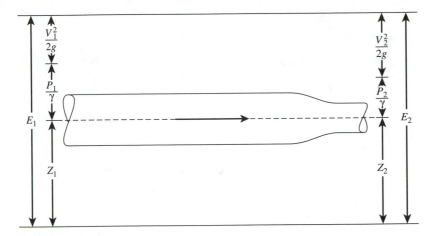

Figure 4–8   Total energy in a constricted pipe.

or

$$E = Z + \frac{p}{\gamma} + \frac{V^2}{2g} \tag{4-8}$$

where E is the total energy head, Z is the height of the water above a reference elevation (ft or m), p is the pressure (psi or kPa), γ is the unit weight of water (specific weight of water, 62.4 lbs/ft^3 or 9.8 kN/m^2), V is the flow velocity (ft/s or m/s), and g is the acceleration from gravity (32.2 ft/s^2 or 9.8 m/s^2).

Because the total energy in a closed system is constant, the total energy head at section 1, or E_1, must equal the total energy at section 2, or E_2. Setting $E_1 = E_2$ and using equation (4-8),

$$Z_1 + \frac{p_1}{\gamma} + \frac{V_1^2}{2g} = Z_2 + \frac{p_2}{\gamma} + \frac{V_2^2}{2g}$$

This equation is called Bernoulli's theorem. It applies to ideal fluids, because viscosity and energy loss because of friction are neglected.

If water is flowing in a horizontal pipeline, $Z_1 = Z_2$. In other words, elevation heads cancel out from both sides, leaving

$$\frac{p_1}{\gamma} + \frac{V_1^2}{2g} = \frac{p_2}{\gamma} + \frac{V_2^2}{2g} \tag{4-9}$$

From the principle of the continuity of flow, the velocity at section 2 must be greater than the velocity at section 1 because of the smaller flow area at section 2. This means the velocity head increases as water flows into the constricted section. Because the total energy must remain constant, however, the pressure head (and, therefore, the pressure) must drop.

Example 4–7

For the pipe system illustrated in Figure 4–7, assume the diameter is 12 in. at section 1 and is 6 in. at section 2. The flow rate through the pipe is 3.0 cfs, and pressure at section 1 is 60 psi. What is the pressure at Section 2?

Solution

First, determine the cross-sectional area (i.e., the flow area) at each section:

$$A_1 = \frac{\pi(1\ ft)^2}{4} = 0.785\ ft^2$$

$$A_2 = \frac{\pi(0.5\ ft)^2}{4} = 0.196\ ft^2$$

Because $V = \dfrac{Q}{A}$, we get

$$V_1 = \frac{3\ ft^3/s}{0.785\ ft^2} = 3.82\ ft/s$$

$$V_2 = \frac{3.0\ ft^3/s}{0.196\ ft^2} = 15.30\ ft/s$$

Applying equation (4-9),

$$\frac{60\ lbs/in^2 \times \dfrac{144\ in^2}{ft^2}}{62.4\ lbs/ft^3} + \frac{3.82\ (ft/s)^2}{2 \times 32.2\ ft/s^2} = \frac{p_2 \times 144\ in^2}{62.4\ lbs/ft^3} + \frac{15.30\ (ft/s)^2}{2 \times 32.2\ ft/s^2}$$

$$138.46 + 0.227 = p_2 \times 2.31 + 3.63$$

$$p_2 = 58.47\ psi$$

Example 4–8

For the pipe system shown in Figure 4–7, the diameter is 0.30 m at section 1 and is 0.15 m at section 2. The flow rate through the pipe is 0.085 m³/s, and the pressure at section 1 is 408 kPa ($408 \times 10^3\ N/m^2$). What is the pressure in the constriction at section 2?

Solution

Compute the flow area at each section:

$$A_1 = \frac{\pi}{4} (0.30 \, m)^2 = 0.07 \, m^2$$

$$A_2 = \frac{\pi}{4} (0.15 \, m)^2 = 0.018 \, m^2$$

Because $V = \frac{Q}{A}$,

$$V_1 = \frac{0.085 \, m^3/s}{0.07 \, m^2} = 1.21 \, m/s$$

$$V_2 = \frac{0.085 \, m^3/s}{0.018 \, m^2} = 4.7 \, m/s$$

Applying equation (4-9),

$$\frac{408 \times 10^3 \, N/m^2}{9800 \, N/m^3} + \frac{(1.21 \, m/s)^2}{2 \times 9.8 \, m/s^2} = \frac{p_2 \, N/m^2}{9800 \, N/m^3} + \frac{(4.7 \, m/s)^2}{2 \times 9.8 \, m/s^2}$$

$$41.63 + 0.0747 = p_2 \times 1.02 \times 10^{-4} + 1.28$$

$$41.63 \, m + 0.0747 \, m = p_2 \times 1.02 \times 10^{-4} \, m + 1.13 \, m$$

$$p_2 = 397.8 \times 10^3 \, N/m^2$$

Flow in Pipes Under Pressure

When water flows in a pipe, there is resistance to that flow because of friction between the flowing water and the pipe wall and between the layers of water moving at different velocities in the pipe due to the viscosity of water. Accordingly, the velocity of water in the pipe varies from almost zero at the wall to maximum along the center line. Therefore, as used in fluid flows, the term *velocity* generally indicates mean velocity unless otherwise specified. Loss of energy in the system results from the frictional resistance to flow, and this loss of energy causes a continuous drop in pressure as the water flows along its path.

Figure 4–9 illustrates a straight pipe attached to a tank full of water. There is no flow in the system; therefore, there is no pressure loss when the valve in the pipe is closed. When the valve is opened, however, flow starts, and the associated energy loss because of friction occurs. This loss can be determined by measuring the pressures along the pipeline. A line connecting the water surface in the tank with the water levels at sections 1 and 2 shows the continuous pressure loss along the pipe. This line is called hydraulic grade line (HGL) or hydraulic gradient.

The HGL indicates the pressure head along the pipe above the center line, and it has a downward gradient unless a pump adds additional energy to the flow in the pipe. The vertical drop in the HGL between two sections (i.e., sections 1 and 2) separated by a distance L is called the head loss (h_L). The ratio $\frac{h_L}{L}$ is the slope S of the HGL, or

$$S = \frac{h_L}{L}$$

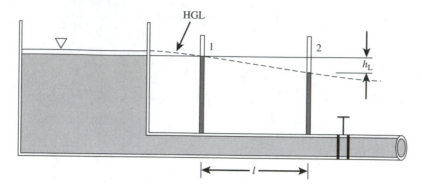

Figure 4–9   Hydraulic grade line (HGL) of water discharging through a pipe.

The HGL passes through the free-water surface of a tank, and this water-surface elevation is the system's pressure head at that point. The greater the flow rate in the pipe, the greater is the head loss and the steeper the slope.

Several equations can be used to solve water flow problems. The most common pipe formula used in the design and evaluation of water distribution systems is the Hazen-Williams equation:

$$V = 1.318CR^{0.63}S^{0.54} \qquad \text{(4-10)}$$

where V is the velocity ft/s; C is the coefficient, which depends on the material and the age of conduit (Table 4–1); R is the hydraulic radius, which is the cross-sectional area divided by the wetted perimeter (ft); and S is the slope of the hydraulic gradient (ft/ft).

By combining the Hazen-Williams formula with equation (4-6) and then substituting the value of $\dfrac{D}{4}$ for R, the formula for the quantity of flow in circular pipes flowing full is

$$Q = 0.281CD^{2.63}S^{0.54} \qquad \text{(4-11)}$$

where Q is the quantity of flow (gpm), C is the coefficient, D is the diameter of the pipe in (in.), and S is the hydraulic gradient (ft/ft).

The nomograph shown in Figure 4–10 is for a C of 100, and it solves equation (4-11) for a 20-year-old, unlined ductile pipe. Given any two of the parameters

Table 4–1   Value of Coefficient C for the Hazen-Williams equation

Pipe Material	C
Asbestos cement	140
Iron	
New, lined	130
20-year-old unlined	100
Concrete	130
Plastic	140–150
Steel	120

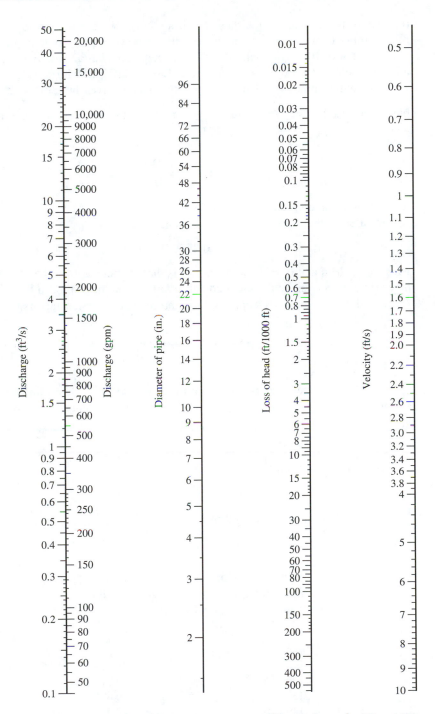

Figure 4–10   Nomogram for the Hazen-Williams formula (C = 100).

(e.g., velocity, discharge, pipe diameter, head loss), the remaining two can be determined from the intersections along a straight line drawn across the nomograph. For example, a flow of 1000 gpm within a 12-in.-diameter pipe has a velocity of 2.8 ft/s, with a head loss of 4 ft/1000ft (0.004 ft/ft). Head losses in pipes with C values other than 100 can be determined by using the correction factors given in Table 4–2. For example, if the head loss at $C = 100$ is $\dfrac{4\,ft}{1000\,ft}$, the head loss for a plastic pipe ($C = 140$) would be

$$0.54 \times 4 = \frac{2.1\,ft}{1000\,ft}$$

where 0.54 is the correction factor K, as explained in Table 4–2.

In SI units, the Hazen-Williams equation is

$$Q = 0.278CD^{2.63}S^{0.54} \tag{4-12}$$

where Q is the quantity of flow (m^3/s), C is the coefficient, D is the diameter of the pipes (m), and S is the hydraulic gradient (m/m). The nomograph can be used to solve for pipe flow in SI units by applying the appropriate conversion factors.

Example 4–9

A 14-in.-diameter pipe is carrying a flow of 1200 gpm. Determine the head loss for (a) $C = 100$ and (b) $C = 140$ in American and SI units.

Solution

From Figure 4–9,

   a. $V = 2.92$ ft/s, and $h_L = 3.5$ ft/1000 ft

Table 4–2   Correction Factors to Determine Head Losses From Figure 4 –10 at Values of C Other Than 100[a]

C	K
100	1
110	0.84
120	0.71
130	0.62
140	0.54

[a] Corrected $h_L = K \times h_L$ at C = 100.

b. $V = 2.92$ ft/s, and using K from Table 4–1,

$$h_L = 0.54 \times 3.5 = 1.89 \text{ ft}/1000 \text{ ft}$$

Example 4–10

A 0.3-m-diameter pipe carries wastewater with a head loss of 1.2 m per 100 m of pipeline. Determine the flow rate in the pipe using (a) equation (4-12) and (b) the nomograph.

Solution

Compute the slope:

$$S = \frac{h_L}{L} = \frac{1.2 \ m}{100 \ m} = 0.012$$

Thus,

$$Q = 0.278 \times 100 \times (0.3)^{2.63} \times (0.012)^{0.54}$$
$$= 0.106 \text{ m}^3/\text{s}$$

Gravity Flow in Pipes

When water flows in a pipe or channel with the water surface exposed to the atmosphere, it is called open-channel or gravity flow. Gravity provides the moving energy, and friction resists motion, thus producing energy loss. Almost all sanitary and storm-water sewers are designed to flow as open channels; stream flow is an example of an open-channel flow. The velocity of flow depends on the steepness of the pipe slope and the frictional resistance.

Manning's formula (equation [4–13]) is used for uniform and steady flow. Uniform flow means that the slope of the water surface and the cross-sectional flow area are constant, and steady flow means that the discharge is constant with time. The coefficient of roughness n depends on the condition of the pipe surface and joints. In a natural streambed, n depends on topographic features and can range from 0.01 for a smooth clay pipe to 0.1 for a small natural stream. A common value of n adopted for sewer design is 0.013. Certain materials (e.g., plastic) in a new pipe may exhibit a lower n, but after pipes are placed in use, they accumulate grease and other solids that change the interior.

Manning's equation can be written as

$$Q = \frac{1.49}{n} AR^{2/3} S^{1/2} \tag{4-13}$$

where Q is the quantity of flow (ft^3/s); n is the coefficient of roughness, which depends on the inside of the pipe; A is the cross-sectional area of flow (ft^2); R is the hydraulic radius, which is the cross-sectional area divided by the wetted perimeter (ft); and S is the slope of the hydraulic gradient (ft/ft).

In SI units, the equation is

$$Q = \frac{1.00}{n} AR^{2/3}S^{1/2} \tag{4-14}$$

where Q is the quantity of flow (m^3/s), n is the coefficient of roughness, A is the cross-sectional area of flow (m^2), R is the hydraulic radius (m), and S is the slope of the hydraulic gradient (m/m).

The nomograph in Figure 4–11 can be used to solve equation (4-13) for circular pipes flowing full based on a roughness coefficient (n) of 0.013. Given any two of the parameters (e.g., quantity of flow, diameter of pipe, slope of pipe, or velocity), the other two can be determined from the intersections along a straight line drawn across the nomograph.

Example 4–11

A 1-m-wide, rectangular channel lined with concrete drains a landfill area into a pond. The slope of the channel is 1%, and the water depth is 0.5 m. Assume n to be 0.013. What is the flow-carrying capacity of the channel flowing full?

Solution

$$A = 1 \text{ m} \times 0.5 \text{ m} = 0.5 \text{ m}^2$$
$$P = 0.5 \text{ m} + 1 + 0.5 \text{ m} = 2 \text{ m}$$
$$R = \frac{A}{P} = \frac{0.5 \text{ m}^2}{2 \text{ m}} = 0.25 \text{ m}$$
$$S = 0.01$$

Thus, from equation (4-14),

$$Q = \frac{1.0}{0.013} \times 0.5 \times (0.25)^{2/3}(0.01)^{1/2} = 1.52 \text{ } m^3/s$$

Example 4–12

An 8-in.-diameter pipeline is built on a slope of 1%. Assuming n to equal 0.013, determine the discharge capacity of the pipeline with full flow, and determine the flow velocity.

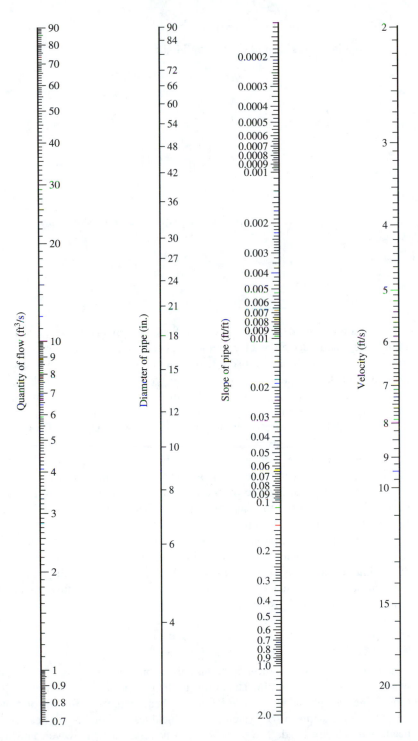

Figure 4–11   Nomogram for the Manning formula (*n* = 0.013).

Solution

Using Manning's formula (equation [4–13]),

$$A = \frac{\pi}{4}(0.67)^2 = 0.353\,ft^2$$

$$R = \frac{A}{2\pi r}$$

$$R = \frac{0.353}{2.10} = 0.168\,ft$$

$$S = (0.01)$$

$$Q = \frac{1.49}{0.013} \times 0.353 \times (0.168)^{2/3} \times (0.01)^{1/2} = 1.22\,ft^3/s$$

$$V = \frac{Q}{A} = \frac{1.22}{0.353} = 3.46\,ft/s$$

PUMPS

A pump is a common mechanical device that adds energy to water or other liquids. Pumps are usually driven by electric motors, but if no electric power is available, gasoline or diesel engines can be used. Pumps are classified according to the mechanical principles of their operation. The two basic types are: positive-displacement pumps, and centrifugal pumps. A positive-displacement pump delivers a fixed quantity of water with each revolution of the pump piston or rotor.

Centrifugal pumps are commonly used in wastewater and water systems. A centrifugal pump adds energy to the water by accelerating it through a fast-rotating impeller. The water moves outward from the vanes, and it passes through a spiral casing. As the velocity drops in the expanding spiral casing, the kinetic energy is converted into a pressure head. Leachate from a landfill is generally collected in a sump, and a centrifugal pump is used to discharge it into a sewer or tanker truck for transportation to a wastewater treatment plant.

The rate at which a pump adds energy to water or some other liquid is the power output of the pump, which is also called the water power. A pump (or any other piece of machinery) can never operate at 100% efficiency. Therefore, the power delivered to the water (P_0) is less than the power put into the pump (P_1). Thus, the efficiency of a pump can be expressed as

$$\frac{P_0}{P_1} \times 100 \qquad\qquad (4\text{-}15)$$

In American units, power is commonly expressed in terms of horsepower (hp), where 1 hp = 550 ft‑lb/min. The power to the pump shaft is called brake horsepower (bhp). The energy added is the total energy head expressed in feet of water (or any other liquid) flowing in the system, which includes the total static head, total friction head, and velocity head. The velocity head is very small, however, and it is usually neglected. The vertical distance between the free-water surfaces on the suction side and on the discharge side of the pump is called the static head (Figure 4–12). The friction head, as explained previously, is the energy lost because of friction between the pipe's inside surface and the flowing water.

Expressing the specific weight (γ) of water in lb/ft^3 and the volume flow rate (Q) in ft^3/s would yield the weight flow rate γQ in lb/s. The power equation thus is

$$P_0 = h_o \gamma Q \qquad \text{(4-16)}$$

Power would be expressed in ft-lb/s or ft-lb/min if Q is expressed in ft^3/min. Horsepower can be converted to kilowatts through multiplying by 0.746 (1 hp = 0.746 kW).

In SI units, power is expressed in terms of kilowatts (kW). Water power (P_0) can be computed with the following formula:

$$P_0 = 9.8 \times Q \times h_0 \qquad \text{(4-17)}$$

where Q is the pump discharge (volume flow rate, m^3/s), h_0 is the total energy head (m), and 9.8 is the unit weight of water kN/m^3.

Power has the units kN-m/s. In the SI system, one kN-m/s is called a kilowatt.

Example 4–13

A leachate collection system operates on a pumping rate of 3 ft^3/m. It includes 1000 ft of 6-in.-diameter pipe ($C = 100$), and the total static head (i.e., distance between suction and delivery) is 15 ft. Compute the required break horsepower (P_1) of the pump if it has an efficiency of 70%.

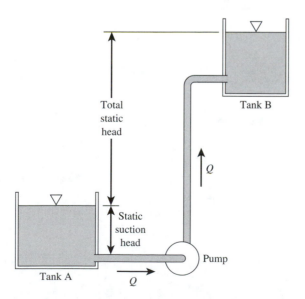

Figure 4–12   Pumping to raise the water level.

Solution

First, determine the total energy head on the pump. The Hazen-Williams nomograph does not go down to a flow of 3 ft^3/m, because this is a very little flow. Consequently, the head loss is also negligible. This fact can be checked by the Hazen-Williams formula (equation [4-11]):

$$Q = 0.281CD^{2.63}S^{0.54}$$

$$= 3\ ft^3/m \times 7.48\ gal/ft^3 = 0.281 \times 100 \times (6)^{2.63}S^{0.54}$$

$$S = 0.0000025\ \text{ft/ft}$$

Thus, for a 1000-ft-long pipe, the head loss is

$$h_L = S \times L = 0.0000025 \times 1000 = 0.0025\ \text{ft}$$

which is very negligible.

The total energy head equals the static head plus the friction head, or 15 ft + 0 = 15.0 ft. Therefore, applying equation (4-16),

$$P_0 = h_0\gamma Q$$

for

$$h_0 = 15\ \text{ft}$$

$$\gamma = 62.4\ \text{lb/ft}^3\ (\text{specific weight of water})$$

$$Q = 3\ \text{ft}^3/m$$

it follows that

$$P_0 = 2,808\ ft\text{-}lb/m$$

and

$$1\ hp = 550\ ft\text{-}lb/m$$

then

$$P_0 = 5.10\ \text{hp}$$

Therefore,

$$P_1(break\ horsepower) = \frac{P_0}{efficiency} \times 100 = \frac{5.24}{70} \times 100 = 7.28\ hp$$

REVIEW QUESTIONS

1. Why is the study of hydraulics relevant to solid and hazardous waste?
2. Define *hydrostatic pressure*.
3. What is the unit weight of water in the American system and the SI system?
4. a. Calculate pressure at the base of a container full of water. The container is 1-m high and has an area of 0.5 m^2.
   b. What would be the pressure at the base of the container if it were 1-m high and had a base area of 1.5 m^2?
5. A closed tank containing leachate is 4 ft deep. The gas from the leachate accumulates in the tank, and a gauge connected to the tank bottom reads 22 psi. Calculate the pressure head of leachate at the tank bottom and the pressure exerted by the trapped gas.
6. Convert the following flow rates:
   a. 10 ft^3/s to gpm and mgd
   b. 10 ft^3/s to 1 m^3/s.
   c. 1 m^3/s to gpm.
7. What is the relationship between flow rate, cross-sectional area, and velocity of flow?
8. Wastewater flows at the rate of 2 m^3/s in a pipe at a velocity of 0.2 m/s. Find the diameter of the pipe.

9. Find the velocities V_1 and V_2 for a pipe section with the following parameters:

$$Q = 5 \text{ ft}^3/\text{s}$$
$$d_1 = 1.0 \text{ ft}$$
$$d_2 = 0.5 \text{ ft}$$

10. Write the equation for the total energy head in a hydraulic system, and explain each term.
11. Find the pressure p_2 in the pipe section given in problem 9 if p_1 is 10 psi.
12. a. Sketch a hydraulic grade line for a flow from a reservoir through a pipe.
    b. Indicate the head loss on the sketch.
13. A 1.0-m-diameter pipe carries wastewater with a head loss of 0.75 m per 100 m of pipeline. Determine the flow rate in the pipe ($C = 100$) using:
    a. The Hazen-Williams equation.
    b. The nomograph.

Chapter 5 Hydrology

This chapter presents fundamental hydrologic concepts relevant to solid and hazardous waste technology. Hydrology is a branch of earth science concerned with the distribution and movement of water both above and below the surface of the earth. To understand the distribution and movement of water, one must have knowledge of the hydrologic cycle.

THE HYDROLOGIC CYCLE

The global system that supplies and removes water from the earth's surface is termed the *hydrologic cycle* (Figure 5–1). Water is in constant motion, such as evaporating or changing into a vapor and moving into the atmosphere. Energy is provided by the sun and by gravity for this circulation of water and water vapor. This natural phenomenon is what the hydrologic cycle represents.

The science of hydrology requires much use of mathematics and statistics. The objective is to measure and analyze the relationships that account for the transformation of water from one phase to another and for its motion from one location to another. Understanding these relationships is very helpful in designing and locating an environmental control system, including a waste landfill.

The major transport mechanisms of water in the hydrologic cycle are precipitation, infiltration (and percolation), surface runoff, and evaporation and transpiration. Because this is a "cycle," a discussion can start anywhere, but it is most appropriate to begin with precipitation.

Precipitation results when moisture in the clouds comes down, according to weather conditions, as rain, snow, ice, sleet, or hail. The annual precipitation for various regions of the United States can be obtained from the U.S. Weather Bureau and from local radio and television stations. As precipitation reaches the ground, it is intercepted by vegetation and depressions through what is called depression storage or ponding. Part of the precipitation that falls on the soil may infiltrate. Infiltration is a process in which precipitation enters under the ground through the surface soil. After satisfaction of the infiltration capacity, depression storage, and interception does surface runoff begins, because any further precipitation has nowhere to go. Runoff eventually flows into lakes, wetlands, rivers, and finally, oceans. Some of the precipitation that infiltrates the ground moves gradually downward because of the pull of gravity and percolates through the unsaturated zone of the earth's crust. The water that remains in the unsaturated zone becomes water in the vadose zone, which is the region from the saturated zone (i.e., groundwater) to the surface. The extra water that cannot be held in the vadose zone

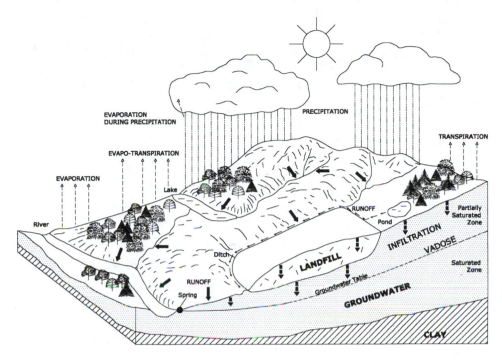

Figure 5–1 The hydrologic cycle.

(i.e., the unsaturated zone) percolates (i.e., continues its movement through pore spaces because of gravity) farther downward into the saturated zone and then forms the ground-water zone.

After reaching the confined aquifer, the water continues to move downward into what are known as recharge areas, because that is where groundwater recharge occurs. The water is in a constant state of flow, and it moves slowly through the soil because of the frictional resistance to flow in the soil and rock openings. The water table is rarely level; it generally follows the shape of the ground surface. The groundwater flows in the downhill direction of the sloping water table and sometimes seeps out into springs, lakes, or streams.

Groundwater is a dependable, long-term water resource, and it is not subject to short-term fluctuations in precipitation. Once an aquifer becomes polluted, however, it takes a long time to flush those contaminants from the system. The direction and velocity of the groundwater flow are important in understanding groundwater contamination.

Some groundwater rises to the surface to form springs and become part of lakes, swamps, rivers, oceans, and so on. From these surface bodies of water, water then evaporates. Water also evaporates from the intercepted precipitation on the wet leaves of plants and trees, buildings, grounds, and so on. In addition, water is lost to the atmosphere through transpiration, which is a process by which plants remove moisture from the subsurface by their roots and then release it to the atmosphere. Because it is sometimes difficult to separate surface evaporation from plant transpiration, these two are frequently combined as one term: *evapotranspiration*. These waters finally evaporate to form clouds, which complete the cycle by again forming precipitation when atmospheric moisture (i.e., water vapor) is cooled and condensed into water droplets.

THE HYDROLOGIC CYCLE FOR A LANDFILL

The hydrologic cycle shown in Figure 5–1 also illustrates the fate of precipitation falling on a landfill. From the previous discussion, it is clear that four things can happen to this precipitation:

1. It can become surface runoff.
2. It may be retained on the surface and evaporate.
3. It can infiltrate the surface layer of the cover material and be removed by plant transpiration.
4. It can infiltrate the waste buried in the landfill, where it may eventually become leachate.

Infiltration

When water is standing on the cover (i.e., surface) of the landfill, it may percolate into the soil cover through gravity. The rate of movement is usually very slow, and in the process, the soil becomes saturated. The rate of percolation or infiltration is determined by the characteristics of the soil covering the site and the vegetation planted on this cover. The flow can be approximated using Darcy's law:

$$Q = KIA \tag{5-1}$$

where Q is the quantity of flow per unit of time; (ft^3/day or cm^3/s), K is the permeability or hydraulic conductivity, (ft/day or cm/s), I is the hydraulic gradient or slope of the water table (ft/ft or cm/cm), and A is the cross-sectional area through which the flow occurs, (ft^2 or cm^2).

Permeability (i.e., hydraulic conductivity) is defined as the flow rate of water through a column of soil at 60°F at a hydraulic gradient of 1 ft/ft. The soil in the cover can significantly affect the amount of water that can infiltrate the landfill. More than one type of soil is required for the landfill cover, because the top layer must absorb some water to maintain a good vegetative cover. (The design of landfill covers is presented in Chapter 15.)

Soil permeability is important in controlling infiltration. A coarse, uniform sand is very permeable (i.e., passes water easily) and would be suitable for a drainage layer, either in the cover or the leachate collection system. On the other hand, clay has very low permeability. A better way to appreciate the significance of the cover material is to compute the time required to penetrate a layer of clay.

Example 5–1

Compute the time required for water to penetrate a layer of clay that is 60-cm deep when exposed to a hydraulic gradient of 1 cm/cm. The permeability K is 1×10^{-8} cm/s.

Solution

The *Depth penetration* over 1 year is

$$(1 \times 10^{-8} \, cm/s) \times 3600 \, s/hour \times 24 \, hours/day \times 365 \, days/year = 0.314 \, cm/year$$

The number of years it would take for the water to penetrate 60 *cm* is

$$\frac{60 \, cm}{0.314 \, cm/year} = 191 \, years$$

With increased depth and a very low-permeability soil, it would take a very long time for any water to pass through the layer. The flow rate increases, however, if there is a static head of water (i.e., standing water) on that layer. This may happen with a liner or a cover if surface grades are not maintained (e.g., a depression where water collects).

Example 5–2

For the previous example, compute the number of years required to penetrate the 60-cm moisture barrier if the static head of water is 20 cm.

Solution

The *hydraulic gradient* is

$$\frac{60 \, cm + 20 \, cm}{60 \, cm} = 1.33 \, cm/cm$$

Thus, the number of years required to penetrate the layer is

$$\frac{1 \, cm/cm}{1.33 \, cm/cm} \times \frac{60 \, cm}{0.314 \, cm/year} = 144 \, years$$

This example shows that a 20-cm static head of water atop a 60-cm moisture barrier layer would reduce the time to penetrate the layer from 191 to 144 years. In other words, the standing water would reduce the time to penetrate the barrier layer in proportion to the depth of the standing water.

Evapotranspiration

The combined process of evaporation and transpiration is called evapotranspiration. It is often defined as the water loss that occurs from a surface fully covered with vegetation if adequate water exists in the soil for that vegetation. Evapotranspiration is a function of solar radiation, differences in vapor pressure between a given water layer and the overlying layer, temperature, wind, atmospheric pressure, and type of vegetation.

During precipitation, the ground surface and the vegetation both receive some moisture. This moisture evaporates, however, soon after the precipitation ends. Soil cover captures the water that infiltrates, and that water then moves down through the soil. If the soil cover has sufficient capacity to retain moisture and is deep enough, most of the water from this precipitation will be held in the cover. Once the storm is over, however, evapotranspiration is usually the main process by which a cover soil loses water and, thus, is a significant factor in leachate generation. Most of the moisture is used by the vegetation on the cover. This moisture is extracted from the soil, and it is returned to the atmosphere by transpiration.

The quantity of water removed by the vegetation (i.e., plants, bushes, grass, and so on) depends on the plant types and the climatic conditions. Plants cannot extract water if the cover material is dry, and a proper selection of plants is essential. Neither deep-rooted nor shallow-rooted plants are suitable for a landfill cover. Deep-rooted plants such as trees will not do well, because the roots will not penetrate the solid waste because of its low pH and high level of dissolved solids. Shallow-rooted plants are limited to the moisture in only the top layer of the soil. Evapotranspiration rates depend on the weather, increasing during warm summer months with good solar radiation and longer daylight hours and decreasing during months with cool temperatures and shorter daylight hours. Wind and low humidity also increase the water loss.

A mix of various types of grasses and broadleaf plants (e.g., clover) can extract moisture by transpiration. Possible transpiration quantities are:

Clover	18–24 in./year
Rye grass	18–24 in./year
Meadow grass	22–60 in./year

Surface Water Runoff

Surface runoff develops after the initial demands of interception, infiltration, and surface storage have been satisfied. Surface runoff quantities are a function of several phenomena, including surface slope, antecedent moisture conditions, vegetative cover, and impervious soil. If the soil is impervious and the surface is sloped for drainage, a greater percentage of precipitation will be surface runoff. This runoff can be estimated by the formula

$$Q = CIA \qquad (5\text{-}2)$$

where Q is the rate of runoff (ft^3/s), C is a dimensionless runoff coefficient, I is the rainfall intensity (in./hour), and A is the drainage basin area (acres).

In American units, rainfall intensity is expressed in terms of inches per hour, and area is expressed in terms of acres. The resulting dimensions for Q become in.-acre/hour, with 1 in.-acre/hour being almost equal to 1 ft^3/s. Therefore, Q is expressed in terms of ft^3/sec (or cfs).

In SI units, rainfall intensity is usually expressed in terms of millimeters per hour. To obtain a correct value of Q, convert millimeters per hour to meters per hour, and then express area A in terms of square meters. It is important to use the appropriate units for I and A.

Table 5–1 presents C values and the total amount of runoff that can be expected for different surface conditions with a rainfall intensity of 1 in./hour for a 6-hour period. The total volume of water as a result of rainfall is

$$\frac{6 \text{ in.}}{12 \text{ in.}} \times 1 \text{ ft} \times 43{,}560 \text{ ft}^2/\text{acre} \times 7.48 \text{ gal/ft}^3 = 162{,}914 \text{ gal/acre}$$

Thus, after comparing this result with the information in Table 5–1, it is clear that the runoff from a tight clay soil with cover and a slope of more than 5% would be approximately 55 percent of the precipitation and approximately 70 percent with no cover. A certain amount of runoff is an important factor in the control of water infiltrating into the waste.

Infiltration into Solid Waste

Water infiltrating a landfill will first be absorbed by solid waste material if that material is below field capacity, which is the upper limit of a porous material's ability to absorb water. Porosity is the percentage of solid waste volume that is occupied by voids or pore spaces. When the waste becomes saturated, water then moves through the waste

Table 5–1   Surface Runoff Potential from Land Areas

Cover (Slope)	Runoff (gal/acre)		
	Sandy Loam	Silt Loam	Tight Clay
Grass			
(0–5%) Flat	16,291 ($C = 0.1$)	48,874 ($C = 0.3$)	65,165 ($C = 0.4$)
(5–10%) Rolling	20,066 ($C = 0.16$)	58,649 ($C = 0.36$)	89,602 ($C = 0.55$)
None			
(0–5%) Flat	48,874 ($C = 0.3$)	81,457 ($C = 0.5$)	97,748 ($C = 0.6$)
(5–10%) Rolling	65,165 ($C = 0.4$)	97,748 ($C = 0.6$)	114,040 ($C = 0.7$)

via gravity, and this water will become contaminated with waste constituents. Leachate quantity depends heavily on precipitation (which is difficult to predict) and the squeezing out of pore liquid in the landfill waste. Decomposition of putrescible waste mass can also release water/liquid, though in negligible amounts.

Example 5–3

Calculate the surface runoff for a 10.0-acre landfill for a 10-year, 6-hour Kansas City storm (i.e., the probability of such a storm is once in 10 years). The land has a good cover surface of sandy loam, vegetative cover, and a top slope of less than 5%.

Solution

The rainfall intensity from the U.S. Weather Bureau's rainfall map is

$$I = 0.75 \text{ in./hour}$$

Then find the appropriate coefficient from Table 5–1:

$$C = 0.1$$

Finally, apply equation (5–2), or

$$Q = CIA$$

and calculate for Q:

$$Q = 0.1 \times 0.75 \text{ in./hour} \times 10 \text{ acres} = 0.75 \text{ in./acre/hour} = 0.75 \text{ ft}^3/s$$

REVIEW QUESTIONS

1. a. What are the similarities between the global hydrologic cycle and a hydrologic cycle of a landfill?
   b. List the components of a landfill hydrologic cycle and their significance.
2. Calculate the infiltration rate for a soil layer with a 100-cm^2 cross-sectional area, a standing water depth of 25 cm, and a hydraulic conductivity of 10^{-7} cm/sec.
3. Which type of soil is preferred for drainage in a landfill cover?

4. What is the relationship between evaporation, transpiration, and weather?
5. What type of vegetative cover is suitable on a landfill cover in the Arizona desert?
6. a. List the factors that affect the surface runoff.
   b. Calculate the surface runoff from a 25-acre landfill with a vegetative cover over sandy loam and a 7-percent slope. Intensity of rainfall is 0.5 in./hour.

Chapter 6 Basic Scientific Concepts of Solid and Hazardous Waste Management

This chapter reviews the basic concepts of chemistry, physics, and biology. This review will be helpful in understanding solid and hazardous waste management technology, because these basic concepts are relevant to the subject.

CHEMISTRY AND PHYSICS

All matter is composed of basic substances called elements. Each element differs in weight, size, and chemical properties. There are more than 100 known elements, and some of the more common elements, along with their symbols and atomic weights, are given in Table 6–1. Symbols for elements are used in writing chemical formulas and equations.

The smallest part of an element is called an atom, and atoms of one element unite with those of another in a definite ratio, which is defined by their valence. Valence is the combining power of an element as compared with that of the hydrogen atom, which has a value of 1. Thus, an element with valence of 2^+ can replace two hydrogen atoms or, in the case of 2^-, can react with two hydrogen atoms. Sodium has a valence of 1^+; and chlorine has a valence of 1^-; therefore, one sodium atom combines with one chlorine atom to form sodium chloride (NaCl). Carbon, which has a valence of 4^-, can combine with four hydrogen atoms to form methane gas (CH_4).

Certain groupings of atoms act together, as a unit, in a number of different molecules. These are called radicals and are given names such as the hydroxyl group (OH^-). Radicals join with other elements to form compounds, and chemistry is concerned with how the elements react and combine with each other to do so.

The smallest part of a chemical compound that still retains the properties of that compound is called a molecule. Molecules are represented using combinations of the symbols for the atoms in that molecule. For example, one molecule of water is composed of two atoms of hydrogen (H) and one atom of oxygen (O); therefore, it's chemical formula is H_2O. The subscript 2 for H indicates two atoms of hydrogen. The formula for methane, which is a landfill gas, is CH_4, thus indicating that one atom of carbon and four atoms of hydrogen combine to form one molecule of this gas.

There are thousands of known compounds, and these compounds are separated into two broad groups; organic compounds, and inorganic compounds. Organic compounds are typically complex molecules of carbon in combination with other elements,

Table 6–1   Basic Information on Common Elements

Name	Symbol	Atomic Weight	Common Valence
Aluminum	Al	27.0	3+
Arsenic	As	74.9	3+
Barium	Ba	137.3	2+
Boron	B	10.8	3+
Bromine	Br	79.9	1−
Cadmium	Cd	112.4	2+
Calcium	Ca	40.1	2+
Carbon	C	12.0	4−
Chlorine	Cl	35.5	1−
Chromium	Cr	52.0	3+
			6+
Copper	Cu	63.5	2+
Fluorine	F	9.0	1−
Hydrogen	H	1.0	1+
Iodine	I	126.9	1−
Iron	Fe	55.8	2+
			3+
Lead	Pb	207.2	2+
Magnesium	Mg	24.3	2+
Manganese	Mn	54.9	2+
			4+
			7+
Mercury	Hg	200.6	2+
Nickel	Ni	58.7	2+
Nitrogen	N	14.0	3−
			5+
Oxygen	O	16.0	2−
Phosphorus	P	31.0	5+
Potassium	K	39.1	1+
Selenium	Se	79.0	6+
Silicon	Si	28.1	4+
Silver	Ag	107.9	1+
Sodium	Na	23.0	1+
Sulphur	S	32.1	2−
Zinc	Zn	65.4	2+

such as hydrogen and oxygen (e.g., CH_4). Inorganic compounds generally do not contain carbon; however, there are some exceptions. Organic compounds are closely associated with living organisms. Inorganic compounds are related to inanimate objects.

Stoichiometry

Stoichiometry is the balancing of chemical equations so that the same number of each kind of atom appears on each side of the equation. It is also the calculation that can be used to determine the amount of each chemical compound involved.

The first step is to balance the equation. For example, when acitic acid (CH_3COOH) is oxidized, carbon dioxide and water are formed through microbial action, as indicated by the following reaction:

$$CH_3COOH + O_2 \rightleftarrows CO_2 + H_2O \qquad \text{(6-1)}$$

Equation (6–1) is not balanced, however, because four hydrogen atoms and four oxygen atoms are on the left and only two hydrogen atoms and three oxygen atoms are on the right. The balanced equation is

$$CH_3COOH + 2O_2 \rightleftarrows 2CO_2 + 2H_2O \qquad \text{(6-2)}$$

Thus, two molecules of oxygen are required to oxidize the acid in carbon dioxide and water. To describe this reaction in terms of the mass of each substance (i.e., how many grams of oxygen are required to react with how many grams of acitic acid), knowledge regarding the mass of atoms and molecules is needed.

The atomic weight of an atom is the mass of that atom as measured in atomic mass units (amu). One amu is defined to be one-twelfth of a carbon atom. The sixth element on the periodic chart, carbon, has an atomic number of 6. The carbon atom contains six neutrons, six protons, and six electrons, and its molecular weight is 12.01115 g. The molecular weight is the sum of the atomic weights of all the constituent subatomic particles. Mass is expressed in moles, which is the mass of a substance divided by its molecular weight. Usually, mass is expressed in grams, so the moles are expressed as g-moles (or lb-moles if mass is expressed in pounds). One g-mole contains 6.02×10^{23} molecules. Expressed mathematically,

$$\text{moles} = \frac{\text{mass}}{\text{molecular weight}} \qquad \text{(6-3)}$$

Now, consider the equation (6-2) again:

$$CH_3COOH + 2O_2 \rightleftarrows 2CO_2 + 2H_2O$$

In this reaction, one acitic acid molecule is oxidized by two oxygen molecules to produce two carbon dioxide molecules and two water molecules. On a larger measure, one mole of acid is oxidized by two moles of oxygen to produce two moles of carbon dioxide and two moles of water.

To express this reaction in grams, first find the number of grams per mole for each substance. Using Table 6–1, the relevant atomic weights are 12 for carbon, 1 for hydrogen, and 16 for oxygen. These values have been rounded slightly, which is a common engineering practice (as discussed in Chapter 3). Thus, molecular weight and, hence, the number of grams per mole are summarized to express the reaction of equation (6-2) or follows:

1 molecule of CH_3COOH + 2 molecules of O_2
$\rightleftarrows$ 2 molecules of CO_2 + 2 molecules of H_2O

or

1 mol of CH_3COOH + 2 mol of O_2 $\rightleftarrows$ 2 mol of CO_2 + 2 mol of H_2O

or

$$60 \; g \; of \; CH_3COOH + 64 \; g \; of \; O_2 \rightleftarrows 88 \; g \; of \; CO_2 + 36 \; g \; of \; H_2O$$

Solutions

A solution is a uniform mixture of two or more substances that are present in a single phase (i.e., as a gas, a liquid, or a solid). A solution of carbon monoxide (CO) dissolved or mixed in the air is a gaseous solution. Solutions in water are called aqueous solutions and are very common in wastewater (e.g., leachate) treatment and disposal. An example of a solid solution is carbon (C) and manganese (Mn) dissolved in iron to form steel.

The largest amount of substance present in a solution is called the solvent; the substances present in smaller amounts are called the solutes. Various chemical changes and reactions occur in solution. An aqueous solution of salt, for example, has the solute (i.e., salt) evenly mixed in water; it does not settle out to the bottom. If one continues adding more salt to the solution, the salt molecules eventually will no longer dissolve. At this point, the solution becomes saturated. Both temperature and the substances already present in the water have a significant effect on the saturation point (i.e., the amount of solute that a solution can hold before becoming saturated). Solid substances (e.g., salt, lime, and so on) are more soluble in warm water than in cold water.

Some substances can dissolve in water without reaching a saturation point. Alcohol, for example, can be mixed with water without limit. Obviously, if there is more alcohol than water, the alcohol is considered to be the solvent (by definition) and the water to be the solute. The solubility of various gases such as carbon dioxide, oxygen, methane, and so on is of interest in both water and wastewater. As with solids, the solubility of gases depends on temperature, but the relationship is the opposite. In other words, the solubility of a gas decreases with increasing temperature. Also, at higher altitudes, the solubility of gas is lower than it is at sea level because of the lower atmospheric pressures at such elevations.

The solubility of oxygen is of great importance for water quality, because dissolved oxygen in water (i.e., aerobic condition) is necessary not only for the survival of beneficial aquatic life in surface waters but also for the treatment of wastewater such as leachate from a landfill (see Chapter 16). Table C–1 in Appendix C lists typical saturation values of dissolved oxygen in fresh water, which illustrates the effect of temperature on solubility.

Suspensions

From the previous discussion, a dissolved substance (i.e., solute) is one that is truly in solution and has the characteristic of its particles not settling, even during a long period of quiescent conditions. The substance is homogeneously dispersed in the liquid, and it cannot be removed without a phase change (e.g., distillation, precipitation, or adsorption). In distillation, either the solute (substance) or the solvent (liquid) itself is changed from a liquid phase to a gas phase to achieve separation. In precipitation, the dissolved substance reacts with another chemical that is added to form a solid phase and, thus, separate the substance from the liquid. Adsorption also involves a phase change in that the dissolved substance reacts with a solid particle to form a solid particle–substance complex. Physical methods such as sedimentation, filtration, or centrifrigation, however, cannot remove dissolved substances.

Large, suspended particles can be separated from the water under quiescent conditions, if given time, by settling. Suspended solids can be removed from water by physical methods such as sedimentation, filtration, and centrifugation. Particles as small as 1 μm (0.001 mm) are likely to settle out eventually; particles smaller than 1 μm are difficult to remove.

Very fine particles less than approximately 0.1 μm are too small to settle out by gravity or to be removed by usual filters. These particles are called colloids, and they occur in liquid as well as in the air. They are in a solid state and can be removed from a liquid by physical means such as filtration through membranes with very small pore spaces. Colloidal particles may have either all positive or all negative charges. Because like charges repel each other, there is a force of repulsion between the colloids; thus, they do not stick together to form larger and heavier particles. This is one reason why colloids are stable and do not settle out of suspension. Certain chemicals can be added to neutralize or reduce the effect of colloidal charges, and by agitation of the liquid, the particles then collide and form larger particles, or flocs, that can be removed from the water and other liquids by sedimentation or filtration.

Concentrations of Solutions and Suspensions.   The properties of solutions and suspensions mainly depend on their concentrations. A concentrated or strong solution has a relatively large amount of solute and a different characteristic than a dilute or weak solution, which has a relatively small amount of solute. Because the solute concentrations are expressed quantitatively instead of qualitatively, weight percent and milligrams per liter are the terms usually used. (Calculations and examples were given in Chapter 3.)

Ionization

When inorganic compounds are placed in water, they dissociate into electrically charged particles called ions, and the process is called ionization. Previously, the chemical bond between Na^+ and Cl^- and the formation of $NaCl$ was described. This compound ionizes in water, as shown by the equation

$$NaCl \rightarrow Na^+ + Cl^- \qquad (6\text{-}4)$$

In other cases, molecules dissociate into charged particles composed of groups of atoms that act together as a unit. These are called radicals, as described previously. For example, a water molecule can dissociate in a hydrogen ion (H^+) and a hydroxyl radical (OH^-) as follows:

$$H_2O \rightleftarrows H^+ + OH^- \qquad (6\text{-}5)$$

When mixed in water, calcium carbonate ($CaCO_3$) will ionize into a Ca^{2+} ion and a CO_3^{2-} radical:

$$CaCO_3 \rightleftarrows Ca^{2+} + CO_3^{2-} \qquad (6\text{-}6)$$

Acid-Base Reactions and pH Levels

A substance that increases the concentration of the hydrogen ions (H^+) in an aqueous solution is called an acid. A base is a substance that causes the concentration of hydroxyls (OH^-) to increase in a solution. Acids and bases may be strong or weak depending on the relative increase in the concentrations of H^+ or OH^-.

Acid-base reactions are a particular type of ionization when H^+ is added or removed from a solution. For example, hydrochloric acid dissociates in water, thus forming H^+ and Cl^- (i.e., chlorine) ions. The reaction can be shown as

$$HCl \rightleftarrows H^+ + Cl^- \tag{6-7}$$

Thus, the hydrogen ion is added to water, and its concentration is increased. A hydrogen ion could also be removed from water by the addition of a base (e.g., NaOH). NaOH dissociates into Na^+ and OH^-, thus increasing the OH^- concentration. A substance like NaOH, which is basic, is also called an alkaline substance.

A basic solution is formed by adding an alkali such as NaOH to water. If both an acid and a base are put in the same water, H^+ combines with OH^- to form water, and if equivalent amounts are added, they neutralize each other, thus forming a salt solution. A solution is said to be acidic if the concentration of H^+ is greater than OH^-, neutral if it is equal, and basic if H^+ is less than OH^-. For neutral solutions, $H^+ = OH^- = 10^{-7}$ moles; the concentration of H^+ is greater than 10^{-7} moles for acidic solutions.

In mathematical terms, a convenient expression for the hydrogen ion concentration is pH, which is given by

$$pH = -\log H^+ \tag{6-8}$$

In a neutral solution (e.g., pure water), the numerical value of the hydrogen ion concentration is 10^{-7}. The logarithm (i.e., exponent) is -7, and the negative of that is 7. Therefore, a neutral solution has a pH of 7 at 25°C. An acidic solution has a pH less than 7, and a basic solution has a pH greater than 7. Figure 6–1 illustrates the idea of relative strength of acidic or basic solutions.

The pH scale is based on logarithms of base 10. Thus, each unit change represents a tenfold change in the degree of acidity or alkalinity of a solution. For example, a solution with a pH of 6 is 10 times more acidic than pure water (pH = 7).

Oxidation and Reduction Reactions

Oxidation and reduction reactions involve changes in valence and the transfer of electrons. Oxidation is the addition of oxygen or the removal of electrons; reduction is the removal of oxygen or the addition of electrons. A classic oxidation reaction is the release of electrons when iron metal corrodes:

$$Fe^0 \rightleftarrows Fe^{2+} + 2e^- \tag{6-9}$$

Soluble ferrous iron (Fe^{2+}) can be removed from a solution by an oxidation or reduction reaction.

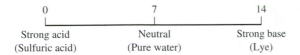

Figure 6–1   Relative strength of acidic and basic solutions.

Organic Compounds

All organic compounds contain carbon atoms, which are connected to each other in chain or ring structures with the other elements attached. These compounds make up much of the hazardous waste that is produced. A basic knowledge regarding the properties and characteristics of organic compounds is essential in the management of hazardous wastes.

Organics are derived from nature (e.g., plant fibers, animal tissues) or produced by synthetic reactions (e.g., rubber, plastics) and fermentation processes (e.g., alcohols, acids). Unlike inorganic compounds, organic substances usually have high molecular weight, have low water solubility, and are usually combustible.

Because most chemical bonds in organic molecules are covalent (i.e., electrons are shared between atoms), these bonds are typically represented by dashes connecting atoms in the molecule. Atoms in a dash structure must be bonded by at least two electrons. A single bond is represented by one dash, a double bond by two parallel dashes, and a triple bond by three parallel dashes. A carbon atom in an organic molecule can have as many as four bonds with other atoms.

Organic compounds that contain only carbon and hydrogen are called hydrocarbons, which are generally divided into two categories: aromatic, and aliphatic. Aromatics are ring or multiple-ring structures containing alternating single and double bonds. Aliphatic compounds are those containing open carbon chains or rings with or without alternating single and double bonds.

The parent compound of aromatic hydrocarbons is benzene (C_6H_6). It is the simplest ring hydrocarbon. The simplest hydrocarbon in aliphatic category is methane (CH_4). Figure 6–2 presents schematic diagrams showing the molecular structure of benzene and methane. Methane is a gas at ordinary temperature and pressure, and it is produced naturally when other organic compounds, such as those found in sewage sludge or municipal solid waste, decompose in an anaerobic environment.

Alcohols are formed from hydrocarbons by replacing one or more hydrogen atoms by hydroxyl groups (OH). Methanol is manufactured synthetically by a catalytic process, and is used extensively in manufacturing organic compounds such as solvents and fuel additives (Figure 6–2).

Partition Coefficients

Partition coefficients are empirical constants that describe how a chemical distributes between two media. Three of these coefficients are important in waste management: the octanol–water partition coefficient, the soil–water partition coefficient, and the vapor–liquid partition coefficient.

Name	Chemical formula	Molecular structure
Methane	CH_4	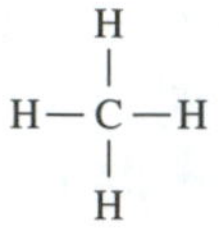
Benzene	C_6H_6	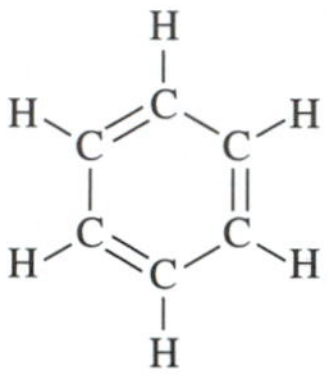
Methanol	CH_3OH	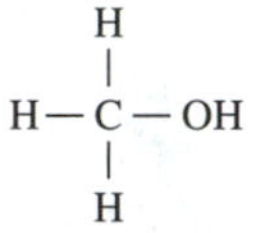

Figure 6–2   Molecular structures of methane, benzene, and methanol.

A partition coefficient is a dimensionless constant that is defined by

$$K = \frac{C_o}{C_w} \tag{6-10}$$

where C_o is the concentration in Octanol (mg/L or μg/L) and C_w is the concentration in water (mg/L or μg/l).

The octanol–water coefficient measures how an organic compound will partition between an organic phase (e.g., chemical taken up by aquatic organisms) and water. Values of K range from 10^{-3} to 10^7. Chemicals with low values of K (<10) are mostly hydrophilic (i.e., "water-loving" or difficult to separate from water) and have both low soil adsorption and a low bioconcentration factor (BCF). The soil–water partition coefficient measures the tendency of a chemical to be adsorbed by soil or sediment. The vapor–liquid partition coefficient measures the ratio of a compound's concentration in the vapor to that in the liquid at equilibrium.

Gases

Knowledge regarding the behavior of gases and gaseous mixtures under various environmental conditions is necessary for solving problems of air, water, and land pollution. For example, the anaerobic decay of organics in a landfill produces CH_4, CO_2, and H_2S, which are, respectively, energy rich and explosive, corrosive, and poisonous gases. Methane (CH_4) from such organic decomposition can cause fires and explosions. In addition, the dissolution of gases in liquids (e.g., air or oxygen for the treatment of wastewater) and the removal of undesirable gases are relevant to the treatment and disposal of solid as well as hazardous wastes.

The Ideal Gas Law.   The ideal gas law states that the product of the absolute pressure and the volume of a gas is proportional to the product of the mass and absolute temperature of that gas. In equation form, this is usually written as

$$PV = nRT \tag{6-11}$$

where P is the absolute pressure (atm), V is the volume (L), n is the mass (mol), T is the absolute temperature (°K), and R is the proportionality constant, which is also called the gas constant or the ideal gas constant.

Equation (6-11) is the simple form of the general equation of state that applies to real gases. The numerical evaluation of R can be obtained from the experimental fact that 1 g/mol of any ideal gas at standard conditions of 0°C (273.15°K) and 101,325 Pa occupies a volume of 22.414 L. Therefore,

$$R = \frac{PV}{nT} = \frac{101{,}325 \; N/m^2 \times (22.414 \times 10^{-3} \; m^3)}{1 \; g \; mol \times 273.15°K} = 8.31 \; N\text{-}m/°K\text{-}mol$$

If P is in atmospheres and V is in liters, then
$R = 0.0821$ L-atm/°K-mol. Other values of R in different systems of units can be obtained from various handbooks.

Example 6–1

In a landfill, anaerobic microorganisms decompose organic matter into methane and carbon dioxide. Estimate the volume of gas produced (at atmospheric pressure and 25°C) from the decomposition of 1 mol of organic matter represented by the formula $C_{16.8}H_{27.0}O_{12.1}N_{0.24}$. Complete decomposition can be described by
$$C_{16.8}H_{27.0}O_{12.1}N_{0.24} + 4.17 \; H_2O \rightarrow 8.65 \; CH_4 + 8.14 \; CO_2 + 0.24 \; NH_3$$

Solution

Each mole of organic matter produces 8.65 mol of CH_4, 8.14 mol of CO_2, and 0.24 mol of NH_3. Thus, from equation (6-11),

$$V = \frac{nRT}{P} = \frac{17.03 \; (0.0821 \; L\text{-}atm/°K\text{-}mol) \times 298°K}{1 \; atm}$$

$$= 416.65 \; or \; 417 \; L$$

Note that the volume of 1 mol of any gas is the same. In other words,

$$0°C = 273°K$$
$$25°C = (273 + 25)°K = 298°K$$

Thus,

$$T_k = 25 + 273 = 298°K$$

Dalton's Law of Partial Pressures.   Dalton's law of partial pressures states that the total pressure for a mixture of several gases is the sum of the partial pressures of the individual gases. In equation form,

$$P_{total} = P_1 + P_2 + P_3 + \ldots = \Sigma P_i$$

where P_i is the partial pressure that gas i would exert if it filled the total volume alone or if

$$P_{total} = \sum_{i=1}^{n} P_i \tag{6-12}$$

Assume that gas A contains n_A moles, gas B contains n_B moles, and gas C contains n_C moles, with a total volume V at temperature T. The partial pressures of three gases are then given by

$$P_A = \frac{n_A RT}{V} = \frac{n_A}{n_{total}} \times P_{total} \tag{6-13}$$

$$P_B = \frac{n_B RT}{V} = \frac{n_B}{n_{total}} \times P_{total} \tag{6-14}$$

$$P_C = \frac{n_C RT}{V} = \frac{n_C}{n_{total}} \times P_{total} \tag{6-15}$$

and the pressure by

$$P_{total} = P_A + P_B + P_C = (n_A + n_B + n_C) \times \frac{RT}{V} = n_{total}\frac{RT}{V} \tag{6-16}$$

From example 6–1,

$$P_A = \frac{8.65 \; mol \times 0.0821 \; L\text{-}atm/°K\text{-}mol \times 298°K}{417 \, L} = 0.508 \; atm$$

$$P_B = \frac{8.14 \; mol \times 0.0821 \; L\text{-}atm/°K\text{-}mol \times 298°K}{417 \, L} = 0.478 \; atm$$

$$P_C = \frac{0.24 \; mol \times 0.0821 \; L\text{-}atm/°K^0\text{-}mol \times 298°K}{417 \, L} = 0.014 \; atm$$

$$P_{total} = 1.0 \; atm$$

Mass Balance

A very familiar scientific concept is that matter can be neither created nor destroyed, but that it can be changed in form. This concept is called a materials balance, or a mass balance, and it is based on the law of conservation of energy. It also serves as a basis for describing and analyzing scientific and technical problems. The general concept of materials balance can be illustrated by equations where accumulation, input, and output refer to the mass quantities of substances either accumulating in the system or, alternatively, coming into or going out of the system.

Applied to a closed system, the first, very basic equation is:

$$\text{input} = \text{output} \qquad \textbf{(6-17)}$$

or simply put, "What comes in must go out." If material accumulates within the system, then

$$\text{accumulation} = \text{input} - \text{output} \qquad \textbf{(6-18)}$$

If material is produced or consumed within the system, the general case can then be expressed as

$$\begin{aligned}\text{(rate of) accumulation} = \;&\text{(rate of) input} - \text{(rate of) output}\\ &+ \text{(rate of) production} - \text{(rate of) consumption}\end{aligned} \qquad \textbf{(6-19)}$$

where (rate of) is used to mean per unit of time.

In the mass balance approach, a sketch or flow diagram of the environmental subsystem is used to analyze and solve the problem. All the known data are calculated and indicated in the same mass units on the diagram as inputs, accumulations, and outputs. The appropriate system boundaries are then selected for the material balance (S) to be determined.

Example (6–2)

In an average week, a family of five purchases and brings into their home 100 kg of consumer goods (e.g., food, magazines, newspapers, furniture, packaging material, junk mail). Of those 100 kg, 40 percent is consumed as food. Approximately 50% of that food is used for body maintenance and released as carbon dioxide; the remainder is discharged to the sewer system. Approximately 2 kg of consumer goods accumulate in the house. The family recycles approximately 30 percent of the solid waste generated (their state mandates a minimum of 25 percent). Estimate the amount of solid waste this family places at the curb each week for collection.

Solution

First, sketch a material flow diagram

Output 1 = body maintenance

↑ 20 kg

Input = 100 kg → | Accumulation 2 kg | → Output 3 = recycle
→ Output 4 = solid waste
for collection

↓ 20 kg

Output 2 = sewer system

Next, write the mass balance equation:

input = accumulation + (output 1 + output 2 + output 3 + output 4)

Third, calculate the known outputs:

Food input = 0.4 (100 kg) = 40 kg

One half of food is used for body maintenance = output 1 = 20 kg

One half of food is discharged into sewer = output 2 = 20 kg

The amount recycled is 30 percent after food and accumulation are taken into account = 0.30 (100 − 20 − 20 − 2) = output 3 = 17.4 kg

Now, determine the solid waste (i.e., output 4) to be collected:

$$\begin{aligned} \text{output 4} &= \text{input} - \text{output 1} - \text{output 2} - \text{output 3} - \text{accumulation} \\ &= 100 - 20 - 20 - 17.4 - 2 \\ &= 40.6 \text{ kg} \end{aligned}$$

Reactors

The tanks used to perform physical, chemical, and biochemical reactions are called reactors, and they are classified based on their flow characteristics and their mixing conditions. In environmental engineering, there are essentially three types of reactors:

1. Batch reactors, in which the materials are added to the tank at the desired conditions and mixed for sufficient time to allow the reaction to occur. The contents are then discharged. The longer the reaction time, the more complete the conversion.

2. Continuously stirred tank reactors, which are also called completely mixed flow reactors, in which the reactants are "continuously" fed to the reactor. Ideally, the contents are uniform throughout the tank, and being well mixed, the concentration of the effluent is the same as the the concentration in the tank. Increasing the detention time in the tank increases the conversion. This reactor is commonly used in wastewater treatment.

3. Plug flow reactors, in which fluid particles pass through the tank in sequence. Those that enter first also leave first. Input is fed at one end of the long reactor, and products are discharged at the other end. Ideally, there is no mixing in the lateral direction. Because the distance traveled along the length of the reactor is a function of time, the extent of conversion depends on that length. This reactor is commonly used in activated sludge systems.

For time-dependent reactions, the time that a fluid particle remains in the reactor affects the degree of completion of the reaction. In ideal reactors, the time in the reactor (i.e., the detention time or flow-through time) is obtained from the following equation:

$$t = \frac{V}{Q} \tag{6-20}$$

where t is the detention time, V is the volume of liquid in the ideal reactor, and Q is the volumetric flow rate.

Real reactors do not behave as ideal reactors do, however, because of turbulence and dead zones within the reactor, short-circuiting as a result of inlet and outlet conditions, and density differences caused by temperature.

MICROBIOLOGY

Microbiology is the study of microorganisms and their activities. Environmental microbiology concerns itself with the microorganisms commonly found in water, wastewater, air, soil, and even in other forms of life, including humans. Many microorganisms perform a useful function; others are pathogenic or disease-causing organisms. Microorganisms play a significant role in the natural environment and in engineered systems.

Microorganisms are divided into two groups: prokaryotic organisms, and eukaryotic organisms. Prokaryotes are single-cell organisms (e.g., bacteria). Eukaryotes may be single cell or multicellular (e.g., algae, fungi, protozoa, rotifers).

Bacteria are the most important and dominant group of organisms in biological wastewater treatment and disposal systems. Figure 6–3 shows the bacteria that are typically found. For this purpose, bacteria are classified as aerobic (i.e., requiring oxygen) or anaerobic (i.e., not requiring oxygen). The growth of bacteria is linked to water, because approximately 80 percent of a bacterial cell consists of water. It is through water that these cells receive their food in dissolved form. Bacteria reproduce through cellular division, which is a process called binary fission.

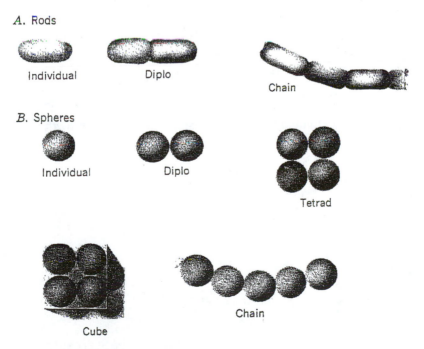

A. Rods

Individual   Diplo   Chain

B. Spheres

Individual   Diplo   Tetrad

Cube   Chain

Figure 6–3   Bacterial forms.

Many factors affect bacterial growth. Bacteria need the proper nutrients, which include fats, proteins, carbohydrates, and trace minerals. Other physical factors are temperature, the gaseous environment, and the pH. Bacteria can be grouped according to the temperature range within which they grow. Psychrophiles grow well at temperatures of 0°C to 20°C, mesophiles from 30°C to 40°C, and thermophiles 40°C or higher. The optimal temperature for most environmental engineered systems is in the range of 30°C to 40°C (Figure 6–4).

The important gases that are directly associated in bacterial growth are oxygen for aerobic biological oxidation and carbon dioxide as a carbon source. Methane is produced during the anaerobic decomposition of municipal solid waste in a landfill.

As mentioned, another important factor in bacterial growth is the pH. Most bacteria exhibit optimum growth in the range of 6.5 to 7.5, with extreme limits for growth between 4.0 and 10.0 (Figure 6–5).

Most of the bacterial processes used in environmental engineering are either aerobic or anaerobic. An aerobic process requires free dissolved oxygen in water, whereas an anaerobic process does not require free dissolved oxygen. The general equation for aerobic process is

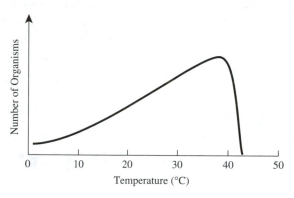

Figure 6–4   Optimum temperature range for bacterial growth.

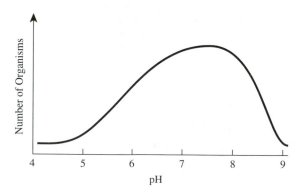

Figure 6–5   Optimum pH range for bacterial growth.

$$\text{organic matter} + HO_2 + \text{aerobes} \rightarrow \text{new cells}$$
$$+ CO_2 + H_2O + NH_4 + NO_3^- + NO_2^- \qquad \textbf{(6-21)}$$

The corresponding equation of the anaerobic process is

$$\text{organic matter} + H_2O + \text{nutrients} + \text{anaerobes} \rightarrow \text{new cells}$$
$$+ CH_4 + CO_2 + NH_3 + H_2S + \text{heat} \qquad \textbf{(6-22)}$$

Facultative bacteria can grow with or without oxygen.

Algae are plantlike, chlorophyll-containing organisms. Most algae are microscopic, but some are large. All algal cells contain photosynthetic pigments; thus, photosynthesis occurs in the presence of sunlight. During periods of light, algae use carbon dioxide and produce oxygen. During periods of darkness, however, no photosynthesis occurs, and algae consume oxygen and produce carbon dioxide.

Algae are important to environmental engineering. They cause taste and odor problems in drinking water supplied from lakes, and they cause problems such as suspended matter or excessive oxygen demand in lagoon wastewater treatment or storage systems. Excessive growth of algae (e.g., algal blooms in lakes) results from nutrients such as phosphorus and nitrogen compounds. Most of the nutrient input comes from wastewater treatment plants and surface runoff from agricultural areas.

Protozoa are special, unicellular organisms. They act as predators of bacteria. Thus, they play an important role in the mixed microbial cultures of aerobic wastewater treatment systems (e.g., activated sludge processes) by consuming large amounts of bacteria (Figure 6–6).

Rotifers are multicellular organisms. They also prey on smaller microorganisms and are important in wastewater treatment systems. They consume large quantities of bacteria and, thus, keep the population of dispersed bacteria under control.

Microorganisms in Engineered Systems

Microorganisms are being used more and more in treating and removing human-generated waste. Because of their special characteristics, specific strains of microorganisms are needed in the production of medicines and alcoholic beverages. Most of the microorganisms used in environmental engineering, however, are not so special. These organisms are found in human and animal digestive systems, and they are present in vegetation, soil, and surface water. They will predominate in an environment conducive to their survival and growth. In engineered systems such as wastewater treatment plants, the environmental controls may include oxygen supply, pH, temperature, and nutrient supply.

Special strains of microorganisms are also being developed, particularly to treat hazardous or toxic substances. These microorganisms are being employed to destroy toxic and hazardous wastes that have leaked into or onto the soil. In many cases, essential nutrients and an oxygen source (e.g., hydrogen peroxide) must be injected into the soil to allow these organisms to thrive. In addition, appropriate moisture, temperature, and pH are required.

Typically, on contact with organic waste, microbes produce extra cellular enzymes. Enzymes are large protein molecules that are composed primarily of amino acids. A myriad of different enzymes (not always from the same species of microbes) act in sequence to degrade organic waste into simpler compounds (e.g.,

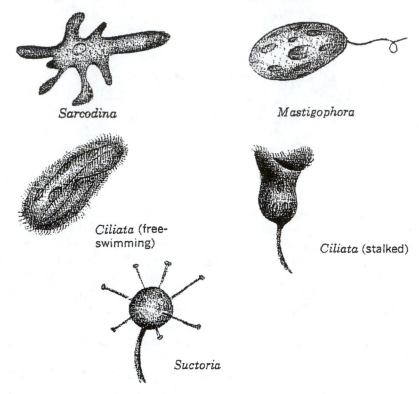

Sarcodina

Mastigophora

Ciliata (free-swimming)

Ciliata (stalked)

Suctoria

Figure 6–6   Basic groups of protozoa.

organic acids to carbon dioxide, hydrocarbons to alcohols). Special strains of oil-degrading organisms have been successful in some of the areas affected by the *Exxon Valdez* spill. Another example of the use of microorganisms is the *in situ* bioremediation of groundwater, as depicted in Figure 6–7. The concept of aerobic *in situ* bioremediation is based on the delivery of oxygen and nutrients to the contaminated subsurface. Water with nutrients and oxygen is injected into the aquifer and simulates the growth of microorganisms, resulting in degradation of the contaminants.

Under the favorable environmental conditions created by landfilling or composting, soil organisms can be used to degrade municipal solid wastes. Microbial activity initially occurs under aerobic and, later, under anaerobic conditions; detailed discussion is provided in Chapters 13 and 16.

Effect of Temperature on the Reaction Rate Constant.   Most reaction rates increase with increasing temperature, approximately doubling for each 10°C rise at lower temperatures (30–40°C). Van't Hoff-Arrhenius proposed a somewhat complex equation for reaction rate constant; a simplified equation is

$$k_2 = k_1 \Theta^{(T_2 - T_1)}$$

(6-23)

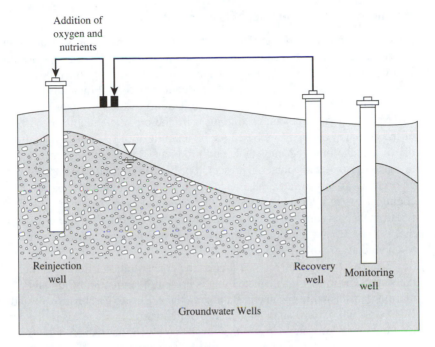

Figure 6–7   In situ bioremediation of groundwater.

where k_2 is the reaction rate constant at temperature T_2, k_1 is the reaction rate constant at temperature T_1, and Θ is the temperature coefficient.

This equation is commonly used in both biochemical and physicochemical reactions to calculate temperature effects. A commonly used temperature coefficient (Θ) for both biological and chemical reactions is 1.072, which doubles or halves the reaction rate for a 10°C temperature range.

Example 6–3

A chemical reaction has a measured reaction rate of 30 per day at 20°C. Using a temperature coefficient of 1.072, calculate k at 23°C.

Solution

From equation (6–23) the reaction rate at 23°C is

$$k_2 = 30 \times 1.072^{(23-20)}$$

$$= 36.95 \text{ per day}$$

Biochemical and Chemical Oxygen Demand

Bacteria and other microorganisms use organic substances as food. During this process, which is called metabolization, they consume oxygen. The organics are decomposed into compounds such as carbon dioxide and water, and the microbes use the energy from this process for the growth and production of cells. Equations (6-21) and (6-22) illustrate the oxygen requirement for stabilization of the oxidizable material.

Various parameters are used to measure the organic strength of wastewater. Most of the common methods are based on the amount of oxygen that is required to convert the oxidizable material into stable end products. Because the oxygen used is proportional to the oxidizable material present, it provides as a relative measure for the strength of organics in wastewater. This need for oxygen is commonly expressed as the biochemical oxygen demand (BOD) and the chemical oxygen demand (COD). The BOD is the measured amount of oxygen the microbes need to oxidize any organic matter that is present, and the COD of wastewater is the measured amount of oxygen needed to chemically oxidize the organics.

The BOD is an indirect measure for the concentration of organic contamination in the water. The more organic matter that is present, the greater the amount of oxygen the microorganisms will consume in stabilizing the wastes to carbon dioxide and water. The BOD does not oxidize all organic matter in the waste. In some cases, the organic waste load on a body of water depletes the oxygen so much that fish are killed. Metabolic activity by bacteria that require oxygen may reduce the level of dissolved oxygen from its normal value. When dissolved oxygen disappears, anaerobic conditions occur, objectionable odors ensue, and poor environmental conditions occur via the formation of gases such as H_2S and methane. Several types of waste, however, undergo anaerobic transformation more efficiently than aerobic transformation (e.g., anaerobic processes to transform halogenated compounds).

Treatment methods for industrial wastewater include chemical and physical processes to remove toxic organics and biological processes to reduce the organic content of high-strength industrial wastewater before discharge. Industry is required to treat wastewater to lower the concentration of suspended solids and organics as measured by the BOD and COD.

The standard conditions for BOD analysis are a temperature of 20°C, an absence of light so that algae will not produce oxygen, and sufficient nutrients so that a lack of nutrients will not limit stabilization. The duration of the analysis is usually 5 days, but the complete decomposition of organic material by microorganisms takes time—usually 20 days (or longer) under ordinary conditions. The time is usually subscripted, as in BOD_5 for the 5-day analysis (i.e., the amount of dissolved oxygen required by microbes over 5 days to decompose organic substances in water at 20°C) or BOD_U for the ultimate test of 20 days or more. In general, the BOD is expressed in terms of milligrams per liter of oxygen. The BOD is a function of time. At the very beginning of the test (time $t = 0$), no oxygen will have been consumed; thus, BOD = 0. Each day thereafter, oxygen is used by the organisms, and the BOD increases. Ultimately, BOD_U is reached, and the organics are completely decomposed. A graph of the BOD versus time, which is called the BOD curve, is shown in Figure 6–8.

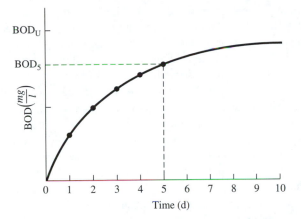

Figure 6–8   Biochemical oxygen demand (BOD) curve.

The mathematical expression for the BOD curve is given by the following equation:

$$BOD_t = BOD_U \times (1 - 10^{-kt}) \tag{6-24}$$

where BOD_t is the BOD at any time t (mg/L), BOD_U is the ultimate BOD (mg/L), k is the rate of BOD reaction (a constant), and t is the time (days).

The rate of oxygen uptake by microbes is represented by the constant k, the value of which depends on the type of microbes, the type of organic matter, and the temperature. For example, a situation of healthy and acclimated microbes, organic matter that is easily biodegradable and does not have material toxic to those microbes, and optimal temperature will have a greater value for k than other situations.

Example 6–4

A sample of a municipal solid waste landfill leachate is found to have a BOD_5 of 1000.00 mg/L. Estimate the BOD_U after 20 days, assuming that k is 0.08 per day for the sample.

Solution

Applying equation (6-24),

$$1000.00 = BOD_U \times (1 - 10^{-0.08 \times 5})$$
$$= BOD_U \times (1 - 10^{-0.4})$$
$$BOD_U = 1{,}667 \text{ mg/L}$$

The BOD_5 of municipal sewage in the United States is generally in the range of 200 to 250 mg/L. The BOD_5 of leachate from a municipal landfill may range from 200 to 1,000 mg/L for old sites (≥ 5 years) and from 4,000 to 20,000 mg/L for young sites. These figures can vary significantly depending on the site characteristics; however, from these figures the reader should get some idea about the strength of leachate.

The COD is the equivalent amount of oxygen that is required to oxidize any organic matter in a water sample by means of a strong, chemical oxidizing agent. COD analysis is performed by using chromic acid, which is a mixture of potassium dichromate and sulfuric acid. Silver sulfate and mercuric sulfate are added to assist in the reaction as a catalyst and as a complexing agent, respectively. The chromic acid mixture is then heated with the water sample and boiled for 2 hours. This process is called chemical digestion. The COD of the sample is the difference between the initial amount of chromic acid and the amount of chromic acid remaining after the digestion. The COD test takes very little time compared with BOD analysis

REVIEW QUESTIONS

1. What is the difference between organic and inorganic chemicals?
2. Write a balanced equation for the oxidation of hydrogen sulfide with chlorine.
3. Calculate the molecular weight of ferric sulfate.
4. a. Express the oxidation (i.e., the addition of oxygen) of methane (CH_4) as an equation.
   b. Express the methane reaction in terms of grams.
5. a. What happens when an inorganic salt is dissolved in water?
   b. Show the dissociation of sodium hydroxide (NaOH) in water as an equation.
6. a. Define an *acid* and a *base*.
   b. What is the significance of pH in waste treatment?
   c. What does a pH of 9 mean when compared with neutral water?
7. What is the difference between methane and methanol?
8. List the primary gases produced during the decomposition of municipal waste in a landfill.
9. Residents of an apartment building bring 1000 lbs of consumer goods into the building per week. This includes food items, furniture, books,

magazines, flowers, newspapers, and so on, 35 percent of which is consumed as food. In addition, 10 lbs per week accumulates in the building. Twenty-five percent of the solid waste generated is sent to a recycling center. How much solid waste per week is deposited in the building's dumpster?
10. Describe the role of microorganisms in waste treatment.
11. What is the role of dissolved oxygen in water during waste biodegradation?
12. List the physical factors suitable for bacterial growth.
13. What is the significance of protozoa in the bacterial metabolism of waste?
14. A chemical reaction has a measured reaction rate of 18 per day at 20°C. Using a temperature coefficient of 1.072, calculate k at 15°C and at 25°C.
15. What is a common parameter used to measure the organic strength of wastewater?
16. Compute the BOD_5 for a sample of leachate whose BOD_U is 1500 mg/L. Assume that $k = 0.1$ per day.

II SOLID WASTE MANAGEMENT TECHNOLOGY

Chapter 7   Overview

This chapter introduces the complex problem of ever-increasing solid waste disposal, defines the term *solid waste management,* and examines its development and functional elements.

Any material that is thrown away or discarded as being useless or unwanted is considered to be solid waste. One of the pressing problems facing municipalities today is the efficient disposal of urban solid wastes on a long-term basis. Once a simple problem, the disposal of solid waste has become a complex dilemma for our modern and wasteful industrial society. More and more, these problems are in the news, discussed in public forums, and argued in courts. Frequently, technical and economic issues have taken a back seat to political and social concerns, and further complicating the problem is the ever-increasing amounts of solid waste being generated by our society and the cost of its processing and disposal. Because these disposal and processing systems are unpopular with the general public, obtaining new sites has become very difficult and costly. Pollution abatement efforts often result in the closing of landfills and public resistance to new landfills or incinerators.

Most municipal solid waste is currently disposed of in a landfill. This method will continue to be employed in the foreseeable future to dispose of a major portion of the urban solid wastes generated in this country. Trends in U.S. solid waste generation, recovery, and discards are discussed in Chapters 10 and 14, but the actual quantities of solid waste being produced nationally remains as elusive today as they did 25 years ago. Numerous estimates have been made, however. On a per-capita-per-day basis, the U.S. Environmental Protection Agency (EPA) estimates approximately 4.4 lbs of municipal solid waste (excluding construction waste, municipal sludge, ash, and automobile bodies) for 1993 (1).

Between 1960 and 1993, the generation of waste increased from approximately 88 million tons to 207 million tons (1). At the same time, waste recovery for recycling and composting increased from 7 percent of the municipal solid waste generated in 1960 to 22 percent in 1993. The total amount of solid waste sent to landfills or incinerators is projected to decline, and in 1993, the proportion of municipal solid waste buried in landfills dropped to approximately 62 percent. A variety of alternative programs and technologies are available, and there is a considerable market for more technologies. Even with extensive energy and materials recovery, however, landfills will be necessary to dispose of the remaining waste material. For example, the volume of total waste may be reduced by approximately 90% in waste-to-energy facilities, but the remaining 10 percent, as well as the materials that cannot be incinerated (e.g., old refrigerators, window air conditioners) must still be disposed of somehow. Only two realistic options are available for the long-term handling of solid waste and residual matter: disposal on or in the earth's mantle, and disposal in the ocean. Federal law in the United States and in most other developed countries prohibits the latter. Therefore, a landfill is needed for disposal of the residue left by waste-to-energy and other material recovery.

DEFINITION OF SOLID WASTE MANAGEMENT

Solid waste is the material arising from human and animal activities that is normally solid and is discarded as being either useless or unwanted. It encompasses the heterogeneous mass of throwaways from urban communities and agricultural, mineral, and industrial wastes as well. In contrast to a liquid, solid material has an angle of repose that allows it to form a pile. The angle that the surface of the material's pile makes to the horizontal is called the angle of repose, and it is a characteristic of the fluidity (i.e., it has the ability to move and change shape). A material that does not form an angle of repose will form a flat, horizontal surface if it is allowed to stand unconstrained.

In the U.S. Resource Conservation and Recovery Act (RCRA) of 1976, solid waste is defined as "garbage, refuse, sludge from a wastewater treatment plant, water supply treatment plant, or air pollution control facility, and other discarded material including solid, liquid, semisolid, or contained gaseous material resulting from industrial, commercial, mining, and agricultural operations, and from community activities" (1). Solid waste does not include solid or dissolved materials in domestic sewage, and it does not include solid or dissolved materials in irrigation return flows or industrial discharges.

Management can be defined as the carefully planned, judicious use of means to achieve an end. In this case, an "end" is the removal and disposal of unwanted material. To achieve this, technical, environmental, administrative, economic, and political problems must be solved. The effort to address these problems is usually referred to as the practice of solid waste management. It encompasses the planning, design, financing, construction, and operation of facilities for the collection, transportation, processing, recycling, and final disposal of residual solid waste materials. All this is based on sound principles of public health, engineering, economics, aesthetics, conservation, and environmental considerations, along with social and ethical issues (2).

DEVELOPMENT OF SOLID WASTE MANAGEMENT

How did the field of solid waste management evolve? A brief review of progress in dealing with waste should answer this question.

From the days of primitive society, humans and animals have used the earth's re-sources to support life and to dispose of wastes. In early times, the disposal of wastes was not a major problem, because the population was small and sparsely distributed. Problems of waste disposal began when humans first started to congregate in relatively small areas. In ancient cities, waste food and other useless items were thrown into the streets, but around 320 BC, in Athens, the first known law prohibiting this practice was put into effect. In ancient Rome, property owners were responsible for keeping the streets in front of their property clean. Disposal methods were very crude, however, such as open pits outside the city. As the population increased, wastes were removed farther away, to an area usually called a dump.

During the 1700s, municipal collection of garbage was started in Boston, New York, and Philadelphia. Again, disposal methods were crude. For example, garbage col-lected in Philadelphia was simply dumped into the Delaware River downstream from the city. The development of a technological society in the United States paralleled the Industrial Revolution in Europe. This resulted in significantly increased solid waste problems, and urban sanitation acts were enacted in many large cities. The primary ob-jective was the removal of waste material for health reasons.

At the beginning of the twentieth century, many of the basic principles and meth-ods underlying what is known today as the field of solid waste management were al-ready well known (3). Obviously, the motor truck has replaced the horsedrawn cart for the collection of solid waste, but the basic methods of such collection remain the same. The commonly recognized methods for the final disposal of solid wastes are:

1. Dumping on land

2. Dumping in surface waters

3. Feeding to swine

4. Mixing into soil

5. Reduction

6. Incineration.

Several U.S. cities started incinerating solid wastes at the beginning of the twen-tieth century. Most large cities, however, were still dumping solid waste on land or in water. (Today, the dumping of waste into any body of water is no longer allowed.) Early incinerators were a source of noticeable air pollution, and the sanitary landfill was developed as a relatively inexpensive alternative to incineration, especially for communities with sufficient land areas. It was better than a city dump, but it was soon determined that without liners, landfills had the potential to contaminate ground and/or surface water. Today, extensive air pollution–control devices on all incinerators, liners for landfills, and other environmental safeguards are required.

In the United States, the responsibility for controlling solid waste was delegated to the "Department of Street and Sanitation." Even today, the collection and disposal in many cities is the responsibility of the Public Works Department. Lately, however, some privatization is occurring.

After World War II, most larger cities were faced with ever-increasing quantities of solid wastes. The cost of hauling these wastes to remote areas also increased. Passage of the Solid Waste Disposal Act of 1965 was the first attempt by the U.S. government to regulate solid waste disposal practices. This law provided funds to states for solid waste management programs, and it established regulations regarding disposal of such waste. The most significant regulations were developed under the RCRA of 1976, which also addressed the management of hazardous waste. This federal statute, which amended the elementary Solid Waste Disposal Act of 1965, reflected the concerns of both the public and Congress. The RCRA emphasized what are appropriately called the "three R's"—*R*educing waste volumes, *R*ecycling materials, and *R*ecovering thermal energy.

Solid waste disposal is now part of an integrated waste management plan; in other words, methods of collection, processing, resource recovery, and disposal should be coordinated to achieve a common objective. Technology associated with resource recovery, reuse, and recycling has evolved since the late 1970s. Most communities now have a solid waste management system that includes some type of recycling plan. Despite these improvements, however, there will always be a need to dispose of some residuals in landfills. Under the RCRA, the EPA has strengthened federal standards for landfills to protect water quality.

FUNCTIONAL ELEMENTS OF A SOLID WASTE MANAGEMENT SYSTEM

Several different activities are associated with a waste management system. Each accomplishes a specific purpose in a sequence of actions. To manage solid waste efficiently, the interrelationships involved should be clearly identified and understood.

In the following discussion, the activities associated with management of solid wastes from the point of generation to the final disposal have been grouped into the six functional elements (Figure 7–1). At least four functional elements are considered to be essential for any satisfactory management of municipal solid waste. These are highlighted here.

The first functional element is the material generated at the source. Materials that are no longer considered to have value are discarded as waste, and the quantity and characteristics of that waste depend on the source. Waste generation is an activity that has almost no controls, but in the future, more controls will probably be exercised through state and federal laws.

The second functional element is waste handling and separation, storage, and on-site processing. Waste handling and separation are the activities that occur before placing waste in storage containers for collection. Wastes are separated into components and stored for collection. Paper, cardboard, glass, plastics, aluminum cans, ferrous metals, and yard wastes are some of these components. Separation at the point of generation is now an important part of waste management, and handling includes the activities associated with separation and with moving the containers to the point of collection.

On-site storage of solid wastes may be in 30- to 60-gallon, conventional trash containers for single-family residences or in a larger container for apartment complexes and other large producers. On-site storage is important for public health and aesthetic considerations. The frequency of collection depends mainly on the production rate and the nature of wastes to be collected. The cost of storage is borne by the owner (or tenant) of the residence or by the commercial establishment who may be generating solid waste.

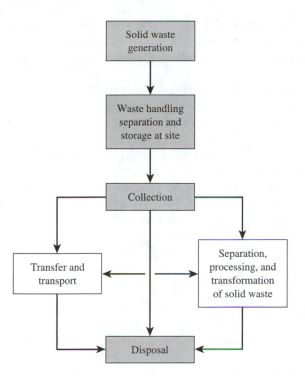

Figure 7–1   Functional elements of a solid waste management system.

The third functional element is collection. This includes picking up solid wastes and emptying containers (most recyclable materials are separated) into a suitable vehicle for storage and transport of these materials to the location where the collection vehicles are emptied. The collection vehicle for a single-family residential area is usually a side- or rear-loading packer truck, into which garbage cans are emptied manually. Large containers servicing large areas require a collection vehicle with mechanisms to transfer solid waste from these containers to the vehicle. These vehicles transport the waste to a processing facility, transfer station, or disposal site (e.g., a landfill or incinerator). Because of economic reasons and the relatively small volume of waste involved, small waste management systems generally use collection vehicles to transport waste directly to the landfill.

For a large or heavily regulated system, additional steps may be needed. The recovery of separated materials, separation and processing of solid waste components, and transformation of waste that occur mainly in locations other than the source of waste generation are included in the fourth functional element. Processing for mass and volume reduction through a resource recovery system includes separation as well as size and volume reduction. Paper, aluminum, plastic, glass, and ferrous metals are all components that are recovered for reuse.

The organic portion of the solid waste can be transformed by chemical and biological processes. For example, the most common chemical transformation is incineration for energy, and biological transformation is used to produce fuel gas or compost for soil conditioner.

The fifth functional element is transfer and transport. This involves the transfer of wastes from a smaller collection vehicle to a larger transport equipment (e.g., large semitrailer, train, or barge). After the transfer, wastes are transported (usually for long distances) to a processing and disposal site.

Disposal is the final functional element. Usually, the final disposition of waste is by landfilling. It does not matter whether these are residential wastes transported directly to a landfill, residuals with no recoverable value, or noncombustible material (e.g., ash) left after energy recovery. A modern sanitary landfill is a well-designed, engineered facility that significantly reduces the nuisance and hazards to public health and the environment.

REVIEW QUESTIONS

1. List the problems involved in the disposal of solid wastes.
2. Define *solid waste* and *solid waste management.*
3. What physical characteristics differentiate solids from liquids?
4. a. What are the sources of municipal solid waste?
   b. List the type of wastes in municipal solid waste.
5. Define *integrated solid waste management.*
6. What is the purpose of the RCRA?
7. a. List the functional elements of solid waste management.
   b. Describe each functional element.

REFERENCES

1. U.S. Environmental Protection Agency. *Characterization of Municipal Solid Waste in the United States.* Washington, D.C.: Office of the Solid Waste, 1994.
2. Tchobanoglous, G.H. Theisen, and S. Vigil. *Integrated Solid Waste Management,* New York: McGraw Hill, 1993.
3. Parsons, H. de B., "The Disposal of Municipal Refuse," *Municipal Solid Waste Magazine,* Issue 2, 1906.

Chapter 8   Environmental Laws and Regulations

This chapter introduces the process that creates and alters environmental laws and briefly describes the current major federal laws. To appreciate environmental compliance and risks, government, industry, and the public should have a through understanding of the various laws and regulations. Environmental regulations have existed for centuries, and controlling environmental pollution has an old and a complicated background. No major environmental legislation existed in any country, however, until the second half of the twentieth century. Before that legislation and the many regulations of the past 40 to 50 years, the basis for most legal actions in the United States was the common law doctrine, which the early U.S. settlers brought from England. Most of the decisions made in those earlier times had their basis in this law. It is a subjective law, however, because it reflects what society considers to be fair and reasonable at the time, which tends to change as society's perception, needs, and scientific and technical knowledge also change.

DEVELOPMENT OF ENVIRONMENTAL REGULATIONS

There was a growing awareness of environmental problems for many years, but not until the 1960s did a the period of heightened environmental awareness crystallize. With the publication of Rachel Carson's *Silent Spring* in 1962, people became aware of the effects of pesticides on birds and other wildlife. Other writings that influenced public concern were Paul Ehrlich's *The Population Bomb,* which described the potentially adverse effect that a rapidly increasing population could have on natural resources, and Barry Commoner's *The Closing Circle,* which explained ecological principles in simple terms.

Many state and federal laws to protect public health and the environment have been enacted over a period of many years. The first significant laws in this area were the U.S. federal statutes passed during the 1970s that dealt with air and surface water quality and hazardous waste. Most of these laws are still in effect in their original form, but they are also in a state of constant transition. In other words, what is legal and accepted practice today may be illegal in the future.

Also influential in generating public concern were events such as the 1969 Santa Barbara oil spill, the Love Canal episode in 1978, and the Times Beach incident in 1979. These events made the entire nation aware of the dangers from hazardous chemicals.

The U.S. Congress writes and passes environmental laws to "empower" regulatory rulemakers and agencies, and it appropriates funds so these regulatory agencies can fulfill their mandates. The Congress consists of the Senate and House of Representatives. It passes laws in response to public interest, public need, or the occurrence of a significant catastrophe. In the United States, a law can be passed by a simple majority of the House of Representatives and the Senate. If the President vetoes the bill, a two-thirds majority of both houses is then required to override. The Clean Water Act of 1972 was passed over President Nixon's veto, and the Resource Conservation and Recovery Act as well as the reauthorization of the Clean Water Act were passed over President Reagan's vetoes in 1987.

The U.S. Environmental Protection Agency (EPA) was formed on December 2, 1970, to protect the environment through coordinated governmental action based on pollution control policy and regulations. When Congress passes environmental legislation, it directs the appropriate federal agency (e.g., the EPA) to promulgate the law into regulations and then publish them. The method used to develop these regulations was established by the Administrative Procedures Act of 1946. Regulatory agency must:

1. Publish an intent to pass a new rule and receive feedback from interested parties, including professional and trade associations.

2. Publish the proposed rule and date of public hearing in the *Federal Register,* which is a chronological publication of proposed and final federal regulations pertaining to almost all subjects, including the environment.

3. Conduct hearings open to the public to resolve any controversy.

4. Publish the final rule in the *Federal Register.*

5. Codify the new rule in the Code of Federal Regulations (CFR).

Figure 8–1 shows a cover from the *Federal Register.*

The CFR is a codification of the rules and regulations published in the Federal Register by the EPA. The CFR is divided into 50 titles, which are further subdivided into parts that reflect the various regulations under a particular department or agency. The CFR is revised each year. EPA regulations are found in 40 *CFR,* Occupational Safety and Health Administration (OSHA) regulations in 29 CFR, and Department of Transportation regulations (DOT) in 49 CFR. Some of the OSHA and DOT regulations are relevant to solid and hazardous waste management. Figure 8–2 shows a cover from the Code of Federal Regulations.

In short, when the EPA first develops and proposes a new or revised regulation, the regulation is published in the *Federal Register* for comment from all the interested parties. After a specified period of time, all the comments are considered by the EPA, and the proposed regulation is then finalized and published in the *Federal Register* and the annual issue of the 40 CFR.

Environmental Protection Agency

Congress created the EPA in 1970 to administer environmental programs and establish standards to protect the environment in a way that is consistent with U.S. goals. The EPA is the primary agency responsible for protecting the environment, but many other agencies are involved in particular areas as well. The EPA's duties include enforcement

Figure 8–1   Cover of the *Federal Register.*
(From Federal Register, *Vol. 55, No. 49, 1990; with permission.)*

of air and drinking water quality standards, stream discharge standards, solid and hazardous waste treatment and disposal standards, and the cleanup of abandoned hazardous waste sites. The agency headquarters is in Washington, D.C., and regional offices are in Atlanta, Boston, Chicago, Dallas, Denver, Kansas City, Philadelphia, San Francisco, Seattle, and New York.

States also have agencies that enforce state environmental laws and regulations and, in many cases, federal regulations as well (with the help and advice of the EPA). State and local governments have the right to adopt environmental laws that are at least as stringent as federal laws. In fact, many state laws are identical to federal laws. Others, such as those in California, are even stricter than federal laws. Interested parties should contact the appropriate EPA regional or state regulatory personnel for up-to-date information on regulatory requirements. Figure 8–3 depicts these regulatory requirements.

Methods of Regulation

A permit is required for each point-source discharger (e.g., city or industry). Acquiring such permits not only necessitates gathering substantial information but also may require significant lead time before such a permit can be issued. The permit must state

Figure 8–2   Cover of the *Code of Federal Regulation.*
(From 40 CFR, 1992; with permission.)

the type of pollutants that may be discharged and the allowable limits of such discharge. The discharger must monitor the various pollutants, keep their discharge below the permitted levels, and submit periodic reports to the EPA or the corresponding state agency. When the discharger violates its permit conditions, it must notify these agencies within a specified period of time that depends on the relevant environmental laws, which include the Clean Air Act, Clean Water Act, and Emergency Planning and Community Right-to-Know Act. The extent of the violation of the act, why it occurred, and what is being done to prevent future violations must be reported.

Rivers and Harbors Act

Federal legislation to control pollution dates back to the Rivers and Harbors Act of 1899, which prohibited disposal into waterways of solid waste that would interfere with navigation. The purpose of this act was to prevent interference with shipping; it had

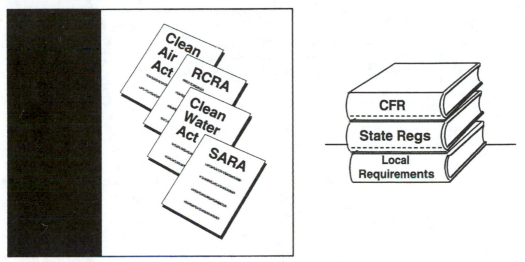

Figure 8–3   U.S. Environmental Protection Agency regulatory requirements.
(From U.S. EPA, Superintendent, U.S. Govt. Printing Office, 1994)

little to do with environmental quality. Even so, the act marked the beginning of solid waste regulation in the United States, and it was the only significant existing environmental law during the 1950s and 1960s.

NATIONAL ENVIRONMENTAL POLICY ACT

On January 1, 1970, President Nixon signed the National Environmental Policy Act (NEPA). Since NEPA was passed, the courts have broadly interpreted its regulations to cover almost every activity that affects the environment and, at the same time, requires a federal permit or uses federal lands or funds. Many states have enacted legislation similar to NEPA for projects not covered by federal regulations. NEPA has been the basis for several environmental lawsuits as well.

The heart of NEPA is the requirement that federal agencies complete an environmental impact statement (EIS) before taking any major action that may have a significant environmental impact. The EIS is a complex, time-consuming, and expensive document to prepare. (Environmental impact analysis was discussed in Chapter 2.)

An EIS is required for any major federal action that significantly affects the human environment. Thus, proper analysis of three components is essential: the action must be federal, must be major, and must significantly affect the quality of the human environment. The first component can be the most misleading. Often, state and even private actions may trigger the EIS requirement if there is some substantial federal connection. For example, in *Maryland Conservation Council v. Gilchrist*, the U.S. Court of Appeals for the Fourth Circuit held that a state highway construction project was a major federal action, because the highway would cross a state park that was established with a substantial federal grant and approval from the Secretary of Interior was necessary to convert the parkland into a highway (1). For an action to be "major," it must be more than merely routine: simply put, projects must involve substantial expenditures of time, money and resources. Because of the statute's failure to define the "human environment," however, the courts have struggled to find a workable formula for determining when the human environment is affected.

Occupational Safety and Health Act

The Occupational Safety and Health Act of 1970 regulates hazards in the workplace, including exposure to hazardous substances, to protect the health and safety of employees. The act is administered by OSHA, which excludes suppliers who fail to meet minimum work conditions from bidding on public contracts. The act also provides penalties, including civil and criminal sanctions, for those suppliers. Among its many provisions, employers are required to keep up-to-date health records for employees exposed to chemical substances.

Clean Air Act

Because air pollution has many causes, a somewhat broad approach has been used to reduce and control such pollutants. Today, these controls include restrictions on fuels, motor vehicles, industries, hazardous and solid waste handling and disposal, and power plants.

The Clean Air Act of 1963, which was a major piece of legislation, set both emission and ambient air-quality limits. This act required the federal government to set ambient air-quality standards for major air pollutants. Matching federal grants for increased research and training were provided to state and local governments as well. States were given until 1975 to meet the act's standards, but many have yet to do so.

The subsequent Clean Air Act of 1970, which was reauthorized (updated) in 1977 and 1990, required progressively stricter controls. Overall, air quality in the United States has gradually improved. The level of lead in the atmosphere has significantly declined, and levels of sulfur oxides, ozone, carbon monoxide, and particulates have also declined.

The 1990 amendments focused on industrial airborne toxic chemicals in addition to motor vehicle emissions. Between 1970 and 1990, airborne emissions of only seven toxic chemicals were regulated. The latest federal air pollution–control regulations (i.e., the 1990 amendments) have taken a broad approach, however, with the goal of a 90 percent reduction in the atmospheric emissions of 189 toxic and cancer-causing chemicals by the year 2003. To comply with this requirement, businesses such as dry cleaners and large manufacturers such as chemical companies must install pollution control equipment. To target acid rain, sulfur dioxide and nitrogen oxide emissions from coal-fired power plants must be significantly reduced by the year 2000.

Clean Water Act

In 1972, the Federal Water Pollution Control Act Amendments of 1972 (Pub L No 92-500) became law after the Congress overrode President Nixon's veto. This act, which was called the Clean Water Act (CWA), established national water-quality goals. Its objective was to "restore and maintain" the physical, chemical, and biological integrity of the nation's waters. The CWA introduced national water-quality goals, technology-based effluent limitations, a national permit system, and federal actions against violators of permit conditions.

The CWA was amended in 1977, 1981, and 1987 and was reauthorized by Congress again in 1990. These acts require the EPA to establish and then monitor the maximum permissible amounts of pollutants that can be discharged into the nation's surface water. Effluent limitations required by the amendments of 1972 and 1977 formed the basis for issuing "National Pollutant Discharge Elimination System Permits."

Safe Drinking Water Act

In 1974, the Congress passed the Safe Drinking Water Act (SDWA), thus giving the EPA power to enforce drinking water quality. This act expanded the coverage of federal regulations, and it set uniform, federal standards for drinking water to guarantee safe public water supplies throughout the country. In addition, it included provisions to protect underground aquifers, which are important sources of drinking water. Any water supply that serves more than 25 people per day for 60 or more days is required to meet the SDWA standards.

The SDWA also required the EPA to determine the maximum contaminant level (i.e., the maximum permissible amount) of any water pollutant that might adversely affect human health. The EPA thus promulgated a more encompassing set of standards, and these regulations were further modified as more information on hazardous substances became available. For example, between 1980 and 1991, standards to prevent corrosion that could cause metals, especially, to dissolve from distribution or consumer piping were established.

Federal Insecticide, Fungicide, and Rodenticide Act

The Federal Insecticide, Fungicide, and Rodenticide Act (FIFRA) was originally passed in 1947 to regulate the effectiveness of pesticides. The FIFRA was amended in 1982 and 1983. Its regulation concerns the registration, use, storage, and disposal of pesticides, and it requires informative and factual labelling. The FIFRA also requires user certification for restricted-use pesticides. In addition, the act specified tolerance levels of certain pesticides and required approval of pesticides before commercial distribution.

Pesticides pose a significant threat to groundwater. In some instances, contamination has occurred. Thus, the FIFRA authorized the EPA to collect detailed information on pesticides suspected of causing pollution and required manufacturers to conduct extensive field-monitoring studies. The EPA also provides educational and technical assistance networks for pesticide users. The purpose is to inform users of the potential problems and of ways to reduce the risk associated with some pesticides.

Solid Waste Disposal Act and Resource Recovery Act

The Solid Waste Disposal Act of 1965 was the first significant step in the federal regulation of solid waste disposal in an effort to improve waste disposal technology. Its main aim was the regulation of municipal waste. Increasing concerns over public health and the environment, however, led to the need for better management of solid waste and supported resource recovery. In turn, this led to the amendments of the 1965 act by Pub L No 95-512, and finally, in 1970, the Resource Recovery Act was passed. This act vastly expanded federal involvement in the management of solid waste, redirecting the emphasis from disposal to reduction, recycling, and energy recovery. By the mid-1970s, both the Congress and the public recognized that additional regulations were needed to ensure proper management of solid waste. These concerns prompted Congress to pass the Resource Conservation and Recovery Act of 1976.

Toxic Substances Control Act

The Toxic Substances Control Act (TSCA) was enacted by the Congress in 1976 to regulate the manufacturing, processing, and distribution of chemical substances that pose a risk to human health or the environment. A "catch-all" statute, the TSCA was intended to cover the gaps between federal statutes regulating particular types of chemicals (e.g., pesticides, food, drugs, cosmetics, nuclear material) by regulating the manufacturers and processors of any chemical not regulated under another statute.

Under the TSCA, the EPA requires chemical manufacturers and processors to provide data about the health and environmental effects of chemicals that may present an "unreasonable" risk and enter the environment or that may present the likelihood of "substantial human exposure." After reviewing the data, the EPA may then issue regulations to reduce the risk to public health or the environment. In the case of imminently hazardous chemicals, the EPA may seize those chemicals through court action or prevent their release through a legal injunction. In addition to enforcement by the EPA, the statute also allows for citizen's lawsuits and contains a "whistleblower" provision. The manufacture, processing, and distribution of products containing polychlorinated biphenyls (PCBs) are banned under TSCA, which thus requires that the EPA promulgate regulations covering the storage and disposal of PCB-contaminated products and wastes.

Resource Conservation and Recovery Act

The Resource Conservation and Recovery Act (RCRA) of 1976 increased controls on nonhazardous solid wastes and began the regulation of hazardous wastes. Revisions to RCRA were made in 1980 and again in 1984. (The 1984 amendments, which are referred to as the Hazardous and Solid Waste Amendments, are discussed later.)

The portions of the RCRA that regulate hazardous waste are contained in a section designated as Subtitle C. It contains expanded provisions for the management of hazardous wastes. Subtitle D covers solid waste disposal, and Subtitle I regulates underground tanks.

The RCRA also established a "cradle-to-grave" system to regulate hazardous waste from its generation, usually in the course of manufacturing or other commercial activities, to its final disposal. This approach has several key elements:

1. A tracking system in which a manifest document is required for every waste that is transported from one location to another.

2. A permit for the operation of facilities that treat, store, or dispose of hazardous waste.

3. A system of controls and restrictions on the disposal of hazardous waste on land.

Hazardous waste is frequently transported from the source to treatment, storage, or disposal facilities. To eliminate any possibility of improper handling during such transport, the waste generators are required by the EPA to complete a hazardous waste manifest form that must accompany the waste when it leaves the site. (The manifest document and cradle-to-grave concept are also discussed in Chapter 18.)

In addition, the RCRA requires that permits be obtained for treatment, storage, and disposal facilities that accept hazardous waste. This permitting system gives the EPA the power to inspect such facilities, enforce standards and requirements, and take legal actions against violators of the RCRA regulations.

Hazardous and Solid Waste Amendments of 1984

During the early 1980s, the EPA was blamed by both the public and the Congress for failing to carry out their mandate. The Congressional response was to pass the Hazardous and Solid Waste Amendments of 1984 (HSWA). The scope of the RCRA regulations was greatly broadened through these amendments.

The HSWA was more detailed than most other pieces of environmental legislation. It prohibited the disposal on land of wastes that did not meet specific treatment standards. Land disposal of bulk and noncontainerized liquid hazardous waste was also prohibited. A major theme of the HSWA was the protection of groundwater. In addition, minimal technical requirements for land disposal facilities were included rather than leaving such technical details to the EPA, which was unusual. The HSWA also established standards for issuing permits, controls on underground storage tanks, and penalties for violating the law.

Comprehensive Environmental Response, Compensation, and Liabilities Act

The RCRA and the HSWA focused on existing and future hazardous waste management, but they failed to deal with older pollution problems such as the many existing, inactive hazardous waste disposal sites, which posed a serious threat to public health and the environment. Therefore, the Comprehensive Environmental Response, Compensation, and Liability Act (CERCLA) of 1980—better known as the "Superfund Act"—became law. This act provided for liability, compensation, cleanup, and emergency response for hazardous substances released into the environment and for the cleanup of inactive hazardous waste disposal sites. CERCLA was generally intended to give the EPA the authority and the funds to clean up abandoned waste sites and to respond to emergencies related to hazardous waste. The law provided for both a response and an enforcement mechanism.

Money for the Superfund came from a trust fund that was created using special taxes on crude oil and chemical companies. In 1986, the CERCLA was reauthorized, and the trust fund was increased from $1.6 to $8.5 billion, with money from taxes on petroleum and chemical companies as well as from a special corporate environmental tax. The CERCLA revisions enacted in 1986 are designated as the Superfund Amendment and Reauthorization Act.

The National Contingency Plan prepared by the EPA pursuant to the CERCLA governs cleanup efforts by establishing procedures and standards for responding to the release of hazardous substances. At the same time, the EPA can sue any responsible parties it can locate for the reimbursement of cleanup costs, thus allowing the federal government to respond while trying to shift the financial responsibility to others. Thus, the Superfund covers cleanup costs if the site has been abandoned, the responsible parties elude detection, or the private resources are inadequate.

Emergency Planning and Community Right-to-Know Act

The Emergency Planning and Community Right-to-Know Act was passed in 1986. It requires written emergency response plans to deal with chemical releases and establishes a Toxic Release Inventory. Employers must report to the EPA and advise employees of any hazardous substances handled and their potential effects. The act also requires that information about any toxic chemicals stored or permitted to be released during the operation of commercial and manufacturing facilities be made available to the public. State and local governments as well as facilities either using or storing hazardous chemicals must have emergency plans for notification and response. These facilities must also report annually the amount and characteristics of certain toxic chemicals both on their premises and released to the environment.

Oil Pollution Act

The Oil Pollution Act was passed in 1924 to prevent oily discharges in coastal waters. Transportation of petroleum products is covered by the Oil Pollution Act of 1990, which can be found in 33 USC §2701-2761. The liability of responsible parties was broadened by this new law. Each responsible party (i.e., owners and operators of vessels, on-shore facilities, or pipelines) is liable for the removal costs, damage to natural resources, damage or injury to or economic losses from the destruction of property, and lost profits.

Pollution Prevention Act of 1990

The Pollution Prevention Act of 1990 calls pollution prevention a "national objective" and establishes an hierarchy of environmental protection priorities as national policy. It also establishes recycling, treatment, and waste minimization as greater priorities than disposal in landfills. Among other provisions, the act directs the EPA to facilitate the adoption of source reduction techniques by businesses and federal agencies and to establish methods of measurement for source reduction.

BROWNFIELDS

Somewhere between very contaminated sites and uncontaminated greenfields is unused industrial and commercial land that is abandoned, idled, or underused, because expansion or redevelopment is complicated by perceived or real environmental contamination. These parcels of land are termed *brownfields,* and developers are unwilling to work under the costly and complex federal, state, and local regulations for environmental remediation at these sites. Public fear of contamination as well as developer and lender concerns for liability are causing developers and companies to leave "brownfields" in urban areas and take their business outside the cities.

To address concerns about the complex regulations and liability issues of the CERCLA and to encourage development in urban areas affected by real or perceived contamination, the EPA initiated the Brownfield Economic Redevelopment Plan in 1995. It is the hope of this initiative to encourage the cleanup of unaddressed contamination in many urban areas (2).

As a part of this initiative, the EPA has removed 27,000 sites from the Superfund tracking system, which is known as the Comprehensive Environmental Response, Compensation, and Liability Information System (CERCLIS). The original inventory tracked 40,000 such sites. Removing sites from the CERCLIS, however, means that the EPA will no longer pursue Superfund action at these sites. Another important aspect of this initiative is that it will not take action against property owners when hazardous substances enter a property through aquifers polluted elsewhere as long as the landowner did not cause the problem or have an agreement or arrangement with the polluter.

There are approximately 30 state voluntary cleanup programs across the country. According to the Assistant Administrator for the EPA's Office of Solid Waste and Emergency Response, the main factor triggering efforts toward the revitalization of brownfield properties is the economy. Most of these cleanups should be overseen by the states. The EPA's role should be to support state and local efforts by providing technical and financial assistance because states should have a say regarding which brownfields are worth developing to boost the economy (3).

The brownfields program involves risk-based corrective action, in which feasible remediation technologies along with good management are expected to provide cost-effective cleanup to benefit inner-city economic development and to protect both public health and the environment.

CASE STUDIES

Pollution Charges Lead to Fines

Two companies and their managers were fined and placed on 3 years' probation Thursday for dumping potentially explosive pollutants in wastewater discharged to a sewage treatment plant.

U.S. District Judge S. Arthur Spiegel sentenced the defendants, who on Jan. 28 had pleaded guilty to having dumped or aided in dumping the wastewater, which wound up in the Ohio River.

Cincinnati's Metropolitan Sewer District said it detected the Oct. 4, 1995, discharge and diverted it into the river to avoid possible damage to the Muddy Creek wastewater plant.

Southside River-Rail Terminal, Inc., a liquid transporting company in Cincinnati, was fined $25,000 for knowingly discharging the polluted wastewater. The company could have been fined as much as $50,000.

Ruetgers-Nease Corp. of Harrison was fined $50,000 for aiding and abetting the unlawful discharge.

Wagner Seed*

In 1986, Wagner Seed Company was held responsible for a lightening bolt that caused an accidental spill from its hazardous waste storage facility. The EPA made the ruling under CERCLA Section 106. Wagner Seed Company argued they should not be considered a responsible party.

In an act of God, the court ruled, there is no third party. In this case, there is only the EPA, God, and Wagner. Reimbursement from the EPA, and God, is doubtful. That leaves only Wagner. (Second Circuit Court)

* This section is reprinted from Lee, K.W. *Practical Management of Chemical and Hazardous Wastes.* Upper Saddle River, NJ: Prentice-Hall, 1995, p. 59; with permission.

REVIEW QUESTIONS

1. A citizen strongly feels that the Federal Insecticide, Fungicide, and Rodenticide Act does not go far enough to protect against groundwater contamination and needs to be amended. How should the citizen seek an amendment to the act?
2. a. List the duties of the EPA.
   b. How does the EPA communicate with the public and seek comments concerning proposed regulations?
   c. How can one find out if a particular statute on hazardous waste storage is in effect?
3. Name the environmental act or acts concerned with the following problems:
   a. Interstate highway construction.
   b. Unsafe working conditions in a steel-manufacturing plant.
   c. A refinery discharging waste in an underground old tank.
   d. A pesticide, which is suspected to cause cancer, being sprayed from an airplane.
   e. A new toxic chemical being sold without proper testing.

4. Discuss the aims and objective of the Resource Conservation and Recovery Act.
5. Trace the route of a manifest document accompanying a waste.
6. Assume there is an old, abandoned hazardous waste disposal site in the county of which you are administrator. How do you go about restoring the site? State your choices under the current law.
7. Consult a book on environmental law, and state what legal action is possible by a citizen in the following cases:
   a. The EPA drags its feet or is very careless in dealing with a corporation accused of contaminating the environment.
   b. A dry-cleaning establishment is discharging toxic fumes.

REFERENCES

1. *Ohio EPA Newsletter,* October 1998.
2. U.S. Environmental Protection Agency. *Brownfields Action Agenda.* EPA/500/F-95/001. Washington, D.C.: U.S. Govt. Printing Office, 1996.
3. Maldonado, M. "Brownfields Boom," *Civil Engineering Magazine,* May 1996, pp. 36–40.

Chapter 9 Environmental Effects and Public Health Aspects

This chapter introduces the role of solid waste management in issues regarding the environment, public health, and worker safety. The problems associated with improper disposal of solid waste and operation of facilities as well as ways to avoid these problems are discussed.

The protection of public health, aesthetics, and the environment are the main reasons for quick removal and disposal of solid wastes. Improperly discarded waste attracts rats, flies, mosquitoes, and many other types of insects. It provides food and a place to breed for rodents and insects, and it has a potential to contaminate both ground and surface water as well as the air. Other components of solid waste such as nails, pointed metal objects, broken glass, household pesticides, solvents, and spray cans can also be hazardous. Increasingly, regulations concerning public health and safety require strict compliance with limiting public access to operational areas and with providing protective devices for worker safety.

SOURCES OF DISEASES

Under warm, moist conditions, most organic or food wastes become a haven for disease-causing organisms (i.e., pathogens). As mentioned, improper storage or disposal of solid waste attracts flies, mosquitoes, and rodents, which are the primary carriers (i.e., vectors) of pathogens. Wastes such as discarded tires and containers also retain water, which provides a breeding place for mosquitoes. The major diseases of concern in the United States associated with flies and mosquitoes are gastroenteritis, hepatitis, dysentery, and encephalitis. Mosquitoes and flies are also responsible for diseases such as malaria, typhoid and paratyphoid fever, cholera, and yellow fever. These diseases are prevalent in most underdeveloped countries because of the unsanitary conditions that exist; so are diseases such as typhus and bubonic plague, which are carried by fleas and lice that live on rodents.

PHYSICAL AND CHEMICAL HAZARDS

The generation of harmful organisms and the transmission of disease are not the only health concerns. Certain solid waste components can also be a source of other public health and environmental hazards.

In the United States, chemicals such as insecticides, herbicides, pesticides, cleaning fluids (mostly ammonia based), solvents, arsenic, or lead-based compounds are used in many households. The containers of these chemicals are not always completely empty before they end up as trash, and even though small in quantity, the chemicals remaining present a danger to both humans and animals. Most of these wastes, including toxic chemicals, asbestos debris, and medical wastes (even though prohibited), may also be present in municipal solid waste. Disposal of such waste has a great potential to contaminate ground and surface water, and sharp and rusty objects such as metals and broken glass can cause serious injuries.

The handlers of waste as well as the public at large also face the hazards of explosive and ignitable solvents and gasoline. When lifting and emptying waste containers in their collection vehicles, a solid waste collection crew is at risk from physical and chemical hazards. Injuries can occur; cuts, bruises, and muscle as well as back pain are the common problems. Road hazards and mechanical equipment related injuries can also be problems. Increasingly, management has turned to training and improved mechanization of collection operations to reduce the rate of worker injuries and increase labor efficiency.

DISEASE AND OTHER HAZARD PREVENTION

Control of flies requires eliminating their breeding places and making food inaccessible. This involves cleanliness, good housekeeping, and application of recommended sanitary control measures: the elimination of food, moisture, and warmth. Organic wastes (e.g., food items) should be placed in tightly closed containers, the outsides of which should be kept clean. The containers should be stored in a cool or airy place, and spillage should be avoided. Soiled solid waste containers should be cleaned with hot water and cleaning compound, and use of a water-resistant lining inside the cans can reduce soiling. Houseflies that develop in landfills can penetrate and emerge through 5 ft of uncompacted cover—and through nearly 6 in. of compacted cover. This emphasizes the importance of prompt collection, disposal, and thorough compaction of solid waste as it is being deposited, followed by a daily earth covering that is compacted in turn.

Mosquito control depends on the elimination of breeding places. If solid waste is not allowed to accumulate and if containers and old tires are broken apart or removed to prevent the accumulation of water, the mosquito problems associated with solid waste can be eliminated.

Rats live off waste food, wherever they can get it. Thus, the rat population can be controlled by keeping organic waste in tightly closed containers made of plastic or metal. In addition, frequency of collection and cleanliness of the on-site storage area is important. A daily, compacted covering of the sanitary landfill will discourage rats as well.

TRANSFER, PROCESSING, RECOVERY, AND DISPOSAL FACILITIES

Two types of public health and safety issues are involved in the design and operation of solid waste facilities. The first relates to the health and safety of the general public. The second relates to the health and safety of workers at those facilities.

Public Access

Because the activities involved with the operation of a transfer station, materials recovery facility (MRF), and disposal facility are potentially dangerous, public access should be controlled and very limited. Convenience stations at an MRF for the deposit of recyclable items by the public should be located away from the main traffic pattern. For reasons of safety, the public should not be allowed to discharge wastes directly into the pit at storage–load transfer stations. Improved practices regarding public safety and site security at landfills have resulted in use of such convenience transfer stations to minimize public contact with the working operations of the landfill. Most sites now have restricted access and are fenced and posted with "No Trespassing" and other warning signs.

Worker Health and Safety

The health and safety of workers at solid waste management facilities is critical. The federal government, through the Occupational Safety and Health Administration, and states have established strict requirements regarding comprehensive health and safety programs for workers. The requirements for these programs change periodically; therefore, the most recent regulations should be consulted. Such requirements, for example, are concerned with:

- Overhead sprays to keep the dust down.
- Dust masks for workers to prevent dust inhalation.
- Enclosed cabs with air conditioning and dust-filtering units.
- Protective clothing, safety shoes, puncture-proof and impermeable gloves, eye protection, and ear plugs for personal safety.

COMPOSTING

Composting is the process of converting most waste organics into useful soil conditioner (see Chapter 13). The public health issue involved is the potential for pathogenic organisms to survive the process. It is essential the finished product be free of pathogenic organisms if it is to be marketed for public use. Generally, most pathogenic organisms found in organic waste are destroyed at the temperatures used in composting operations (typically 55°C for 2 to 3 weeks).

Heavy metals may also be a problem in composting, especially when mechanical shredders are used. When metals in solid waste are shredded, metal dust particles may become attached to the organic material that, after composting, is applied as a soil conditioner. Metals such as cadmium, lead, and mercury are of concern because of their toxicity.

AIR POLLUTION AND WATER POLLUTION FROM SOLID WASTE MANAGEMENT

One of the problems associated with sanitary landfills is the production of methane gas by microorganisms that decompose organic material anaerobically. The gas may seep through the trash and accumulate in underground pockets, and if it is not properly vented, there can be a potential for explosion. It is even possible for methane to seep into the basements of nearby homes and become a potential hazard there as well.

Another problem is water pollution from an improperly designed and operated landfill. (Landfill design and operation are discussed in Chapters 15 and 16, respectively.) Without proper control of the infiltration that results from precipitation and the surface runoff from adjacent land areas, water will slowly flow through the solid waste. This liquid (i.e., leachate) absorbs the organics and heavy metals in the landfill, which can endanger the quality of ground or surface water supplies.

Incinerators also produce large quantities of ash, which are currently disposed of in special sanitary landfills or hazardous waste sites. This ash contains toxic materials, including heavy metals and possibly dioxins, and there is a concern these materials could contaminate groundwater.

Improper disposal of solid waste can pollute the atmosphere through odors from decaying garbage, smoke and fumes from fires at the site, or the products of open burning. During the 1960s, uncontrolled burning of solid waste through on-site incineration at large housing complexes, shopping centers, and commercial buildings caused severe air pollution. The ban on open burning in urban areas, however, has shifted the focus on air pollution to the location of commercial incinerators. Without proper air pollution–control devices, these units can cause significant air pollution. Air pollution caused by particulates and gaseous pollutants from landfill sites—and especially from municipal incinerators—are another environmental problem related to solid waste disposal. Gaseous pollutants from an uncontrolled or poorly operated solid waste incinerator may include carbon monoxide, sulfur dioxide, nitrogen oxide, hydrogen chloride, dioxin, and furan. These gases can cause serious health and environmental problems, such as respiratory illness, cancer, and smog or acid rain.

Early land disposal sites were considered to be "dumps," because the waste was discarded without any engineering and scientific consideration. For example, liners were not provided, and soil cover was used only when it was convenient or thought to be needed. Landfilling of solid wastes or the residues from incineration can endanger the quality of ground or surface water supplies. Sanitary landfills, however, are engineered disposal systems. Thus, proper design and careful operation are needed to minimize the risk associated with contaminated drainage (i.e., leachate) coming from decomposing organic waste in the landfill.

Other hazards to public health and the environment, such as toxic, explosive, or ignitable wastes, can be minimized through careful separation of such wastes and their disposal as hazardous wastes (see Part III).

ENVIRONMENTAL IMPACT STATEMENTS

The National Environmental Policy Act of 1969, Section 102 (2)(C), requires federal agencies to prepare a "detailed statement" of environmental impacts for "major federal actions significantly affecting the quality of the human environment and health" (1). A principal task of those preparing an environmental impact statement (EIS) is to forecast the environmental effects of alternative actions. For any particular action, an *impact* is defined as the difference between the future state of the environment if the action occurred and the future state if no action occurred (2). When a new project such as a solid waste disposal facility is planned, an assessment is undertaken, which leads to an EIS. The EIS includes information concerning effects on human health and the environment; the contents and method of assigning relative weights for comparison were given in Chapter 2.

REVIEW QUESTIONS

1. List the reasons for quick removal of solid waste from residential areas.
2. What climate is most conducive for the growth of disease-causing organisms in a heap of solid waste?
3. a. List the physical and chemical hazards associated with solid waste.
   b. How can these hazards be minimized?

4. Briefly discuss the management of solid waste to prevent disease.
5. a. What is an environmental impact statement?
   b. Write an environmental impact statement for a proposed solid waste landfill in your town after obtaining the pertinent information from your state environmental protection agency.

REFERENCES

1. U.S. Environmental Protection Agency. *Comprehensive Studies of Solid Waste Management*. Bureau of Solid Waste Management, Publication 2039. Washington, D.C.: U.S. Govt. Printing Office, 1970.

2 Ortolano, L. *Environmental Planning and Decision Making*. New York: John Wiley & Sons, 1984.

Chapter 10   Sources, Composition, and Characteristics of Solid Waste

This chapter focuses on the sources, composition, and characteristics of municipal solid waste.

Solid wastes include all solid or semisolid materials that no longer have any use or value for their possessor. The study of sources, composition, and characteristics is essential for an effective solid waste management system.

SOURCES

The sources of solid waste in a community are generally related to land use and zoning. These wastes can be grouped or classified in several ways, but classifications are necessary to address effectively the complex challenges of solid waste management. As a basis for later discussion, solid waste is divided into four general categories:

1. Municipal waste.
2. Industrial waste.
3. Agricultural waste.
4. Hazardous waste.

Hazardous waste is discussed in Part III, but it should be noted here that the classifications vary depending on the profession and geographical area.

Municipal Waste

Municipal solid waste includes waste from household, institutional, commercial, municipal, and industrial sources (excluding process wastes).

Residential Waste.   This category of waste includes the rejected solid material that originates from single-family, multifamily, and high-rise dwellings. These wastes are often called household wastes, and they consist of garbage, rubbish and trash, bulky waste, and ash.

   Garbage. This type of waste results from food preparation, packaging, consumption, and associated activities. Most of this waste is putrescible. Quick removal from the place of generation, careful storage, and disposal are necessary, because such waste tends to attract rats and flies and to produce strong odors.

   Rubbish and Trash. This category consists of paper and paper products, cans, bottles, plastics, old clothes, leather products, metal products, glass, ceramics, dirt, dust, garden wastes, and so on. In this category, however, only the garden wastes are putrescible.

Bulky Waste. This category includes heavy and large wastes, such as appliances, furniture, mattresses, toys, tires, consumer electronics (e.g., computers, television sets, stereos), and similar items. Because of the size, weight, and irregular generation of these items, special handling and collection techniques are needed.

Ash. This waste is the end product from burning firewood, coal, and so on for the heating of residential units.

Municipal Services.   Municipal solid waste also includes the solid residue from municipal functions and services.

Water and Wastewater Plant Sludge. Sludge that is generated by water and wastewater treatment plants (especially wastewater sludge) needs to be disposed of properly to prevent ground or surface water contamination. These treatment systems are operated either by the municipality, public utility companies, or a sanitary district.

Street Refuse. This type of waste results from the collection of street sweepings and debris that are primarily inorganic in nature (e.g., sand, dirt, grit). The quantity and content of this solid waste depends on the season and the frequency of cleaning operations. For example, during the fall, it may consist of leaves, whereas during the winter, it may consist of sand or cinders used for snow and ice control. When storms strike, there may be branches and leaves from fallen trees.

Public Park and Beach Refuse. People using these facilities also generate refuse, which includes bottles, cans, paper products, food products, ashes, and so on. In addition, waste results from the maintenance of trees, lawns, bushes, and the debris from storm damage. In most cities, an arbor department has the responsibility of maintaining trees and other vegetation in the parkways of streets and other public lands. This may be the responsibility of the park district as well.

Dead Animals. This is a major problem in areas close to habitats with a large population of wild animals (e.g., deer, gophers) or in communities that either do not have animal control laws or do not enforce them. The municipality is responsible for the removal and disposal of dead animals.

"Abandoned" Waste. Even though it is against the law, people still pitch bottles, cans, and paper products into streets, drainage ditches, and parks. Occasionally, junked appliances are also abandoned on public grounds and old automobiles abandoned on streets after removing their license plates. Most cities have a towing and disposal system for such vehicles, which includes legal steps such as title clearance.

Commercial and Institutional Waste. Solid waste also originates from stores, restaurants, markets, offices, hotels, printing businesses, service stations, repair shops, educational and research institutions, hospitals, prisons, and so on and is subdivided into garbage and rubbish. Garbage is generated in restaurants, cafeterias, and fast-food establishments. Rubbish is generated in commercial establishments because of packaging material; in offices and institutions because of paper, and in stores because of plastic, wood, and metal. In addition, there are special wastes generated by hospitals and research laboratories which are solid and semisolid materials. These may include toxic chemicals, radioactive materials, explosive materials or pathological materials.

Pathological materials may include used surgical items, and materials (blood vials, bags, etc) associated with humans and animals in hospitals and research laboratories. Because of the hazardous nature of wastes, they require special collection, handling, and disposal, depending on the characteristics of the material. More discussion on hazardous wastes is in part III of this book. Commercial waste is generated by stores, restaurants, hotels, motels, markets, offices, service stations, and so on. It includes paper, cardboard, wood, plastics, metals, glass, and bulky items. Institutional waste is generated by schools, universities, hospitals, government centers, and so on. It includes items similar to those in commercial waste.

Demolition and Construction Waste. This class of solid waste includes wood, metal, concrete, bricks, glass, plastics, plumbing, wiring, and so on. These are the materials used in the construction of buildings and pavement and that are removed by the destruction of such structures. The quantity and components of such waste can be highly variable.

Industrial Waste

Industrial process wastes are excluded from the category of municipal wastes (though wastes such as corrugated boxes, office papers, cafeteria wastes, wood pallets, and so on are included). There are two general sources of solid waste at industrial sites: the commercial and institutional components, and the process solid waste. The quantities and characteristics of these types of waste are different.

Commercial and Institutional Waste.   This type of waste is generated by office, cafeteria, and other personnel-related activities. These wastes are included in the category of municipal waste.

Process Waste.   These wastes are generated by various industrial processes. These industries include chemical plants, refineries, electrical, printing, wood, and many others. Some of this waste may be hazardous and is handled accordingly (see Part III). Much of this waste is managed on site by the generating industries, but some is landfilled. The end product (i.e., output) from an industrial process may be less than the input, but the balance of the material may end up as solid waste. Because of economic considerations and waste reduction efforts, secondary products are manufactured by recovering a certain portion of this waste.

Agricultural Waste.   In rural areas, the removal and disposal of solid waste resulting from agricultural activities present a significant problem. These wastes are not discussed in any detail in this text, except to indicate that such wastes are generated from animal feedlots and crops. Substantial quantities of manure are generated from feeding operations of cattle, hogs, chicken, or turkeys, and managing these large quantities is a major cost to feedlot owners.

Solid waste associated with raising vineyards and orchards results from prunings and cuttings. The large volume of this material has the potential to harbor insects and plant diseases. Many other disposal problems are also associated with agricultural production. For example, empty containers and bags of fertilizer, insecticide, or pesticide may have some residuals and chemicals, which have the potential for water and land contamination during their disposal.

COMPOSITION

The term *composition* is used to describe the individual components making up the solid waste stream and their relative distribution, which is usually based on percentage by weight. This information is important when evaluating equipment requirements and management plans. For example, separation of recyclable components may be needed.

The composition of solid waste in the United States has changed considerably over the years. Until the 1940s, the bulk of municipal solid waste was ash from coal burning and food waste; scrap materials such as metals and wood products were recovered by scavengers. Even the garbage was not wasted. It was fed to livestock or swine, or it was plowed under fields for nutrients. Scavenging of solid waste is still practiced in many Third World countries, either at the source or the disposal site. In the United States after World War II, however, urbanization (i.e., the influx of people to metropolitan areas) and industrialization changed the citizens' lifestyles, and this was reflected in solid waste generation and composition. During the 1960s, new products appeared in abundance and at affordable prices. Since then, items such as cans, glass bottles, plastic containers, tires, appliances, clothes, and so on are considered to be more economical to discard than to reclaim, and the packaging for many items, and throwaway dinnerware create a vast array of waste material.

Table 10–1 presents historical changes in the U.S. municipal waste stream and composition as generated and discarded by weight (1). Clearly, the percentage of certain waste components has changed over the years. For example, the percentage of paper products and plastics has increased. Table 10–2 lists the breakdown of municipal solid waste in a typical U.S. community (2). Next to residential and commercial waste, construction and demolition waste are the most prevalent.

Significant events during the last 50 years have made a major impact on the composition of solid waste in the United States. These events resulted from technological advances, lifestyle changes, and regulatory restrictions. Lifestyle changes are tied to technological changes as well.

Technological Changes

Soon after World War II, home heating fuel changed from coal to liquid or natural gas. Since then, the ash content of solid waste has decreased markedly. Oil fields were initially developed to supply the large quantity of fuel needed for war machines. Along with liquid fuel, abundant quantities of natural gas were produced as well. After World War II, fuel producers found a big market for space heating. Pipelines and other equipment were developed to transport the fuel to cities, along with the convenience of automatic burners, and this relatively low-priced, clean fuel resulted in a rapid switch from coal to gas or liquid. The ash produced by burning coal is no longer a significant

Table 10–1 Historical Changes in the U.S. Municipal Solid Waste Stream and Its Composition

Component	1970 Generated		1970 Discarded		1980 Generated		1980 Discarded		1990 Generated		1990 Discarded	
	million tons	% composition	million tons	% composition	million tons	% composition	million tons	% composition	million tons	% composition	million tons	% composition
Paper and paperboard	44.2	36.3	36.8	32.5	54.65	36.1	42.75	31.3	73.28	37.5	52.34	32.3
Yard waste	23.14	19	23.25	20.5	27.44	18.2	27.55	20.1	34.9	17.9	30.74	19
Food	12.78	10.5	12.78	11.3	13.22	8.7	13.22	9.6	13.22	6.7	13.22	8.1
Glass	12.67	10.4	12.45	11	14.98	10	14.21	10.4	13.22	6.7	10.57	6.5
Ferrous metals	11.56	10.3	12.45	11	11.57	7.7	11.24	8.2	12.34	6.3	10.35	6.4
Nonferrous metals	1.54	1.3	0.66	1.1	2.86	1.9	1.54	1.5	3.85	2	1.98	1.2
Plastics	3.1	2.5	3.1	2.7	7.82	5.2	7.82	5.7	16.19	8.3	15.86	9.8
Wood	3.96	3.3	3.96	3.5	6.72	4.4	6.72	4.9	12.34	6.3	11.9	7.3
Rubber and leather	3.19	2.6	2.86	2.6	4.29	2.8	4.18	3.1	4.62	2.4	4.4	2.7
Textiles	1.98	1.6	1.98	1.8	2.64	1.7	2.64	1.9	5.62	2.9	5.28	3.3
Miscellaneous	2.64	2.2	2.31	2	5.06	3.3	4.62	3.3	6.06	3	5.28	3.4
Total	110.5	100	102.2	100	137.3	100	123.9	100	177.6	100	147	100

Source: U.S. Environmental Protection Agency. *Characterization of Municipal Waste in the U.S.* Washington, D.C.: U.S. EPA, 1992; with permission.

Table 10–2   Breakdown of Municipal Solid Waste in a Typical U.S. Community

Waste Source	Range (% of wt)	Typical (% of wt)
Residential and commercial (nonhazardous)	50–75	62
Special waste (e.g., bulky)	3–12	5
Hazardous	0–0.99	0.1
Institutional	3–5	4
Construction and demolition	8–20	14
Street sweepings	2–5	4
Landscaping	4–9	6
Treatment plant sludge	3–8	5

component of municipal solid waste. Table 10–3 describes the typical composition of domestic waste in several countries.

Modern technological advances in packaging, encouraged by a general economic prosperity, also caused a significant change in the composition of solid waste (2–4). Increasing use of plastics in the beverage industry and of paper, plastic film, and aluminum foil in the food industry are examples. As the technology of packaging and freezing various food items improved, the industry began marketing different types of prepared foods, such as the "TV dinner" (i.e., a meal that needs a few minutes of thawing and heating before it can be eaten). After the meal, the empty package (e.g., paper, foil) is thrown away. The development of microwave ovens further added to the convenience. The marketing of prepared or semiprepared food requires processing to a full or a partial extent, removing nonedible or unwanted portions of the food (e.g., carrot tops, potato skins, corn cobs and husks, bean pods, bones) that would have become household garbage, which instead remain at the processing plant and become

Table 10–3   Typical Composition of Domestic Waste

Component	United States (%)	United Kingdom (%)	Poland (%)	China (%)
Food wastes	9	25	24	36
Paper, cardboard	40	29	11	2
Plastics	7	7	2	1.5
Glass	8	10	6	1
Metals	9.5	8	2	1
Clothing/textiles	2	3	10	1.5
Ashes, dust	3	14	45	57
Unclassified (e.g., garden, yard, wood)	21.5	4	—	—

Source: Data from Tchobanoglous, G.H., Theisen, H., Vigil, S. *Integrated Solid Waste Management.* New York: McGraw-Hill, 1993; Mortensen, E. "Introduction to Solid Waste," Lectures Notes for Graduate Diploma in Environmental Engineering, University, College Cork, Ireland, 1990–1993; and World Health Organization. *Urban Solid Waste Management.* Firenze, Italy: WHO, 1991.

part of the industrial solid waste. These biodegradable food wastes are replaced by plastic, paper, and aluminum foil, which being synthetic materials are mostly non-biodegradable. Introduction of the garbage grinder as an appliance for most new residential units (and some old ones) has significantly decreased the garbage content of solid waste (Table 10–1). The food waste instead reaches the sewer through the garbage grinder and adds to the organic load of the wastewater treatment plant.

Another example of the effect of new products on the composition of solid waste is the beverage-container industry. As late as the 1960s, almost all beverages were marketed in returnable glass bottles. The wide acceptance of cans and nonreturnable bottles followed the development of pop-top cans and manufacture of such bottles. Consumption of beverages increased from approximately 53 billion units in 1958 to approximately 80 billion units in 1976, and during this period, throwaway containers increased from 10 billion to 58 billion units and returnable bottles remained constant at approximately 1.6 billion units (5).

Recently, however, this same industry's efforts to decrease use of disposable containers has affected the amount of refuse entering the waste stream. As a percentage of municipal solid waste, containers and packaging have been declining. This has resulted from increasing use of relatively lightweight aluminum and plastic and a decreasing use of heavier steel and glass containers (6). The market share of aluminum cans for disposable containers has increased from approximately 1 percent in 1964 to more than 90 percent in 1984. In 1985, more than 66 billion aluminum cans were used. For large beverage containers (≥ 1 L), including water, milk, juice, and soft drinks, plastic containers continue to be popular, and this trend may even eliminate aluminum and glass from the beverage market altogether (7).

Because of marketing needs and for protection from damage, a significant amount of packaging material (e.g., cardboard, styrofoam, paper, plastic, aluminum foil) is used to display items in the shop. Often, more packaging material is used than is really needed. Estimates for the quantities of packaging material added to the municipal solid waste stream in the United States (Table 10–4) indicate an increase in the quantity of such material (8, 9). The percentage of the solid waste stream accounted for for pack-

Table 10–4   Estimated Quantities of Packaging Material in the Municipal Solid Waste Stream

	1971[a]	1980[b]	1990[b]
Million tons	35.1	38.2	45.1
% Of total	36.9	36.6	35.5

[a] Data from U.S. Environmental Protection Agency. *Recycling Works! State and Local Solutions to Solid Waste Problems.* Washington, D.C.: U.S. Govt. Printing Office, 1989.

[b] Data from U.S. Environmental Protection Agency. *Assessing the Consumer Market.* 21P-1003. Washington, D.C.: U.S. Govt. Printing Office, 1991.

aging materials, however, has decreased slightly. Use of lighter plastic material instead of heavier paper has decreased the percentage by weight of the packaging material.

The printing industry also adds a significant amount of paper to the solid waste stream. Both the technological advances in printing and cheap mailing rates for advertising material have caused an avalanche of printed paper in the mail, that eventually becomes a part of the solid waste stream. Table 10–5 indicates the quantities of paper used for different printing purposes and shows an increase from 1971 to 1993 (8, 10).

Data for the solid waste collection rate in other countries is shown in Table 10–6 (11). By comparing Table 10–6 with Table 10–3, the following observations are made:

1. Food waste constitutes approximately 30 percent of all waste, except in the United States. The low U.S. figure of 9 percent results from kitchen garbage grinders and food-processing methods. Garbage grinders increase the organic load on wastewater treatment plants, and food processing increases the organic solid waste of the processing industry. In both cases, domestic organics—and, thus food wastes—are reduced.

2. The ash content is approximately 50 percent in China and Poland because of the greater use of coal in less-developed countries. The ash content of solid waste in developed countries is less than 5 percent because of the use of other energy sources (e.g., gas, electricity, oil) for home heating and cooking.

Table 10–5   Quantities of Paper Used for Various Printing Purposes

	1971 (million tons)	1990 (million tons)	1993 (million tons)
Newspapers	6.62	9.11	12.9
Books, periodicals	4.43	6.96	7.5
Office paper	2.4	5.71	6.8
Magazines	—	—	2.2
Corrugated boxes	—	—	26.35
Total	13.45	21.78	55.75

Source: Data from *Waste Age* April 1980: Raven P., et al. *Environment.* Sander College Publication, 1995; and U.S. Environmental Protection Agency. *Generation and Recovery of Major Categories of Paper and paperboard in U.S. in 1993.* Washington, D.C.: U.S. Govt. Printing Office, 1993.

Table 10–6   Domestic and Commercial Solid Waste Output (1959–1961)[a]

	Population	Range	Median	
Country	Range	lbs/cap/year	lbs/cap/year	lbs/cap/day
U.S.A.	50,000–8,500,000	1,100–1,700	1400	3.83
England	70,000–8,600,000	450–1,080	650	1.78
France	50,000–4,500,000	400–900	575	1.58

[a] This comparison indicates that excessive solid waste is produced in the United States.

Source: Rogus, C.A. "Refuse Collection and Disposal in Western Europe," *Public Works*, April 1962.

3. Paper, cardboard, and plastics form a very low percentage of solid waste in China and Poland compared with the United States and United Kingdom. Again, economic conditions and level of development are the reasons.

Lifestyles

Technological advances and industrialization have allowed residents of the United States to enjoy one of the highest standards of living in the world. Unfortunately, the standard of living is inevitably tied to the generation of solid waste, to the philosophy of wastefulness and the quick obsolescence of products.

The behavior of U.S. consumers today constitutes an aberration in both time and space. A typical New Yorker, for example, discards approximately 4 to 5 lbs of solid waste each day, whereas a resident of Hamburg or Rome throws out only about half of that (12). Per-capita waste generation in Calcutta, India, is approximately 1.12 lbs/day (Indian Association of Environmental Engineers, personal communication, 1993). In the United States, cheap products such as appliances, furniture, tires, and so on are relatively cheap, but the high cost and inconvenience of repairing them encourages consumers to buy new products rather than to repair the old ones. Thus, these old products become a part of solid waste. Table 10–7 provides a few comparisons (13, 14).

People like to consume food and drinks from easy-to-serve, easy-to-discard containers, paper cups, and plates. Soon after use, these items join the solid waste stream in large quantities. Americans use many household chemicals (e.g., personal hygiene products, cleansers, insecticides, pesticides, automobile oil) in easy-to-use containers. These empty or partially full containers are then discarded as solid waste.

Recently, the growth of the computer and video industry has impacted the solid waste stream by adding broken and used equipment, tapes, printer paper, and so on. Modern technology has made computer and entertainment equipment affordable, thus increasing the amount of waste.

Table 10–7   Per-Capita Generation of Municipal Solid Waste in Selected Countries

Country	Year of Estimate	Per Capita Generation (lbs/year)
Germany	1987	700
Italy	1989	663
Japan	1988	878
United Kingdom	1989	786
United States	1993	1608

Sources: Data from U.S. Environmental Protection Agency. *Materials Generated in Municipal Solid Waste by Weight, 1993.* Washington, D.C.: U.S. Govt. Printing Office, 1994.

Regulatory Changes

Federal and state regulations to control air pollution have impacted the composition and quantities of municipal solid waste. Not enough data are available on the characteristics or quantities of such waste generated before the ban on burning to create a good estimate for that period. As an approximation of the legislation's impact, however, consider a typical solid waste stream of approximately 25 percent ash, 25 percent water, and 50 percent organic material. Assuming that combustion is approximately 70 percent efficient in "burning" the 50 percent organic material and in evaporating almost all of the water, only approximately 35 percent of the weight of the original waste (0.50 × 0.70) remained. This material would be approximately 60 percent inorganic (0.25 + 0.35), such as ash, glass, metals, and dirt. It has little or no value for recycling, is inert, and would be suitable only for landfilling.

This inert material must be hauled to the land disposal site. After the "burn," the density of the inert material is almost twice that of the unburned material. With this density increase and accompanying volume reduction, the number of trips to the disposal site might be fewer (e.g., one-fourth) than that required with unburned refuse. The ban on open burning has significantly impacted the composition of solid waste as well.

Recycling

Recycling is an important practice of solid waste management, and interest in recycling has revived during the last 15 years. Many state and local governments have passed mandatory recycling legislation. For example, in 1987, New Jersey passed the first statewide recycling legislation. Under this law, residents were required to recycle 25 percent of all solid waste generated and towns to compost leaves by 1989 (14). This legislation is significantly affecting the composition of solid waste.

The returnable bottle deposit law in many states aims to reduce highway litter, conserve resources, and reduce the volume of solid waste for disposal. The bottle law is usually applicable to all types of beverage containers, including glass, metal, and plastic, but not to other types, such as food jars, plastic and paper cups, wine and liquor bottles, and so on. Some returnable bottle laws are being amended, however, to include liquor, wine, and possibly other containers.

Quantities and Characteristics

The U.S. Environmental Protection Agency (EPA) estimates the United States generated 207 million tons of municipal solid waste (MSW) in 1993. This is approximately 4.4 lbs (92 kg) per person per day (15). This figure includes solid waste from residential, commercial, institutional, and industrial sources, but it does not include such things as construction waste, automobile bodies, municipal sludge, combustion ash, and industrial process waste. When items such as construction and demolition waste and municipal plant sludge are included, this estimate compares favorably with that by Tchobanoglous *et al.* (2) of 6.1 lbs/person/day. Chapter 7 indicated that the actual quantities of MSW being generated remain elusive, and it should be noted that average values are subject to wide variation from city to city, season to season, and with respect to the methodology used.

Table 10–8   Typical Per-Capita Solid Waste Generation Rates from Other Municipal Services

Municipal Service	Generation Rate (lbs/person/day)
Construction debris	0.86
Street sweepings	0.25
Trees and landscaping	0.19
Parks and recreational areas	0.12
Catch basins	0.04
Sewage treatment plant sludge	0.37
Total	1.83

Sources: Data from U.S. Environmental Protection Agency. *Materials Generated in Municipal Solid Waste by Weight, 1993.* Washington, D.C.: U.S. Govt. Printing Office, 1994; Tchobanoglous, G.H., H.L. Theisen, and S. Vigil. *Integrating Solid Waste Management.* New York: McGraw-Hill, 1993; and Schwarz, S. *Energy and Resource Recovery from Waste.* Park Ridge, NJ: Noyes Data Corporation, 1983.

The best estimate for the residential fraction of MSW probably lies between 2.3 to 2.7 lbs/person/day (15). When these figures are added to the fraction of waste from other municipal services given in Table 10–8, the MSW generation rate is close to 6.1 lbs/person/day. Typical numbers for estimating solid waste generation should be used only when other data are not available, however, and then only for preliminary evaluation.

Figure 10–1 shows a materials breakdown for the EPA's estimate of 207 million tons of waste (14). Paper and paperboard products are the largest component, or 37.6 percent by weight. The second largest component is yard trimmings, at 15.9 percent. Figure 10–2 shows the historical growth in generation of waste by source (16). Over

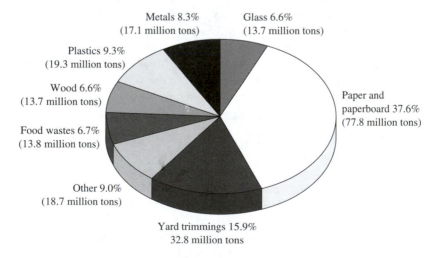

Figure 10–1   Components of municipal solid waste by weight in 1993.

(From U.S. Environmental Protection Agency. Materials Generated in Municipal Solid Waste by Weight, 1993. *Washington, D.C.: U.S. Govt. Printing Office, 1994; with permission.)*

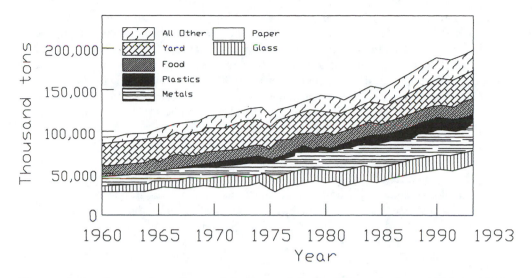

Figure 10–2   Generation of materials in municipal solid waste from 1960 to 1993.

(From U.S. Environmental Protection Agency. Generation of Materials in Municipal Solid Waste, 1960–1993. Washington, D.C.: U.S. Govt. Printing Office, 1994; with permission.)

the years, the paper and paperboard products component has grown steadily. In contrast, the yard trimmings component has recently been gradually declining as backyard composting and mulching have become more prevalent. Many regulating agencies also now require that yard trimmings not be landfilled.

Figure 10–3 shows trends in U.S. solid waste generation, recovery, and discards from 1960 to 1993, with projections to the year 2000 (16). Between 1960 and 1993, the

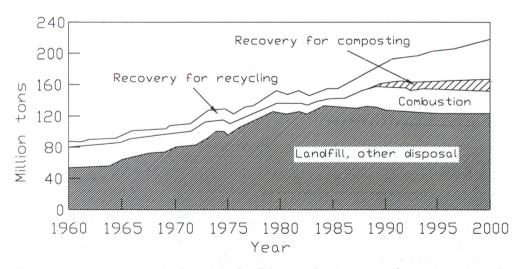

Figure 10–3   Management of municipal solid waste from 1960 to the year 2000 (projected).

(From U.S. Environmental Protection Agency. Generation of Materials in Municipal Solid Waste, 1960–2000. Washington, D.C.: U.S. Govt. Printing Office, 1994; with permission.)

generation of waste more than doubled, from 88 million to 207 million tons. Things are not so bad as these numbers might indicate, however. Combined materials and energy recovery increased from 7 percent of solid waste generated in 1960 to 22 percent in 1993. In addition, the total amount of solid waste sent to landfills or incinerators is projected to decline from its peak of 162 million tons in 1990 to 152 million tons in the year 2000— even though total solid waste generation is projected to increase to 218 million tons.

Figure 10–3 also indicates that in 1985, 83 percent of MSW was buried in landfills, but in 1993, that figure had dropped to 62 percent. In addition, in 1960, approximately 30 percent of MSW was burned in incinerators, which did not have air pollution controls or energy recovery. As people became aware of the environmental problems associated with air quality during late 1960s and 1970s, the old incinerators were closed, and the fraction of total waste that was burned dropped to a low of 10 percent in 1980. The need for energy recovery and air emission controls resulted in new and modified incinerators, and the fraction of waste burned has since gone up to approximately 15 or 16 percent.

Earlier discussion identified the changes in solid waste quantities and characteristics that are more universal in character. Additional variations may be area specific, and must be considered in solid waste management as well.

Household Variations

The quantity and characteristics of solid waste may vary from household to household within the same community depending on the size of the household, the income of its members, and their attitudes. A study of solid waste generation from households of different sizes and characteristics showed that this variation is significantly influenced by the number of people in a household (17). The results of this study are presented in Figures 10–4 through 10–6. For various reasons, average household size has decreased more than 20 percent during the last 33 years, from 3.42 people in 1960 to 2.67 in

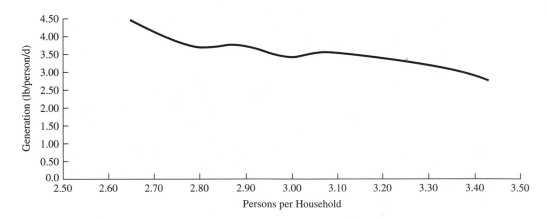

Figure 10–4   Generation of municipal solid waste as a function of household size.

(From U.S. Environmental Protection Agency, Generation of Materials in Municipal Solid Waste, 1960–2000. Washington, D.C.: U.S. Govt. Printing Office, 1994; with permission.)

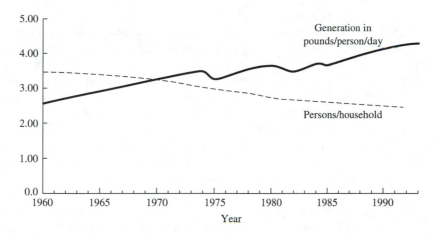

Figure 10–5   Generation of municipal solid waste as a function of household size over time.

(From U.S. Environmental Protection Agency. Generation of Materials in Municipal Solid Waste, 1960–2000. Washington, D.C.: U.S. Govt. Printing Office, 1996; with permission.)

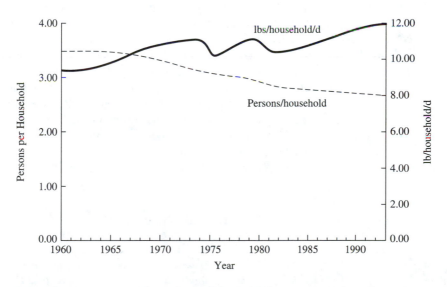

Figure 10–6   Generation of municipal solid waste as a function of persons/household and lbs/household/day.

(From U.S. Environmental Protection Agency. Generation of Materials in Municipal Solid Waste, 1960–2000. Washington, D.C.: U.S. Govt. Printing Office, 1999; with permission.)

1993. The U.S. Bureau of the Census and the American Demographics project the average household size will drop to 2.4 by the year 2000 before starting to increase again.

Smaller households have resulted in higher per-capita consumption rates for several products that end up in the waste stream. Items such as newspapers, appliances, furniture, and yard trimmings would be expected to relate more closely to the number of households than to the overall population. For example, households with one or two people generally read the same number of newspapers and have the same number of appliances as larger households.

Figure 10–4 shows how per-capita MSW generation decreases with increasing household size. Generation and average household size are shown as a function of time in Figure 10–5, and as expected, per capita generation is seen to rise as the household size decreases. Figure 10–6 shows that despite decreasing household size, the amount of MSW generated per household has continued to increase, but not as rapidly as per-capita generation has. (Note that the correlation between generation in pounds per household per day and people per household is weaker than the correlation between generation and pounds per person per day and between people per day and people per household.)

The effect of socioeconomic strata on the waste generation (18) and on domestic waste composition (19) was studied for short terms in Berkeley, California, in 1967; Cincinnati, Ohio, in 1970 to 1971; and in Boston, Massachusetts, in 1969 to 1970. Based on these limited data (Table 10–9), solid waste from low-income domestic sources appears to contain more glass, metal, and food waste and less paper, textiles, plastics, leather, and rubber than average waste. These conclusions are consistent with lesser use of heavily packaged prepared foods, enhanced use of canned versus frozen foods, and the consumption of less clothing and shoes by individuals with lower incomes. In addition, fewer low-income households have food grinders, so waste food is discarded as solid waste to be collected rather than to sewers through the grinders. This explains, in part, the larger amount of food waste from low-income households.

Attitudes of people play a significant role as well. People who are willing to change their habits and lifestyles to conserve natural resources generate less solid waste in their homes, and the wastes they do generate likely has different characteristics. For example, people take old magazines to retirement homes, use more returnable or reusable containers, give unwanted clothes and appliances to the Salvation Army, and separate recyclable material before discarding the solid waste for final disposal.

Table 10–9   Composition of Solid Waste from Low-Income Urban Areas

	Component (% of wt)						
Income level	Glass	Textile rubber	Metal waste	Leather, food, and plastics	Food	Paper, food, and yard waste	Paper
Low-income	14.5	1.9	9.7	2.3	22.2	66.4	42.7
Average income	8.8	2.2	8.6	2.9	17.1	70.5	44

Seasonal Variations

Because of seasonal contribution to waste quantity (e.g. yard wastes, recreational area waste), significant variations occur in the per-capita generation rate. During the spring and summer, quantities of grass clippings as well as tree and bush trimmings, especially in low-density areas, increase. In the fall, these garden trimmings give way to leaves. Many states restrict the use of landfills for disposal of this material. Burning is also prohibited in metropolitan areas. During the Christmas season, there is an increase of packaging material, batteries, and of course, Christmas trees after the holiday. Figure 10–7 shows an estimate of the expected pattern of seasonal variation in waste load (20).

Any solid waste collection, processing, and disposal system must be designed to deal with variations in the production of such wastes. For example, collection is significantly affected by these variations. When large increases in the quantity of solid waste occur, additional crews and vehicles are needed, or overtime must be paid. Both drive up the cost of collection. During seasonal storms, additional workers or overtime may also be required to remove damaged trees and shrubs. If this material is to be disposed of in a landfill, then the problem becomes capacity. For incineration, garden trimmings and grass usually have a high moisture content and, thus, an impact on the energy content of the solid waste. Such waste may require auxiliary fuel to start and to maintain combustion.

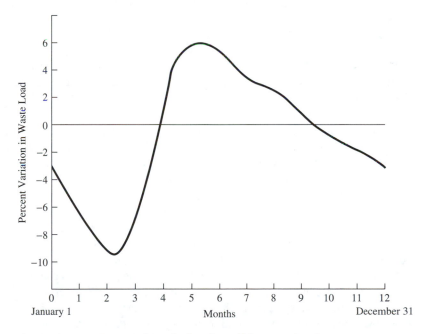

Figure 10–7   Seasonal variation in solid waste load.

(From Davidson, G.R., Jr. A Study of Residential Solid Waste Generation in Low-Income Areas. Report SW-83. Washington, D.C.: U.S. Govt. Printing Office, 1972; with permission.)

Regional Variations

Seasonal variation itself can differ in different regions. In areas with high precipitation, the moisture content of solid waste will be considerably higher than in arid regions. For example, in wet areas such as Sao Paulo, Brazil, the moisture content of solid waste is typically 50 percent (20). This would mean a combination of more yard and garden waste as well as more precipitation to wet the additional refuse. In contrast, the low humidity common in arid regions may significantly reduce the moisture originally present in the waste. The amount of such waste in arid regions also would be less than in regions with high rainfall.

There are other regional variations, but these are mostly minor. They may become more pronounced, however, because of legislation at the state level. In states having a beverage-container deposit law, the proportion of such containers in the solid waste is likely to be less than that in the states without such a law. If all beverages are sold in disposable (e.g., plastic) containers, the glass and aluminum content would be substantially less as well.

HAZARDOUS WASTE IN MUNICIPAL SOLID WASTE

People use a host of products containing hazardous chemicals and then discard those products and/or their containers without realizing the consequences. This section discusses the problems associated with hazardous waste components in MSW and the ways to eliminate it.

Many products used each day inside the home (e.g., furniture polish, wood preservatives, paint and paint thinner, stain remover, household cleaners, automotive products, batteries) as well as lawn and garden products (e.g., herbicides, pesticides, fungicides) contain hazardous chemicals. They can be found in the garage, in the basement, under the kitchen sink, and in the bathroom. The hazardous materials in these products can be harmful to public health and the environment when improperly used or disposed. Figure 10–8 shows a few such household products. Studies have not been made to estimate the amount of waste stored in the home, disposed of improperly, or put out for collection. The residential waste stream, however, includes everything put out in the trash, poured down the drain, or dumped on the ground. The relative distribution of hazardous wastes will vary as well, depending on the time of year (e.g., greater during the spring cleanup) (1).

Another source of hazardous waste is commercial establishments. These establishments are mostly small-quantity generators. Typical examples include solvents from auto-repair shops and dry cleaners, paints and thinners from painting contractors, inks from print shops, and synthetic material from appliance repair shops, and herbicides and fungicides from plant nurseries and crop-growing establishments.

Because these hazardous wastes from homes and small commercial operations contribute to the hazardous waste in MSW entering sanitary landfills, the EPA is concerned about these wastes entering the soil, water, or air after disposal. The amount of hazardous material found in residential and commercial MSW is in the range of 0.075 to 0.2 percent by weight, of which approximately 75 to 85 percent is from residential sources (2). Table 10–10 presents selected examples of such hazardous materials and the products in which they are found.

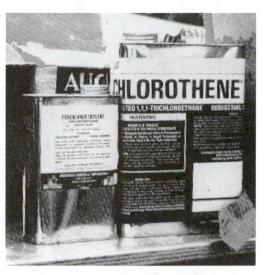

Figure 10–8   (a) Old pesticides now either banned or restricted by the U.S. Environmental Protection Agency. (b) Chlorinated solvents commonly found in households.

(From Ostler, N., ed. Introduction to Environmental Technology. *Upper Saddle River, N.J.: Prentice-Hall, 1996; with permission.)*

Problems from Hazardous Waste

Even the small amounts of hazardous waste found in MSW can cause problems by their presence in solid waste landfill or combustion units and their persistence when discharged in the environment. These hazardous wastes affect the recovery of materials, combustion and conversion products (e.g., compost), and landfills. Unless an effective source separation is done, trace amounts of hazardous organic constituents will be found in the MSW and compost produced from such MSW is not desirable as a soil conditioner. Combustion of hazardous MSW results in gaseous emissions and ash that are toxic because of heavy metals such as barium, cadmium, chromium, lead, and so on.

Trace organic constituents have also been found in leachate and gaseous emissions from MSW landfills. The source of these organics is hazardous waste in the MSW or biological and chemical reactions in the landfill. Again, most hazardous waste found in MSW tends to persist over a long period of time. Examples of persistent organic wastes include chlorinated pesticides as well as high-molecular-weight chlorinated and aromatic hydrocarbons. Examples of nonpersistent organic wastes include oils, biodegradable pesticides (mostly nonchlorinated), most detergents, and low-molecular-weight solvents.

Physical, Chemical, and Biological Transformation of Constituents

The mechanisms and processes associated with physical, chemical, and biological transformation of hazardous waste are discussed in detail in Chapters 13 and 16.

Table 10–10   Selected Examples of Hazardous Ingredients Found in Common Household Products

Product	Ingredient
Automotive Products	
Antifreeze	Ethylene glycol
Auto battery	Sulfuric acid
Gasoline	Hydrocarbons, benzene
Motor oil	Trichloroethylene
Used motor oil	Lead, hydrocarbons, trichloroethylene
Household Cleaners and Similar Products	
Adhesives	Acrylic acid
Cosmetic (e.g., perfume), wood stain	Aniline
Paint	Arsenic oxide
Household cleaners (e.g., spot remover, open cleaner), stain, varnish, adhesives cosmetics (e.g., nail polish remover)	Benzene
Ni-Cd battteries, paints, photographic chemicals	Cadmium
Flea powders	Chlordane
Latex paint	Chlorinated phenols
Household degreasers	Chlorobenzene
Insect repellents	Hexachloroethane
Paint, varnish, auto batteries	Lead
Paint, batteries, lamps	Mercury
Household cleaners, paint strippers, adhesives	Methylene chloride
Shoe polish	Nitrobenzene
Batteries, photographic chemicals	Silver
Rat and mice control	Warfarin
Transmission fluid, degreassers, paint, fabric, cosmetics, microfilm	Xylene

Source: U.S. Congress, Office of Technology Assessment. *Facing America's Trash*, Washington, D.C.: U.S. Government Printing Office, 1994; with permission.

Management of Hazardous Waste

The most effective way to eliminate the small quantities of hazardous waste found in MSW is to separate them at the point of generation. In addition, the citizens should be educated about the proper use, storage, and disposal of household hazardous wastes. To minimize improper disposal, a number of communities have established product exchange programs, special collection days, and collection sites.

Reduce, Reuse, and Recycle.   The concept behind these three words may encompass a variety of actions and practices that collectively are known as source reduction and are applicable to hazardous waste in MSW. The "Three R's" are discussed in detail in Chapter 14.

A process that uses excessive amounts of raw materials or produces excessive amounts of hazardous waste is environmentally and economically undesirable. If the materials use or waste generation itself cannot be reduced, then an item or material should be reused, because reuse provides a significant benefit. Recycling means that waste is converted into a reusable product; for example, lead from lead–acid batteries is reclaimed for new products. In addition, used oil, a very large waste stream of which only a relatively small percentage is collected at auto-service stations and even at curbside as part of waste disposal services, can be partially re-refined and used as a fuel after blending.

Product Exchange Programs.   Items of interest can be exchanged for use by individuals, such as in paint exchange programs. A buyer of a particular color or type of paint may no longer wish to use it, but another person may find use for it.

Specific Collection Days.   A common approach to the management of household hazardous waste is to hold community collection day or days. On these collection days, community members can bring their waste, at little or no charge, to a specific location for reuse, recycling, treatment, or disposal by persons in the business of handling such wastes. The cost of such programs are paid by local governments, businesses, or industries.

Permanent Collection Sites.   For convenience, many communities establish permanent collection sites (e.g., city warehouses, landfill sites, fire stations). Permanent collection sites are popular, because citizens can drop off wastes according to their own schedules.

REVIEW QUESTIONS

1. a. List the sources of solid waste.
   b. Briefly discuss the characteristics of each.
2. a. Define the composition of solid waste.
   b. How has the composition of plastics and paper products changed in the last 30 years? State possible reasons for such changes.
3. a. List the possible causes of changes in the composition of municipal solid waste.
   b. How does climate affect the composition?
4. Why do the estimates of municipal solid waste generation from various sources differ?
5. a. Estimate the municipal solid waste generation per four-person and per six-person household. (*Hint*: Use the graphs in this chapter.)
6. Check your (or any other) house, in the garage and under the bathroom and kitchen sinks, and then list:

   a. Name of the products you find.
   b. Chemicals listed on the product label.
   c. Any caution or warning on the label.
   d. Intended use.
   e. Adverse effects of listed chemicals on public health and the environment.
      (*Hint*: You can consult a book on chemicals or contact an EPA District Office).
7. a. What problems may arise from mixing household hazardous waste with other domestic waste?
   b. Discuss ways to minimize these problems.
8. a. Does your town or county have a special collection program for household harzardous waste?
   b. If so, discuss the operation of the program.

REFERENCES

1. U.S. Environmental Protection Agency. *Characterization of Municipal Waste in the U.S.* Washington, D.C.: U.S. EPA, 1992.
2. Tchobanoglous, G.H., H.L. Theisen, and S. Vigil. *Integrated Solid Waste Management.* New York: McGraw-Hill, 1993.
3. Mortensen, E. "Introduction to Solid Waste," Lectures Notes for Graduate Diploma in Environmental Engineering, University College, Ireland, 1990–1993.
4. World Health Organization. *Urban Solid Waste Management.* Firenze, Italy: WHO, 1991.
5. Midwest Research Institute. *The Role of Packaging in Solid Waste Management, 1966 to 1976.* Kansas City, MO: Midwest Research Institute 1980.
6. Council for Solid Waste Solutions. *The Solid Waste Problem.* Chicago: Public Works Dept., 1989.
7. Pfeffer, J.T. *Solid Waste Management Engineering.* Upper Saddle River, N.J.: Prentice-Hall, 1992.
8. *Waste Age*, April 1980.
9. Raven, P., et al. *Environment.* Sander College Publication, 1995.
10. U.S. Environmental Protection Agency. *Generation and Recovery of Major Categories of Paper and Paperboard in U.S. in 1993.* Washington, D.C.: U.S. EPA, 1994.
11. Rogus, C.A. "Refuse Collection and Disposal in Western Europe," *Public Works*, April 1962, pp. 17–23.
12. Worldwatch Institute Publication, 1987, pp. 7–14.
13. World Resources Institute, 1992.
14. U.S. Environmental Protection Agency. *Materials Generated in Municipal Solid Waste by Weight, 1993.* Washington, D.C.; U.S. EPA, 1994.
15. Schwarz, S. *Energy and Resource Recovery from Waste.* Park Ridge, N.J.: Noyes Data Corporation, 1983.
16. U.S. Environmental Protection Agency. *Generation of Materials in Municipal Solid Waste, 1960–1993.* Washington, D.C.: U.S. EPA, 1994.
17. U.S. Environmental Protection Agency. *Characterization of Municipal Solid Waste in the United States.* Washington, D.C.: U.S. Govt. Printing Office, 1994.
18. Davidson, G.R., Jr. *A Study of Residential Solid Waste Generated in Low-Income Areas.* Report SW-83. Washington, D.C.: U.S. EPA, 1972.
19. "Analysis of composition of Rubbish in the United States," *Solid Waste Management*, September 1972, pp. 62–64.
20. Kennedy, J.C. "Seasonal Variations in Municipal Solid Waste Output," Paper presented at the Engineering Foundation Research Conference on Solid Waste R & D, Philadelphia, July 24–28, 1967.

Chapter 11 Physical, Chemical, and Biological Properties of Municipal Solid Waste

This chapter introduces the physical, chemical, and biological properties of municipal solid waste (MSW). Many of these properties can be quantified through calculations and data collection, and these properties must be known to develop and design a solid waste management system.

In the past, waste handlers needed to know little about the physical, chemical, and biological properties of solid waste, because almost all of it was disposed in landfills. Today, however, proper solid waste management techniques involve recycling, reuse, transformation, and highly regulated disposal, so more knowledge regarding the properties of MSW is necessary.

PHYSICAL PROPERTIES

Important physical properties of MSW include specific weight, moisture content, particle size distribution, field capacity, and hydraulic conductivity. Because MSW is a heterogeneous mass of constantly changing composition, these physical characteristics must be viewed on an aggregate and an individual-component basis. In addition, the various waste processes will alter the physical character of the waste, which thus requires recharacterization after each processing state. For example, shredding homogenizes solid waste, alters the shape of the waste components, and reduces the particle size distribution.

Specific Weight

Specific weight is defined as the weight of a material per unit of volume (lb/ft^3, lb/yd^3). It is important here to distinguish between compacted and uncompacted waste. Specific weight data are needed to get an idea of the total mass and volume of the waste that must be managed. Table 11–1 presents specific weights for a few typical categories of MSW.

Specific components of MSW vary with location, season, and length of time in storage, so the selection of "typical" values should be carefully made. As delivered in a compaction vehicle, MSW varies from 300 to 800 lbs/yd^3. A typical value is approximately 500 lbs/yd^3. In a noncompacting collection truck, the specific weight is 150 lbs/yd^3, whereas in an individual pickup truck, it is 100 lbs/ft^3.

These specific weights assume that each collection vehicle is full. If this is not the case, specific weights should be confirmed with an on-site weighing program. For

153

Table 11–1   Typical Uncompacted Specific Weights for Municipal Waste Components

Components	Specific Weight (lbs/yd^3)[a]
Broad Categories	
Paper, cardboard, plastics	135
Food wastes	500
Miscellaneous rubbish[b]	270
Ash, dirt, brick, ferrous metal	800
Municipal solid waste	250
Specific Categories	
Plastic	110
Aluminum	270
Tin cans	150
Yard waste	170

[a] The actual specific weight may vary by as much as 50% from the values shown depending on the nature and moisture content of the constituents.

[b] Miscellaneous rubbish includes glass, nonferrous metal, wood, rubber, leather, and textiles.

Source: Adapted from Glynn, H., and G. Heinke. *Environmental Science and Engineering.* Upper Saddle River, N.J.: Prentice-Hall, 1996; with permission.

example, a compaction truck with a volume of 20 yd^3 and a net weight (i.e., the weight of the loaded truck less the weight of the compactor truck) of 10,500 lbs will have a specific weight of 10,500 lbs/20 yd^3, or 525 lbs/yd^3.

The volume occupied by solid waste under certain conditions determines the number and size or type of refuse containers, collection vehicles, transfer stations, and land requirements for disposal. For example, Kaiser (1) found that loose solid waste weighed 108 to 135 lbs/yd^3. Under a load equivalent to 30 ft of solid waste, the weight was 349 lbs/yd^3 when the moisture content was 27 percent and 480 lbs/yd^3 when the moisture content was 42 percent. Moisture increases the specific weight of solid waste until the material becomes saturated with water. As water replaces air in the voids, the specific weight increases, but additional water may decrease the specific weight by replacing the solids.

Reduction in the volume of solid waste is an important consideration in collection, transport, and disposal. Volume reduction may be expressed in terms of a compaction ratio as well as a percentage. The performance of baling and compaction equipment is rated by the percentage of volume reduction and the compaction ratio as well. Expressed as an equation:

$$\text{volume reduction } (\%) = 100\left(1 - \frac{V_f}{V_i}\right) \tag{11-1}$$

where V_i is the initial volume of wastes before compaction (yd^3) and V_f is the final volume of wastes after compaction (yd^3). Thus, the compaction ratio is defined as

$$\frac{V_f}{V_i} \tag{11-2}$$

Example 11–1

The initial volume of a mass of solid waste is 20 yd³. After compaction, the volume is 5 yd³. Compute the percentage of volume reduction and the compaction ratio.

Solution

Applying equation (11-1),

$$V_i = 20 \text{ yd}^3 \quad V_f = 5 \text{ yd}^3$$

Thus,

$$\text{volume reduction } (\%) = \frac{20 - 5}{20} \times 100 = 75\%$$

and the compaction ratio = 20/5, or 4.

Moisture Content

The moisture content of solid waste is expressed in the wet weight method or the dry weight method of measurement. In the wet weight method, moisture is expressed as a percentage of the wet weight of the material. In the dry weight method, it is expressed as a percentage of the dry weight of the material. In solid waste management, the wet weight method is most commonly used, expressed as an equation:

$$m = \frac{w - d}{w} \times 100 \tag{11-3}$$

where m is the moisture content (%), w is the initial weight of the sample as delivered (lb or kg), and d is the weight of the sample after drying at 105°C (lb or kg).

For most MSW in the United States, the typical moisture content varies from 15 to 40 percent depending on composition, season of the year, and weather conditions (e.g., rain, draught). The lower percentage represents solid waste from an arid region; the higher percentage represents solid waste from a region with high precipitation. The moisture content of solid waste is an important parameter in determining the heating value, density, and leachate from sanitary landfill. Table 11–2 provides typical moisture contents of MSW components.

Example 11–2

Solid waste with a dry specific weight of 200 lbs/yd³ can absorb a significant amount of water, which fills the voids by replacing air. If this waste has a moisture content of 20 percent, compute the wet specific weight increase from equation (11-3).

Table 11–2   Typical Moisture Content of Municipal Solid Waste Components

Component	Moisture Content (% by wt)	
	Range	*Typical*
Food wastes	50–80	70
Paper	4–10	6
Cardboard	4–8	5
Plastics	1–4	2
Textiles	6–15	10
Rubber	1–4	2
Leather	8–12	10
Garden trimmings	30–80	60
Wood	15–40	20
Glass	1–4	2
Tin cans	2–4	3
Nonferrous metals	2–4	2
Ferrous metals	2–6	3
Dirt, ashes, bricks, and so on	6–12	8
Municipal solid wastes	15–40	20

Source: Adapted from Tchobanoglous, Theisen, and Vigil. *Integrated Solid Waste Management.* New York: McGraw-Hill, 1994; with permission.

Solution

Because $m = 20\%$ and $d = 200$ lbs/yd^3,

$$0.20 = \frac{w \text{ lb/yd}^3 - 200 \text{ lb/yd}^3}{w \text{ lb/yd}^3}$$

or

$$w \text{ (wet specific weight)} = 250 \text{ lbs/yd}^3$$

Thus, at 50-percent moisture, the wet specific weight is 400 lbs/yd^3.

Particle Size Distribution

The particle size and the size distribution of various component materials in solid waste are considered in biological transformation, recovery of materials, and incineration. The size distribution and the shape characteristics of solid waste vary with the composition. The size range can extend from grains of sand and dirt to large, bulky items of furniture and household appliances. Most components are either cylindrical, spherical, or plateline in shape.

Particle size is also relevant for recycling and reuse and for sizing equipment used in processing waste. For example, aluminum soft-drink cans are typically

5.9 in. in height and 2.36 in. in diameter, and they are categorized by an effective size of

$$(l \times d)^{1/2} = (5.9 \times 2.36)^{1/2} = 3.73 \text{ in.}$$

Waste components are usually described by length $\times$ breadth $\times$ height ($l \times b \times h$). Knowledge of the largest dimension is needed for sizing equipment (e.g., sieves, conveyor belts, grinders). Table 11–3 provides typical particle size distributions. Shredders and separators are used to reduce the MSW components to desirable particle sizes for further processing or treatment.

Field Capacity

The field capacity of solid waste is the total amount of moisture that a sample of that waste will hold freely against the downward pull of gravity. Field capacity has great significance in determining the formation of leachate in landfills; the water in excess of the field capacity will drain away as leachate.

   Both the waste material and the cover material can retain water against gravity. The field capacity varies with the overburden weight. For solid waste in a landfill, the field capacity decreases with the overburden weight and can be estimated by using the following empirically derived equation:

$$FC = 0.6 - 0.55 \left(\frac{W}{10,000 + W} \right) \tag{11-4}$$

where FC is the field capacity (% of dry weight of waste) and W is the overburden weight calculated at the midheight of the waste in the lift in question (i.e., average lift lb) (2). Lift is the height of the solid waste pile with cover from its base. Overburden weight is the sum of the weight of the cover material, of the solid waste, and of moisture (including rainfall weight entering the waste).

Table 11–3   Typical Particle Size Distribution of MSW

Component	Size Range (in.)
Food	0–8
Paper, cardboard	4–20
Plastics	0–8
Glass	0–8
Metals	0–8
Clothing/textiles	0–12
Ashes, dust	0–4

Source: Adapted from Kiely, G. *Environmental Engineering.* New York: McGraw-Hill, 1997; with permission.

Example 11–3

Calculate the field capacity of a landfill site with the previously described formula and the following data:

> Compacted specific weight for the waste of 1000 lbs/yd^3 (includes moisture)
> Soil specific weight of 2500 lbs/yd^3 (including moisture)
> 100 lbs of rainfall entering the landfill in 1 year
> Landfill volume of 1 yd^3

Solution

Assume waste to cover ratio of 5:1 (by volume)
First, determine W:

$$W = \frac{1}{2}\left(\frac{1000\ lbs}{yd^3} \times \frac{5}{6}\ yd^3 + 100\ lbs\right) + \left(\frac{2500\ lbs}{yd^3} \times \frac{1}{6}\ yd^3\right)$$

$$= 884\ lbs$$

Next, determine the field capacity factor using equation (11-4):

$$FC = 0.6 - 0.55\left(\frac{W}{10,000 + W}\right)$$

$$= 0.6 - 0.55\left(\frac{884}{10,000 + 884}\right)$$

$$= 0.55$$

Assuming a 20-percent moisture content of the waste (Table 11–2),

$$\text{dry weight of waste} = 1000\ lbs/yd^3 \times yd^3 \times 0.80 = 800\ lbs$$

Thus, the amount of water that can be held in the solid waste is

$$0.55 \times 800\ lbs = 440\ lbs$$

Permeability of Compacted Waste

The hydraulic conductivity of compacted waste governs the transport rate of leachate and other contaminants (e.g., gases, microbiological contaminants) within the solid waste fill. Solid waste is usually not homogeneous, and hydraulic conductivities are not isotropic. Typically, the hydraulic conductivity of solid waste is approximately 3.3×10^{-5} ft/s, but it depends on the density. Table 11–4 presents selected data. The quantity of leachate transmitted is determined by Darcy's law (see Chapter 5).

Table 11–4   Permeability of Municipal Solid Waste

Processing	Hydraulic Conductivity (ft/s)
Dense baled waste	23×10^{-6}
Loose waste	49×10^{-5}
Shredded waste	3.3×10^{-4} to 3.3×10^{-6}
Typical solid waste[a]	3.3×10^{-5}

[a] Depends on density.

Example 11–4

Determine the quantity of leachate (Q) transmitted through a layer of solid waste ($A = 0.075$ ft^2) with a hydraulic conductivity (K) of 3.3×10^{-5} ft/s and the slope of 1 percent.

Solution

Recall from Chapter 5 that

$$Q = KA \times h/l$$

Thus, if A is 0.075 ft^2, K is 3.3×10^{-5} ft/s, and S is 1%, then

$$Q = 3.3 \times 10^{-5} \text{ ft/s} \times 0.075 \text{ ft}^2 \times 0.01 = 2.475 \times 10^{-8} \text{ ft}^3/\text{s}$$

CHEMICAL PROPERTIES

Information on the chemical composition of MSW is needed to evaluate alternative processing and recovery options. For example, the feasibility of using MSW for compost or combustion depends on the chemical composition of that waste. MSW can be separated into combustible and noncombustible materials, but if the solid waste is to be used as fuel, it is important to consider the following properties:

1. Proximate analysis.
2. Ultimate analysis.
3. Energy content.

When the organic fraction of MSW is to be composted or used in other biological conversion products, information on both the major and trace elements is needed.

Proximate Analysis

Proximate analysis is a chemical characteristic that determines the amount of surrogate parameters in place of the true chemical content. These parameters are normally moisture (i.e., loss of moisture when heated to 105°C for 1 hour), volatile matter (i.e., ignition at 600°C to 950°C), fixed carbon (i.e., carbon not burned or combustible residue left after the volatile matter is removed), and ash (i.e., weight of residue after combustion). As Table 11–5 shows, proximate analysis is important in evaluating the combustion properties of a fuel. For example, moisture adds weight and evaporation of water will reduce the heat release from the fuel. In other words, the higher the moisture content, the lower the material's value as a fuel.

Ash also adds weight, but it does not generate heat. Even so, the heat loss with ash is small.

Volatile matter and fixed carbon are important combustion characteristics of a fuel. Volatile matter changes to gases as the temperature increases. These gases then enter a second chamber, where combustion occurs. Heat release is rapid and mostly complete. Fixed carbon is essentially the residue that remains after the volatile matter has been removed. Combustion occurs on the surface of this residue. Fuels with a high percentage of fixed carbon require more time on the furnace grates for complete combustion than fuels with a low percentage of fixed carbon.

Ultimate Analysis

Ultimate analysis determines the percentage of each element in solid waste. The percentages of the five primary elements, moisture, and ash for solid waste are shown in Table 11–6. The results of ultimate analysis are used to characterize the chemical composition of organic matter in the waste. They are also used to obtain a proper mix of waste materials for a suitable carbon:nitrogen ratio, which is essential for a satisfactory biological conversion processes because both carbon and nitrogen are needed for bacterial degradation.

Carbon, hydrogen, and oxygen constitute the majority of the mass in solid waste (Table 11–7). The sulfur content is important when considering a combustion process, however, because a low sulfur content causes less air pollution during combustion. Because of concern over emissions of chlorinated compounds during combustion, the halogen content is often included in an ultimate analysis as well. The ash fraction contains the residues left after combustion of the organic material as well as in the inorganic environment. If a good recovery system is used to process the waste before com-

Table 11–5 Proximate Analysis

| | Solid Waste (% of wt) | |
	Range	Typical
Moisture	15–40	20
Volatile matter	40–60	53
Fixed carbon	5–10	7
Ash	10–30	20

Source: Kaiser, E. "Chemical Analysis of Refuse Components," *Proceedings of the National Incinerator Conference*, New York: ASME, 1966, pp. 1–5; with permission.

Table 11–6   Ultimate Analysis

Element	Solid Waste (% of dry wt)
Moisture	20
Carbon	30
Hydrogen	4
Oxygen	25.5
Nitrogen	0.37
Sulfur	0.13
Ash and metal	20

Source: Kaiser, E. "Chemical Analysis of Refuse Components," *Proceedings of the 1968 National Incineration Conference.* New York: ASME, 1968; with permission.

Table 11–7   Typical Data on the Ultimate Analysis of the Combustible Components in Residual Municipal Solid Waste

Component	Element (% of dry wt)					
	C	H	O	N	S	Ash
Paper	45.4	6.1	42.1	0.3	0.12	6.0
Cardboard	43.70	5.70	44.90	0.09	0.21	5.30
Plastics	60	7.2	22.8	—	—	10
Textiles	46.2	6.4	41.8	2.2	0.2	3.2
Rubber	77.7	10.3	—	—	2	10
Leather	60	8	11.50	10	0.4	10
Food wastes	41.7	5.8	27.6	2.8	0.25	21.9
Yard wastes	49	6.5	36	3.0	0.35	5.0
Wood	48	6	42.4	0.3	0.11	2.9

bustion, the ash content (especially ceramic, glass, and metal) can be reduced to less than 10% of that from conventional incineration systems (3). Many of the heavy metals are also present in the organic components (e.g., plastics, paper, leather). Heavy metals in the ash require disposal as a hazardous waste; leachability of these metals is determined using a procedure specified by the U.S. Environmental Protection Agency. Other metals, such as iron, manganese, magnesium, and so on, are also present in the ash but are not considered to be toxic.

Example 11–5

Determine the chemical composition of the organic fraction, with and without water, for residential MSW with the typical composition shown in Table 11–8.

Table 11–8   Typical Composition of a 100-lb Solid Waste Sample

Component	% of dry wt	Moisture (%)	Dry Weight (lbs)
Organic			
Paper	40	6	37.6
Cardboard	7	5	6.65
Plastics	9	2	8.82
Textiles	2	10	1.8
Rubber	1	2	0.98
Leather	0.5	10	0.45
Food wastes	8	70	2.4
Garden trimming	14.5	60	5.8
Wood	2	20	1.6
Total	84		66.1
Inorganic			
Glass	6	2	5.88
Nonferrous metals	5	2	4.9
Ferrous metals	3	3	2.91
Dirt, ash, and so on	2	8	1.84
Total	16		15.53

Solution

First, determine the percentage distribution of the major elements composing the waste. The necessary information is given in Table 11–9.

Also, from Table 11–8, moisture = 84 − 66.1 = 17.9 lbs

Next, convert the moisture content to hydrogen and oxygen. The molecular weight of water is 18 g; that is,

$$2 \text{ mol of H} = \frac{2 \text{ g}}{18 \text{ g}}$$

$$1 \text{ mol of O} = \frac{16 \text{ g}}{18 \text{ g}}$$

From Table 11–9, the moisture content of the waste is 17.9 lbs. Thus,

$$H = \frac{2 \text{ g}}{18 \text{ g}} (17.9) \text{ lbs} = 1.99 \text{ lbs}$$

$$O = \frac{16 \text{ g}}{18 \text{ g}} (17.9) \text{ lbs} = 15.91 \text{ lbs}$$

Table 11–10 summarizes the values to this point.

Now compute the molar distribution of the elements, neglecting ash. Your figures should match those in Table 11–11.

At this point, determine an approximate chemical formula with and without water. After computing the normalized mole ratios, confirm they match those in Table 11–12.

Table 11–9   Computations for Example 11–5.

Component	Wet Weight (lbs)	Dry Weight (lbs)	Composition (% dry wt)						
			C	H	O	N	S	Ash	Total
Paper	40	37.6	17	2.3	15.82	0.112	0.045	2.25	37.53
Cardboard	7	6.65	2.90	0.38	2.98	0.006	0.014	0.35	6.63
Plastics	9	8.82	5.29	0.63	2.01	—	—	0.882	8.81
Textiles	2	1.8	0.83	0.11	0.75	0.04	0.003	0.057	1.79
Rubber	1	0.98	0.761	0.1	—	—	0.0196	0.098	0.98
Leather	0.5	0.45	0.27	0.036	0.051	0.045	0.0018	0.045	0.45
Food wastes	8	2.4	1	0.14	0.66	0.067	0.006	0.52	2.39
Yard wastes	14.5	5.8	2.84	0.377	2.0	0.17	0.020	0.29	5.69
Wood	2	1.6	0.768	0.096	0.678	0.0048	0.0017	0.046	1.59
Total	84	66.1	31.65	4.16	24.94	0.44	0.11	4.53	

Table 11–10   Summary for Example 11–5

Component	Weight (lbs)	
	With H_2O	Without H_2O
Moisture	17.9	—
Carbon	31.65	31.65
Hydrogen	4.16 + 1.99 = 6.15	4.16
Oxygen	24.94 + 15.91 = 40.85	24.94
Nitrogen	0.44	0.44
Sulfur	0.11	0.11
Ash	4.53	4.53

Table 11–11   Molar Distribution for Example 11–5

Components	Atomic Weight (lb/mol)	Moles	
		With H_2O	Without H_2O
Carbon	12.01	2.635	2.635
Hydrogen	1.01	6.089	4.119
Oxygen	16.00	2.553	1.559
Nitrogen	14.01	0.0314	0.0314
Sulfur	32.07	0.0034	0.0034

Thus, the chemical formulas are

With water: $C_{775}H_{1790.8}O_{750.8}N_{9.23}S_1$

Without water: $C_{775}H_{1211.0}O_{458.2}N_{9.23}S_1$

Note that the calculations here have been rounded.

**Table 11–12   Mole Ratios
for Example 11–5**

Component	With H_2O	Without H_2O
Carbon	775	775
Hydrogen	1790.8	1211.0
Oxygen	750.8	458.2
Nitrogen	9.23	9.23
Sulfur	1	1

Energy Content

Knowledge regarding the heating value of any material is essential to evaluate its potential as fuel in a combustion system. This parameter is a function of the composition of the solid waste (i.e., of the percentage of materials with higher Btu values). The energy content of the organic components in MSW can be determined by using a boiler as a caloric meter, by calculation (if the elemental composition is known) and by experimental determination using a bomb calorimeter test. Because of the difficulty of instrumenting a full-scale boiler, most data on the energy content of organic components are based on the results of bomb calorimeter tests, which measure the heat release at a constant temperature of 25°C (77°F) from the combustion of a dry sample. The energy content of MSW depends on its mix of materials and its moisture content. The results obtained are known as the higher heat or gross heat value. Included in this value is the energy contained in the vaporized water that is produced. The following chemical reaction occurs during combustion of a fuel, say, cellulose:

$$(C_6H_{10}O_5)_n + 6nO_2 \rightarrow 6nCO_2 + 5nH_2O \text{ (vapor)} \tag{11-5}$$

Because this vapor is not usually condensed, the energy is lost, and a better estimate of the energy that can be recovered during combustion, which is called the lower heat value, or net energy, results. Table 11–13 lists the higher heat values of several MSW components. As shown, the energy content values are on an as-received basis, which means that waste has undergone no processing. These values may be converted to a dry basis using equation (11-6):

$$\text{Btu/lb (dry basis)} = \text{Btu/lb (as discarded)} \times \left(\frac{100}{100 - \% \text{ moisture}} \right) \tag{11-6}$$

Similarly, the energy content values may be converted to an ash-free dry basis:

$$\text{Btu/lb (dry ash-free basis)} = \text{Btu/lb (as discarded)}$$
$$\times \left(\frac{100}{100 - \% \text{ ash} - \% \text{ moisture}} \right)$$

The heating value of solid waste on an as-received basis can be greatly reduced as the ash and moisture content increases.

Table 11–13   Energy Content of Various Components in Municipal Solid Waste

Material	Energy Content (Btu/lb)
Mixed paper	6,800
Mixed food waste	2,400
Mixed green yard waste	2,700
Mixed plastics	14,000
Rubber	11,200
Leather	8,000
Textiles	8,100
Demolition softwood	7,300
Waste hardwood	6,500

Source: Niessen, W. "Properties of Waste Material," *Handbook of Solid Waste Management.* New York: Van Nostrand-Reinhold, 1977; with permission.

Example 11–6

Estimate the energy contained in a pound of as-received, discarded solid waste.

Solution

First, using data in Table 11–14 for discarded material and in Table 11–13 for energy content, create a table based on 100 lbs of solid waste. Some adjustments can be made for insufficient data. Percentages for rubber and leather are not given separately, so use the average energy:

$$\frac{11,200 + 8,000}{2} = 9600 \; Btu$$

Table 11–14   Data for Example 11–6

Material	lbs	Btu/lb	Btu
Paper and paperboard	31.7	6,800	215,560
Glass	6.6	0	0
Metals	7.4	0	0
Plastics	11.5	14,000	161,000
Rubber and leather	3.6	9,600	34,560
Textiles	3.3	8,100	26,730
Wood	7.6	6,900	52,440
Food wastes	8.5	2,400	20,400
Yard trimmings	16.2	2,700	43,740
Other	3.6	0	0
Total	100		554,430

For wood, use the average energy for two kinds of wood:

$$\frac{7,300 + 6,500}{2} = 6,900 \ Btu$$

For other wastes, assume an energy value of zero. Your table should look like Table 11–14.

Next, determine the as-discarded energy content per pound of waste:

$$\frac{554,430}{100} = 5,544 \ Btu/lb$$

If the Btu values are not available, approximate Btu values for individual materials can be determined using the Dulong equation:

$$Btu/lb = 145C + 610(H_2 - 1/8O_2) + 40S + 10N \qquad \textbf{(11-7)}$$

where C is carbon (% wt), H_2 is hydrogen, (% wt), O_2 is oxygen (% wt), S is sulfur (% wt), and N is nitrogen (% wt).

Example 11–7

Estimate the energy value of typical, residential MSW with the average composition determined in Example 11–6 and with water and sulfur.

Solution

The chemical composition of the waste is $C_{775}H_{1790.8}O_{750.8}N_{9.23}S_1$. Determine the percentage distribution by weight of the elements composing the waste, round off the coefficients, and create a table. When completed, your table should look like Table 11–15.

Table 11–15   Data for Example 11–7

Component	Atoms per Mole (n)	Atomic Weight	lbs	% of wt
Carbon	775	12	9,300	39.97
Hydrogen	1,791	1	1,791	7.70
Oxygen	751	16	12,016	51.65
Nitrogen	9	14	126	0.54
Sulfur	1	32	32	0.14
Total			23,265	100.00

Using equation (11-7), the energy content of the waste is

$$Btu/lb = 145(39.97) + 610(7.70 - [1/8 \times 51.65])$$
$$+ 40(0.14) + 10(0.54) = 6,565 \ Btu/lb$$

Note that Btu/lb $\times$ 2.326 converts to kJ/kg. Also the computed energy content of solid waste is higher in Example 11–7 than the value in Example 11–6, because only the organic fraction of the MSW was considered in Example 11–6.

BIOLOGICAL PROPERTIES

The biological properties of MSW are relevant, because aerobic/anaerobic digestion is used to transform waste into both energy and beneficial end products. Anaerobic decomposition of "food wastes," with methane, carbon dioxide, and other end products, have been used. A few organic MSW components, such as rubber, leather, plastic, and wood, are undesirable for biological conversion. The relevant fractions for biological transformation include proteins, lipids, and carbohydrates. The fraction of MSW conducive to bidegradion consists of materials with proteins, lipids, and carbohydrates.

Proteins are made of carbon, hydrogen, oxygen, nitrogen, and sulfur, with basic building blocks of amino acids (e.g., $C_5H_7NO_2$). Bacteria are mainly protein. Proteins are nitrogenous compounds that consist of organic acid with a substituted amine group (NH_2). Food wastes and yard wastes are protein sources and provide a good source of nutrients for the biodegradation of solid waste. Incomplete decomposition of proteins can produce amino acids, which have odors.

Fats, waxes, and oils are insoluble in water but are soluble in some organic chemical solvents and are slowly biodegradable. The primary sources of lipids are garbage, cooking oils, and fats. Fats and oils have the general formula $C_nH_{2n+1}COOH$. Lightweight compounds are oily, and heavyweight compounds are waxlike. Many of these compounds are sparingly soluble in water but are biodegradable. They also have a high energy value of approximately 16,000 to 18,000 Btu/lb and, thus, can be very suitable for an energy-recovery process.

Carbohydrates (i.e., hydrated carbons) contain carbon, hydrogen, and oxygen exclusively. Their general formula is $(CH_2O)n$. They include cellulose, hemicellulose, starch, and lignin, all of which (except for lignin) are readily biodegradable. The starch polymers hydrolyze to glucose and sugars, are water soluble, and readily biodegrade. These are abundant in potatoes, rice, corn, and other edible plants. They also attract flies and rats. The main source of carbohydrates in residential solid waste is food and yard wastes.

Crude Fibers

The category of crude fibers includes natural fibers. The major polymers are cellulose and lignin; both occur together in many fibers and result in material that does not readily biodegrade. Natural fibers in paper products and in food and yard wastes are the main source of these polymers. Textiles made from natural fibers, cotton, wool, and leather would also be classified in this category.

Synthetic Organic Materials

The remaining organic material in solid waste is classified as plastic, which is composed of synthetic compounds. Recently, synthetic material has become a significant component, accounting for between 4 and 10 percent of MSW.

Resistance to biodegradation makes these materials environmentally objectionable. Several processes to develop biodegradable plastics are being researched, however. One approach uses starch as a natural polymer. Another uses microorganisms, and a third involves use of wood fibers.

Even so, there is resistance to the idea of degradable plastic. For example, degradable plastic would defeat the purpose of the plastic. There are also safety-related issues, including the potential toxicity of chemicals leaching from degradable plastics into the environment.

Another factor of interest is the high energy content of plastic, which is approximately 14,000 to 20,000 Btu/lb. An increase in plastic content increases the potential for energy recovery. A problem associated with the combustion of plastic, however, is that some of it contains polyvinyl chloride, which is a source of chlorine and dioxin. Acid gas (from chlorine) causes corrosion, and along with dioxin, it may cause air pollution as well.

Biodegradables

An important characteristic of the organic component in MSW is that most of it can be biodegraded into gases and relatively inert and inorganic solids. The putrescible nature of organics has the potential of producing odors and attracting flies.

Biodegradability of the food fraction of MSW is given by the equation

$$BF = 0.83 - 0.028 LC \qquad \text{(11-8)}$$

where BF is the biodegradable fraction expressed on a volatile solids basis, 0.83 and 0.028 are empirical constants, and LC is the lignin content of volatile solids (% of dry wt). Table 11–16 lists the biodegradable fraction for several organic MSW components.

Odors

The organic component of MSW (i.e., food and yard wastes) is putrescible. Odors can develop when organic wastes are stored for long periods, especially in warm climates, because of anaerobic decomposition of the readily decomposable organics. Under

Table 11–16   Data on the Biodegradable Fraction of Selected Organic Waste Components

Component	Volatile Solids (% of total solids)	Lignin Content (% of volatile solids)	Biodegradable Fraction
Food wastes	7–15	0.4	0.82
News print	94.0	21.9	0.22
Yard wastes	50–90	4.1	0.72

anaerobic conditions, sulfate can be reduced to sulfide, which combines with hydrogen to form hydrogen sulfide, which is highly odorous and toxic. Anaerobic decomposition can be represented as

$$\text{organic matter} + \text{microorganisms} \rightarrow CO_2 + H_2O + \text{new cells}$$
$$+ \text{ unstable products } (H_2S, NH_3, CH_4)$$

These end products are objectionable, because they are not stable and because odors are produced.

Nonbiodegradables

The remaining materials in MSW are usually noncombustibles: glass, ceramics, dust, dirt, and metals. These usually account for 12 to 20 percent of the dry solids. They are the residue left after combustion.

REVIEW QUESTIONS

1. List the physical properties of municipal solid waste.
2. What is the specific weight of solid waste in a compactor truck with a volume of 22 yd^3 and a net weight of 12,000 lbs?
3. a. What is the significance of volume when handling waste?
   b. Compute the percentage volume reduction and compaction ratio when a waste volume of 22 yd^3 is reduced to 10 yd^3.
4. What is the significance of moisture content in municipal solid waste?
5. A sample of municipal solid waste has a dry specific weight of 150 lbs/yd^3. Calculate the weight if moisture is added so that the moisture content is 25 percent.
6. a. What is the relevance of waste particle size in waste management?
   b. List the equipment used for reducing particle waste size.
7. a. Define *field capacity*.
   b. What happens to water in excess of the field capacity of a landfill?
   c. Calculate the field capacity factor of a landfill site given the following:

   Compacted specific weight of waste = 800 lbs/yd^3

   Soil specific weight = 2000 lbs/yd^3

   Weight of rain water entering landfill = 70 lbs/year

   Landfill volume = 10 yd^3

   (Use Table 11–2 for moisture content.)

8. a. Why is hydraulic conductivity of waste relevant?
   b. Compute the quantity of leachate through a loose layer of 1 ft^2 of solid waste when its slope is 1.5 percent.
9. a. What is the difference between proximate and ultimate analysis?
   b. Why are carbon and volatile matter of solid waste important?
   c. List the main elements that constitute the majority of solid waste material mass.
   d. Why is ash content a problem in solid waste management?
10. Determine the chemical composition of the organic fraction of municipal solid waste without sulfur and water. Use Table 11–8 with the following changes: plastic, 11 percent; and garden trimmings, 12.5 percent.
11. Use the chemical composition in problem 10, and estimate the energy value of typical residential municipal solid waste.
12. a. Describe the relevant fractions of municipal solid waste in biological transformation.
    b. Why is there a problem with synthetic materials in this area?

REFERENCES

1. Kaiser, E. "Refuse Is the Sweetest Fuel," *American City*, May 1967, pp. 32–37.
2. Tchobanoglous, G., H. Theisen, and S. Vigil. *Integrated Solid Waste Management*. New York: McGraw-Hill, 1994.
3. Pfeffer, J. *Solid Waste Management Engineering*. Upper Saddle River, N.J.: Prentice-Hall, 1992.

Chapter 12   Storage, Collection, and Transportation of Solid Waste

This chapter discusses various aspects of solid waste handling, such as storage, collection, and transportation. The economic aspects of collection and transportation are emphasized through cost calculations.

Handling, separation, and processing of solid waste at the source—before collection—follows the generation of waste. Because waste diversion is mandated by law in several states, separation of waste components, including waste paper, cardboard, aluminum cans, glass, and plastic containers, at the source is done for recovery and reuse of materials. After separation, homeowners store the separated components in a container. These separated components are then collected by the municipality or the waste collection agency, or they are taken to a local buy-back or recycling center by the homeowner. (Source reduction, reuse, recycling, and recovery are discussed in Chapter 14.)

STORAGE AT THE SOURCE

Factors to consider in the storage of solid wastes include the effects of storage on the waste, the type of container to be used, the location of the container, and public health and aesthetics. Solid waste is generated at the source more or less continuously, but collection occurs intermittently. Therefore, on-site storage of solid waste is needed until collection day. When the solid waste is temporarily stored on the premises (i.e., accumulated between collections), an adequate number of suitable containers should be provided.

Single-Family Residential Sources

The most common method of solid waste storage is in one or more "garbage" cans with a capacity of 20 to 32 gal each. These containers are designed with tight-fitting lids and are made from galvanized metal or plastic. They are manually lifted and emptied in the collection vehicle. Recently, roll-out containers have become popular as well. These containers on wheels have a capacity of 50 to 55 gal each and are made of plastic. On collection day, the resident rolls the container to the street, where it is mechanically loaded onto the collection vehicle. Figures 12–1 and 12–2 show typical storage containers and collection vehicles. If the collection crew uses a "shoulder barrel," which is a plastic container with a capacity of approximately 50 gal, then the cans are emptied into it and set out is not required. However, if the weight of the waste exceeds the crew's ability (i.e., approximately 60 lbs) to carry it to the vehicle, then an extra trip is needed.

Figure 12–1 Typical containers used for storage of commingled wastes and a bin for recyclables.

Figure 12–2 Loading solid waste from storage containers.

Any oversize or bulky solid wastes (e.g., furniture, appliances) that are generated at a single-family residence require special handling. Several options exist for collecting this type of waste, such as collect with the other waste, pickup at the homeowner's request, or periodic or seasonal pickup along a defined route. There is usually an additional charge for this service.

In general, the size and number of containers as well as their location and the types of waste within them are defined by municipal ordinance or regulation or by the contract for a private collection company. The point of collection affects the crew size and storage method, and it also controls the cost of collection. Curbside and alley collection require residents to place the waste containers at the curb or alley; residents must then retrieve the empty containers after collection. When alley collection is possible, the containers are placed at the back of the property. Only older developments have alleys, however, and the alley width may not allow the passage of modern collection vehicles.

Multifamily Residential, Commercial, and Institutional Sources

Other sources generate much larger quantities of solid waste than single-family residential sources. Obviously, it is inefficient to use individual containers for storage in this case. In addition, because some users are not careful in depositing the waste, it overflows, spills over the area, and causes a nuisance. Thus, for better aesthetics and hygienic considerations, large containers (i.e., dumpsters) are used for multifamily residential, commercial, and institutional sources. The capacity of such containers ranges from 0.5 to 40 yd^3. These containers are built from sheet steel and have significant weight of their own. The larger containers are located to permit access by the collection vehicle; smaller containers may be mounted on rollers and moved from the storage area to the collection vehicle. Figure 12–3 shows a dumpster for large quantities of waste. The solid waste in containers with a capacity of 6 yd^3 or less is emptied into the collection vehicle by a mechanical loading system and is then compacted. Containers with a capacity between 6 and 10 yd^3 may be dumped into a truck or hauled directly to

 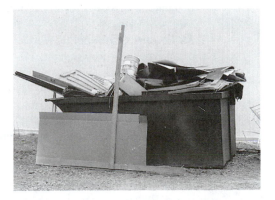

Figure 12–3   Dumpsters for large quantities of solid wastes. (a) Tilt-frame, hauled container. (b) Smaller dumpster.

a processing facility or disposal site; again, the waste is compacted with a stationary compactor. Containers with a capacity of greater than 10 yd^3 are generally included in the hauled container system, in which they are loaded on as part of the truck, hauled to the disposal site, emptied, and returned to their original or other location (1).

COLLECTION

Solid waste collection refers to the gathering of waste from places such as residential, commercial, institutional, and industrial areas and public parks. Collection may occur once or twice a week or, in some cases, daily depending on the quantity generated and type of waste. Both storage and collection operations must be coordinated.

The collection and transport of solid waste to processing and disposal sites accounts for roughly three-fourths of the total cost of the service; therefore, this is a very important function of solid waste management. Before the push for recycling and recovery, collection and transport decisions were focused on selecting the proper number and size of vehicles, the most efficient collection routes and schedules, and the location and design of transfer stations (if needed). With the push for recycling and recovery, these operations have become more complicated. Now, a municipality or private collection agency may require separate storage containers, vehicles, routes, schedules, and destinations for recyclable and recoverable material, which must be coordinated with the existing collection systems. Figure 12–1 illustrates the bin and container for separate storage.

Types of Collection Systems and Equipment

During the past 15 years, a variety of systems and equipment have been used in the collection of solid wastes. These systems may be classified in several ways, such as by types of waste collected, mode of operation, and equipment used. In this text, collection systems are classified according to their mode of operation into two categories: stationary container systems, and hauled container systems. In the former, the container is emptied into a collection vehicle at the point of collection; in the latter, the container is hauled away from the collection point.

Stationary Container Systems. In this system, the containers used for the storage of all types of waste remain at the point of generation, except when they are moved to the curb or other location to be emptied. These systems vary according to the type and quantity of waste to be handled and according to the number of generation points. There are two main types: systems in which mechanically loaded collection vehicles are used, and systems in which manually loaded collection vehicles are used. Large containers are mechanically loaded; small containers generally are manually loaded on to a collection vehicle.

The usual collection vehicle for residential areas is a packer truck, which is a manually rear- or side-loaded compaction vehicle. Figure 12–4 shows a rear-loader packer truck. Because of the economic advantage involved, almost all collection vehicles in use today are equipped with an internal compaction mechanism. Larger, self-loading compactor vehicles needing only one driver-operator are also available and can auto-

Figure 12–4 Rear end–loader garbage truck.

matically unload full-storage containers at apartments, shopping centers, and similar locations, replace the empty containers for reuse, and then deliver the compacted contents along with those of other containers to the disposal site.

Typical municipal solid waste at curbside has a density of approximately 200 to 250 lbs/yd^3. At that low density, collection vehicles fill too quickly, which means additional trips. Most compactor trucks can compact the waste by a factor of two or more. The solid waste is fed through a hopper, the hydraulic ram is activated, and the density of the waste is increased by the action of the ram. In the transport of solid waste, volume is a very important factor.

Hauled Container Systems. These container systems are suited for the collection from sources with a high waste generation rate. The collection vehicle delivers a large, empty storage container to an institution, construction or demolition site, or commercial operation, and it picks up a full one, which is then hauled to a processing or disposal facility. One driver can perform all the loading and unloading of these containers (unless regulations require that the driver have a helper). The containers may be carried by a hoist truck, which lifts relatively small containers into place, or by a tilt-frame truck, which can handle larger containers called drop boxes that hold loose or compacted refuse. With the advent of self-loading packer trucks and the tilt-frame hauled container system, both of which can carry much larger loads, the hoist truck has been replaced in most applications. Figure 12–5 shows a hauled container.

Figure 12–5   Dumpster being hauled.

Methods of Collection

Single-Family Residences.   The manner by which solid waste is transferred from storage containers to the collection vehicle depends on several factors, including the type and size of the storage containers, the size of the collection crew, and the type of collection vehicle. Municipal solid waste collection is a labor-intensive activity, and labor is usually one of the most costly aspects of a municipal collection system. Most collection crews have three—or fewer—members. Under certain conditions, such as a curbside location for containers and only one crew member (who drives and loads), a cost-efficient collection can usually be accomplished. As previously stated, there are two main types of collection vehicles: side-loaders, and rear-loaders. The side-loading truck can be operated by one member crew.

There are several ways by which a collection crew can operate. Figure 12–6 illustrates techniques for one-person, two-person, and three-person crews. In all three cases, waste is collected from both sides of the street. For a one-person crew, a typical side-loading vehicle stops with the front end between two residences to be serviced. The driver gets out, picks up the can, empties it into the side of truck, puts the can back, walks to the other side, and does the procedure again. The driver then moves the vehicle to the next location, and the operation is repeated.

For a two-person crew, the truck stops between four residences to be serviced. The driver goes to collect waste from location 2, returns to the truck to empty the shoulder barrel, and then services location 1. During this time, the second crew member is ser-

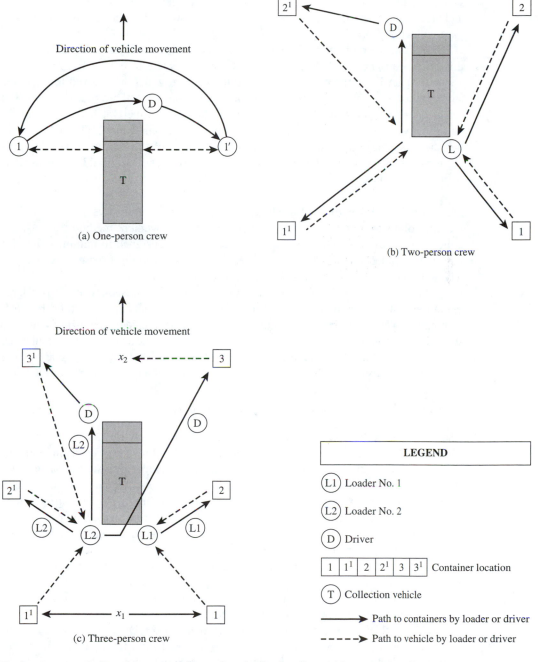

Figure 12–6 Collection procedures for single-family residences. (a) One-person crew. (b) Two-person crew. (c) Three-person crew.

vicing the residences on the other side of the street. The driver then returns to the cab and drives forward to the next location. This method works for both curbside and back-yard collection (1).

For a three-person crew, six locations are serviced for each stop. As the vehicle is driven forward, the two loaders get down from the vehicle to collect from residences 1 and 1'. They bring the collected waste to the vehicle, which has moved forward and stops almost in the middle of the six locations being serviced (three on each side). The driver goes from the cab to service location 3. Meanwhile the two loaders empty the barrels of waste picked up earlier and then service locations 2 and 2'. The driver emp-ties the waste barrel and moves on to location 3'. One of the loaders drives the vehi-cle forward and picks up the driver, who has now serviced location 3'. This cycle is then repeated (1).

Methods-Time-Measurement Techniques.

There are several ways the crew can col-lect the solid waste and still achieve an efficient and convenient method. Because the collection process is labor intensive, the procedures are designed to minimize the time required to collect the solid waste. Numerous studies involving time-motion analysis have been conducted to develop efficient procedures.

Solid waste collection from containers has economic consequences. It is a large operation and has many variables. The financial aspect can be understood by envi-sioning that a 6-second reduction in time spent servicing a residence translates into 900 seconds (15 minutes) if 150 residences are served with one truck. If 15 minutes are saved per truck, then 1500 minutes are saved if there are 100 trucks and 3000 minutes if each truck hauls two loads. This equals 50 hours, and for a two-person crew receiving $18.00 per hour, the reduction in cost is

$$18 \times 2 \times 50 = \$1800.00$$

The reduction in the operating cost should also add to the savings as well.

Methods-time-measurement (MTM) is a system developed by experts in industrial engineering to predict or measure the time needed to perform almost any manual task. In this case, MTM involves studying the actions of a crew while collecting solid waste. Expert analysts apply appropriate MTM values to determine the basic expected time to perform a specific waste collection task with different crew sizes. Comparisons are then made with the values obtained through actual field studies, and "standard time" for the task is determined. This is the actual time required to complete the task. The MTM system, however, does not include all time delays for factors that can not be standardized (e.g., fatigue) personal delays (e.g., taking a drink of water) equipment problems, and so on. The standard times measured for each procedure can be com-pared, and the most efficient procedure can be found.

Table 12–1 presents the results of such an analysis (2). Data in this table indicate an alley location of the garbage cans requires the least time (i.e., person-minute per stop) for collection. Only older developments have alleys, however; subdivisions de-veloped after the early 1950s rarely provided land for alleys. In addition, because of their size, modern collection vehicles may not be able to operate in alleys, but if a smaller collection vehicle is used, the cost of collection will be higher.

Table 12–1   Standard Times for Solid Waste Collection

Crew Size (n)	Can Location	Standard Time (min)[a]	Services per Stop (n)	Person-Minutes per Service Stop[b]
1	Backyard	2.40	2	1.35
2	Backyard	2.38	4	1.33
3	Backyard	2.35	6	1.33
1	Alley	0.910	2	0.606
2	Alley	0.490	2	0.79
3	Alley	0.350	2	0.97
1	Curbside	1.58	2	0.88
2	Curbside	2.11	4	1.20
3	Curbside	2.06	6	1.18

[a] Standard time per collection stop for both sides of the street or alley; two containers per service stop.

[b] Includes travel time between collection stops on the route; using shoulder barrels (except when cans are located in the alley).

Source: U.S. Public Health Service. *A Study of Solid Waste Collection Systems.* Publication 1892 (SW-96). Washington, D.C.: U.S. PHS, 1969; with permission.

The curbside collection of solid waste from single-family residences offers the second best location. Approximately 0.3 person-minutes can be saved per service stop by reducing the crew size from three to one with this location. The time required for collection from the backyard is not sensitive to crew size and is greater than that for alley or curbside collection.

Table 12–2 presents the results from a field survey of a collection crew in Southern California to evaluate the time required to service residences (2). Four are municipal collection agencies (A, B, C, D) and two are private agencies (X, Y) were involved. From the time required to collect the residential solid waste and the weight of the solid waste collected, the labor required per ton of waste can be computed. Collection costs can be computed by multiplying the person-minute per ton by the cost of labor and then adding the equipment cost.

Table 12–2   Time Required for Residential Service Based on Crew Size, Cans, and Items per Service.

Agency	Crew Size (n)	Cans per Service (n)	Items per Service (n)[a]	Time (min/service) Collect	Time (min/service) Travel	Refuse per Service (lb)	Person-Minutes per Ton
A	1	2.45	3.40	0.68	0.15	77.1	26.3
B	2	2.65	3.55	0.59	0.17	81.2	43.0
C	3	2.70	4.01	0.58	0.17	73.2	63.5
D	1	1.79	2.60	0.57	—	56.9	37.6
X	1	2.74	4.04	0.99	—	88.1	33.8
Y	1	2.07	2.87	0.59	—	60.5	39.0

[a] Number of items per service includes cans as well as other types of containers or bundles.

Source: U.S. Public Health Service. *A Study of Solid Waste Collection Systems.* Publication 1892 (SW-96). Washington, D.C.: U.S. PHS, 1969; with permission.

Table 12–2 also presents the collection times for an increasing number of containers and for different crew sizes. An increase in the number of containers will increase the time required for collection. A larger crew will reduce the time required to service a residence. This reduction in time is not in proportion to the increase in crew size, however. A crew of three does not collect solid waste three times as fast as a crew of one. Again, the collection times in Table 12–2 do not include time loss because of nonproductive activities, which may be predictable or unpredictable.

Predictable unproductive time is associated with activities that are part of the collection system. These include the travel time for driving the collection vehicle and crew from the garage or storage or other base at the beginning of the work to collection route and the return time from the disposal site at the end of the day. Also included is the time allowed for lunch and breaks, route retracing, relief time, and time to go to another route after one is finished. Traveling over certain streets that have already been serviced, which is called route retracing, cannot be avoided. The relief time is time at which there is not enough time to collect additional solid waste and take it to a disposal site within the 8-hour workday. The collection vehicle is then driven back to the central base or storage area, and the crew is paid for the full day.

Unpredictable nonproductive events include accidents, equipment malfunction or failure, traffic congestion, and crew activities. Crew activities may include talking, drinking beverages, smoking, using the restrooms, and so on. Fatigue also adds to the time.

Certain delays cannot be controlled, but proper hiring practices followed by training, supervision, and occasional activities associated with morale-boosting can help to save valuable time. Careful planning of vehicle routes and times can avoid streets with traffic congestion. Good training of crew and maintenance of equipment can reduce the possibility of accidents and equipment failure.

Waste Collection From Large Producers.   Collection of solid waste from single-family homes is more labor intensive than collection of waste from large producers, which is more mechanized. As indicated previously, waste from large producers is stored in containers made of sheet metal with varying capacities. These containers are very heavy, and even the smaller containers cannot be loaded manually. The solid waste stored in the containers is either emptied into the collection vehicle, or the container is loaded onto the vehicle and is hauled to the disposal site.

Smaller containers are engaged with mechanical arms to lift and transfer their contents to the collection vehicle. The empty container is then put back, and the vehicle is driven to the next location. This sequence is repeated until the vehicle is filled and then unloaded at the disposal site.

The larger containers are not emptied into the collection vehicles. The filled larger containers are instead loaded onto the truck chassis and hauled to the disposal site. The container is then emptied at the disposal site and returned to its location. Figure 12–7 shows a truck with a container.

Estimates of Collection Time and Cost.   Planning of collection patterns and truck capacities requires estimates of the times associated with the collection system. Figure 12–8 shows a basic collection system, consisting of an overnight parking lo-

Figure 12–7   Container being emptied before return to its location.

cation or garage for the vehicles, a collection route, and the disposal site. If the distance from the collection area to the disposal site is large, a transfer station may be built between the collection route and the disposal site to save time; incorporation of such a transfer station is discussed later.

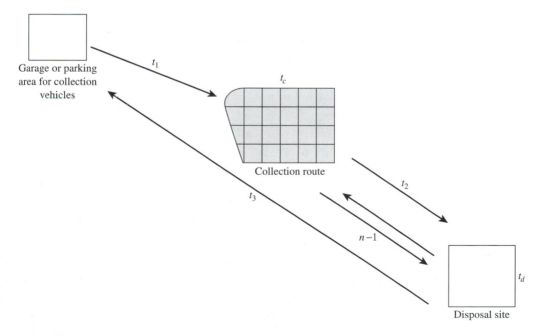

Figure 12–8   Timing for solid waste collection.

A simple equation can be developed to calculate time and derive a basic plan for solid waste collection (3). The following notation relates to time:

T = total time for 1 day's collection of solid waste

t_1 = time to drive from vehicle location to beginning of collection route

t_r = total time for solid waste collection

t_2 = time to drive between collection route and disposal site

t_d = time at disposal site dropping vehicle load

t_3 = time to drive from disposal site back to vehicle location at end of day

t_b = time spent on breaks and so on per workday plus unexpected delays

n = number of trips from collection route to disposal site

In other words, each day's collection requires the following:

- Time for driving from the location where the collection vehicle is parked to the collection route.
- Time for the collection route.
- Time for a run to the disposal site.
- Time for unloading at the disposal site.
- Time for driving back from the disposal site to the vehicle's parking location.
- Time for (possibly) more than one run between the collection route and the disposal site depending on the capacity of the collection vehicle, number of locations to be serviced, and the amount of waste to be picked up.

Thus, the total time for a day's collection of solid waste is given by

$$T = t_1 + t_r + t_2 + t_d + (n - 1)(2t_2 + t_d) + t_3 + t_b$$

$$T = t_1 + (2n - 1)t_2 + nt_d + t_3 + t_b + t_r \qquad \text{(12-1)}$$

Example 12–1

It takes 20 minutes to drive a collection vehicle from its location to the beginning of the route, 35 minutes to drive between the route and the disposal site, and 15 minutes to return the empty vehicle from the disposal site to its original location. It takes 14 minutes to offload a vehicle at the disposal site. The crew takes two 15-minute breaks per day, and another 30 minutes are allowed for unexpected delays. If two runs are made to the disposal site per day, compute the time left in an 8-hour workday for actual solid waste collection. In addition, compute the time left if only one run is made per day.

Solution

Rearranging equation (12-1), we have

$$t_r = T - t_1 - (2n - 1)t_2 - nt_d - t_3 - t_b$$

where T is 480 minutes, t_1 is 20 minutes, t_2 is 35 minutes, t_d is 14 minutes, n is 2, t_3 is 15 minutes, and t_b is 60 minutes. Thus,

$$t_r = 480 - 20 - ([2 \times 2] - 1)35 - 2(14) - 15 - 60 = 252 \text{ minutes}$$

In other words, 4.2 hours are left for actual waste collection with two runs.
     With one run,

$$t_r = T - t_1 - (2n - 1)t_2 - nt_d - t_3 - t_b$$

where T is 480 minutes, t_1 is 20 minutes, t_2 is 35 minutes, t_d is 14 minutes, n is 1, t_3 is 15 minutes, and t_b is 60 minutes. Thus,

$$t_r = 480 - 20 - ([2 \times 1] - 1)35 - 1(14) - 15 - 60 = 336 \text{ minutes}$$

In other words, the time available for actual collection with one run is 5.6 hours.

     To estimate the number of households that can be served per day by one collection truck, we need the average length of time to service each stop. The time per stop is equal to the average time it takes to drive from one stop to another plus the time it takes at each stop to empty the containers. For example, assume that the speed of the collection vehicle averages 5 ft/s and that it takes 13 seconds to empty a container (these values are fairly realistic). Thus, the average time per stop is

$$t_s = 0.2D + 13N_x \qquad\qquad \textbf{(12-2)}$$

where t_s is the average time for one stop (seconds), D is the average distance between stops (ft), and N_x is the number of containers to empty per stop (3).
     To find the volume of a collection vehicle, we need to know the vehicle compaction ratio, the number of stops, and the volume of refuse per stop, or

$$V = \frac{vN_s}{r} \qquad\qquad \textbf{(12-3)}$$

where V is the collection vehicle volume (ft^3), v is the average volume of solid waste at each collection (ft^3/stop), r is the compaction ratio $\left(\dfrac{\text{curbside volume of the waste}}{\text{compacted volume of waste in truck}}\right)$, and N_s is the number of stops per truckload.

Example 12–2

Assume that the average distance between stops along the route is 500 ft and that there are two containers per stop (a typical situation). The collection vehicle stops at each residence, which puts out 9 ft^3 of solid waste at the curbside each week. The density of this waste at curbside is 210 lb/ft^3 (7.8 lb/yd^3). How many stops would be made per vehicle load, and what should be the vehicle capacity if the compactor reduces the waste volume by a factor of three?

Solution

Using equation (12-2), the time needed to service a stop would be

$$t_s = 0.2D + 13N_x$$
$$= (0.2 \times 500\,ft) + (13 \times 2\,containers/stop) = 126\,s/stop$$

In the example 12–1, the computed collection time is 5.6 hours per day. At 126 s/stop and one vehicle load being collected, the number of stops are

$$N_s = \frac{5.6\,hours/day \times 3600\,s/hour}{126\,s/stop \times 1\,load/day} = 160\,stops/load$$

Substituting N_s in equation (12-3), the volume is

$$V = \frac{9\,ft^3/stop \times 160\,stops/load}{\dfrac{9\,ft^3\,at\,curb}{3\,ft^3\,in\,vehicle}} = 480\,ft^3\,or\,18\,yd^3$$

where 3 is the compaction ratio. Thus, a vehicle with at least 18 yd^3 of hauling capacity should be satisfactory.

Economic Analysis.   The size of the collection vehicle in the example 12–2 is such that the vehicle can make only one run to the disposal site per day. If a vehicle of 18 yd^3 or slightly more is not available, two runs may be needed. Thus, examples 12–1 and 12–2 would need to be reworked using two or three runs. With more runs, however, time is also spent making trips back and forth to the disposal site. The decision makers should attempt to answer if it is more economical to design a collection system for a large collection vehicle resulting in fewer trips to the disposal site or a system for smaller collection vehicles resulting in more back and forth runs to the disposal site each day. Obviously, larger vehicles cost more than smaller vehicles.

Economic analysis was discussed in Chapter 2. To come up with a vehicle size that will provide the most economical waste transport, the annual cost of owning and operating the vehicle and of paying its crew should be determined. The annual cost of owning a collection vehicle can be described as

$$A = P\left[\frac{i(1 + i)^n}{(1 + i)^n - 1}\right] \tag{12-4}$$

where A is the annual cost, P is the purchase price, i is the interest rate, and n is the amortization period. Assume the cost of a mechanized collection vehicle is $127,000 in 1997 dollars (4), which is amortized over a 6-year period using an 8 percent interest rate. This vehicle would have an annualized cost of

$$A = 127,000 \left(\frac{0.08\,[1 + 0.08]^6}{[1 + 0.08]^6 - 1} \right)$$

$$= 127,000 \left(\frac{0.1269}{0.5868} \right)$$

$$= \$27,465 \text{ per year.}$$

These collection vehicles are large and heavy. Constant starting and stopping is hard on the vehicles and also on their mileage per gallon. A midsize collection vehicle driven approximately 30 miles per day for 5 days a week (260 days/year) for 8 hours a day would cost annually to operate approximately, in 1997 dollars as follows:

Gasoline	7,228
Maintenance, repair, and insurance	25,219
Total	32,447

Including the amortized cost determined earlier, the total annualized cost of purchasing, fueling, and maintaining a midsize compactor collection vehicle is $59,912. In this approach, the volume of the vehicle has not been considered. The true annualized cost will, in fact, depend on the size of the vehicle, and one approach is to estimate by a simple regression equation as follows:

$$\text{annualized cost} = a + bV \tag{12-5}$$

where a and b are empirically determined estimates based on a cost survey of available vehicles to determine a relationship between the cost of a collection vehicle and its volume V.

Example 12–3

A collection vehicle with a two-person crew with labor charges of $18 per hour per person (benefits included) and a capacity of 18 yd^3 is used to collect solid waste from 160 households in a single trip per day for transport to the disposal site. The curbside pickup is provided once a week for each house; collection is done 5 days a week. What is the cost per ton of solid waste collected if each household puts out 9 ft^3 of waste per week with a curbside density of $\dfrac{7.8 \text{ lbs}}{\text{ft}^3} \left(\dfrac{210 \text{ lbs}}{\text{yd}^3} \right)$?

Calculate the cost per year per household by:

a. Estimating separately the cost of ownership, operation, and labor.
b. Using a cost model with volume and labor.

Solution

For part (a), the total number of houses served over a 5-day week is

$$5 \text{ days} \times 160 \text{ houses/day} = 800$$

Thus, the total amount of solid waste collected by vehicle per year is

$$\frac{\dfrac{9\,ft^3}{residence \ per \ week} \times 7.8 \ lbs/ft^3 \times 800 \ residences \times 52 \ weeks/year}{\dfrac{2000 \ lbs}{ton}}$$

$$= 1460.16 \ tons$$

Recall that the annual cost of waste collection equals the cost of owning and operating the vehicle plus the labor cost. As shown earlier, the cost of owning (purchasing) and operating the collection vehicle is $59,912.00 per year. The labor cost is

$$2 \ persons \times \$18/hour \times 8 \ hours/day \times 5 \ days/week$$
$$\times 52 \ week/year = \$74,880.00$$

Thus, the total annual cost per ton of waste collection is

$$\frac{(\$59,912 + \$74,880)}{1460.16 \ ton/year} = \$92.31$$

and the annual cost per household for this collection service is

$$\frac{\$92.31/ton \times 1460.16 \ ton/year}{800 \ residences} = \$168.48$$

For part (b), the annualized cost of the collection vehicle (including fuel, maintenance, insurance) based on the vehicle is, as shown earlier,

$$\text{annualized cost} = a + bV$$

If a and b are determined to be 10,000 and 3,000, respectively, for a 18-yd^3 vehicle, then

$$\text{annualized cost} = 10,000 + 3,000(18) = \$64,000$$

As determined earlier, the labor cost is $74,880. Thus, the total annual cost per ton of waste collection is

$$\frac{\$64,000 + \$74,880}{1460.16 \ ton/year} = \$95.11$$

and the annual cost per household for this collection service is

$$\frac{\$95.11/ton \times 1460.16 \ ton/year}{800 \ residences} = \$173.59$$

Tables 12–3 and 12–4 extend the analyses of these examples to trucks making one, two, or three runs to the disposal site. Larger trucks make fewer runs, but their capital cost is much greater than that of smaller trucks. Use of smaller trucks, however, offsets the capital cost disadvantage of larger trucks because the labor costs are less with smaller trucks.

The solution represented by Table 12–3 is sensitive to the distance between the collection area and the disposal site because of time. With increasing distance, fewer trucks driving a long distance once per day begins to become economical. This is clearly shown in Table 12–4, which describes the annual cost based on drive time between the collection route and the disposal site and on the number of trips.

Layout of Collection Routes.   The general steps in laying out a collection route are:

1. On a reasonably large-scale (i.e., easily readable) map of the area to be served, data such as location, collection frequency, and number of containers should be plotted for each solid waste pickup point. For commercial and industrial sources, the estimated quantity of waste to be collected at each location should also be shown. For residential sources, only the number of homes per block need be shown, because the average quantity of solid waste per residence generally is approximately the same.

2. Estimate the total quantity of wastes to be picked up from the locations being serviced. Determine the average number of residences from which wastes are to be collected during each trip based on the collection vehicle volume and compaction ratio.

3. Beginning from the dispatch point (or garage), lay out collection routes that include all the pickup locations to be serviced. The route should be planned so that the last of these locations is nearest the disposal site.

4. After the collection routes are determined, the container density and haul distance for each route should be determined. Labor needs versus the available time per day should be determined as well. In some cases, it may be desirable to adjust the collection routes to balance the workload. The finalized routes should be drawn on the map.

Table 12–3   Comparison of Costs for One, Two, or Three Trips per Day to the Disposal Facility[a]

Number of Trips per Day (n)	Residences per Vehicle (n)	Minimum Vehicle Size (yd³)	Annual Costs		Cost per Ton	Annual Cost per Household
			Vehicle	Labor		
1	800	18	$64,000	$74,880	$95.11	$173.59
2	600	6.67 (6.70)	$30,100	$74,880	$95.87	$174.96
3	400	2.96 (3)	$19,000	$74,880	$128.6	$234.70

[a] Based on example 12–3.

Table 12–4   Annual Cost Based on Drive Time (t) between Collection Route and Disposal Site and on Number of Trips (n)

t (minutes)	$n = 1$	$n = 2$	$n = 3$
0	t_r = 6.18 hours N = 177 $N \times n \times 5$ = 885 solid waste = 1615 tons/year annual cost = $\dfrac{\$74{,}880 + \$64{,}000}{1{,}615}$ = \$85.99/ton	t_r = 5.95 hours N = 85 $N \times n \times 5$ = 850 solid waste = 1551.42 tons/year annual cost = $\dfrac{\$74{,}880 + \$30{,}100}{1551.42}$ = \$67.67/ton	t_r = 5.71 hours N = 54.5 $N \times n \times 5$ = 817.50 solid waste = 1492 tons/year annual cost = $\dfrac{\$74{,}880 + 19{,}000}{1492}$ = \$62.92/ton
12	t_r = 5.98 hours N = 170.95 $N \times n \times 5$ = 854.75 solid waste = 1560 tons/year annual cost = \$89.03/ton	t_r = 5.35 hours N = 76.42 $N \times n \times 5$ = 764.20 solid waste = 1394.81 tons/year annual cost = \$75.26/ton	t_r = 4.71 hours N = 44.85 $N \times n \times 5$ = 672.75 solid waste = 1227.90 tons/year annual cost = \$76.46/year
24	t_r = 5.78 hours N = 165.14 $N \times n \times 5$ = 825.70 solid waste = 1507 tons/year annual cost = \$92.16/ton	t_r = 4.75 hours N = 67.85 $N \times n \times 5$ = 678.50 solid waste = 1238.39 tons/year annual cost = \$84.77/ton	t_r = 3.71 hours N = 35.34 $N \times n \times 5$ = 530.10 solid waste = 967.13 tons/year annual cost = \$97.07/ton
36	t_r = 5.58 hours N = 159.43 $N \times n \times 5$ = 797.15 solid waste = 1454.95 tons/year annual cost = \$95.45/ton	t_r = 4.15 hours N = 59.28 $N \times n \times 5$ = 592.70 solid waste = 1081.97 tons/year annual cost = \$97.03/ton	t_r = 2.71 hours N = 25.81 $N \times n \times 5$ = 387.15 solid waste = 706.62 tons/year annual cost = \$132.86/ton
48	t_r = 5.38 hours N = 153.71 $N \times n \times 5$ = 768.55 solid waste = 1402.75 tons/year annual cost = \$99.01/ton	t_r = 3.55 hours N = 50.71 $N \times n \times 5$ = 507.10 solid waste = 925.55 tons/year annual cost = \$113.42/ton	t_r = 1.71 hours N = 16.28 $N \times n \times 5$ = 244.20 solid waste = 445.71 tons/year annual cost = \$210.63/ton
60	t_r = 5.18 hours N = 147.99 $N \times n \times 5$ = 739.95 solid waste = 1350.55 tons/year annual cost = \$102.83/ton	t_r = 295 hours N = 42.14 $N \times n \times 5$ = 421.40 solid waste = 769.13 tons/year annual cost = \$136.49/ton	—
84	t_r = 4.78 hours N = 136.57 $N \times n \times 5$ = 682.85 solid waste = 1246.33 tons/year annual cost = \$111.43/ton	t_r = 1.75 hours N = 25 $N \times n \times 5$ = 250 solid waste = 456.30 tons/year annual cost = \$230.07/ton	—

Collection Vehicle Routing Methods. Routing of the collection vehicle may be managed by one of the following methods:

1. *Daily route method:* This scheme requires the crew to finish its assigned route before calling it a day. If the route has not been finished, the crew must work overtime.

2. *Large route method:* In this method, the crew has sufficient work for the entire week, and the collection along the route must be completed in 1 week. The crew decides when to pick up. This method works for backyard collection, because the residents do not know when pickup will occur.

3. *Single-load method:* According to this method, routes are planned to get a full collection vehicle and crew assigned to as many loads as it can collect during a day's work. This method considers crew size, length of travel, vehicle capacity, waste generated, and other variables.

4. *Definite working day method:* In this scheme, the crew works for its assigned number of hours (or an agreed-on number of hours based on contract negotiations). This usually is the case where labor unions are involved.

The single-load method is the preferred method, but the definite working day method is common in communities where unions are strong.

After determining a scheme for managing the collection vehicle, a route must be selected. The selection of a proper route, which is known as route optimization, can produce considerable savings for the waste collection agency or the city. Computer programs can develop an optimal route but are seldom used, because such programs take time and expertise to write and debug for each specific situation, because a relatively efficient route can be selected by experience and common sense, and because collection crews may change the routes to suit themselves. The purpose of routing is to subdivide the community into units that will permit collection crews to work efficiently. The community can be divided into districts, with each district constituting 1 day's work for the crew. The route is the planned path of travel for the collection vehicle.

Heuristic Routing. Common sense routing is sometimes called heuristic routing. The Office of Solid Waste Management Programs of the U.S. Environmental Protection Agency (EPA) has developed a simple, noncomputerized, "heuristic" (i.e., rule-of-thumb) approach to routing based on logical principles. Some sensible rules of thumb, when followed, will go a long way toward producing the best collection solution. The goal is to minimize deadheading, delay, and left turns by developing, recognizing, and using certain patterns that repeat in every municipality.

As indicated earlier, routing skills depend on certain rules and experience. According to the EPA (6), these rules include:

1. Routes should not be fragmented or overlap. Each route should be compact, consisting of street segments clustered in the same geographical area.

2. Total collection plus haul times should be reasonably constant for each route (i.e., equalized workloads).

3. The collection route should be started as close to the garage or motor pool as possible and take into account heavily traveled and one-way streets.

4. Heavily traveled streets should not undergo collection during rush hours.

5. For one-way streets, it is best to start the route near the upper end of the street and to work down it through the looping process.

6. Services on dead-end streets can be considered as being services on the street segment they intersect, because they can only undergo collection by passing down that street segment. To keep left turns at a minimum, collect the dead end-streets when they are to the right of the truck; they must be collected by walking down, backing down, or making a U-turn.

7. When practical, service stops on steep hills should undergo collection on both sides of the street while the vehicle is moving downhill for safety, ease, speed of collection, lower wear on the vehicle, and conservation of gas and oil.

8. Higher elevations should be at the start of the route.

9. For collection from one side of the street at a time, it is generally best to route with many clockwise turns around blocks. (*Note:* Heuristic rules 8 and 9 emphasize the development of a series of clockwise loops to minimize left turns, which generally are more difficult and time-consuming than right turns. Especially for right hand–drive vehicles, right turns are safer.)

10. For collection from both sides of the street at the same time, it is generally best to route with long, straight paths across the grid before looping clockwise.

11. For certain block configurations within the route, specific routing patterns should be applied.

Example 12–4

Plan a collection route (i.e., lay out a residential collection route) for the section of a city shown in Figure 12–9. Assume that the following conditions apply:

1. Streets

    a. All streets are two way on all four sides of the pattern

    b. Collection is on one side of the street at a time

    c. No U-turns are allowed in the streets

2. General

    a. Occupants per residence, 3.5

    b. Solid waste collection rate, 4 lbs/person/day

    c. Collection frequency, once per week

    d. Type of collection service, curbside

    e. Collection crew size, one person

    f. Collection vehicle capacity, 15 yd^3

    g. Compacted specific weight of solid waste in collection vehicle, 500 lbs/yd^3

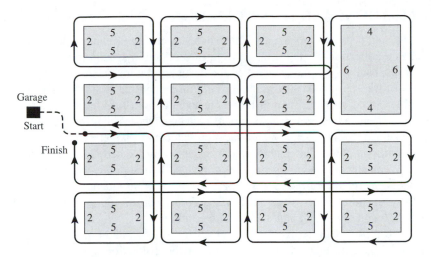

Figure 12–9   Plan of a collection route showing the number of residences to be serviced.

Solution

First, develop the data needed to establish collection routes (i.e., step 2 in the layout of collection routes).

1. Determine the total number of residences from which waste is to be collected:

$$\text{residences} = 14(14) + 1(20) = 216$$

2. Determine the compacted volume of solid waste to be collected per week:

$$Volume/wk$$
$$= \frac{216 \; residences \times 3.5 \; persons/residence \times 4 \; lbs/person/day \times 7 \; days/week}{500 \; lbs/yd^3}$$
$$= 42.34 \; yd^3$$

3. Determine the number of trips required per week:

$$trips/week = \frac{42.34 \; yd^3}{15 \; yd^3/trip} = 2.823 \; \approx 3$$

4. Determine the average number of residences from which waste is to be collected on each trip:

$$Residences/trip = \frac{216}{3} = 72$$

Now, lay out collection routes by trial and error, as indicated previously, using the route constraints listed earlier.

TRANSFER STATIONS AND TRANSPORTATION

The haul time to a processing and/or disposal site significantly affects the cost of solid waste collection. As the distance from the collection system to the processing facility or disposal site (collectively called the destination point) increases, the cost of hauling (or of transportation) also increases, until direct hauling is no longer economically feasible. At a certain transport distance, management must decide whether to build a transfer station, which is a facility where collected wastes may be stored temporarily or transferred from several small collection vehicles to larger vehicles for hauling to the destination point. The transfer station is generally located close to town.

Transfer stations are sometimes perceived by the public as a source of noise, dust, odors, increased traffic, rodents, flies, and litter. Consequently, property values suffer. Most of these concerns can be alleviated, however, through proper site selection, good design and operation, and good public relations.

Careful planning is critical, and several criteria determine where a transfer station should be located. It should be near the collection area to minimize the crew time for hauling to the transfer station, which is nonproductive time. In addition to proximity to the collection routes, access to major haul routes is also important to optimize transfer vehicle productivity. Access roads must be able to handle heavy truck traffic, and routes must be designed to minimize the impact of such vehicles on neighborhoods. Aside from routing issues, the land on which the facility is built should have adequate area for future expansion and be zoned for industrial purposes. This area should also provide for isolation of the facility. Availability of utilities such as water, sanitary and storm sewers, electricity, and fuel for heating should be considered in site selection as well. The cost to obtain the necessary utilities is an important part of overall site development costs, but sizing procedures for transfer stations are not well established. For a transfer station with a capacity of 250 tons/day, approximately 5.5 to 6.5 acres may be needed depending on the sizing procedures and design (4).

A recent trend in waste management is development of large, integrated materials recovery/transfer facilities. Private facilities receiving loads with large quantities of recyclables have taken advantage of this by selling easily separated materials. Developing a good, comprehensive recycling program at a transfer station may involve significant planning; a case study is provided later in this chapter.

Cost

The cost of a transfer station depends on many factors, such as its size, the price of land, local construction costs, and a variety of technological factors. Simple and relatively inexpensive transfer stations may consist of a building shell with a thick, concrete slab that is used as a tipping floor. Collection vehicles drop the solid waste onto this floor. A front-end loader is then used to scoop up the waste and to load the transfer vehicle. These facilities also may include hoppers for direct transfer of waste from collection vehicles to transfer vehicles, and they may have compaction equipment to compress waste for loading onto the transfer vehicles. Some transfer stations have equipment for baling as well. Figure 12–10 shows a simple transfer station.

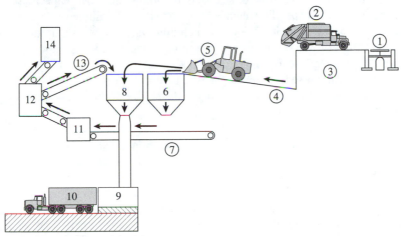

Transfer station with storage, processing, and compaction facilities

1. Scale to weigh the incoming solid waste load.
2. Collection vehicle.
3. Unloading platform for collection vehicle to deposit waste into the storage pit (4).
4. Storage pit for incoming waste before processing.
5. Front-end loader to feed waste to the hoppers (6 and 7).
6. Hopper to discharge the waste onto the conveyor (7).
7. Conveyor to convey the waste from the hopper (6) to the shredder (11).
8. Hopper to feed the waste to the compactor (9).
9. Compactor for the compaction of waste before loading into the transfer trailer (10).
10. Transfer trailer to transport the waste to MRF or disposal facility.
11. Shedder to shred the waste and convey it to the separator (12).
12. Separator to separate the waste into nonferrous waste (13) and ferrous scrap (14).
13. Conveyor to move the waste to the hopper (8) for compaction (9).
14. Ferrous scrap for further processing.

Figure 12–10   (a) Transfer station with a depressed ramp where transfer trailers are located. (b) Transfer station with storage, processing, and compaction facilities.

Transfer Vehicles and Equipment

Road transport is the most common mode of moving solid waste to distant disposal sites. Rail and water transport are also used in rare cases. Two kinds of transfer trailers are used; open-top trailers, and enclosed trailers.

Open-top, noncompaction trailers are lighter than their compactor counterparts; consequently, a larger payload usually can be loaded in these vehicles. Specific weights achieved in noncompaction trailers range from 275 to 400 lbs/yd^3 for municipal solid waste.

Compaction trailers are enclosed vehicles that are loaded by a stationary compactor. A hydraulically operated, push-out blade is used to unload the trailer. Specific weights achieved in compaction trailers range from 500 to 600 lbs/yd^3.

These trailers are made of steel and aluminum, and their capacities range from 65 to 96 yd^3. With a specific weight of 600 lbs/yd^3 at a volume of 96 yd^3, the weight of municipal solid waste is 57,600 lbs. Including the vehicle weight, this is less than the weight limit of 80,000 lbs for most highways. Other reasons, however, may limit the load these vehicles carry. For example, secondary and access roads to a disposal site generally will not support trailer loads, especially during wet weather; therefore, vehicles must be loaded or sized according to the ability of the access roads to support their weight.

At the site, the trailer can be unloaded using various devices. Many trailers have a hydraulically operated, push-out blade or a cable winch that pulls a panel in the front of the trailer to the rear, thus forcing out the solid waste. In other cases, the cable is attached to a tractor, and the waste is pulled out of the trailer.

Baling municipal solid waste has several potential advantages, including lower haul cost and lower cover material requirements. Most stations using baling equipment have floor storage for the waste. Bales are loaded onto flatbed trucks, barges, or rail cars for transport to the disposal site. After the bales are loaded, they are covered by a tarpaulin to keep the waste contained and to prevent littering. Loading and unloading of baled waste is done with a forklift or crane.

The specific weight of bales may be as high as 2000 lbs/yd^3 (7), and a single bale may weigh more than 5 tons. With rail and barge hauling, the waste is transferred to a vehicle at the landfill site, which then moves the waste to its burial site. In some cases, containers are used for convenience in loading and unloading barges or rail cars. These containers are then transferred to trucks for hauling to the landfill, and mechanical unloaders dump the contents of the containers and place them back on the truck for reuse.

Economics

Selection of a transfer system depends on the site, type of station, available transportation links between the generation source and disposal site, length of haul to the disposal site, quantity of waste, and weight limits (4). An economic comparison is made of the unit cost associated with using the collection vehicle as the haul vehicle versus the cost of constructing and operating a transfer station and of using a transfer vehicle. One can reasonably conclude that the ton-mile cost of hauling with the collection vehicle is much less than the ton-mile cost of hauling with the transfer vehicle, so the distance must be large enough to necessitate the investment in the transfer station.

There are two general types of transfer stations: direct-discharge transfer stations, and storage transfer stations. In the former, collection vehicles dump their loads directly into the larger transfer vehicles; in the latter, the loads are emptied into storage pits or platforms. Later, the waste is loaded into big transport vehicles for hauling to the destination point. Figure 12–10 shows the two types of transfer facilities.

To decide whether to build a transfer station, management should consider at what point the distance to the disposal site justifies the added cost of a transfer station. Figure 12–11 indicates the break-even point for a transfer system. Examination of the sketch indicates that beyond a certain distance, direct hauling is uneconomical compared with the cost of a transfer station.

Example 12–5

A transfer station with a capacity of 250 tons/day operates 5 days per week. It cost approximately $6 million to build. Another $700,000 per year is needed to operate the station. A tractor trailer costs $209,256 and carries 17.5 tons of waste per trip. The annual operation and maintenance (including fuel) cost for the truck is $35,000. The annual salary of the driver (including benefits) is $50,000. The capital cost of the building is to be amortized over a 20-year period and the capital cost of the tractor trailer over a 10-year period. The interest rate is 10 percent.

Assume that it takes 1 hour to make a roundtrip from the transfer station to the disposal facility and that five trips are made each day. Calculate the hauling costs from the transfer station.

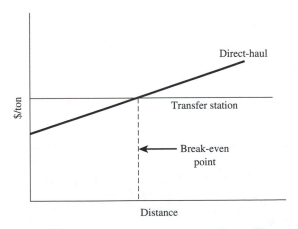

Figure 12–11   Break-even point for a transfer station.

Solution

The capital recovery factor is

$$CRF = \frac{i(1 + i)n}{(1 + i)^n - 1} \tag{12-6}$$

For a building, i is 10% and n is 20 years. Thus,

$$CRF = \frac{0.10 \, (1 + 0.10)^{20}}{(1 + 0.10)^{20} - 1} = 0.1175 \; per \; year$$

For a tractor trailer, i is 10% and n is 10 years. Thus,

$$CRF = \frac{0.10 \, (1 + 0.10)^{10}}{(1 + 0.10)^{10} - 1} = 0.1627 \; per \; year$$

The annual cost of the transfer station is

$$(\$6{,}000{,}000 \times 0.1175) + \$700{,}000 = 1{,}404{,}400 \; per \; year$$

and the annual cost per ton of waste is

$$\frac{\$1{,}404{,}400/year}{250 \; tons/day \times 260 \; days/year} = \$21.60 \; per \; ton$$

The annual cost of a truck and its driver is

$$(\$209{,}256 \times 0.1627) + \$35{,}000 + \$50{,}000 = \$119{,}046$$

and the cost per ton of waste hauled is

$$\frac{\$119{,}046}{17.5 \; tons/trip \times 5 \; trips/day \times 260 \; days/year} = \$5.23$$

Therefore, the total cost per ton of the transfer station and vehicle is $21.60 + $5.23 = $26.83 per ton

In the example 12–5, the calculation of cost was based on a roundtrip from the transfer station to the disposal facility of 1 hour. Figure 12–12 shows how cost changes with trip length. The fixed cost for the 1-hour roundtrip (including unloading time) to the transfer station is $21.60 per ton, and the variable cost of the vehicle (30 minutes each way) is $5.23 per ton.

Suppose a good site for a transfer station is located 30 minutes from the last collection stop. Continuing with the previous example, Figure 12–13 indicates that the most economical waste collection system involves two runs per day from the last collection point to the transfer station at a cost of $92 per ton. The figure also shows the added cost of a transfer system to carry the waste the rest of the way to the disposal site. For a one-way drive time between the collection route and the disposal facility of less than 52 minutes, no transfer station can be economically justified. It is more eco-

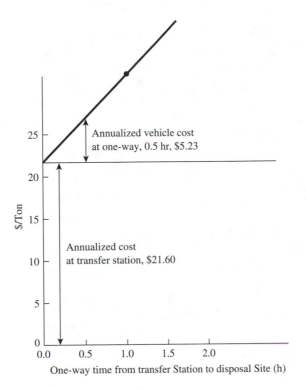

Figure 12–12 shows: y-axis labeled $/Ton with values 0, 5, 10, 15, 20, 25. x-axis labeled "One-way time from transfer Station to disposal Site (h)" with values 0.0, 0.5, 1.0, 1.5, 2.0. Annotations: "Annualized vehicle cost at one-way, 0.5 hr, $5.23" and "Annualized cost at transfer station, $21.60".

Figure 12–12   Effect of distance on the cost of including a transfer station in a waste collection system. (Data for graphs are provided in example 12–5.)

nomical to make two runs per day from the last collection point to the disposal site. If the one-way drive exceeds 52 minutes, however, a transfer station should be considered, because the total cost is lower. (These results are only for the example 12–5, but they provide a method to evaluate various tradeoffs in making a good decision.)

Site Selection Based on Operational Control

In situations where two or more transfer stations and disposal sites (landfill or incinerator) are available, it is desirable to determine the optimum, most cost-effective allocation of waste to each facility. Given that most waste management agencies have many external constraints on the location of waste processing and disposal facilities, however, cost consideration becomes less important.

Consider the locations of transfer station (T) and disposal site (D). Figure 12–14 illustrates the links between T and D.

The primary objective of waste distribution among the various options is to obtain the least-cost disposal system. The costs considered are those of each disposal option,

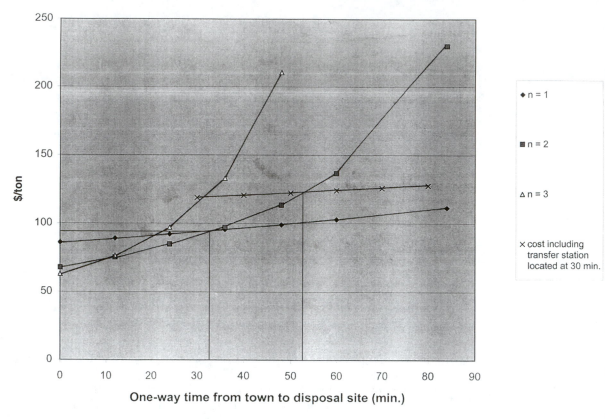

Figure 12–13   Optimum number of trips per day to the disposal site.

which may be different for each disposal site, and those of hauling the waste from each transfer station to each disposal site. The assumptions are:

1. The total amount of waste to be hauled to all disposal sites must equal the amount available at a transfer station.
2. Each site can only accept specified amounts of wastes.
3. The amount of waste hauled from each transfer station is equal to or greater than zero.

In the symbolic form, the allocation problem is set as follows:

1. Designate transfer station sites by i.
2. Designate disposal sites by j.
3. X_{ij} is the tons of waste transported from transfer station i to disposal site j.
4. S_{ij} is the cost of hauling waste from transfer station i to disposal site j.
5. Q_j is the total amount of waste that can be accepted at the disposal site.

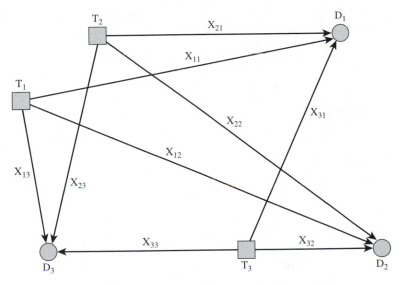

Figure 12–14   Allocation of solid waste to disposal sites.

6. L_j is the cost of waste disposal at site j.

7. R_i is the the total amount of waste delivered to transfer station i.

8. Thus, the objective function can be written as the sum of the following terms:

$$X_{11}(S_{11} + L_1) + X_{12}(S_{12} + L_2) + X_{21}(S_{21} + L_1) + X_{22}(S_{22} + L_2)$$
$$+ X_{23}(S_{23} + L_3) + X_{31}(S_{31} + L_1) + X_{32}(S_{32} + L_2)$$
$$+ X_{33}(S_{33} + L_3) + \ldots + X_{ij}(S_{ij} + D_j) \quad \textbf{(12-7)}$$

The problem is to minimize the cost. In mathematical terms,

$$\sum_{i=1}^{3} \sum_{j=1}^{3} X_{ij}(S_{ij} + L_j); \, i = 1 \text{ to } 3; \, j = 1 \text{ to } 3 \quad \textbf{(12-8)}$$

subject to the following constraints:

1. *Total waste available:* The total amount of waste shipped from the transfer station must equal the amount available, which is a material balance problem:

$$\sum_{j=1}^{3} X_{ij} = R_i \quad \textbf{(12-9)}$$

2. *Total capacity at the disposal site:* Each disposal site has a specific capacity that cannot be exceeded. This may be a result of the site's capacity and/or limited road access. Expressed mathematically,

$$\sum_{i=1}^{3} \leq Q_j \quad \textbf{(12-10)}$$

3. *Waste flow:* The amount of waste hauled from the transfer station must have a positive flow; it is absurd to haul waste back from the disposal site to the transfer station. Expressed mathematically,

$$\sum_{i=1}^{3} X_{ij} \geq 0 \qquad\qquad (12\text{-}11)$$

Several solution methods are available. Most require computers. Several approximate solution methods have also been developed that provide results fairly close to the optimum solution. The optimum solution itself may be obtained with some of the computer programs.

CASE STUDY

*A Prescription for Reviving a Transfer Station Gives a Boost to Recycling Program**

In 1973, Ohio's Miami County solid waste incinerator was shut down when the Ohio EPA required additional costly air emissions–control facilities. The county decided to continue solid waste disposal services by transporting the waste to two landfills, which had been closed.

To provide better and economical waste disposal services, a transfer station was needed. Part of the county incinerator (now shutdown) building was converted into a transfer station by installing compactors adjacent to the incinerator's receiving pit. The old overhead crane and grapple bucket, which were used to feed the incinerator boilers, were fixed for use to feed the compactors. The trash would be loaded into the compactors by the crane and grapple bucket, and the compacted trash would be pushed into the transfer trailer.

Unfortunately, the aging crane frequently broke down, because electrical components malfunctioned and grapple welds were coming apart. At such times, there was no way of loading the trash from the pit. During these periods, the trash would be dumped on the ground in another area and loaded into the trailers with front-end loaders. In addition to the crane, there was damage to the compactors and trailers.

Defining the Problems

Faced with the frequent problems, the county hired a new sanitary engineer and a new transfer station manager. The county health department asked the sanitary engineer to prepare plans for a new transfer station, but meanwhile, he and the

---

**Source:* U.S. Public Health Service. *A Study of Solid Waste Collection Systems.* Publication 1892 (SW-96). Washington, D.C.: U.S. Govt. Printing Office, 1969; with permission.

station manager should keep the present station in working order. The manager, after a careful study, identified the following problems:

- The equipment was underdesigned, overused, misused, and poorly maintained. For example, the crane was running 14 hours per day, even though it was designed to run on a 2-hour duty cycle. The compactors were also being overloaded to three times their capacity and being used to compact large objects for which they were not designed.
- There was no preventive maintenance program.
- Workers' morale was low.

Solutions

Obviously, the priority was to fix the equipment. Instead of the patchwork repairs being made by the current crane maintenance company, a new firm was asked to do what was needed so that the equipment would work well. After a careful evaluation, the firm completely overhauled the crane and the bucket and made them fully functional by repairing and changing both parts and controls. A preventive maintenance program was also put into place (e.g., selecting the right lubricant and changing the gear oil in the crane, which had not been changed for the last 23 years). This resulted in less wear and tear of the equipment and in better efficiency.

Now that the equipment was fixed, the next step was to prevent the overuse and misuse of that equipment. There was a widespread practice among the employees of salvaging materials such as aluminum, brass, and other items of value for themselves. Employer were depositing the waste on to the upper floor, removing what they needed, and then using the crane to push the remainder back into the pit again, to pick it up, and to load it into the compactor. This caused overuse of the equipment. This practice had been going on for last several years, and no one had discouraged the employees from it. The new manager took a strong stand against this practice by telling employees that salvaging of material for personal use would be considered theft and, thus, carry the possibility of prosecution and job termination.

The excessive damage due to misuse was tackled by excluding large objects, such as appliances, metal piping, car parts, and steel decking, and other items, such as construction and demolition debris. This action eliminated monthly repairs costing about $23,500 and improved the cranes' efficiency.

To achieve efficiency and reliability, changes were also made in the operation of the compactors. When fed together in the hopper, large objects would cause damage to the hopper and floor and cost $3000 to $4000 to fix. The problem was solved either by eliminating the large items or by feeding those items in one by one. Also, the pressure on the compactor was set at 3800 psi, whereas the correct setting should have been 1800 psi or lower. Obviously, the pressure setting was reduced.

Several steps were taken to boost the morale of the workers as well. Upward revision of pay scales and biweekly meetings by the manager to review work, obtain feedback from workers, and keep workers informed of policies and programs were some of the steps taken. The occupational health of the employees was monitored by periodic checkups. Employees could see that their efforts resulted in a good facility, thus giving them a sense of pride in their accomplishment.

Recycling Center

The revived transfer station gave a boost to the county's recycling program. The transfer station began removing valuable materials from the incoming waste and selling them to scrap dealers and other recyclers. Revenues generated were in the range of $15,000 to $20,000 per year. In addition, the facility removed wooden poles, leaves, brush, and tires from the incoming waste. Except for the tires, this material was shredded and distributed as mulch: waste tires were sent to tire recyclers.

In May 1990, the county hired a solid waste resource coordinator to plan solid waste recycling activities and to help develop the county's waste management program as mandated by Ohio's new solid waste management law—House Bill 592. Among the recycling activities was a large, public drop-off recycling center at the transfer station, which started with a trailer to collect old newspapers. This replaced a local center that collected approximately 8 tons of newspapers per week as part of the fund-raising activities but closed when the bottom fell out of the market for newspapers.

The county wanted to provide an outlet for committed recyclers, who were driving miles to deposit newspapers for recycling. (The county's newspapers go to a nearby mill that makes corrugated cardboard.) Next, the drop-off facility started accepting aluminum and bimetal beverage cans, steel food cans, all colors of glass containers, and recyclable plastics, cardboard, and other office paper.

The drop-off center is now collecting more than 70,000 lbs of recyclables per month. According to the county officials and consulting engineers, a major advantage of the transfer station drop-off recycling center over any other drop-off program in the county is the full-time recycling monitor, who helps the recycling program by providing on-the-spot education and by guiding the public to properly prepare and place the recyclable material. He also does some processing of the waste to be recycled. The recyclables are picked up by a recycling company to be bailed, loaded, and transported to a mill. The company pays the county for metal cans and glass containers and does not charge to transport the recyclables. The county's only cost after the initial expense of the recycling building, collection bins, and signs is the monitor's pay.

The drop-off program at the transfer station costs the county approximately $35 to $50 per ton. Compared with most curbside programs, which range in cost from $120 to $250 per ton, this is low. This cost can be even lower if the market value of recyclable newspaper and cardboard goes up, because approximately 75 percent of the recyclables is newspaper and cardboard.

Present Situation and Future Planning

The transfer station drop-off recycling center is meeting the recycling needs of the Miami County residents. It has provided efficient transfer and disposal service, and it has provided environmental benefits by diverting significant quantities of solid waste from landfills and preserved valuable natural resources through recycling. By implementing sound management practices, the operation has improved. The repair and maintenance costs of the transfer station have been reduced from $278,000 in 1988 to $74,500 in 1990. Plans are also underway to replace the current facility with a modern one, which is now almost complete.

REVIEW QUESTIONS

1. a. What is the purpose of on-site storage for solid waste?
   b. List the precautions needed during such storage.
2. a. How has the practice of storage and collection changed in the last 25 years?
   b. On the solid waste pickup day in your town, check at least 10 residential curbside locations and report the number of containers and the types of recyclables at each location. Are the recyclable separated from the rest of the waste?
3. a. Why are the collection and transportation of solid waste considered an important function of solid waste management?
   b. What is the difference between collection from single-family residences and from multiple-family dwellings?
4. a. What is the purpose of waste compaction?
   b. If the density of waste at curbside is 250 lbs/yd^3 and the waste is then picked up and compacted by a compactor vehicle with a factor of 2.5, what is the density of the waste as received at the landfill?
5. Sketch and explain the municipal waste collection process with a two-person crew.
6. A city crew of two collects waste of 180 residences per week. The city engineer devises a new system to save 5 seconds per residential collection. If the labor cost is $18/hour, compute the annual savings based on 260 service days.

7. a. List the productive and nonproductive events that occur during solid waste collection.
   b. Can some of these delays be avoided?
8. It takes 30 minutes to drive a collection vehicle from the city garage to the beginning of the route, 40 minutes to drive between the route and the disposal site, and 20 minutes to return the empty vehicle from the disposal site to the garage. It takes 15 minutes to off-load the vehicle at the disposal site. The crew takes two 15-minute breaks per day and allows 30 minutes for unexpected delays. If two runs are made to the disposal site per day, calculate the time left in an 8-hour workday for the actual solid waste collection.
9. Based on the data in question 8, find the capacity of the vehicle using the following information:

   - Distance between the stops along the route, 400 ft
   - Number of containers per stop, 2
   - Volume of waste generated per residence, 10 ft^3/week
   - Vehicle compaction factor, 2.5

10. Using the vehicle capacity from problem 9, calculate the collection cost per ton. Assume a two-person crew, labor charges of $16.00 per hour per person, and 150 households serviced on a single trip. The curbside pickup service is provided once a week for each house, and the waste collection service is provided 5 days a week.

11. a. List the common sense routing rules for the solid waste collection.

b. Sketch a collection route for the town (or a section of the town) in which you live. Justify the proposed route.

REFERENCES

1. Techobanoglous, G.H. Theisen, and S. Vigil. *Integrated Solid Waste Management.* New York: McGraw-Hill, 1994.

2. U.S. Public Health Service. *A Study of Solid Waste Collection Systems.* Publication 1892 (SW-96). Washington, D.C.: U.S. Govt. Printing Office, 1969.

3. Masters, G. *Introduction to Environment Engineering and Science,* 2nd ed. Upper Saddle River, N.J.: Prentice Hall, 1991.

4. U.S. Environmental Protection Agency. *Decision Maker's Guide to Solid Waste Management.* Washington, D.C.: U.S. Govt. Printing Office, 1989.

5. U.S. Environmental Protection Agency. *Decision Maker's Guide in Solid Waste Management.* Washington, D.C.: U.S. Govt. Printing Office, 1976.

6. Shuster, K.A., and D.A. Schur. *Heuristic Routing for Solid Waste Collection Vehicles.* U.S. EPA Publication No. SW-113. Washington, D.C.: U.S. Govt. Printing Office, 1974.

7. Robinson, W. *Solid Waste Handbook.* New York: Wiley-Interscience, 1986.

8. McGarry, M. Bhatt, and T. York. "Prescription for Reviving a Transfer Station," Paper Presented at the Waste Equipment and Recycling Exposition, in Dayton, OH, 1991.

Chapter 13   Solid Waste Basic Processing Technologies

This chapter introduces the recovery of materials, the separation and processing of solid waste components, and the transformation processes that convert waste to recover useful products. The methods used to recover source-separated materials include curbside collection and homeowners taking such materials to drop-off and buy-back centers. The segregation of solid waste components may be done at the point of generation (i.e., on-site processing) or at a central processing facility.

On-site processing requires the education and cooperation of the waste producer: homes, commercial establishments, industries, and the like. Because of regulations, an increasing number of waste collection agencies, require source separation for recycling. During on-site processing, wastes are segregated into types at the point of generation; for example, cans are put into one container, plastics in another, paper in another, and so on. The collection crew can then pick up and deposit the separated wastes in correspondingly separate compartments of the vehicle. In Ada, Ohio, a special vehicle collects segregated items on a weekly basis. These items are left at the curbside by homeowners in large yellow or red plastic baskets supplied by the collection agencies. If on-site processing is not done, segregation into components may occur at a central facility.

Further separation and processing of wastes that have been source separated, as well as the separation of commingled wastes, usually occur at materials recovery facilities or combined materials recovery/transfer facilities. The latter may include the functions of a drop-off center for separated wastes, a separation facility, composting, and bioconversion of wastes as well as a facility for the production of fuel. Figure 13–1 shows handsorting of municipal solid waste components.

PREPROCESSING

Preprocessing of solid waste is done to produce a waste stream with a greater homogeneity and to permit recovery of materials such as aluminum, glass, and ferrous metals. These activities, which are normally used at a resource recovery facility, include weighing, receiving and storage, screening, shredding, and air classification.

Weigh Stations

Weighing incoming waste provides accurate information on the quantity received and allows equitable fees for processing to be established. The station consists of platform scales of a size to accommodate the largest vehicle. Figure 13–2 shows a

Figure 13–1   Handsorting of municipal solid waste.

weigh station at the entrance to a processing facility. Three main types of scales are used for weighing solid waste: the beam weight, the loan cell, and a combination mechanical–electronic scale.

The information recorded when vehicles are weighed includes the date and time, vehicle identification, tare weight, gross weight, and net weight. The tare weight (i.e., weight of the truck) is deducted from the gross weight, which includes the weight of the solid waste, to obtain the net weight (i.e., weight of the waste). This information

Figure 13–2   Weighing station at the entrance to a solid waste processing facility.

(Courtesy of Miami County, Ohio.)

is used to bill the owner of the vehicle for the amount (tons) of waste received. This is termed the *tip fee.* Information on the overall amount of waste received for processing is also recorded. For a sanitary landfill, the accumulated weight received indicates the rate at which the landfill capacity is being used and provides data for determining the total solid waste production of the area served by the facility. Coded vehicles provide data on the waste production along the specific routes served by a specific vehicle. These data can be used to establish a more efficient route collection system.

Weigh stations can be operated manually by inputing the data with a keyboard. This requires an operator in the scale house. Most stations also control access to the facility by having an operator at the checkpoint. Stations with no need for the access control are automated. In this case, vehicles have magnetic cards that are inserted into a card reader, and the information is collected and computed automatically. The degree of automation of the weigh station depends on the needs of the facility and cost. A scale system should be able to handle the maximum expected number of vehicles without excessive delays.

Receiving and Storage Areas

Receiving and storage areas receive incoming vehicles, provide space for them to unload, and allow storage of waste material before processing. Receiving and storage areas also buffer the processing system from variations in the rate of incoming waste. Most resource-recovery facilities operate continuously, whereas solid waste is generally collected only 5 days a week, in one shift per day.

Example 13–1

A 600 ton/day resource-recovery facility receives municipal solid waste in an 18-yd^3 compactor vehicle at 400 lbs/yd^3.

The nominal processing rate is

$$\frac{600\ ton/day}{24\ hours/day} = 25\ ton/hour$$

The vehicle capacity is

$$\frac{18\ yd^3}{2000\ lbs/ton} \times 400\ lbs/yd^3 = 3.6\ tons\ per\ vehicle$$

or

$$25\ tons/hour \times \frac{1\ vehicle}{3.6\ tons} = 7\ truckloads\ per\ hour\ arriving\ to\ unload$$

Collection occurs 5 days per week in one shift per day, whereas the facility processes waste 24 hours a day, 7 days a week. In addition, actual work is done only 6 hours per shift, because 2 hours are used for lunch, breaks, and washing up. Calculate the actual receiving rate and the peak receiving rate for this facility.

Solution

The actual receiving rate is

$$600 \ tons/day \times \frac{7 \ days}{5 \ days} \times \frac{1}{6 \ hours/day} = 140 \ tons \ per \ hour$$

and

$$\frac{140 \ tons/hour}{3.6 \ tons/truck} \approx 39 \ truckloads \ per \ hour$$

Allowing for an hourly peaking factor of approximately 150 percent and a seasonal peaking factor of 125 percent, the peak receiving rate is

$$140 \times 1.5 \times 1.25 = 262.5 \ \text{tons per hour}$$

or

$$\frac{262.5 \ tons/hour}{3.6 \ tons/vehicle} \approx 73 \ truckloads \ per \ hour$$

As shown in example 13–1, for a 600 ton/day resource-recovery facility, the peak number of trucks arriving is 73 per hour, or one truck arriving every 0.82 minutes. A 10-minute unloading time, which is fairly representative, would require approximately 12 to 13 spaces for unloading. This should allow for the peak receiving rate to be achieved. Without considering the peaking factor, one truck would arrive every 1.53 minutes, and a 10-minute unloading time would require approximately 7 spaces. It is not unusual for several loaded vehicles to arrive at the same time as well, because when similar routes are started at the same time, collection is finished for the day at about the same time. The time differential between the end of a route and the tip area will produce the distribution of truck arrivals at the tip site.

Proper planning of collection routes can assist in distributing arrival times over the workday. Usually, a critical time is the end of the workday. If the workday ends at the same time for all collection crews, then all vehicles are likely to arrive at the tip floor at approximately the same time. This can result in considerable waiting to unload (i.e., a long queue) and, consequently, in loss of productive time. For economical consideration, providing too many spaces is not desirable, because the facility may not get enough use. Therefore, another option is to stagger the workday starting time of the crews.

Solid waste processing facilities can rarely match the rate at which the waste is received with the rate at which it is processed. As discussed, solid waste is not received uniformly by the facility over the workday. There are periods of low receipts and periods of high receipts. Thus, the solid waste is fed into a processing line at a constant rate, there are times when not enough waste is being received to meet the demand rate and times when too much waste is being received. Therefore, a "surge" or storage facility is needed to temporarily store the waste. In addition, if the solid waste is processed 24 hours a day, sufficient waste must be available in the storage area at the end of the collection day for processing until the vehicles bring in more waste. For facilities processing continuously and without other storage, a 3-day storage capacity generally must be provided to handle long weekends. Common methods of solid waste storage at resource-recovery facilities are:

- Pit and crane
- Live bottom pit
- Tipping floor
- Atlas storage and retrieval system

Pit and Crane.   This is the oldest type of system and remains a common method of storage in a waste combustion facility, as shown in Figure 13–3. The tip floor may be 20 to 40 ft above the bottom of the pit. Trucks back up to the edge of the pit and then dump the waste into it. The waste is picked up by an overhead crane with a claw and is delivered to a hopper, through which it is fed to the furnace. The crane operator controls the feed stream's quality (e.g., alternating wet and dry waste) to obtain a balanced heat load in the furnace. This method has the advantage of proven technology and requires relatively little area. Disadvantages are the high construction and maintenance costs and the difficulty of controlling fires in the pit.

Live Bottom Pit.   Another technique for reclaiming solid waste includes "live bottom" pits. As the name indicates, the bottom of the pit moves. This type of storage facility consists of a large pit with conveyors at the bottom. Advantages include

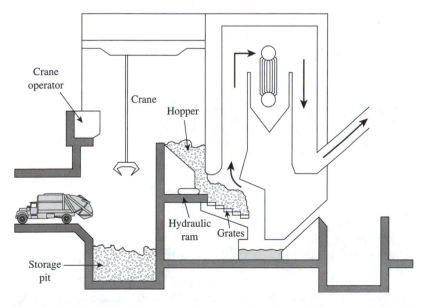

Figure 13–3   Typical incineration system using pit storage with retrieval of solid waste by an overhead crane.

automation and a large storage capacity in a reduced area. Disadvantages include the high construction cost and the difficulty of separating items that cannot be processed. This system has had limited success, because many waste items tend to stick together on the conveyor.

Tipping Floor.   This is an alternative to pit storage and is often used where front-end processing is intended. A tipping floor consists of a concrete slab on which the collection truck unloads waste. The slab is surrounded by a reinforced concrete wall (i.e., a push wall) designed to withstand the force of the front-end loader pushing against the wall to load a bucket. Advantages include a lower cost than pit storage and the ability to clean the floor regularly and to presort the waste. Figure 13–4 shows a tipping floor and a crane with a claw for a municipal solid waste incinerator.

Figure 13–4   (a) Vehicle unloading its contents onto a tipping floor. (b) Operator in the balcony manipulating the claw to pick up waste for processing.

Example 13–2

Size a pit-and-crane storage system for a 600 tons/day resource-recovery facility to handle 73 truckloads per hour if unloading takes 10 minutes.

Solution

Assume a 50-ft width, a density of 400 lbs/yd^3, and a storage capacity of 3 days. Thus,

$$\frac{60 \; minutes}{10 \; minutes} = 6 \; truckloads \; per \; bay \; per \; hour$$

and

$$73 \; truckloads/hour \times 6 \; truckloads/bay/hour \approx 12 \; bays$$

Assuming a 13-ft length per bay, the total system length is

$$12 \; bays \times 13 \; ft/bay = 156 \, ft$$

Atlas Storage and Retrieval System.   Atlas bins have had a long history in the storage and retrieval of waste products. This equipment is used extensively for storing wood shavings, bark, sawdust, and wood chips.

The concept behind the Atlas bin is fairly simple. The bin consists of a metal silo in the shape of an inverted cone. The processed solid waste (i.e., fuel) is fed into the bin through an opening in the top and onto a rotating slide, which distributes the material around the bin. The waste then forms a pile in the shape of a cone (Figure 13–5). The waste pile grows from the center, because the incoming stream drops down from the center of the inverted cone. Recovery of the bulk material from storage is accomplished by chains of sweep buckets, three to six of which are used, depending on the bin diameter and the volume of flow required. Each sweep chain is fixed at one end to a powered, rotating "pull ring" that encircles the storage area; the other end is free or trailing. As the pull ring rotates around the periphery of the bin, the sweep chains automatically trail toward the center. They drag along the bottom of the pile and pull the refuse toward the exit conveyor, which is located in a channel under the floor. The waste pulled by the chain then drops on to this conveyor and is conveyed to the desired process.

One problem with this system is damage to the floor because of the abrasive qualities of the sand and glass found in the feed. This increases the maintenance costs. Another problem occurs when the waste is allowed to accumulate in the bin. If the bin is not completely emptied on a frequent basis, some of the waste may remain in it for extended period of time, and if the moisture content is adequate, biodegradation will result in odors.

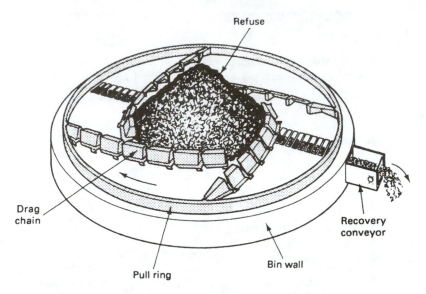

Figure 13–5   Solid waste retrieval from an Atlas bin.

(From Waste Age, *Vol. 16, No. 7, July 1985; with permission.)*

Refuse Conveying

A variety of transport processes are available to move solid waste into, away from, and between processing stages. At least four separate systems are applicable. The equipment selected depends on the processing requirements and transport properties of the material to be conveyed. Particle shape and size distribution, density, abrasiveness, moisture content, and angle of repose are some of the important properties to consider.

Flight Conveyor.   The flight conveyor is also called a pan or an apron conveyor. It consists of a continuous loop of steel plates that are pinned together and overlapped and supported by rollers. It is enclosed with stationary, channel-like vertical plates, and the conveyor serves as the bottom of a receiving bin for feeding solid waste to the processing facilities. The steel plates are ruggedly built to withstand the impact of large, dense objects. The conveyor is driven up the incline by a variable-speed drive that enables the operator to adjust the feed rate. Figure 13–6 shows a typical flight conveyor.

Belt Conveyor.   A belt conveyor is a continuous belt constructed with several plys of rubber canvas fabric stretched around two pulleys. The belt is carried by a series of rollers that are supported by the structural frame. The rollers are spaced to support the load being carried by the belt and are positioned to create a concave shape for the belt so that it can carry a greater load. Placing vertical metal plates on the sides of the belt can increase the depth of the material being carried. Belts are commonly used conveyors. They can convey materials or waste for long distances, up slopes as great as 30°, at a normal speed of 70 to 150 ft/min and with loads as great as 5000 tons per

Figure 13–6   Close-up view of a conveyor.

hour. Problems associated with this type of conveyor include waste falling from the belt because of backsliding and belt damage from sharp or ragged objects in the waste. Figure 13–7 shows typical belt conveyors.

Bucket Conveyor.   The bucket conveyor is designed for lifting materials vertically. It is also used for transporting solid waste. Bucket elevators are primarily used for shredded and separated solid waste. The buckets can be attached to a chain and lifted up a metal channel, and a type of rubberized belt with built-in buckets is common conveyors of solid waste.

Pneumatic Conveyor.   As the name implies, the pneumatic conveyor uses a stream of air to transport solid waste. These systems can be operated in either a positive- or negative-pressure mode. Positive pressure is used to blow material through the conveyor; this is suited for transporting material from a single site to multiple sites. Negative pressure is best suited for transporting material from multiple sites to a single site. Pneumatic transport provides a closed, compact means of moving solid waste for long distances at a high speed. Air-velocity requirements are in the range of 3000 to 5000 ft/min. Power requirements are higher than those for other types of conveyors. Figure 13–8 describes a typical pneumatic conveyor.

Air locks are used to feed and discharge the solid waste transported through the closed system. The solid waste is recovered by passing it through a cyclone to remove the larger particles. Pneumatic conveyors generally are used in conjunction with air-classification systems in which air is used for other purposes and, because the waste is already suspended in air, it becomes practical to use a pneumatic conveyor.

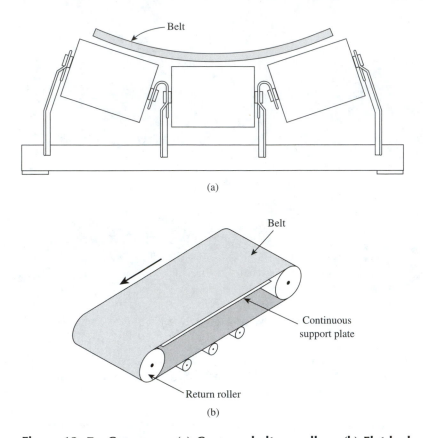

Figure 13–7   Conveyors. (a) Concave belt on rollers. (b) Flat bed on rollers.

(From U.S. Environmental Protection Agency. Handbook of Materials Recovery Facilities for Municipal Solid Waste. *Washington, D.C.: U.S. Govt. Printing Office, 1990; with permission.)*

PHYSICAL PROCESSING

Particle Size Distribution of Unprocessed Solid Waste

Raw solid waste is a mixture of boxes, bags, newspapers, bottles, cans, clippings, tires, and so on, all with varying sizes and shapes. Therefore, all solid waste received at a processing facility is checked for its condition. As a first step in processing, the waste contained in boxes and bags is removed by opening these containers, which are passed through a rotating, drumlike device (with knives) that breaks them open or through a shredder that reduces the solid waste to a certain particle size.

The size distribution and shape characteristics of the solid waste will vary with the composition of that waste. The constituent size can range from grains of sand and dirt to large, bulky items of furniture and appliances. There is variation in shape, but most waste components are either cylindrical, spherical, or platelike. It should be recognized that most solid waste is composed of a diversity of components with a wide range of shapes, which can affect processing.

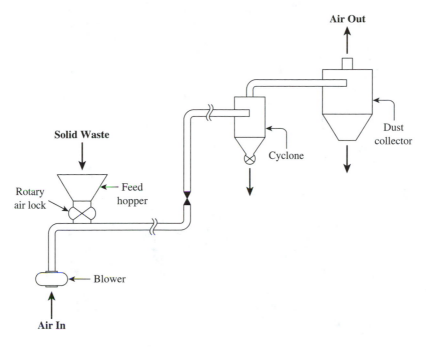

Figure 13–8   Pneumatic conveyor.

An extensive study of the size distribution and morphology of raw solid waste has been reported by Ruf (2). The cumulative and frequency distribution curves (Figure 13–9) developed in that report provide an evaluation of size distribution. From these curves, it is possible to get an idea of the size distribution of such waste to provide an approach for finding streams that are rich in certain waste components. The separation of waste according to size alone can provide clues for recovering specific materials by applying a variety of available technologies.

Ruf has further investigated the size distribution of primary shredded waste components. The cumulative and frequency distribution curves for the range of components in primary shredded waste are also shown in Figure 13–9. These curves characterize primary shredded waste and can be used to predict the performance of a separation process.

Particle size differs for different mills. The rating of a mill is based on a "nominal" particle size, which is a screen that will pass 90% (by weight) of the shredded waste. For example, a mill that produces a nominal particle size of 4 in. will have a processed waste in which 90% of the weight consists of particles less than 4 in. in size. Commercial solid waste falls primarily in the range of 2 to 20 in., whereas residential solid waste ranges from 1 to 10 in. After shredding, the particle sizes of mixed waste, which is mainly from commercial and residential sources, are reduced to a nominal size of between 2 to 4 in. When a large portion of the input contains oversized or bulky material and the required particle size of the output is small, two-stage shredding may be required.

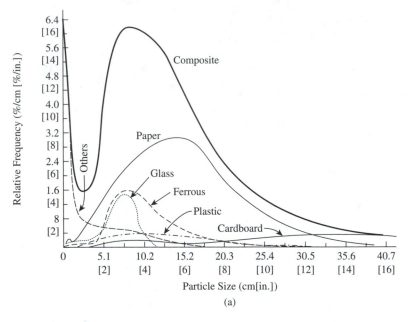

(a)

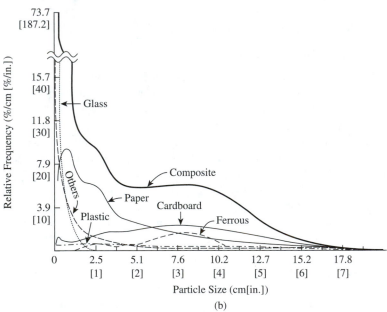

(b)

Figure 13–9   Particle size distributions. (a) Raw solid waste components. (b) Shredded waste.

(From Ruf. J. Particle Size Spectrum and Compressibility of Raw and Shredded Municipal Solid Waste. PhD thesis, University of Florida, 1974; with permission.)

Shredding and Size Reduction

Shredding is done to reduce the size of waste and to produce a relatively uniform material. Over time, the emphasis on massive size reduction has changed, with the focus shifting to finding more efficient separation and recovery processes. Forces such as tension, compression, and shear are employed to tear, cut, crush, grind, pulverize, and shred the waste. *Shredding,* however, is the general term applied to the mechanical processes used to reduce and homogenize waste. This section describes shredders and indicates some of the advantages and disadvantages of the various systems.

Processing rates of 500 to 1000 tons per hour can be achieved (3). In most cases, primary shredding will reduce solid waste to particles of less than 6 in. in size. The size distribution of the material exiting a hammermill is quite variable, however, and is a function of solid waste composition, moisture content, feed rate, exit grate, rotor speed, and size of the opening.

Hammer Mills.   Hammer mills are the most commonly employed equipment for solid waste reduction. The main element of the hammer mill is the rotor, and the mills are classified according to the orientation of the rotor or shaft. The rotor is encased in a heavy-duty housing. Two basic types of hammer mills are available: the vertical rotor shaft, and the horizontal rotor shaft. The latter is more common. Figure 13–10 shows both types.

Vertical Shaft Mill.   The vertical shaft mill has a mounted vertically drive shaft. Hammers are attached to this shaft, and when rotating at high speed, these hammers strike the solid waste being fed into the mill. The hammers are enclosed in a steel housing that serves as a counterforce to the hammers. Waste input is fed at the top of the rotating, vertical shaft. Size reduction is controlled by the spacing between the side walls and the hammers on the rotary shaft; the mill is tapered so that the bottom is smaller than the top.

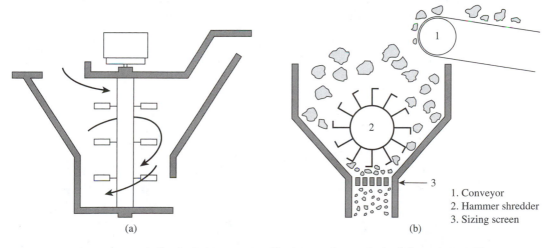

(a)   (b)

1. Conveyor
2. Hammer shredder
3. Sizing screen

Figure 13–10   (a) Verticle shaft hammer mill. (b) Horizontal shaft hammer mill.

The fan action of the rotating hammers causes an air current that moves from the top toward the bottom, which along with gravity pulls the waste material into the mill. Material that has not been reduced in size will not pass, and the impact from the hammers will impart a centrifugal motion, thus causing the object to be reduced. Very large items will not be fed in to prevent damage to the mill. The retention time in the mill and the number of hammer impacts are important factors that control size reduction. The spacing between the hammer tips and the housing in the lower part of the mill regulate the time of passage through the mill. The number of hammers in the mill determines the number of impacts, which in turn affects the particle size. The desired particle size can be obtained by changing the number and location of the hammers in the mill.

Horizontal Shaft Mill. The horizontal shaft hammer mill is more common than the vertical shaft hammer mill. As the name implies, the rotating shaft in this case is horizontal. Steel hammers are pinned to this rotor, and when spinning, they impact the waste in the mill. In this type of mill, the size of the existing refuse is controlled by the size of the openings in the floor grate. Only those particles smaller than the opening will be passed. These mills are used extensively for the crushing of ore and stone, but they do not work well for shredding solid waste. Because of the grate, this mill cannot pass these materials and is subject to damage from oversized and difficult material.

Flail Mill. The flail mill can also be used for shredding solid waste. In this mill, two sets of hammers (i.e., flails) mounted on parallel shafts are rotated in opposite directions. This type of mill has a low power requirement, which is commensurate with its limited ability to reduce particle size. As solid waste passes through the mill, the hammers strike it and knock it against the anvil plate. Small waste will go through the mill without being reduced in size, but the flails tear large containers such as bags, boxes, cans, and so on. Figure 13–11 describes a flail mill.

Low power and maintenance requirements as well as minimum size reduction are the primary advantages of flail mills. Low power and maintenance mean a lower operational cost than the hammer mill. There is also an advantage from its limited ability for particle size reduction, which allows the various constituents of waste to retain

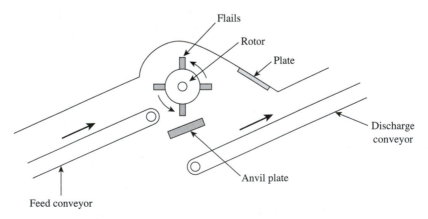

Figure 13–11   Flail mill.

their original size. For example, inorganics such as can metal and glass do not break into small particles and mix with organics in the waste, and can recovery is easier when cans are whole rather than in small pieces. In addition, the organic fraction is easier to process and use when it is not mixed with inorganics.

Shear Shredder. Figure 13–12 illustrates a shear shredder. This type of shredder operates with a scissorlike action, in which two counter-rotating shafts support cutters that literally cut the material as it becomes entrapped between the shredder's teeth. The waste material to be shredded is directed to the center of the counter-rotating shafts. The size of the material is then reduced by the shearing or tearing action of the cutter disks, and the shredded material drops or is pulled through the unit. Figure 13–13 shows a close-up view of this process. These shredders have also been used as bag breakers. Compared to hammer mills, however, these shredders are low-speed machines, ranging from 60 to 190 rpm. The shafts are hydraulically driven, and power cost is less than that of hammer mills because of the slow speeds. Rotation of the cutters automatically reverses to free jammed material. Large objects such as tires can be shredded without any difficulty as well. Size of the waste output is controlled by the spacing and orientation of the cutters and by the spacing between the shafts. This mechanism provides an advantage for the subsequent separation.

It should be noted that shredding is a very energy-intensive item in the processing of solid waste. Energy consumption directly relates to the size reduction requirement (4). For example, a nominal particle size of 5 in. will require approximately 4 to 6 $\frac{\text{hp-hr}}{\text{ton}}$ of dry weight processed. As the nominal particle size is reduced to 2 in., the power demand increases to 19 to 21 $\frac{\text{hp-hr}}{\text{ton}}$, and at 1 in., the power demand is 22 to 24 $\frac{\text{hp-hr}}{\text{ton}}$ (1).

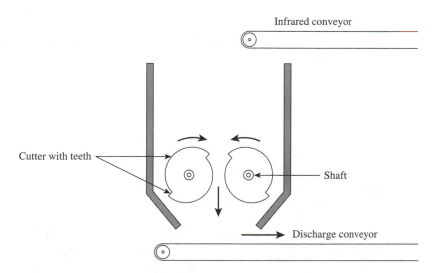

Figure 13–12   Shear shredder.

Figure 13–13   Shredder for metal items.

Volume Reduction.   Compaction of materials into bales reduces their volume and increases their density. Recyclables such as paper, cardboard, and aluminum cans are baled for ease in handling and reduced transportation costs. Unprocessed municipal solid waste is sometimes baled before landfilling.

Baling equipment compresses the material in one or more directions by applying pressure of 2000 to 3500 lbs/in^2 from hydraulically operated pistons. The density of a bale of municipal solid waste is approximately 2000 lbs/yd^3. The density of compacted waste in a landfill is usually less than 1000 lbs/yd^3. Factors that affect the baling process include the characteristics of the material, the time during which the pressure is applied, and the volume of the bale.

Operational Problems.   Several problems may arise during the operation of shredding equipment; these include fires, explosions, and material jams. Workers may also be exposed to high levels of dust in the area. Potentially explosive items in solid waste include gas cans, propane containers, ammunition, cigarette lighters, and so on. The possibility of explosion can be significantly reduced, however, by providing an escape channel that diverts gases away from the equipment and to the outside of the building that houses the shredder.

Friction in a shredder can cause a fire. Normally, waste is in and out of the shredder quickly enough to avoid igniting. If a shredder is stopped or jammed, however, a fire may occur. Fire extinguishers should be available to combat such a fire. Jamming occurs when the shredder cannot break waste such as a steel beam or large tire and, thus, cannot expel it through the chute. The jammed object can cause damage to the machine. Dust can be controlled by spraying a mist of water on the waste, but the moisture content of the waste will therefore increase and possibly negate its value as a fuel.

Separation of Waste Components

After the solid waste has been subjected to some kind of size reduction, which may simply be opening the bags and boxes, it is subjected to several processes to recover components of interest. Usually, each process recovers a spcific component. A sequence of processes can be designed and used to recover all the components of interest. The sequence may vary depending on the composition of the waste being processed. Economic considerations play an important role in material recovery.

Separation by Particle Size. Screening is the operation of separating a feed into oversize and undersize products. Oversize products do not pass through the openings of the screen; undersize products pass through the openings of the screen. Screens may be classified as primary, secondary, and tertiary, depending on their location in the waste-processing flow. The screen with the largest opening is the primary screen (e.g., a trommel). These screens are put ahead of all separation units in a recovery facility.

Both raw and shredded solid waste are processed through the screen. With proper screen sizes, a variety of waste streams, each with specific components of raw waste, can be obtained. Overlap in the size range does not always produce complete separation for some of the waste, however. Additional processing may be needed to achieve better results. Removal of many small and unwanted particles by the initial screening will reduce the load on the shredder and remove those components that cause the greatest wear on that equipment.

For efficient screening, the size of the openings must be appropriate for the particles to be separated and exposed to those openings. To determine the right opening size for waste separation, a particle-size analysis of the feed material is needed. A rotary screen or trommel is effective for processing raw waste before shredding by removing dirt, rocks, glass, metal objects, and oversized plastic as well as paper wastes.

Trommel. The trommel or rotary screen has been an effective processing unit at many resource-recovery plants. Trommels are revolving screens of perforated, cylindrical tubes that are mounted on drive units and usually on an incline (Figure 13–14). The feed is introduced at the upper end of the tube, and the material is screened as it tumbles down the drum. As the drum rotates, particles are carried up the side of the drum until they reach a certain height, where they fall to the bottom to repeat the cycle. The more cycles, the better the separation. Figure 13–15 shows a trommel.

The inside face of the trommel drum may have protruding knives or cutters, which facilitate breaking open plastic bags. The material tends to follow a helical path through the length of the tube. The size of the perforations varies depending on the type of material being screened. Perforations in the trommel can be of a single size, or there can be two or three zones of different-size perforations.

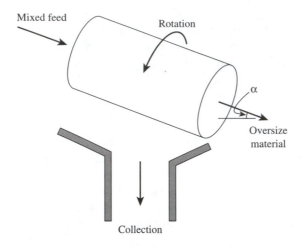

Figure 13–14   Trommel.

Figure 13–15   Close-up view of a trommel.

Compound trommels consist of more than one concentric screening tube on the same axis and are used for multifraction separation. They can be effective in concentrating certain components of municipal solid waste.

The requirements of a trommel are generally set by the type of waste being screened. Design parameters include trommel length, diameter, and slope. Operating parameters include rotational speed and feed rate. Trommel diameter and rotating speed must be selected with care.

Particles should not be allowed to simply cascade down the length of a trommel screen, nor should they be allowed to centrifuge on the side. Cascading results from a slow rotation, whereas centrifuging results from a fast rotation. The centrifugal force will hold the particles to the drumside until the gravitational force exceeds it. The ideal operation would be for particles to climb up the sides of the screen and then drop when after reaching the highest point in rotation.

The critical speed of rotation can be determined by

$$\omega = \sqrt{\frac{g \cos\alpha}{r}} \qquad (13\text{-}1)$$

where ω is the angular speed (in radians per unit of time), r is the radius of the trommel screen, and α is the angle of inclination of the trommel, and g is the acceleration because of gravity.

It is clear from equation (13-1) that larger the drum, lower is the critical speed. Typical rotational speeds are in the range of 10 to 30 rpm.

Example 13–3

Calculate the critical speed (rpm) of a 6-ft-diameter trommel inclined at 3°.

Solution

Using equation (13-1),

$$\omega = \sqrt{\frac{g \cos\alpha}{r}}$$

$$\sqrt{\frac{32.2 \times \cos 3°}{3}} = 3.27 \ rad/s$$

Thus,

$$\frac{3.27}{2\pi}(60) = 31.24 \ rpm$$

Disk Screens. A disk screen consists of lobed or star-shaped disks mounted on rotating shafts perpendicular to the direction of material flow. The shafts and disks all rotate in the same direction, so the particles move down the length of the bed. Disk screens are used in many of the same applications as trommels. As the disks turn, the lobes carry material across the surface. Figure 13–16 illustrates a disk screen.

These screens can be built to provide a range of particle-size cuts, from as small as 0.25 in. to larger than 8 in. Only large particles remain on the screen. Disk screens provide a reasonably distinct separation of particles according to size. For example, cans and bottles are isolated easily.

Separation by Density. Common methods of separation by density are liquid flotation and air classification. Both are discussed here.

Liquid flotation separates materials with a density lower than the liquid (i.e., floating fraction being used) from those materials with a density greater than the liquid (i.e., sinking fraction). For example, a recycled soft-drink container (including the cap) is fed into a grinder. The constituents are then separated by water flotation. The polymer polyethylene terepthalate (PET or PETE) forms the body and the cap, has a higher density than water, and sinks. High-density polyethylene (HDPE) forms the base cup, has a lower density than water, and floats.

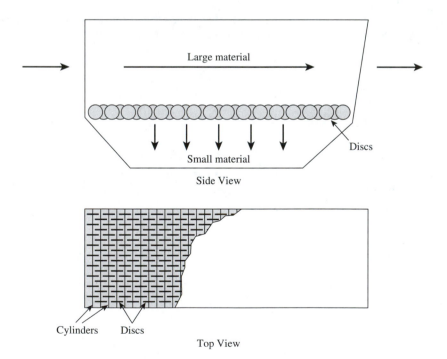

Figure 13–16   Disk screen.

(From U.S. Environmental Protection Agency. Handbook of Materials Recovery Facilities for Municipal Solid Waste. *Washington, D.C.: U.S. Govt. Printing Office, 1990; with permission.)*

Air classification is a commonly used, dry-density separation method for processing municipal solid waste. This method uses gravity and air currents to separate waste that is introduced into a stream of moving air. Shredded solid waste is separated into a light fraction, such as paper, plastic, and other light material, and a heavy fraction consisting of organic and inorganic material, such as gravel, heavy metal pieces, and so on.

The separation process in an air classifier is a function of air velocity and flow. Three main factors affect the separation of particles in an air stream: particle size, particle shape, and particle density. Along with density, particle diameter determines the particle mass, and the downward force exerted by gravity is determined by the mass of the particle. The upward air velocity creates a drag force that counters the gravitational force. This drag force is affected by the particle shape and the Reynolds number, but it is impossible to perform a theoretical design analysis of an air-classification unit for shredded solid waste. Manufacturers base their design of commercial units on pilot-scale units.

Figure 13–17 illustrates a vertical air classifier that separates inorganic from organic particles (5). Components of low density and high air resistance are called the light fraction; components of high density and low air resistance are called the heavy fraction. The light fraction consists mostly of combustible components, whereas the heavy fraction consists mostly of noncombustible materials. In most municipal solid waste, the light fraction is 60 to 80 percent of the total. The basic principle used in the air-classification method is that low-density material tends to move with the air stream but heavy-density material tends to move very little or remain stationary.

Air classification of shredded waste presents some problems. The moisture content in municipal solid waste is variable, and wet waste particles tend to agglomerate

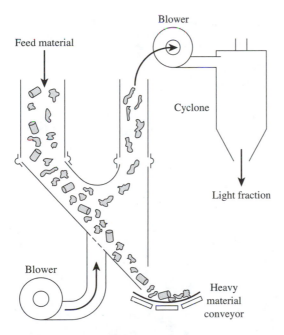

Figure 13–17   Air classifier.

and form larger particles. Wet paper also tends to pick up dirt, which is then carried over to the light fraction. Carryover glass fines are also a problem, because they become embedded in the paper and other portions of the waste.

Air-knife application of the air-classification concept is used for a waste stream that has undergone a separation process. Figure 13–18 illustrates the concept of an air knife. The waste is introduced into a horizontal air stream, and the shredded components drop out of that air stream at different points along the flow path as a function of their weight and shape. The air current will carry the plastic and paper with it, but dense material will fall more rapidly. Two to five fractions can be recovered depending on the design. These air-knife units work well and have almost no problem with dust (because the air is contained in a hood), a low initial cost, and a low operating cost.

Magnetic Separation. Magnetic forces of attraction or repulsion can separate metal objects and provide another means of recovering solid waste components. A variety of equipment is available for separating the ferrous materials in solid waste, and there are several types of magnetic separators. Most make use of a magnetic end pulley on a conveyor belt or a drum located at the end of the conveyor belt. The velocity of the belt imparts inertia to the material as it is discharged, thus causing it to move in the direction of the belt's motion. The magnet in the head pulley binds the ferrous metal to the belt surface, and the metal then moves back under the conveyor as the belt returns. Soon after, the metal drops out of the magnetic field and into a recovery bin. Figure 13–19 illustrates the workings of this separation process. Figure 13–20 shows a different type magnetic separation system, with a simple, belt-type magnet.

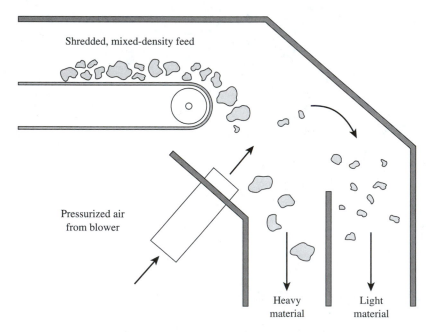

Figure 13–18   Air knife.

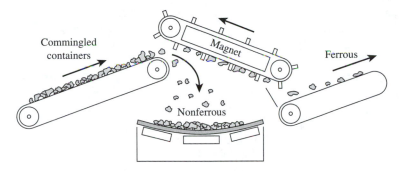

Figure 13–19 Magnetic belt.

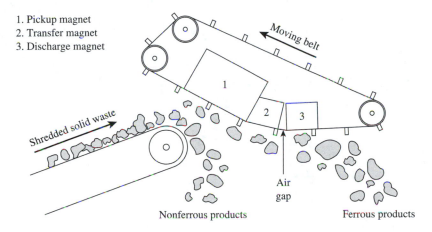

Figure 13–20 Magnetic belt for the recovery of ferrous metal.

This belt is suspended above a conveyor belt that transports processed solid waste. This suspended, self-cleaning type of strong electromagnet has the power to recover relatively heavy pieces of ferrous metal. The magnets are covered by a belt that transports the metal to a recovery bin. The magnetic fraction in the solid waste is lifted by the first magnet and then conveyed to the second magnet, which is reversed in polarity, thus causing the metal to rotate. In turn, this agitates the material collected and causes entrapped nonmagnetics (e.g., paper, plastics) to be released. The ferrous material attracted to the third magnet is then carried to the discharge end of the belt. Available data indicate an efficiency rate of approximately 80 to 85 percent (6).

The market value of the ferrous fraction (i.e., scrap) is relatively low. Transportation and processing costs may equal or exceed the scrap value. Removal of this metal during the earlier stages of the processing sequence is very desirable, however, because of the problems this metal may cause.

Aluminum is the principal nonferrous metal contained in municipal solid waste and it may account for as much as 85 percent of the total nonmagnetic metal content. Unlike ferrous metal, however, the value of recovered aluminum is high. It generates approximately 20 times as much revenue per ton as glass, steel cans, or newspapers. National Recovery Technologies, Inc., has developed a recovery system for aluminum (Figure 13–21[c]) (6).

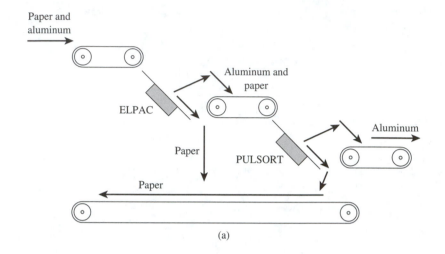

(a)

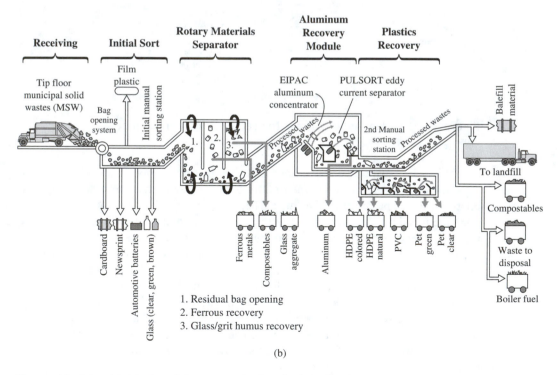

1. Residual bag opening
2. Ferrous recovery
3. Glass/grit humus recovery

(b)

Figure 13–21   (a) National Recovery Technologies aluminum recovery system. (b) National Recovery Technologies automated recovery system.

(Courtesy of National Recovery Technologies.)

Initially, shredded municipal solid waste is separated to facilitate greater concentration of aluminum in the waste stream. The removal of glass, plastic, organic and inorganic particles, other metals, and paper also increases the efficiency. The aluminum-rich stream then travels down an inclined plane with a series of metal detectors, each of which is connected to a high-pressure air jet underneath it. When the detector finds metal, a small blast of air from the jet blows the metal and other impurities onto another conveyor. This second stream has a good amount of aluminum and some paper and plastic, and this aluminum is then separated from the impurities. This technique is based on the principle that when passing through an electromagnetic field, nonferrous metals will have induced eddy currents within them. In turn, these eddy currents will produce magnetic fields that repel an applied magnetic field. A repelling force is thus generated that facilitates separation of the metal. This oppositely directed force ejects the aluminum metal from the stream.

National Resources Technology Group. The U.S. government, the public, and industry have pushed to find alternatives to the conventional disposal of waste in landfills. Recently, great emphasis has been placed on diverting recyclable materials from landfills. The National Resources Technology (NRT) group has been able to accomplish this economically by using its own proprietary separation equipment and its own operating experience with the best available supporting technology.

The NRT has developed a processing system incorporating several technologies to produce refuse-derived fuel and to recover various materials from solid waste. This system includes a bag-opening system, rotary material separator, aluminum separator, Vinylcycle, and autosort plastic separation systems. Figure 13–12(b) shows a typical configuration of the NRT's automated recovery system.

Waste is first unloaded from collection vehicles onto the tipping floor, where large, bulky items are removed. The waste is then transported by a conveyor to a bag-opening system, where the bags are mechanically opened and emptied. Paper, cardboard, sheet plastics, automotive batteries, and glass products are manually removed, and the waste enters a large, rotary, material-separator drum. The bags and boxes are opened by the cutters (i.e., sharp, metal, knifelike protrusions) located in the first third of this drum, and their contents are released. The second third of the drum has magnets attached to the surface, which attract the ferrous metals and carry them to the top of the drum, where they are then swept from the drum surface and collected onto a conveyor belt. In the last third of the drum, lifting bars are attached to the drum wall. Here, waste moves with the drum rotation and is thus elevated. The paper, plastics, textiles, and so on, fall down, but the yard waste, food scraps, grit, broken glass, and so on are retained by the lifting bars and fall onto a conveyor near the top of the drum.

Three streams thus exit the drum: one containing paper, organics, and nonferrous metals; one containing broken glass, grit, food scraps, and so on; and one containing ferrous metals and some organics. The first stream is sent for aluminum recovery (discussed earlier). The part of the first stream intended for RDF may need to be refined and shredded further. The other two streams are passed through an air knife to remove fine organic material. The ferrous metal and unshattered glass can be recovered; the grit, shattered glass, and so on are sent to the landfill.

Approximately 85 to 90 percent of glass, metals, and noncombustibles are removed in this processing (7). Of course, aluminum can recovery is almost 100 percent. This recycling of aluminum, ferrous metals, and glass can also help mass burn incinerators to run well.

Examples of Innovative Technologies

Vinylcycler is a proprietary electromagnetic screening process that enables vinyl containers to be separated from mixtures of whole or crushed postconsumer containers with both speed and accuracy. The presence of chlorine atoms in the vinyl triggers a computer-timed air burst that separates the vinyl containers from the mixed plastic stream.

A multisort unit sorts plastic bottles from mixed streams of as much as 7000 lbs/hour. The sort capabilities are PETE from HDPE, green PETE from clear PETE, and natural HDPE from mixed-color HDPE.

Robosort is another automated sorting device. An employee operates it by touching selected items on a video screen. The operator can be located away from the dirt and noise of the conveyor line but still see the items on the conveyor as they scroll on a computer monitor. The operator then simply touches an object on the screen to eject it from the waste stream.

CHEMICAL TRANSFORMATION (COMBUSTION)

Definition of Combustion

Combustion is an engineered process that employs thermal decomposition via thermal oxidation at high temperatures ($\geq 1400°F$) to convert a waste to a lower-volume, nonhazardous material or to energy. This is a chemical reaction in which the elements of a fuel are oxidized to a stable state.

The major elements in fuel are carbon, hydrogen, and oxygen. Some fuels have a significant sulfur content, and some have a significant nitrogen content. When adequate oxygen is available, carbon is oxidized to carbon dioxide (CO_2), hydrogen to water (H_2O), sulfur to sulfur dioxide (SO_2), and nitrogen to nitrogen oxide (NO).

Necessary Conditions for Combustion

Combustion is a chemical reaction and, therefore, follows the laws of chemical equilibrium, chemical kinetics, and thermodynamics. Combustion reactions are a function of oxygen, time, temperature, and turbulence.

Thermal processing of solid waste is done by combustion. If done with the same amount of oxygen or air needed for complete combustion, it is known as stoichiometric combustion. Combustion with oxygen in excess of the stoichiometric requirements is termed excess air combustion. Partial combustion of solid waste under substoichiometric conditions to generate a combustible gas containing carbon dioxide, hydrogen, and hydrocarbons is termed gasification. Pyrolysis is the thermal decomposition of waste in the complete absence of oxygen.

Having oxygen in excess will drive the reaction to completion more rapidly. If a stoichiometric quantity of oxygen is available, the reaction will eventually reach completion. Sufficient time must be available for the combustion reactions to proceed to completion. Many compounds also have an ignition temperature that must be exceeded to initiate the reaction (i.e., to ignite the waste). A higher temperature results in a higher reaction rate but also higher emissions of nitrogen oxide, a drawback in destroying waste by forming air pollutants. Turbulence is important as well, because it causes mixing of the combustion gases in the furnace, which is essential for completion of the reaction. The flame zone is the area where the volatized gases mix with the oxygen in the air. With excess air and turbulence, the reaction proceeds to completion within 1 or 2 seconds. The practice of waste combustion has resulted in the need for higher temperatures to increase the opportunity to use the waste as an energy and to destroy toxic compounds.

The basic reactions for stoichiometric combustion in the organic fraction of waste are:

Carbon: $\frac{C}{12} + \frac{O_2}{32} \rightarrow CO_2$ (13-2)

Hydrogen: $\frac{2H_2}{4} + \frac{O_2}{32} \rightarrow 2H_2O$ (13-3)

Sulfur: $\frac{S}{32.1} + \frac{O_2}{32} \rightarrow SO_2$ (13-4)

Assuming that dry air contains 23.15-percent oxygen by weight and, as equation (13-2) indicates, that each mole of carbon needs two moles of oxygen, then the amount of air required to oxidize 1 lb of carbon would be equal to

$$\frac{32}{12} \times \frac{1}{0.2315} = 11.52 \text{ lb}$$

Similarly amounts of hydrogen and sulfur can be calculated and are 34.56 and 4.31, respectively.

As discussed, incineration is fundamentally a form of chemical processing that involves the rapid oxidation of materials. Simply stated, the combustion process involves several stages in which the waste is dried (i.e., moisture is evaporated) as it enters the furnace, the organic compounds are volatilized, and the volatile compounds are ignited in the presence of oxygen.

Incineration of Solid Waste

The percentages of carbon, hydrogen, oxygen, nitrogen, sulfur, halogens, and phosphorus as well as the moisture content in the waste must be known to determine the air requirements for stoichiometric combustion. The elements and percentages vary according to the characteristics of the waste. The byproducts of incineration are gases, ash, and heat energy.

Waste must be combustible in to be destroyed. Primary products from the combustion of organic waste are carbon dioxide, water vapor, and ash; several other products (e.g., NO) are also formed. The fundamentals of incineration are the same for both solid waste and hazardous waste, but there are some significant differences in the combustion requirements. For example, for hazardous wastes, an efficiency of destruction and removal of 99.999 percent is typically required. The emissions are also more toxic than those from solid waste. (Thermal destruction of hazardous waste is discussed in Chapter 20.)

As the solid waste enters the combustion chamber and its temperature increases, the volatile materials escape as gases. A continued increase in temperature will cause many of the organic components to thermally "crack" and form gases as well. When the volatile compounds are driven off, fixed carbon and ash remain. When the temperature reaches the ignition temperature of carbon (>1300°F), then the carbon is also ignited. To achieve complete destruction (i.e., burnout) of all combustible material, the temperature must be greater than 1300°F throughout the bed, and there must be sufficient oxygen. This oxygen is supplied by blowing air under the grates, which then passes up through the solid waste on those grates. Sufficient time should be allowed for a good burnout.

Heat Value. A major consideration in the selection or design of an incinerator is the fuel value of the material being burned. Fuel value is described in terms of the gross heat or the higher heat value (HHV) and of the net heat or lower heat value (LHV). The gross heat value can be determined by using a laboratory bomb calorimeter and completely burning a weighed sample in the presence of oxygen in the calorimeter and then calculating the liberated heat by measuring the temperature rise of the surrounding water bath. The American Society of Testing and Materials specifies a procedure for measuring the gross calorific value of solid waste–derived fuels; these values for various materials in municipal solid waste are listed in Table 13–1 (7). If the elemental composition of municipal solid waste is known, the heat value can be calculated from that as well.

In addition to the heat value, other important fuel properties are the moisture and the combustible and noncombustible (i.e., ash) contents. Fuel with a high heat value, moisture content of less than 50 percent, and ash content of less than 60 percent can be burned without supplementary fuel. Waste material with a low heat value, high moisture content, and high ash content require supplementary fuel.

A waste analysis to determine the levels of moisture, volatile combustibles, fixed carbon, and ash is called a proximate analysis. An elemental or ultimate analysis of a waste component typically involves determining the percentage of carbon, hydrogen, oxygen, nitrogen, sulfur, and ash. Because of concern regarding the emission of chlorinated compounds, determination of the halogen level may be included in the ultimate analysis in some cases. The results of the ultimate analysis are used to characterize the chemical composition of the organic matter in municipal solid waste.

As stated earlier, without calorimetric data, the heat value of combustible material may be estimated by using one of the many equations developed for this purpose, which are based on the ultimate analysis of waste composition. To calculate the (HHV)

$$145.7(C) + 619(H) - 59.8(O) + 45.1(S) \qquad \textbf{(13-5)}$$

where C, H, O, and S are the weight percentages of carbon, hydrogen, oxygen, and sulfur, respectively, and the answer is in terms of Btu/lb (8).

Table 13–1 Key Features of New Federal Regulations for Municipal Waste Combustors

Good Combustion Practices

Load must not exceed 110% of maximum level (4-hour average) as demonstrated during a dioxin/furan test. Maximum inlet temperature of the particulate matter control device must not exceed 30°F beyond the maximum demonstrated temperature during the dioxin/furan performance test.

CO level[a] (block averaging time) and the concentration in the flue gas must not exceed the following:

Modular starved and excess-air combustors	50 ppmv (4 hour)
Mass burn waterwall and refractory combustors	100 ppmv (4 hour)
Combustors using fluidized bed combustion	100 ppmv (4 hour)
Mass burn, rotary waterwall combustors	100 ppmv (24 hours)
Refuse-derived fuel stokers	150 ppmv (4 hour)
Coal/RDF mixed fuel–fired combustors	150 ppmv (24 hours)

American Society of Mechanical Engineers or State certification for combustor supervisors and training for plant operators.

Annual stack testing for particulates, dioxins/furans, and hydrogen chloride.

MWC Metal Emissions Limits[a,b]

Particulate matter	34 mg/dry standard m^2
Opacity	10% (6-minute average)
Best demonstrated technology	Fabric filter

Combustor Organic Emissions[a,b,c]

Dioxins/furans	30 ng/dry standard m^2
Best demonstrated technology	Fabric filter

Combustor Acid-Gas Emissions[a]

SO_2	80% or 30 ppmv (24-hours average)
HcL	95% reduction or 25 ppmv
Best demonstrated technology	Fabric filter and spray dryer

Nitrogen Oxides Emissions[a]

NO_x	180 ppmv (24 hour daily)
Best demonstrated technology	Noncatalytic reduction

Monitoring Requirements

SO_2	Continuous emission-monitoring system, 24-hour geometric mean
NO_x	Continuous emission-monitoring system, 24-hour arithmetic average
Opacity	Continuous emission-monitoring system, 6-minute average
CO, load, and temperature	Continuous emission-monitoring system, 4- or 24-hour average

[a] All emission levels are at 7% O_2 (dry basis).

[b] Verified at annual stack compliance test.

[c] Dioxins/furans measured as total tetra- through octachlorinated dibenzo-*p*-dioxins and dibenzofurans.

To account for the effect of moisture, it is often useful to speak of the net heat value or the LHV:

$$HHV - 10.50(W + 9H) \qquad \text{(13-6)}$$

where W is the mass or weight of the moisture and H is the percentage weight or mass of hydrogen in dry waste (9). Again, the answer is expressed in terms of Btu/lb.

Example 13–4

Assume the proximate and ultimate analyses of a solid waste are:

Proximate Analysis	(% wt)	Ultimate Analysis	(% wt)
Moisture	20	Moisture	20
Volatile matter	60	Carbon	30
Ash	20	Hydrogen	4
Total	100	Oxygen	25.5
		Nitrogen	0.37
		Sulfur	0.13
		Ash and metal (inerts)	20
		Total	100

Also assume the following conditions are applicable:

1. The as-fired heating value of the solid waste is represented by equation (13-5).
2. The grate residue contains 10 percent unburned carbon.
3. The entering air is at 77°F; the residue is at 700°F.
4. The radiation loss is 0.006 Btu/Btu.
5. The specific heat of residue is 0.3 BTU/lb°F
6. The latent heat of water is 1040 Btu/lb.
7. All the oxygen in the waste is bound as water.
8. The theoretical air requirements based on stoichiometry are:

Element	Reaction with Oxygen	Air Requirement
C	$C + O \rightarrow CO_2$	11.52 lb/lb
H	$2H_2 + O_2 \rightarrow 2H_2O$	34.56 lb/lb
O	$S + O_2 \rightarrow O_2$	4.31 lb/lb

9. The moisture in combustion air is negligible.
10. The net hydrogen available for combustion is equal to the fraction $\left(H - \dfrac{O}{8} \right)$.

   In other words, this is the percentage of hydrogen minus $\frac{1}{8}$th the percentage oxygen. This accounts for "bound water" in the dry combustible material.

11. The heating value for carbon is 14,000 Btu/lb.

Now:

a. Estimate the HHV and LHV of this waste as received.

b. Calculate the air requirement for complete combustion of 100 tons of this waste.

c. Determine the heat available in the exhaust gases from combustion of 100 tons/day of this waste by heat balance.

Solution

a. From equation (13-5),

$$HHV = 145.7(C) + 619(H) - 59.8(O) + 45.1(S)$$
$$= 145.7(30) + 619(4) - 59.8(25.5) + 45.1(0.13)$$
$$= 4371 + 2476 - 1525 + 5.86$$
$$= 5327.86 = 5328 \text{ Btu/lb}$$

and from equation (13.6),

$$LHV = HHV - 10.50(W + 9H)$$
$$= 5328 - 10.50 (20 + [9 \times 4])$$
$$= 5328 - 588$$
$$= 4740 \text{ Btu/lb}$$

b. First, compute the weights of the elements of the waste:

Element	Computation	lbs/day
Carbon	0.30 × 200,000 lbs	60,000
Hydrogen	0.04 × 200,000 lbs	8,000
Oxygen	0.255 × 200,000 lbs	51,000
Nitrogen	0.0037 × 200,000 lbs	740
Sulfur	0.0013 × 200,000 lbs	260
Inerts	0.20 × 200.000 lbs	40,000
Moisture	0.20 × 200,000 lbs	40,000

Second, compute the residue:

$$\text{inerts} = 40,000 \text{ lbs/day}$$

Because the grate residue contains 10-percent unburned carbon (an assumption made for this problem),

$$\text{total residue} = \frac{40,000}{0.90} = 44,444 \left(\text{or} \ \frac{40,000}{[100 - 10]\%} \right)$$

$$\text{Carbon in residue} = 44,444 - 40,000 = 4,444 \text{ lbs/day}$$

Third, compute the available hydrogen and bound water:

$$\text{net available hydrogen} = \left(4.0\% - \frac{25.5\%}{8}\right) = 0.8125\%$$

$$= 0.008125 \times 200,000 \text{ lbs}$$

$$= 1,625 \text{ lbs/day}$$

$$\text{hydrogen in bound water} = 4\% - 0.8125\% = 3.18\%$$

$$= 0.031875 \times 200,000 \text{ lbs}$$

$$= 6,375 \text{ lbs/day}$$

Therefore,

$$\text{bound water} = \text{oxygen} + \text{hydrogen in bound water}$$
$$= 51,000 + 6,375 = 57,375 \text{ lbs/day}$$

Fourth, calculate the air required:

Element	Air Requirement (lbs/day)
Carbon = $(60,000 - 4,444) \times 11.52$	640,005
Hydrogen = $1,625(34.56)$	56,160
Sulfur = $260(4.31)$	1,120
Total dry theoretical air (C + H + S)	697,285
Total dry air (100-percent excess)	1,394,570

c. First, determine the amount of water produced by combustion of the available hydrogen. Because it takes 2 lbs of H (or two molecules) to form 18 lbs of H_2O (or one molecule),

$$H_2O = \frac{18 \text{ lbs of } H_2O}{2 \text{ lbs of } H} (1,625 \text{ lbs/day}) = 14,625 \text{ lbs/day}$$

The remaining computations of heat balance for the process are as follows:

Item	Heat Value (10^6 Btu/day)
Gross heat input	
$2 \times 10^5 \dfrac{lb}{d}\left(5,328 \dfrac{Btu}{lb}\right)$	1,065.6
Heat lost in unburned carbon:	
$4,444 \dfrac{lb}{d}\left(14,000 \dfrac{Btu}{lb}\right)$	62.21
Radiation lost:	
$0.006 \dfrac{Btu}{Btu}\left(1,0655 \times 10^6 \dfrac{Btu}{d}\right)$	6.39
Inherent moisture:	
$40,000 \dfrac{lb}{d}\left(1,040 \dfrac{Btu}{lb}\right)$	41.60

Moisture in bound water:

$$57{,}375 \ \frac{lb}{d}\left(1{,}040 \ \frac{Btu}{lb}\right) \qquad\qquad 59.65$$

Moisture from the combustion of available hydrogen:

$$14{,}625 \ \frac{lb}{d}\left(1{,}040 \ \frac{Btu}{lb}\right) \qquad\qquad 15.21$$

Sensible heat in residue:

$$44{,}444 \ \frac{lb}{d}\left(0.3 \ \frac{Btu}{lb}\,(700°F - 77°F)\right) \qquad\qquad 8.30$$

Total losses   193.38

Net heat available in flue glasses:

$$(1{,}065.6 - 193.38) \times 10^6 \ \frac{Btu}{d} \qquad\qquad 872.22$$

Combustion efficiency:

$$\left(\frac{872.22 \times 10^6 \ \dfrac{Btu}{d}}{1{,}065.5 \times 10^6 \ \dfrac{Btu}{d}}\right) \times 100\% \qquad\qquad 81.85\%$$

Types of Incinerators.   Incinerators can be designed to operate with unseparated (i.e., mass burn) or with processed municipal solid waste refuse (i.e., derived fuel).

Mass burn. In this type of incinerator, the solid waste is processed only minimally before being fed to the incinerator. Wastes are deposited in a pit or tipping floor, where it is visually checked by the crane operator, who then removes objects that are unacceptable for incineration (Figure 13–3). Despite this check by the crane operator, however, anything in the municipal solid waste stream, including large, noncombustible items (e.g., damaged or broken bicycles, appliances, furniture) and even hazardous waste either deliberately or accidently discarded, may enter the incinerator. Therefore, the system must be designed to handle these wastes without damage. The energy content of the mass fixed waste can be extremely variable, depending on the climate, season, and source. Despite the possible drawbacks, mass-fired combustion has become the technology of choice.

After the check, the waste is pushed into a pit or onto conveyors that lead to the feed hoppers. An overhead crane or hydraulic ram feeds the waste into the furnace at a regulated rate. Mass-burning systems use water-tube-wall (often called waterwall) furnaces for combustion. Waterwall units are enclosed by closely spaced, water-filled, steel tubes that recover heat by both radiation and convection. This heat recovery enables them to operate at lower temperatures than refractory-lined units, which in turn makes them smaller and cheaper to build and more efficient in energy recovery. Figure 13–3 shows a mass-burn incinerator.

Incinerators are also used to burn hazardous waste. In addition to the fluidized bed incinerator, liquid-injection and rotary kiln incinerators are often used. Rotary kiln incinerators can be used to burn solid wastes, slurries, and liquids.

Refuse-Derived Fuel. Refuse-derived fuel (RDF) is the combustible fraction of municipal solid waste that is processed to remove metal, glass, and other noncombustible materials and, thus, to produce a more homogeneous product. The combustible material that is obtained may be further processed to increase its density by compression into pallets or briquettes, or it may be shredded to a fluff.

In RDF-fired incinerators, the RDF is typically burned on a traveling grate stoker. Because of its movement, the traveling grate introduces the fire from below, and along with turbulence, this provides for uniform combustion.

Because of the higher energy content of RDF compared with unprocessed municipal solid waste, RDF combustion systems can be physically smaller than similarly rated, mass-fired systems. Because of the more homogeneous nature of RDF, an RDF-fired system can also be controlled better than a mass-fired system. The highest thermal values with the least amount of ash content are found in RDF, which represents only 40 to 50 percent of the amount of MSW after initial processing. The heat value of unprocessed municipal solid waste in the United States usually averages between 3500 and 6500 Btu/lb. When the combustible fraction is processed into RDF, the heat value may range from 6550 to 8000 Btu/lb. The largest U.S. RDF facility currently in operation is in Dade County, Florida; it has a design capacity of 3000 tons/day.

Fluid Bed Combustors.   A simple form of fluid bed combustor system (Figure 13–22) consists of a vertical steel cylinder with a sand or limestone bed (usually refractory lined), a supporting grid plate, and air-injection nozzles. When air is forced upward, the bed fluidizes and expands to as much as twice its original volume. These inciner-

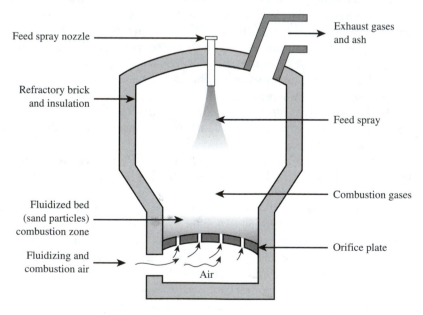

Figure 13–22   Fluid bed incinerator.

ators can burn a variety of fuels, including municipal solid waste, biomass, sewage sludge, and numerous chemical wastes. Use of limestone bed material allows the combustion of high-sulfur fuel (e.g., high-sulfur coal) with very little sulfur dioxide (SO_2) emissions. Limestone ($CaCO_3$) reacts with oxygen and sulfur dioxide is formed by the sulfur-containing wastes to form calcium sulfate ($CaSO_4$) and carbon dioxide (CO_2). Expressed as an equation:

$$2CaCO_3 + 2SO_2 + O_2 \rightarrow 2CaSO_4 + 2CO_2 \qquad \text{(13-7)}$$

Multiple-Chamber Incinerators. Figure 13–23 shows a simple, in-line, rectangular incinerator (10). The three major compartments are the ignition or primary combustion chamber, the mixing chamber, and the secondary combustion chamber. Two parameters dictate the geometry of these chambers; gas flow velocities, and retention times. Gas velocities control the mixing, and retention times are associated with the combustion rate. An adequate retention time is necessary for good combustion. Air pollution–control regulations specify both the velocities and the retention times.

Incinerator design is still being improved through research and development. Some helpful guidelines have been developed by the Incinerator Institute of America (11). For example, the size of the primary combustion chamber depends on the arch height and the

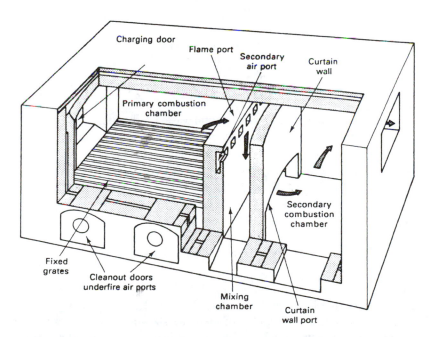

Figure 13–23 In-line, multichamber incinerator.

(From U.S. Department of Health, Education, and Welfare–Public Health Service. Air Pollution Engineering Manual. Publication No. 999-AP-40. Washington, D.C.: U.S. Govt. Printing Office, 1967; with permission.)

grate loading. The grate loading is an empirical relationship that is unique for the type of grate that is employed. The loading on stationary grate (Figure 13–23) is given by

$$L_G = 10\log R_C \tag{13-8}$$

where L_G is the grate-loading rate in $\dfrac{\text{lb}}{\text{hr-ft}^2}$ and R_C is the solid waste combustion rate in lbs/hour. Grate area (A_G) is obtained by dividing the solid waste combustion rate by the allowable loading rate $\left(\dfrac{R_C}{L_G}\right)$. This area determines the horizontal dimensions of the primary combustion chamber for the many different types of grates. Several grates are

mechanical, and as indicated earlier, each has a different allowable loading rate based on evaluations of grate performance.

Incinerators are generally rated in accordance with the estimated weight of solid waste they can burn in 24 hours. Loadings can range to as much as just slightly more than 100 lbs of solid waste per hour per square foot of grate area. Small incinerators for apartment buildings and institutions are loaded at much lower rates. Incinerator Institute of America standards suggest grate loading rates for municipal solid waste of 20 lbs/hour/ft^2 in 100-lb/hour burning units to 30 lbs/hour/ft^2 in 1000-lb/hour burning units. Using equation (13-8),

$$L_G = 10\log R_C$$

$$L_G = 10\log \times 100 \, \frac{\text{lb}}{\text{hour}}$$

$$= 20 \, \frac{\text{lb}}{\text{hr-ft}^2} \text{ of as-received waste}$$

The heating value of as-received solid waste would range from 4000 to 5500 Btu/lb depending on the moisture and ash content. The heat release rate for these grates would be for a loading rate of

$$20 \, \frac{lb}{hr\text{-}ft^2} \, (4,000 \text{ to } 5,500 \, Btu/lb) = 80,000 \text{ to } 110,000 \, \frac{Btu}{hr\text{-}ft^2}$$

and

$$30 \, \frac{lb}{hr\text{-}ft^2} \, (4,000 \text{ to } 5,500 \, Btu/lb) = 120,000 \text{ to } 165,000 \, \frac{Btu}{hr\text{-}ft^2}$$

This is a typical rate for a fixed grate. Mechanical grates can be loaded at higher rates, however, resulting in higher heat values. Many different types of grates are used solid waste incineration, and a number of these are shown in Figure 13–24. Most incinerators provide necessary turbulence by burning fuel on sloping grates that move in some fashion to agitate the waste. This movement causes the waste to tumble forward through the combustion chamber. There is no significant advantage of one grate over another. Grate systems allow ash to fall through the grates as the solid waste is transported from the point of introduction to a final collection bin. The grates control the speed with

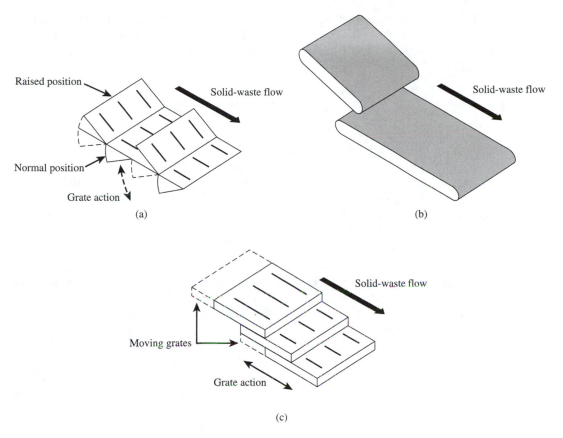

Figure 13-24 Types of grates used in solid waste incinerators. (a) Rocking grates. (b) Traveling grates. (c) Reciprocating grates.

which the materials move through the incinerator and the turbulence for complete combustion. As a rule of thumb, mechanical grate area and the corresponding grate loading can be found on the assumption of a heat release value for solid waste equivalent to approximately 300,000 $\dfrac{\text{Btu}}{\text{hr-ft}^2}$ of grate area.

All moving grates can be considered for use in combustion of RDF, but the characteristics of unprocessed solid waste are so variable that these grates may not work well. Because of the nonhomogeneity and bulky nature of the solid waste, grate action is required to mix and expose the waste to the high temperature and combustion air within the primary combustion chamber. These grates are built as stair steps, with the alternate step being stationary. The movable steps slowly move back and forth, gradually pushing the waste with a tumbling action as it moves from one step to the next. Thus, the waste is mixed and exposed to the combustion air.

The method for determining the horizontal dimensions of primary chambers was discussed earlier. The other dimension needed, however, is the height (or the arch height)

of the primary chamber. The retention time of the combustion gases, which is associated with the combustion rate, is a factor that determines the chamber height by an empirical relationship that is unique for each furnace. The arch height (H_A) can be calculated as

$$H_A = \frac{4}{3} (A_G)^{\frac{4}{11}} \tag{13-9}$$

Another factor to consider is the length-to-width ratio of this chamber. This ratio
is based on the waste combustion rate. Combustion gases exiting the primary chamber
pass through a flame port into a mixing chamber, where the secondary air is mixed
with the incoming combustion gases. In the flame port and mixing chamber, gas velocity is significant in generating turbulence. The gas flow rate and the cross-sectional
area determine the velocity; that is,

$$V = \frac{Q}{A}$$

where V is the velocity (ft/s), Q is the flow rate (ft^3/s), and A is the cross-sectional area
(ft^2). In a multiple-chamber incinerator, the recommended velocity is 55 ft/s at 1000°F.
It must be noted that the volume of the gas is temperature dependent, with the base
temperature being 1000°F. Thus, if a lower temperature is encountered, the gas velocity will decrease, but at a higher temperature, the gas velocity will increase. The recommended downward velocity in the mixing chamber is 25 ft/s, and the velocity
through the wall port (at 950°F) is approximately 70% of the mixing chamber velocity. The height and width of the mixing chamber are based on the dimensions of the
primary chamber.

The gas velocity in the secondary chamber is much lower. The preferred velocity
range is 5 to 6 ft/s and always less than 10 ft/s. The length and width of the primary
chamber need to be adjusted, because the height and width of the secondary chamber
are fixed by the primary combustion chamber.

Combustion Air.   Combustion air, in proper quantities, is essential to complete burning. Theoretical combustion air represents the amount of air required for a stoichio-
metrically balanced reaction. This reaction is impossible to achieve in an incinerator,
however, so an excess of combustion air must be supplied, which is referred to as actual combustion air. The air supplied for combustion is usually 2 to 4 times the stoichiometric requirement to ensure complete combustion. As an equation,

(2.0 to 4.0) × (theoretical combustion air lbs/min) = actual combustion air lbs/min
(actual combustion air lbs/min) 70°F) = actual combustion air ft^3/min

where 13.35 ft^3 of dry air at 70°F weighs 1 lb.

Combustion air is introduced at three locations: underfire air ports, overfire air
ports, and mixing chamber air ports. As the air passes through the hot grates and ash,
it picks up some of the heat and, thus, cools the grates and ash. In addition, it provides the oxygen to oxidize the residual organics in the ash. The underfire air is approximately 10 percent of the total combustion air. Too much air will suspend large

quantities of ash in the combustion gases and, consequently, increase the air pollution problem resulting from particulates. Approximately 60 to 70 percent of the air required is added to the primary chamber as overfire air to obtain high temperatures and good burning. This overfire air reacts with the solid waste in the chamber to form a flue gas. Lower air velocities tend to reduce the amount of ash particles in flue gas. Flue gas entering the mixing chamber usually contains some incomplete products of combustion, however, which require the secondary air reaction. The balance of the air is added to provide excess oxygen in the secondary combustion chamber. This excess supply of oxygen, high temperatures, and a required retention time of 1 to 2 seconds bring the combustion to completion. Primary heat release occurs in the secondary combustion chamber. Volumes of such chambers range from 10 to 25 ft^3 per ton of rated capacity.

Air distribution is controlled by dampers on the air ducts or parts where the air enters the furnace. Air-handling systems are overdesigned by more than 50 percent to provide for the flexibility of air distribution. With induced air or forced draft, head loss of approximately 0.1 in. of water gage or less is maintained.

Because of the disadvantages associated with high chimneys (e.g., high construction costs, need for strong foundations), the current trend is toward relatively shorter stacks. Because of the lower heights of these stacks, the natural draft created within is not sufficient, in general, to provide the desirable air supply pressures for the incinerator operations. Therefore, where such stacks are used, draft fans must also be used. The design of incineration systems includes fans to ensure positive control over the airflow. The temperature of the combustion gases is reduced to recover the energy, but the lower temperature differential causes a reduced draft. Thus, the discharge of combustion products into the atmosphere is unsatisfactory. The fans help to move the gases in the stack and to ensure satisfactory discharge into the atmosphere.

Stacks should be designed for gas velocities of approximately 25 ft/s with maximum air. A rough approximation would be 0.3 ft^2 of stack area per ton of rated capacity. Stack heights usually range from 100 to 180 ft to create a natural draft for the dispersion of exhaust gases into the atmosphere.

Design Guidelines and Objectives.

No set design standards are either recommended or required by the U.S. Environmental Protection Agency (EPA) for incinerators. Each furnace is unique, and each design is based on the experience of the designer, manufacturer, and operator to achieve the objectives—that is, to meet the guidelines set by the EPA or state government. Each incinerator requires a discharge permit from the state government and most guidelines are incorporated into that permit.

On December 19, 1995, pursuant to Sections 111 and 129 of the Clean Air Act, the EPA promulgated emission guidelines applicable to existing municipal waste combustor (MWC) units and new source performance standards applicable to new MWC units (12). These guidelines and standards are codified in the CFR, Part 60, Subparts C_b and E_b. These subparts only apply to combustors with a capacity of greater than 250 tons per day.

The key features of the new federal incinerator regulations are given in Table 13–1. Even so, state and local air pollution–control districts can issue emission limits that are stricter than federal regulations after providing the due process of public hearings. Because federal, state, and local regulations are in a constant state of flux, the reader should consult directly with regulatory officials for the most current standards and regulations.

Air Pollution.   Incineration is a resource-recovery process with significant potential for creating air pollution. Other resource-recovery processes also generate pollutants that are emitted into the atmosphere, but the potential for air quality problems is greatest with incineration. Because municipal solid waste is a heterogenous mixture, uniform combustion conditions are difficult to achieve. Careful operation with the proper combination of turbulence, temperature, retention time, and treatment of exhaust gases, however, can achieve substantial reduction of pollutants.

Particles and Heavy Metals.   Solids in the form of ash contain bits of glass, metal, unburned carbon particles, and inert substances (e.g., sand). This accumulates as bottom ash, which falls through grates in the combustion chamber.

Fly ash is made of light particles that are carried out by combustion gases. These particles are made of noncombustible materials that consist of inorganic oxides, including heavy metals that are emitted over a wide range of particle sizes. Condensation of particles occurs during partial combustion, producing vapors. When the combustion gas cools, these vapors condense to form particles that are organic compounds. If the temperature in the combustion chamber is high, the metals will be present as a vapor. At a temperature of 1800°F or higher, all of the metals except lead will be present as a vapor. Salts of some of the metals have lower boiling points; for example, lead chloride will boil at a temperature lower than 1800°F. Mercury is also a problem, because it boils and forms vapor at 673°F. Whereas most vaporized metals return to a solid state when the combustion gas cools, mercury remains in the vapor state. A wet scrubber is used to remove mercury. Studies indicate that very small particles in the respirable fraction (approximately 0.1 to 15 μm) are emitted from solid waste incinerators and include heavy metals and trace elements.

Other Pollutants.   Emissions from municipal solid waste incinerators include sulfur dioxide, nitrogen oxide, hydrocarbons, and carbon monoxide. These emissions vary, depending on the combustion techniques, operating practices, and combustibility of the waste.

Hazardous Air Pollutants.   Emissions of potentially hazardous air pollutants from the incineration of municipal solid waste include heavy metals (discussed earlier), dioxins, polychlorinated biphenyls, and aromatic hydrocarbons. Combustion flue gases typically contain corrosive hydrogen chloride (HCl) as a result of burning polyvinyl chloride or other chlorinated plastics found in the waste stream. Compared with fossil fuels, the flue gas of incinerated solid waste has higher amounts of chlorine compounds. The organic chlorine found in plastics and solvents such as polyvinyl chloride or methylene chloride is converted to HCl. Some of the HCl (possibly 20 to 40 percent) is absorbed by alkaline particulates. The average concentration in a U.S. incinerator falls between 100 to 150 ppm (13).

Studies indicate that dioxins are toxic and carcinogenic. Incinerators are not the only source of dioxins, however. Residential fireplaces, diesel engines, pesticides, wood preservative manufacturing, paper mills, and many other industrial processes also produce these harmful compounds.

Air Pollution–Control Systems. Although incinerator design, operating procedures, regulations, and fuel cleaning can significantly reduce the amount of pollutants produced in waste-to-energy plants, some pollutants are inevitable. The quantity and composition of air emissions depend on the composition of the refuse, the incinerator design, and the completeness of combustion. Different types of pollutants require different control devices. Modern, well-maintained, and well-operated air pollution–control systems provide good emission control. Even so, significant environmental impacts can be associated with the solids and gases that are produced during combustion.

Acid Gas and Dioxin Control. Acid gases such as hydrogen chloride (HCl), hydrogen fluoride (HF), sulfur dioxide (SO_2), and nitrogen oxide (NO_x) have limits imposed on their discharge by some permits. The combustion temperatures are generally less than 2000°F which limit the production of NO_x. In sunlight, NO_x contributes to the formation of photochemical smog. Because municipal solid waste has very low sulfur concentrations of approximately 0.1 to 0.2 percent, SO_2 emissions are very low. Production of a mixture of acid gases, however, creates both the operational and environmental problems. These gases must be converted to a solid form by precipitation or absorbed by solid or liquid particles, with the resulting particles then being removed by an appropriate device. The three processes applicable to acid gas removal are: wet scrubbers, semiwet scrubbers, and dry scrubbers (discussed later).

Dioxins and furans are chlorinated compounds with somewhat similar chemical structures. Both consist of two benzene rings that are linked together by oxygen bridges. Highly toxic substances dioxins are produced as byproducts of incineration, chemical processing, chlorine bleaching of paper and pulp, and burning of diesel fuel. In general, control of dioxins and furans is very dependent on the combustion efficiency; there is a correlation between combustion temperature and residence time and dioxin and furan emissions (14). The California Air Resources Board recommends minimum temperatures in thermal processing systems of 1800°F $\pm$ 190°F with a residence time of 1 second or longer. In addition, the conditions that minimize generation of carbon monoxide also minimize generation of dioxins and furans (15).

Scrubbers. Scrubbers (followed by a particulate control device) are considered to be current technologies for controlling acid emissions. Greater removal efficiencies are usually achieved by greater condensation; devices that lower gas temperature and, thus, increase condensation can enhance scrubber effectiveness. The lower temperatures allow mercury, dioxins, and furans to condense and then be captured by a particulate device. For all scrubbers, temperature and, for dry and semiwet (i.e., spray-dry) scrubbers, the amount of lime to neutralize acids are the important factors in scrubber efficiency. To maximize control of emissions, the scrubber should generally be adequately sized, operate at temperatures less than 270°F, and permit gas circulation through the scrubber for at least 10 to 20 seconds.

Wet scrubbers are used for removing particulates from the gas. Sometimes, lime or other alkaline agents are added in small amounts to neutralize or reduce the acidic environment. Improved designs can remove 99 percent of HCl and SO_2 and more than 80 percent of dioxin, lead, and mercury (16). Figure 13–25 shows a wet scrubber.

Advantages of a wet scrubber include simultaneous removal of particulate matter, acid, and other components of exhaust gases. In addition, high temperature flue gas streams do not cause any problem. Disadvantages include high energy input to remove small particles, high maintenance costs, and requirements for wastewater treatment.

Semiwet and dry scrubbers control acidic gases (e.g., SO_2, HCl) by neutralization. When flue gases pass through an alkaline mist of a calcium- or sodium-based slurry, the droplets neutralize the acids as the gases evaporate. The larger, dried particles settle to the floor, and the smaller particles exit with the flue gas to be collected in an electrostatic precipitator or bag house. This method eliminates the scrubber water, which must be treated or disposed. Dry scrubbers can neutralize 99 percent or more of HCl and SO_2 under optimal conditions, such as a temperature less than 300°F, high lime/acid ratios, and a long gas residence time in the scrubber. Dioxin emissions are also considerably reduced (16).

Compared with wet scrubbers, the advantage of semiwet and dry scrubbers is that the residue is dry and has less volume. Dry residue is less expensive to dispose because of the lesser volume. In addition, the absence of water reduces the corrosion problem associated with a wet system.

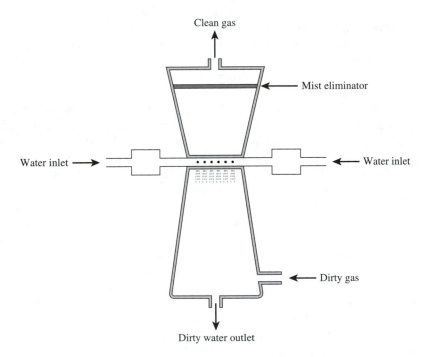

Figure 13–25   Wet vertical venturi scrubber with throat injection.

Particulate Control Devices. Emissions of particulates and heavy metals are best reduced by collection in one of two basic types of add-on particulate control devices: fabric filters, and electrostatic precipitators (ESPs). Heavy metals are captured as they condense out of flue gas and into the particles at temperature of 450°F.

An ESP is a device that captures particles by electrostatic attraction. These devices consist of one or more pairs of electrically charged plates or fields. Charge plates consist of negative electrodes (to increase the voltage potential) and a grounded, positive plate where particles are collected and neutralized. Particulates ranging from 0.1 to 50μm are effectively removed. Figure 13–26 shows a typical ESP. The advantages of ESPs include high particulate collection efficiency, low operating costs, and operation at high temperature. Disadvantages include high capital costs, large space requirements, and explosion hazards when treating combustible particulates and gases.

Baghouse filters, as the name implies, are structures containing fabric filter bags through which flue gases pass for the removal of particulate matter. These bags are made of a variety of natural or synthetic fabrics. A typical baghouse is shown in Figure 13–27. Fabric filters are a state-of-the-art particulate control technology, with a 99-percent removal efficiency over the range of particulate sizes, specified for the particular fabric filter. Flue gases are directed through ducts with the fabric bags attached to the ends. The bags then filter the particulates from the air as it passes through them, and the bags are occasionally shaken to release particles to a collection tank below. These filters remove particles with diameters larger than approximately 0.5 μm. Advantages include high particulate collection efficiencies, dry collection of solids, insensitivity to gas stream fluctuations, and simple operation. Disadvantages include the need for special fabrics at temperatures higher than 550°F, fire or explosion hazard because of concentrations of some dusts in the collector, high maintenance requirement (i.e., bag replacement), and possible caking or plugging of the fabric.

A third type of particulate control device is the cyclone, which is a device that funnels flue gases into a spiral, thus creating a centrifugal force that removes large particles. Flue gases enter a cylindrical chamber tangentially, and they swirl around the chamber. Inertia carries particulates to the wall. These particulates move downward in the conical section and are collected in a hopper. Figure 13–28 shows a cyclone. Particles of 10 μm or larger are effectively removed. Cyclones have low capital and operating costs.

The metals in batteries, appliances, electronic items, and so on are mostly removed during the various separation processes used to prepare the RDF. The contaminants contained in other components of municipal solid waste, however, such as paper, plastic, textiles, leather, and rubber, may end up in the fuel. The benefits of separation are evident from the results of the NRT separation process. In addition to recovering items such as aluminum, ferrous metals, glass, and other inert materials, this process reduces stack emissions by 52 percent for lead, 64 percent for chromium, and 73 percent for cadmium (compared with burning unprocessed refuse) (17). Carbon monoxide was also reduced by 63 percent. This reduction in emissions results from uniformity of waste and good combustion control. Clearly, preparation to provide a homogeneous fuel with reduced material can greatly improve operation.

Significant reduction in emissions is also possible by careful operation of the incinerator. The system should be designed and built for operational flexibility. Sufficient

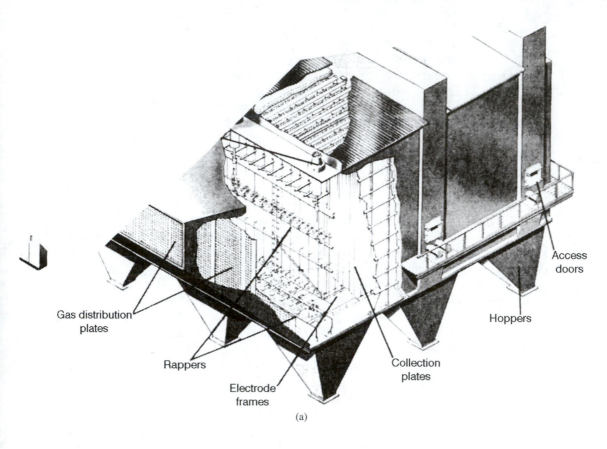

Gas distribution plates

Rappers

Electrode frames

Collection plates

Access doors

Hoppers

(a)

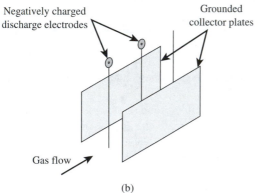

Negatively charged discharge electrodes

Grounded collector plates

Gas flow

(b)

Figure 13–26   Flat surface–type electrostatic precipitator. (a) Cutaway view. (b) Schematic.

(U.S. Department of Health, Education, and Welfare–Public Health Service. Air Pollution Engineering Manual. *Washington, D.C.: U.S. Govt. Printing Office, 1967; with permission.)*

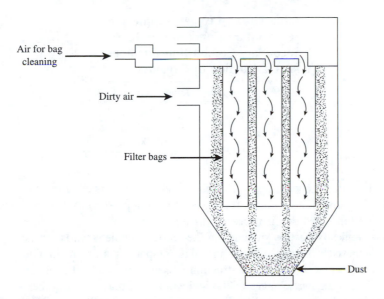

Figure 13–27   Baghouse collector. Note that this type of baghouse collects dust on the outside of tubular filters. Reverse air (under pressure) is used for cleaning the bags.

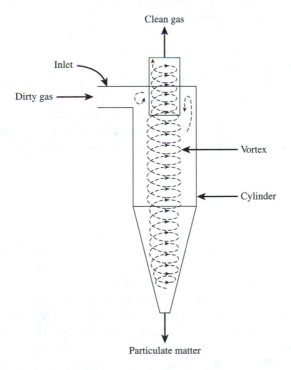

Figure 13–28   Cyclonic dust collector.

Table 13–2   Oxygen Supply and
Temperature Conditions in the
Secondary Combustion Chamber
for Destruction of Organics

O_2 (%)	Temperature (°F)
20	1350
2.5	1500
0	1800

air supply with the capability to distribute required air at needed rates to various sections of the furnace is very important. To maintain a proper oxygen level and temperature, a uniform feed rate is necessary as well.

To chemically destroy organics in the waste, complete oxidation is required and can be achieved if excess oxygen and high temperatures are maintained in the secondary chamber. A 99-percent destruction rate of pentachlorophenol (i.e., a precursor formed during the early stages of the incineration of plastics and other similar products) was noted with the conditions given in Table 13–2 (18). To minimize formation of NO_x, the temperature should not be more than 2000°F.

Disposal of Ash.   An incinerator produces two types of ash: bottom ash, and fly ash. Bottom ash is the residue left from the burned waste, and it often contains partially burned materials. Bottom ash accounts for approximately 90 percent of the total ash produced. Fly ash is more uniform in composition than bottom ash. Both bottom ash and fly ash, however, will contain all of the metals present in the waste, which are mostly oxides. Leachate from the incineration tends to be acidic and could release heavy metal ions.

Most states consider ash as being nonhazardous; however, studies indicate this may not be true (19). Consequently, the EPA categorizes these ashes as hazardous if they fail the toxicity test. This test is a procedure that passes a weak acid solution through the ash to determine what chemicals leach from that ash. If any chemicals exceed the regulatory limit, the ash qualifies as being hazardous and must be treated in accordance with regulations requiring disposal at a hazardous waste site. (Chapters 19 and 20 provide further discussion.)

BIOLOGICAL TRANSFORMATION

The organic matter in municipal solid waste presents a significant disposal problem, but it also has the potential to be converted into useful chemicals and fuels. Microorganisms biodegrade the organics into gases, solids, and energy. To continue to reproduce and function properly, however, these organisms must have a source of energy; carbon to synthesize new cells, inorganic elements (i.e., nutrients), proper pH and temperature, and a nontoxic substrate. In addition, the types of microorganisms (i.e., aerobic and anaerobic transformation) used during biodegradation should be considered. One important application of biological principles is in the process of composting solid waste (e.g., yard wastes, food wastes).

Composting

Composting is becoming a popular waste management option as communities find ways to divert portions of their waste from landfills and meet mandated waste diversion goals. Composting can be applied to transform yard waste and the organic fraction of municipal solid waste, and co-composting can be applied with wastewater sludge.

Yard trimmings and food waste account for one-fourth of the municipal solid waste generated in the United States. Before the 1990s, almost all waste was disposed of in a landfill or an incinerator. As current landfills run out of space, sites for new disposal facilities become increasingly difficult to find, more restrictions are put into place by regulating agencies, and public opposition to such facilities grows, it becomes apparent that source reduction and recycling programs must be implemented for wastes.

One such program is composting. In fact, composting of yard and garden waste, particularly leaves and grass clippings, is now mandatory in several U.S. states and has been adopted voluntarily by others to meet legislated waste diversion goals. In countries such as Holland, solid waste composting was practiced as early as the 1930s (20). According to the EPA (21), recovery of yard trimmings has increased from 12 percent of the yard-trimming mass generated in 1960 to 19.8 percent of that generated in 1993.

Microbiology.   Composting is the biochemical degradation of organic waste. The end product of composting is a humuslike material that can be used primarily as a soil conditioner.

Most composting processes are designed for aerobic operation. In an aerobic process, microorganisms (in an oxygen environment) decompose the organic food wastes as follows:

$$\text{organic matter} + O_2 \xrightarrow[\text{bacteria}]{\text{aerobic}} \text{new cells} + CO_2 + H_2O + NH_3 + SO_4 \quad \textbf{(13-10)}$$

The aerobic composting organisms are bacteria, fungi, and protozoa.

The process of composting has several parameters of relevance. These are moisture content, temperature, oxygen, carbon/nitrogen ratios, and pH.

Moisture content. The optimum moisture content is 50 to 60 percent. At less than 50-percent moisture content, the metabolic activity slows. At greater than 60 percent, the moisture inhibits oxygen access by filling the voids between particles and causes a temperature reduction.

Temperature. Composting is an exothermic process and undergoes temperature changes throughout the decomposition process as follows:

Psychrophilic	59°F to 68°F
Mesophilic	77°F to 95°F
Thermophilic	122°F to 140°F

Best results are achieved if the thermophilic stage can be reached within a few weeks and then maintained. Temperatures greater than the thermophilic range may inhibit biological activity, but at the same time, pathogenic bacterial kill is achieved.

Oxygen. Oxygen is essential for aerobic decomposition. With a low oxygen level, the decomposition process becomes anaerobic, which is a much slower process and also generates odors. Turning and ventilating the compost will keep the oxygen at a sufficient level.

Carbon/nitrogen Ratio. This ratio is a measure of the optimum biochemical conditions, which occur in the range of 20:1 to 40:1. If the ratio is less than 20:1, carbon-rich material is needed; if the ratio is greater than 40:1, nitrogen-rich compounds should be added.

pH. The optimum pH range is from 6 to 8. In the initial days of the biodegradation process, however, the pH is as low as 5 as organic acids are formed. Then, the pH rises as these acids are decomposed during the thermophilic stage.

Biochemical Composition.   The biochemical composition of the waste significantly influences how the process develops. Materials such as plants, human waste, and food waste are easily degradable, but materials with high lignin content, such as tree bark, wood, yard waste, and certain paper products, are slow to biodegrade.

Composting is a developing municipal solid waste management technology in the United States. Unlike yard waste compositing, a large amount of preprocessing of incoming waste is required before composting. Preprocessing is done to isolate the compostable portion of the waste stream from the noncompostable portion (e.g., glass, plastics, metals). The preprocessing of waste before composting is largely a separation task, as addressed earlier in this chapter.

Co-composting refers to simultaneous composting of two or more diverse waste streams with sewage sludge or other nitrogen-rich material. The sludge provides moisture and nutrients for the compost. Combining sludge with municipal solid waste for composting is done to provide balanced nutrients and to use the sludge.

Backyard Composting.   Backyard composting involves homeowners starting a compost pile of yard and degradable household wastes on their own property. Backyard composting is a source-reduction activity, because materials composted do not end up in the municipal waste stream. Commonly used methods of backyard composting include windrows and Pen.

Windrows are elongated piles from 2 to 5 ft high and constructed by layering the waste. The piles are turned periodically. To protect the material from excessive moisture during rainy season, the piles are sometimes covered with tarp.

The pen method involves building a compost pile within a pen of woven wire (e.g., chicken wire). This type of system is easily moved, and the wire allows for good air circulation.

Composting Techniques.   The two main methods of composting now used in the United States may be classified as agitated and static. In the agitated method, the composting waste is agitated periodically to aerate (i.e., to provide oxygen), to control the temperature, and to mix the waste. In the static method, air is blown through a static pile of composting waste. The common agitated and static methods are known as the windrow and static pile methods, respectively.

Windrows. Windrow composting has been practiced in many locations throughout the United States, including the joint U.S. Public Health Service–Tennessee Valley Authority project. As described earlier, composting waste is piled into heaps or windrows (Figure 13–29[a]). The piles are typically elongated and dome shaped to shed rain and snow. A windrow system can be constructed by forming the organic waste to be composted into windrows of 6 to 8 ft high and 15 to 17 ft wide at the base (22). The windrows are turned once or twice each week during the composting period of approximately 5 to 6 weeks to provide aeration and mixing. Turning can be accomplished with either a front-end loader or a clamshell bucket on a crane. Before the windrows are formed, organic material is processed by shredding and screening it to a size of approximately 1 to 3 in., and the moisture content is adjusted to 50 to 60 percent. The temperature is approximately 130°F to 135°F. Turning the piles is accompanied by odors. After the turning period, the compost is cured (i.e., stabilized) for another 4 weeks.

Windrow composting requires extensive land space. A generally accepted rule of thumb is that 1 acre of land is required for every 3000 to 3500 yd^3 of leaves collected (22).

Static Piles. In static pile composting, aeration is provided to the waste piles by blowing or drawing air through a system of perforated, flexible drainage pipes. Figure 13–29(b) illustrates the static pile process. The aerated static pile system consists

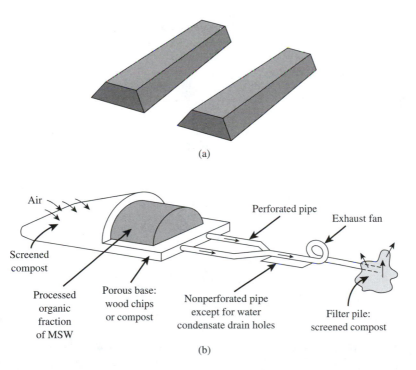

(a)

(b)

Figure 13–29   (a) Windrow piles. (b) Aerated static pile.

of a grid of aeration or exhaust piping, over which the processed organic fraction of municipal solid waste is placed. Typical pile heights range from 7 to 8 ft, and a layer of fine compost soil is often placed on top of the newly formed pile for insulation and odor control.

Air is introduced to provide the oxygen needed for biological conversion and to control the temperature within the pile. With use of temperature and oxygen sensors, the system can be made relatively easy to operate in terms of optimizing composting activity and maintaining sufficiently high temperatures to kill pathogens.

In-Vessel Composting. In-vessel composting, which is also called closed composting, is a closed reactor vessel. The biology of this process is the same as in the open process. Closed composting, however, permits better control of the environment, (e.g., temperature, moisture, aeration) during composting. An in-vessel system takes approximately 14 days for composting and 20 days for curing, whereas an open system takes a minimum of 21 days for composting and 30 days for curing. A closed system also requires less land area.

Mechanical systems for in-vessel composting are designed to minimize odors and the processing time by controlling the environment within the reactor. It is a capital-intensive operation, however, and maintenance costs are high. There are many variations of in-vessel composting designs. Figure 13–30 illustrates a typical reactor. The performance and end result are almost the same as in other composting techniques but have varying levels of control.

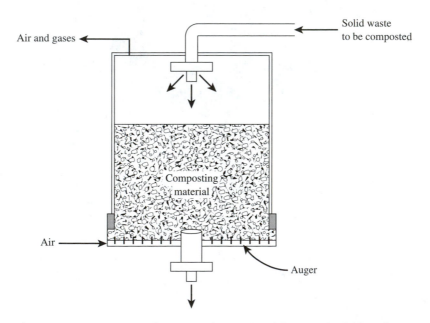

Figure 13–30   In-vessel composting unit with a vertical plug flow reactor.
(From Tchobanoglous, G., H. Theisen, and S. Vigil. Integrated Solid Waste Management. *New York: McGraw-Hill, 1993; with permission.)*

Costs of Centralized Composting.   Costs related to the development of a centralized yard waste–composting facility include land, labor, site preparation, windrow formation, water source, and turning equipment, storage pile formation, separation/ shredding/screening, disposal of noncompostable and unacceptable materials, contingencies (e.g., additional costs from bad weather, equipment breakdown), and overhead. Costs at medium-sized facilities ($\leq$30,000 yd^3 of leaves) have been estimated at $5.50/yd^3. Costs at larger facilities ($\geq$80,000 yd^3 of leaves) using modern mechanical equipment have been estimated at $8.00/yd^3 (22).

These costs do not include those of collecting and transporting the yard waste. Larger facilities are more expensive on a dollar per ton or a volume basis, because larger facilities require increased capital expenditures for equipment and labor. The cost of composting equipment is given in Table 13–3 (22).

Pyrolysis

The term *pyrolysis* refers to the act of decomposing organic compounds through application of heat. It is closely related to combustion, a thermal process in which organic materials are heated to high temperatures in the oxygen-free environment (i.e., in a closed container). Pyrolysis, however, is a distillation process. It is a reaction in which heat must be supplied for the reaction to occur (i.e., endothermic), whereas in incineration, heat is produced (i.e., exothermic).

A typical pyrolytic reaction using cellulose is

$$C_6H_{10}O_5 \rightarrow CH_4 + 2CO + 3H_2O + 3C$$

in which a gas is produced containing methane (CH_4), carbon monoxide (CO), and moisture. The CO and CH_4 are combustible, providing the produced gas with a positive heating value. The carbon residual (3C), a char, also has a heating value. The char is a carbon-rich solid.

Table 13–3   Composting Equipment and Approximate Cost in 1997 Dollars

Equipment	Cost ($)
Vacuum leaf collection	
Trailer mounted	18,000–28,000
Truck	65,000–78,000
Front-end loaders	97,000–194,000
Water tank truck	65,000–84,000
Aerating and turning equipment	39,000–207,000
Separating and shredding equipment	22,000–194,000
Tub grinder	142,000

Source: Data from U.S. Environmental Protection Agency. *Decision-Makers Guide to Solid Waste Management.* (U.S. EPA/530-SW.89-072.) Washington, D.C.: U.S. Govt. Printing Office, 1989.

The composition and yield of the products of pyrolysis can be varied by controlling the operating parameters (e.g., pressure, temperature, time, feedstock size, auxiliary fuels). High temperatures in excess of 1400°F (760°C) favor the production of gases such as hydrogen, methane, carbon monoxide, and carbon dioxide. Temperatures from 850°F to 1350°F (450°C to 730°C) produce tar, charcoal, and liquids such as oils, methanol, and acetic acid.

Municipal solid waste has a high heat value, with a range of 10 to 12 million Btu per ton (see example 13–4). Such waste is not only heterogeneous but may differ greatly from one batch to the next. Therefore, considerable research and pilot work are needed before pyrolysis can be applied as an efficient process to recover energy from solid waste. Figure 13–31 illustrates an experimental unit that uses the pyrolysis process from the Battelle Northwest final report. Solid waste is shredded to a size of less than 4 in. and is passed through a magnetic separator. It is then fed into the reactor. As it enters the reactor, the waste passes through the drying zone, the pyrolysis zone, the char gasification zone, and then settles in the ash bed in the bottom. A mechanism removes the ash from the reactor, passes it through a crusher, and then loads it onto trucks for landfill. The air-and-steam mixture injected into the bottom of the reactor reacts with

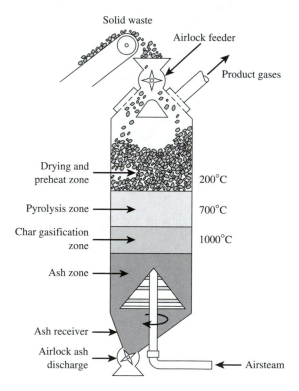

Figure 13–31 Battelle pyrolysis reactor.

(From U.S. Environmental Protection Agency. Battelle Northwest Final Report, City of Kennewick, Washington. Washington, D.C.: U.S. Govt. Printing Office, 1972; with permission.)

the char residue to produce the heat required to pyrolyze the incoming waste. Half the steam decomposes during the reaction; the other half heats the solid waste stream. Water from the condensed steam is then piped for treatment, and the gas from the reactor is compressed and piped to a turbine for power generation.

LIFE CYCLE ASSESSMENT

The life cycle assessment (LCA) of a product is a systems approach for examining pollution sources in a process. It is a new and developing environmental management technique. In considering the environmental effects of trash, it is important to consider the entire life cycle of a product, from synthesis or manufacture through distribution and use to waste disposal. Such LCAs have been used in solid waste management.

It is evident that obtaining and processing raw materials and eventually disposing of the product itself may be as much of a problem as the product manufacture. An LCA of material and product use can be helpful in dealing with such problems. This analysis uses an energy and materials balance approach at every stage in the life cycle of a product, process, and associated activity, which includes raw materials, raw materials refining, manufacturing, transportation, distribution, use, maintenance, retirement, recycling, and final disposal. Such a holistic approach should indicate how the relative contributions of the life cycle stages impact the environment, and Figure 13–32 shows a possible flow chart.

A LCA has great potential for product improvement—and for abuse as a marketing tool by manufacturers and trade associations seeking to promote their products. Different users have different purposes and demands, and an industry trying to promote a product has a different outlook and purpose from others. One problem is that this type of study is often conducted by groups representing the industry or the individual corporations involved. Obviously, most outcomes of such studies tend to favor the industry or corporation conducting or sponsoring the study.

Many environmental studies compare the attributes of competing products, such as disposable versus reusable cloth diapers or polystyrene versus paper coffee cups. In a market economy, environmental comparisons between different products can often be complex and difficult to assess. For example, take the study of disposable diapers manufactured by Procter & Gamble versus reusable cloth diapers (23). Procter & Gamble found in a study done for their company that cloth diapers consume three times more energy than disposable diapers. According to the National Association of Diaper Services, however, disposable diapers consume 70 percent more energy than cloth diapers. The difference arose from the accounting procedures and viewpoint. For example, if the energy contained in a disposable diaper is recoverable energy in a waste-to-energy facility, then the disposable diapers are more energy efficient. Also, will the cloth diapers be washed in hot or lukewarm water, and will a diaper service be considered? Answers to these questions should provide further insight into this matter.

For an example of the difficulties in performing an LCA, consider a comparison of the environmental impact of a single-use, hot-drink cup made from polystyrene foam versus a similar cup made from paper. Will this analysis provide a definite answer as to which cup is better for the environment? A simple and obvious answer would be

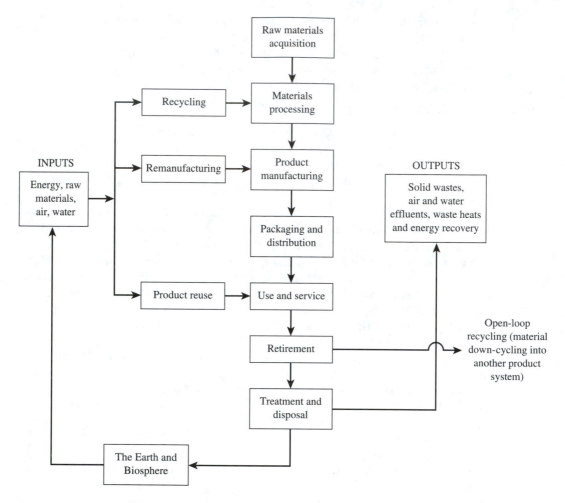

Figure 13–32   Life cycle stages of a product.

neither cup (i.e., use a reusable mug). Disposable cups are used in social gatherings, however, and they are necessary in hospitals. So, try the life cycle approach to answer the question posed.

The paper for the cup comes from trees, and the removal of trees harms the environment. The foam for the polystyrene comes from hydrocarbons such as oil and gas, and this results in environmental damage and the use of valuable and nonrenewable resources. The production of paper for the cup, however, results in higher air and water effluents. Particulates, chlorine, carbon dioxide, and sulfur compounds are emitted into the air, and the biochemical oxygen demand and suspended solids run into water. Production of the foam cup produces none of these pollutants, but its production emits significant quantities of pentane. Paper production does not. Therefore, it is difficult to determine whether, for example, air pollutants emitted during paper production are more significant than pentane emitted during polystyrene manufacture. In terms of the energy requirement, processing raw wood to a paper cup uses approximately 3.8 g of fuel, but a polystyrene cup uses 4.5 g of fuel oil (24). A paper cup's poor structural integrity limits its use to one time, but the polystyrene cup is strong, holds hot beverages well, and can be reused at a gathering. When incinerated, both cups burn clearly, but a paper cup yields approximately half as much heat recovery per kilogram as polystyrene. Paper biodegrades slowly in a landfill, however, and its degradation in the landfill produces carbon dioxide, methane, and leachate with a high BOD. A polystyrene cup is fairly inert, though, and will remain in the landfill for a considerable time before eventually biodegrading.

Therefore, at this stage, the question of which cup is better for the environment is difficult to answer. Perhaps in the future, more detailed and specific life cycle studies may help to answer this question.

An LCA needs good data and facts, but such data are often unavailable. Data uncertainties and lack of an accepted methodology for LCAs are also of concern and have led to calls for peer review, use of valid data, and need for complete data presentation (including limitations). If conclusions are based on reliable data obtained using an accepted methodology, the LCA technique can be a good tool for determining the environmental impact associated with a product.

For example, compare the total environmental effect of returnable glass bottles with the environmental effect of nonreturnable (i.e., nonrefillable) bottles. Note from Table 13–4 that refillable glass bottles cause the least environmental impact. A comparison of refillable and nonrefillable containers with plastic bottles and aluminum cans indicates that the latter have an environmental impact but one that is less than that of nonrefillable bottles.

Unfortunately, returnable bottles either are no longer available or are being phased out in favor of throwaway containers. In the United States, plastic bottles and aluminum cans are very popular. Consumers like the convenience of not having to keep and return

Table 13–4   Environmental Impacts Associated with Containers[a]

Container	Energy (Btu/L)	Air Emissions (g/L)	Waterborne Emissions (g/L)	Solid Waste Mass (kg)	Solid Waste Volume ($\times 10^{-4} m^3$)
16-oz. refillable glass bottle used 8 times	4032	6.5	1	0.18	2.2
16-oz. non refillable glass bottle	9118	18.2	2	0.52	6.7
16-oz. PET bottle	8281	11.1	1.9	0.1	3.4
12-oz. aluminum can	8620	11	3.2	0.63	1.6

[a] For 1987 recycling rates.

Source: Date from Sellers, V.R., and J.D. Sellers. *Comparative Energy and Environmental Impacts for Drink Delivery Systems*. Prairie Village, KS: Franklin Associates, 1989.

bottles, and in general, stores do not like to accept the returnable bottles. From their viewpoint, it is a waste of time. Even in relatively poor countries like India, where resources are limited, people like the convenience of throwaway beverage containers.

REVIEW QUESTIONS

1. a. What is on-site processing of solid waste?
   b. What is the function of a materials recovery facility?
2. a. What is the basic function of a receiving and storage area?
   b. A 500-ton/day recovery facility receives waste in a 20-yd^3 compactor vehicle at 350 lbs/yd^3. Collection service is provided 5 days per week, one shift per day. The facility processes waste 24 hours per day, 6 days a week. Calculate the peak receiving rate in terms of tons and truckloads per hour.
3. Size a storage system for the truckloads in question 2. Assume a bay that is 12-ft long and 48-ft wide.
4. a. What are the methods used in the physical processing of solid waste?
   b. List the equipment used for such physical processing.
5. Discuss the innovative technologies used by NRT.
6. What is a chemical transformation of waste?
7. Write the equations for the basic stoichiometric combustion of the organic fraction of waste.
8. a. Why is excess air needed for combustion?
   b. What is the difference between low heat value (LHV) and high heat value (HHV)?
9. Calculate the LHV and HHV for the following ultimate analysis (% w): moisture = 25, C = 25, H = 5, O = 26.5, N = 0.35, S = 0.15, and ash = 18.
10. What is the difference between a mass burn and a refuse-derived fuel incinerator?
11. Discuss a fluid bed combustor.
12. a. List the problems associated with incineration of solid waste.
    b. Discuss ways to reduce these problems.
13. What are the common air pollution–control devices?
14. Sketch a baghouse collector, and explain how it removes pollutants from the air.
15. a. Why has the popularity of composting increased in recent years?
    b. Assume that a town of 60,000 persons is planning to compost 30 percent of its organic waste. What advice would you give regarding the location, design, and operation of such a facility?

REFERENCES

1. Schwartz, S., and C. Brunner. *Energy and Resource Recovery from Waste.* Park Ridge, NJ: Noyes Data Corp, 1983.
2. Ruf, J. *Particle Size Spectrum and Compressibility of Raw and Shredded Municipal Solid Waste.* PhD Thesis, University of Florida, 1974.
3. Hecht, N. *Design Principles in Resource Recovery Engineering.* New York, NY: Butterworth, 1983.
4. Trezek, G., G. Savage, and L. Diaz. "Solid Waste Management," May 1980.
5. National Center for Resource Recovery. *New Orleans Resource Recovery Facility: Implementation Study of Equipment, Economics, and Environment.* September 1977.
6. National Recovery Technologies, Inc., Nashville, Tennessee. *Journal of the Air and Waste Management Association,* April 1989, Vol. 39.
7. Kiser, J.V.L., and B.K. Burton. Energy from Municipal Waste: Picking up Where Recycling Leaves Off," *Waste Age,* Vol. 23, No. 11, November 1992, pp. 39–46.

8. Hazome R, et al. "Investigation of Elemental Analysis of Refuse for Calorific Value," Nippon K.E.S. Shoho, 1979.

9. Hougan, et al. *"Chemical Process Principles. Part I: Material and Energy Balances,* 2nd ed. New York: 1954. Wiley,

10. U.S. Department of Health, Education and Welfare–Public Health Service. *Air Pollution Engineering Manual.* Washington, D.C.: U.S. Govt. Printing Office, 1967.

11. *Incinerator Standards.* Chicago: Incinerator Institute of America, 1968.

12. U.S. Environmental Protection Agency. *Municipal Waste Combustion: Summary of the Requirements for Section 111(d)/129–State Plans for Combustor Emission Guidelines.* Washington, D.C.: U.S. Govt. Printing Office, 1996.

13. Robinson, W.D. *The Solid Waste Handbook: A Practical Guide.* New York: Wiley, 1986.

14. Hasselris, F. "Optimization of Combustion Conditions to Minimize Dioxin, Furan, and Combustion: Gas Data from Test Programs at Three MSW Incinerators," *Journal APCA,* Vol. 37, No. 12, December 1987.

15. *Air Pollution Control at Resource Recovery Facilities, 1991.* Sacramento, CA: California Air Resources Board, 1991.

16. Hershkowitz, D. "Garbage Burning: Lessons from Europe: Consensus and Controversy in Four European States," Paper Presented at the New York Waste Conference, 1988.

17. *Waste Age,* Vol 19, No. 2, February 1988, p. 55.

18. *Waste Age,* Vol 18, No. 12, December 1987, p. 162.

19. Liptak, B.A. *Municipal Waste Disposal in the 1990s.* Radnor, PA: Chilton 1991.

20. Rhyner, C.R., L.J. Schwarz, R.B. Wegner, and M.G. Kohrell. *Waste Management and Resource Recovery.* New York, NY: Lewis, 1995.

21. U.S. Environmental Protection Agency. *Characterization of Municipal Solid Waste in the United States: 1994 Update.* Washington D.C.: Office of Solid Waste, 1994.

22. U.S. Environmental Protection Agency. *Decision-Makers Guide to Solid Waste Management.* (U.S. EPA/530-SW.89-072.) Washington, D.C.: U.S. Govt. Printing Office, 1989.

23. "Life Cycle Analysis Measures Greenness, but Results May Not Be Black and White," *Wall Street Journal,* 28 February 1991.

24. Hocking, M.B. "Relative Merits of Polystyrene Foam and Paper in Hot Drink Cups: Implications for Packaging," *Environmental Management,* Vol. 15, No. 6, June 1991, pp. 731–747.

Chapter 14 Source Reduction, Reuse, Recycling, and Recovery of Municipal Solid Waste

This chapter introduces the issues involved in source reduction and recovery of materials from municipal solid waste (MSW). According to the principles described in the U. S. Environmental Protection Agency's (EPA's) *Agenda for Action* (1), states, municipalities, and the waste management industry should use the following hierarchy to reduce the solid waste management problem most effectively:

1. Source Reduction
2. Reuse
3. Recycling
4. Treatment
5. Disposal

The first three elements address waste prevention and diversion of materials from the waste stream, and they are discussed in this chapter. The last two involve the transformation, destruction, or disposal of materials once they enter the waste stream. They are discussed in subsequent chapters.

"Waste that is not produced does not have to be collected" is a very good concept; consequently, preventing waste and pollution has become a major issue. Source reduction programs may include the design, manufacture, and packaging of products with little or no toxic content, a minimum volume of material, and/or a longer product lifetime. This involves changing the way that products are made and marketed. Through source reduction, landfill capacity and natural resources are conserved, less energy is used in manufacturing, and land, air, and water pollution are reduced. Elements of source reduction activities include product reuse, reduced material volume, reduced toxicity, increased product lifetime, and decreased consumption.

A few examples of product reuse are reusable shopping bags, clothes and other items at the Salvation Army, retreaded tires, and recharged batteries. Reduced material volume is possible by using concentrates, lighter-metal cans, and glass containers. Non-refillable (i.e., one-way), 16-oz. glass bottles, for example, are 37 percent lighter today than they were in the 1970s (2). Larger food containers can reduce the amount of metal; for example, a single 16-oz. can uses 68 g of metal, or approximately 40 percent less than the 95.4 g used in two 8-oz. cans. Obviously, however, use of larger containers may lead to food spoilage.

On average, each U.S. resident generated approximately 515 lbs of packaging waste in 1990 (3). Packing comprises one-third of the national waste stream by weight, and it is a potential target for waste reduction. Some packaging is essential for protecting, transporting, and marketing, but waste reduction is possible by eliminating unnecessary packaging, designing better packages, and reusing and refilling.

In Europe, dramatic steps have reduced packaging waste. These strategies are based on the principle that "the polluter pays." This makes the producers of packaging responsible for the packaging waste, in effect "internalizing" the cost of waste management and providing incentives for source reduction. For example, Germany passed the Packaging Ordinance in May 1991. This ordinance almost eliminates packaging materials from the public waste stream. It requires that retailers provide a container so that consumers can remove and leave secondary packaging (i.e., packaging to prevent theft or promote marketing) at the store. These are powerful incentives for source reduction. Germany has also passed strong recycling laws as well.

REDUCING QUANTITY AND TOXICITY

Reductions in the quantity and toxicity of waste and reuse of materials before they enter the waste stream are practices implemented by manufacturers and consumers. Manufacturers should be aware of the possibilities for waste reduction in various processes and how their products and packaging will affect the waste stream. Consumers should be aware of the environmental consequences of their purchasing and consumption choices. Knowledge of environmental and waste management issues may encourage both manufacturers and consumers to examine waste reduction and reuse practices.

Products can be designed and formulated before manufacturing to contain less—or even none—of the substances that pose risks when these products become part of the waste stream. Toxic substances used in products, for example, can create environmental risks when those products are disposed of in landfills and incinerators.

Toxic materials have always appeared in household wastes, but in the middle of this century, as synthetics replaced many traditional materials, the synthetically derived toxic materials in such waste have increased appreciably. These toxic constituents in solid waste include heavy metals, chlorinated hydrocarbons, and used motor oil.

Reduction in toxicity can be achieved by using less (or no) problematic substitutes for toxic constituents. For example, battery manufacturers eliminated mercury from zinc-carbon cell batteries during the early 1990s (4). They also reduced the amount of mercury in alkaline cells and developed cadmium-free, rechargeable batteries for certain uses. Substitution for lead in paint and elimination of lead in gasoline have significantly reduced lead toxicity. Products with longer lifetimes should be used over shorter-lived alternatives that are designed to be discarded after use, and examples here include longer-lasting tires, light bulbs, and so on. Source reduction policies should encourage repairs and continued use rather than disposal. Consumption of materials that, when disposed, are harmful to the environment (e.g., certain types of plastics and other oil-based products) should be avoided as well.

Source reduction activities vary widely, and many factors need consideration when evaluating, based on careful analysis and common sense, the economic and environmental effects. For example, source reduction can save disposal costs, because a smaller waste stream may reduce pollution, (e.g., less landfill leachate).

RECYCLING

Recycling is the separation of a given waste material from the waste stream for reuse or processing to be suitable for use as a raw material for manufacturing. Since 1970, recycling programs have increased. Initially, it was viewed as a singular activity (i.e., segregation and collection of glass, papers, and so on) rather than as a holistic activity that "closes the loop," as symbolized by the familiar recycling logo shown in Figure 14–1.

After source reduction, which is given top priority in the solid waste management hierarchy, recovery of materials for recycling and composting is the next important activity. The commonly accepted definition of solid waste recycling is to use one or more components in a way that they are not deposited in a sanitary landfill and that conserves natural resources. Most recycling programs are subsidized financially, because the collection and transport of waste for recycling require substantial amounts of labor and energy. The recycling process itself includes separating recyclables by type, collecting them, processing them into new forms, manufacturing them into products, and marketing them as goods made from reprocessed materials.

Usually, separation comes before collection. Separation is generally done by the generators (i.e., citizens, businesses, industries, institutions). Recyclables are then delivered or picked up for delivery to a materials-processing center or a scrap processor. In a few instances, mixed materials are collected and then separated. Source separation, however, provides the cleanest, most well-defined fractions of waste suitable for subsequent recycling or reuse. Mechanical or manual sorting at the destination is not always well defined.

Source-separated wastes may be collected at the curbside or delivered to a drop off center. In some areas, drop-off centers are in convenient locations, such as supermarket parking lots or a town's municipal warehouse, and they accept recyclables that the consumer delivers in person. In rural areas, these centers may provide the only recycling service in a community or may supplement curbside collection. Curbside col-

Recyclable   Recycled   Recycled

Figure 14–1   Labels designating paper products as being recyclable or recycled.

(From U.S. Environmental Protection Agency. Handbook of RCRA. *Washington, D.C.: U.S. Govt. Printing Office, 1994; with permission.)*

lection systems vary, however. Usually, a separate vehicle for recyclables makes its own collection run, but in co-collection systems, a single vehicle picks up both recyclables and residential solid waste. On the collection day, the residential owners set out the recyclables in a bin. Commercial establishments and industries usually make separate arrangements with waste management companies to pick up their recyclables. Figure 14–2 shows bins used for leaving recyclable materials at the curbside; Figure 14–3 shows a large drop-off container with separate cells for each type of recyclable material (e.g., plastic, aluminum cans, glass).

According to the EPA (5), recovery for recycling and composting had little effect on the total waste stream until the 1980s, when that portion was less than 10 percent of generation. Because of the pressures from decreasing landfill capacity, environmental

Figure 14–2   Curbside bins for recyclable materials.

Figure 14–3   Container with separate compartments for recyclable materials.

impacts, improving markets, economic incentives, and political support, a strong emphasis on recovery for recycling (including composting) developed during the latter part of the 1980s. In 1987, New Jersey passed the first statewide, mandatory recycling legislation. Under this law, residents in the state's 567 communities were required to recycle 25% of the solid waste generated, and all towns were required to compost leaves, by 1989 (6). Since then, 42 states have passed such laws. As an incentive, the EPA has set a national goal that 25 percent or more of all solid waste should be recycled; many states have moved to enact this goal requirement as well. For example, California has mandated that 50 percent of all solid waste be recycled by the year 2000. In the United States, overall recovery is projected to reach approximately 30 percent of total generation by the year 2000.

Table 14–1 presents the generation, recovery, and percentage of MSW recovered for the years 1980, 1990, and 1993. As shown, the largest category of waste generation and recovery is paper and paperboard. The next largest is yard waste.

Home hazardous waste (HHW) is collected separately. These collection programs began out of a concern for the environment and public health. These programs are locally financed, and the most common are:

- *One-day:* Citizens drop off their HHW on a certain day at a certain location.
- *Curbside collection:* HHW is collected from each household by the program's staff.
- Door-to-door pickup: Allows housebound individuals to participate.
- *Permanent dropoff:* Citizens drop off their HHW at a permanent collection site.

These programs are expensive, however, averaging $100 per participant (7).

Uses for Recycled Materials

Historically, manufacturers have resisted using recycled materials in product manufacturing. Clearly, processing virgin raw materials is more economical than processing recycled materials, but there are other reasons for using virgin materials as well. Manufacturers have come to rely on the known quality of the raw material, and the composition of the virgin material is consistent. Continued reuse, however, may increase the level of unwanted materials, which may affect the quality of the consumer goods. Tax benefits and lack of advances in recycling technology are other reasons for industries preferring virgin materials. Table 14–2 indicates the products that can be recycled.

Paper and Paper Products.   Recycled paper is used to make newsprint, paperboard for various types of boxes, container board, and construction products (e.g., cellulose insulation, fiberboard). Other applications include garden mulch, garden pots, and animal bedding. In 1993, 77.8 million tons of paper products were generated, and 26.46 million tons (34 percent) of that were recovered (Table 14–1).

The various paper products recovered from solid waste can be repulped and made into new products. For example, newsprint is pulped and then made into new newsprint, and recycled wastepaper has long been used as a source of fiber to augment pulp made

Table 14–1 Generation and Recovery of Materials in MSW in the United States

Materials	1980 (thousands of tons)			1990 (thousands of tons)			1993 (thousands of tons)		
	Generation	Recovery	% Recovered	Generation	Recovery	% Recovered	Generation	Recovery	% Recovered
Paper and paperboard	54,730	11,850	21.7	72,680	20,250	27.9	77,840	26,460	34
Glass	14,950	750	5	13,180	2,630	20	13,670	3,010	22
Metals									
Ferrous	11,580	370	3.2	12,440	1,710	13.7	12,930	3,370	26.1
Aluminum	1,760	340	19.3	2,860	1010	35.3	2,970	1,050	35.4
Other non-ferrous metals	1,120	540	48.2	1,100	730	66.4	1,240	780	62.9
Total metals	14,460	1,250	8.6	16,400	3,450	21	17,140	5,200	30.3
Plastics	7,870	20	0.3	16,820	370	2.2	19,300	680	3.5
Rubber and leather	4,290	130	3	5,930	330	5.6	6,220	370	5.9
Textiles	2,610	20	0.8	6,450	580	9	6,130	720	11.7
Wood	6,760	Negligible	Negligible	12,310	390	3.2	13,690	1,320	9.6
Other materials	2,870	500	17.4	3,150	680	21.6	3,300	730	22.1
Total Materials and Products	108,540	14,520	13.4	146,920	28,680	19.5	157,290	38,490	24.5
Other wastes									
Food wastes	13,200	Negligible		13,200	Negligible		13,800	Negligible	
Yard trimmings	27,500	Negligible		35,000	4,200	12	32,800	6,500	19.8
Miscellaneous inorganic wastes	2,250	Negligible		2,900	Negligible		3,050	Negligible	
Total other wastes	42,950	Negligible		51,100	4,200		49,650	6,500	
Total MSW	151,490	14,520		198,020	32,880		206,940	44,990	

Source: U.S. Environmental Protection Agency. *Generation and Recovery of Materials in the U.S.A.* Washington, D.C.: U.S. Govt. Printing Office, 1994; with permission.

Table 14–2   Recyclable Products and Applications

Recyclable Products	Applications
Paper	Reprocessed as newsprint, roofing material, paperboard, insulation, egg cartons, and paper masha.
Plastics	Reprocessed for auto parts, playground equipment, fiberfill, strapping, toys, mats, clothes, and shoes.
Glass	Containers, construction material, and planters.
Aluminum	Reprocessed for cans and other types of containers.
Other metals	Reprocessed for containers, garden tools, and structural products.
Construction waste	Reprocessed for pressed boards, pavements, and other construction purposes.
Tires	Playground equipment, floor mats, artificial reefs, erosion barriers, boat docks, pier bumpers, and sandals and belts.
Yard waste and other organics	Composted for use as soil conditioners for gardens and wood chips for landscaping
HHW	
Latex paint	Reprocessed and used for antigraffiti paint.
Used oil	Refined for use as lubricant.
Car batteries	Melted, with lead recovered for use.
Household batteries	Processed to recover mercury and silver.

Source: U.S. Environmental Protection Agency. *Household Hazardous Waste Management: A Manual for 1-Day Community Collection Programs.* Washington, D.C.: U.S. Govt. Printing Office, 1993; with permission.

from virgin materials. The proportion of recycled paper blended with the virgin fibers depends on the quality of the recycled material. Each recycling of wastepaper, however, results in shortening of the paper fiber, which soon reaches a size at which it is not possible to use anymore.

Economic considerations continue to drive the recycled paper market. Legislated dictates seem to have increased the quantity of paper available, but waste paper processors are reluctant to make large capital investments to increase plant capacity. The price of waste paper has fluctuated considerably as well. Only recently, communities had to pay someone to haul old paper away. In 1987, the price received by communities for old paper ranged from $5 to $80 per ton (8). The economic value of recycled paper depends on the characteristics of the paper collected. Better recycling technologies, limited supply of paper, and President Clinton's executive order to federal agencies to use at least 20 percent postconsumer material by the end of 1994 and 30 percent by the end of 1998, led to increased prices in 1994, but later, the prices dropped again.

Plastics.   Plastics are the bane of environmentalists. As Table 14–1 shows, even though plastics only make up approximately 9.3 percent of U.S. MSW by weight, they compose about one-fourth of the waste stream by volume because of their bulk. In 1993, 19.3 million tons of plastics were produced, but only 0.68 million tons (3.5 percent) were recovered.

Table 14–3 Common Types of Recyclable Plastics

Code Number	Chemical Name	Nickname	Typical Uses
1	Polyethylene terephthalate	PETE	Soft-drink bottles
2	High-density polyethylene	HDPE	Milk cartons
3	Polyvinyl chloride	PVC	Food packaging, wire insulation, and pipe
4	Low-density polyethylene	LDPE	Plastic film used for food wrapping, trash bags, grocery bags, and baby diapers
5	Polypropylene	PP	Automobile battery casings and bottle caps
6	Polystyrene	PS	Food packaging, foam cups and plates, and eating utensils
7	Mixed plastic		Fence posts, benches, and pallets

Most plastics are synthetic compounds composed of polymers containing hydrogen, carbon, and oxygen, and they are usually manufactured from petroleum and its derivatives. Table 14–3 lists common types of plastics that may be recycled, along with code number, nickname, chemical name, and uses. Figure 14–4 shows the identifiers for various types of plastics. Most common types of plastics that are recycled to any appreciable extent are polyethylene terephthalate (PETE) and high-density polyethylene (HDPE). The amount of plastic that is recycled continues to increase, and there is a significant demand for good, recyclable plastic. PETE and HDPE together account for 79 percent by volume of all plastics recovered from the waste stream, most in the form of beverage bottles (3).

Plastic recycling requires great care because of possible contamination by the products the plastic once contained or even by a small quantity of different types of plastic with different resins. Different resins have different melting points, so if a batch of mixed plastic is being heated and made into products, some resins may not melt—and some may end up burning. Most recycled plastic is used to manufacture other plastic products, but recycled PETE bottles can be used as the starting material for several other products (e.g., fiberfill, fibers for outdoor clothing). Recycled PETE is also used to make surfboards, carpets, and soft-drink bottles.

For several other applications, sorting is not necessary. Comingled plastics or mixtures can be shredded, melted, and extruded into useful forms. For example, plastic lumber can be used for fenceposts, siding, park furniture, play equipment, and so on.

Figure 14–4 Plastic identification markings.

The primary market for HDPE is the production of extrusion products such as flower-pots, toys, car components, tiles for landscape, containers, and so on.

Aluminum.   The recycling of aluminum, especially aluminum cans, has been very successful. Both social and economic pressures initially forced an unwilling industry to participate in this recycling. The economic incentive is directly attributable to the fact that recycled aluminum uses only 2 to 3 percent of the energy required to make new aluminum from bauxite ore (5). Industry has recognized the marketing advantage of recyclable aluminum containers, and the major aluminum companies use this in their advertising to induce people to use more aluminum containers.

Table 14–1 shows the amount of aluminum generated and recovered. In 1993, almost 63 percent of aluminum beer and soft drink cans (approximately 60 billion cans per year) were recovered from the waste stream and recycled back into production of aluminum. This recycling helps to reduce the demand placed on landfills and saves significant amounts of energy.

Other Metals.   Metals can be classified into two categories: ferrous, and nonferrous. As the name indicates, ferrous metals contain iron, but nonferrous metals (e.g., lead, zinc, copper, aluminum, silver) contain almost no iron. The only metals considered here are those found in MSW such as metal appliances, water heaters, garden tools and equipment, and so on.

The concern over "metal contamination" of steel has reduced the market for steel cans, but recycling activities have increased the availability of tin-coated steel cans. In 1993, 17.1 million tons of metal products were generated (5). Ferrous metals accounted for 75.5 percent of total metal production and aluminum and other nonferrous metals for 17.3 percent and 7.2 percent, respectively, that same year.

Glass.   Glassmakers return glass that is broken during manufacturing to the glass furnace. Many manufacturing plants also have a buy-back program for broken glass when their own supply is inadequate. Broken glass mixed at a ratio of 15% with raw materials is used for new product manufacture. As long as the recycled glass is the same color, it can be used without additional refining.

It is necessary to separate glass according to color and then run it through a crusher to break the bottles and remove the caps and labels. This reduces the volume of the glass and, consequently, the handling and shipping costs. The cost of long-distance shipping may make the glass recycling prohibitive.

Glass has been losing its share of the beverage-container market. As Table 14–1 shows, approximately 14.9 million tons of glass entered the waste stream in 1980, of which only 5 percent was recovered. In 1993, approximately 13.6 million tons entered the waste stream, and 22 percent was recovered. This decrease indicates the impact of lighter-weight aluminum cans and plastic bottles.

Some cullet (i.e., scraps of glass) is mixed with asphalt to form a new road-paving material called glasphalt. Cullet is also used as a road base material in place of gravel. Other products formed from cullet include glass-wool insulation, fiberglass, abrasives, light-weight aggregate for concrete, and so on. In Asia, broken glass is embedded on top of cement concrete walls built around property to provide privacy and security. Would-be intruders are discouraged from climbing over the wall, because the sharp glass pieces can cause injuries.

Construction and Demolition Wastes

Construction and demolition (C&D) wastes result from the construction, renovation, and demolition of buildings and from bridge and road repair. Wastes in this category include concrete, asphalt, bricks, stone, dirt, wood and associated products, painted or treated lumber, metals, glass, asbestos, plaster, insulation materials, and plumbing and heating items.

Only a small percentage of C&D wastes are recovered. Greater amounts will probably be recovered in the future, however, because of the increasing costs of landfills and because of mandatory legislation. Materials recovered from C&D wastes include asphalt, concrete, wood, and metals.

Asphalt. Asphalt waste comes from repairs and repaving. Old pavement material is processed with concrete and stones or by itself. It is crushed and screened, and this product is then used as a road base or is mixed with an asphalt binder and used as paving material.

Concrete. Most concrete is crushed and screened for use as a road base, or it is mixed with new concrete or used as aggregate in asphalt pavement.

Glass. Glass is crushed and graded. It is used as a road fill or is transported to a glass facility for use in making new products.

Wood. Wood wastes consist of lumber, treated wood, plywood, particle board, and asbestos. Clean wood is processed for fuel and landscaping. The remaining wood is shredded and passed through a classifier, where large pieces are separated. Magnets are used to remove any ferrous metals (e.g., nails). Smaller wood fragments are ground for mulch.

Metals. Steel used as reinforcement in foundations, slabs, and pavement is recovered and sold as scrap metal. Nonferrous scrap metal in gutters, siding, copper piping, aluminum frames, and so on can also be reclaimed.

Used Tires. Though only approximately 1 percent of U.S. MSW by weight, car tires, bus tires, truck tires, and tractor tires pose a major disposal problem (9). More than 2 billion used tires are stockpiled in the United States.. Every year, an estimated 237 million tires are discarded, 10 million are used, and 33.5 million are retreaded. Only approximately 14 percent of the discarded tires are used as a fuel. An estimated 5 percent of the discarded tires are used for rubber-modified asphalt and other uses, and 4 percent are exported.

Piles of tires are eyesores and cause environmental and public health problems. They are also a fire hazard and, when burned, can produce noxious black smoke and fumes. Piles of tires provide a breeding ground for disease-spreading pests as well. Water can collect in tires, thus providing an incubation site for mosquitoes. Rats and other rodents also make their homes in tire piles.

Many communities no longer allow tires to be discarded with other residential waste. Tires are classified as "Special Waste," and disposing of them may cost $3.00 or more a piece. Tire dealers, auto garages, and many landfill operators pay independent contractors to pick up used tires. Whole tires are no longer accepted at most landfills, because they take up a large volume of the landfill and also tend to rise to the surface. In rising to the top, they can disrupt and destabilize portions of landfill and break the clay caps that are placed over landfills.

Tire collectors separate the tires that can be reused after retreading. The remainder are stored or shredded for disposal in a landfill. Tires are also used for generating energy through incineration. Unfortunately, few practical uses have been devised for used tires. The most obvious way to reduce the number of tires entering the waste stream is to buy a better quality and make them last longer.

Because of their initial expense, many tires are retreaded or remanufactured. More than 800 U.S. companies retread or remanufacture. Retreading means replacing the tread, whereas remanufacturing means replacing both the tread and the sidewall rubber. Replacement tires for very large vehicles (e.g., earth-moving equipment, farm equipment, large trucks) are very expensive; therefore, many of these tires are retreaded several times.

As mentioned, tires can be incinerated or burned as a fuel in specially designed power plants. Approximately 33 million tires are burned annually in waste-to-energy and manufacturing plants. Tires can also be burned with other materials in more conventional power-generating incinerators or as a source of heat and energy for industrial operations. Tire burning has the potential for toxic air pollution, however, and has not found favor with the public.

Approximately 10 million used tires are utilized annually for miscellaneous purposes. These include use as playground equipment, as artificial reefs (especially in New Jersey and Florida), as bumpers on boat docks and piers, as sandals, and as floormats. Tires have also been used as a primary building material for breakwaters and erosion barriers on beaches. Split and punched tires can be used to make belts, gaskets, roofing, and molded items.

Asphalt Rubber.   A promising, large-scale, secondary use for tires is in "asphalt rubber," in which tires are ground up and blended with asphalt at 400°F to form a chemical bond. Paving contractors and state highway and city road departments generally do not like to use this material because of uncertainty regarding performance and concern about increased cost.

Oils, Solvents, Acids, and Metals

Oil Recovery.   Used lubricating oils can be recovered to a quality essentially equal to that of virgin lubricating oils. The dirt and sludge that build up in these oils make decontaminating and reclaiming the lubricant a complex chemical process, which is typically called oil re-refining. Used lubricants collected from automobile service stations and industrial plants contain a wide range of materials, including detergents, metal particles from the parts being lubricated, unburned fuel constituents, and combustion byproducts. The preferred method for re-refining used oil is distillation. Distillation of oil separates impurities, such as detergents, through vaporization and condensation.

Solvent Recovery.   Recovery facilities separate contaminants from waste solvents, thus restoring the solvent to its original quality or a to lower-grade solvent. Distillation is used by most commercial solvent processors, but other separation methods include evaporation, filtration, centrifugation, and stripping.

Acid Regeneration.   Acid recovery usually involves separation of unreacted acid from an acid waste (e.g., spent pickle liquor generated by steel mills). Impurities such as oxidized iron are removed as a precipitate by cooling the acid.

Metals Recovery.   Metals can be recovered by using differences in the melting and boiling properties to separate them at high temperatures. Another technology removes and concentrates metals from liquid waste by using processes such as precipitation, ion exchange, membrane filtration, electrodialysis, adsorption, reverse osmosis, and solvent stripping.

Yard Waste and Other Organics

Table 14–1 indicates that yard trimmings and food waste account for almost one-fourth of all MSW generated in the United States. Before the 1990s, most of this waste was disposed of in a landfill or incinerator. As current landfills run out of space and suitable sites for new landfills become more and more difficult to find, source reduction and recycling programs must be implemented. To reduce the amount of material going into a landfill, many communities now collect and process yard waste separately. The impact of this practice is evident from the increased recovery of yard trimmings, from negligible in 1980 to 12 percent in 1990 to 19.8 percent in 1993 (Table 14–1). More households now use yard waste for mulching or backyard composting, and municipalities use their own composting programs. Figure 14–5 shows bins for storing grass in a park.

Compost is a humuslike material that results from the aerobic, biological stabilization of organic materials in solid waste. It is useful as a soil conditioner, and it contains approximately 1 percent or less of the major nutrients. Brush and woody materials generally require shredding or chipping. Most major composting facilities are located near a landfill, where waste brought to the landfill serves as a major source. The science and methods of composing were discussed in Chapter 13.

Mulch can be produced by mixing leaves with shredded brush and tree prunings. It can be used for residential and commercial landscaping and to enhance plant growth to revegetate a landfill for its closure.

Figure 14–5   Bins for storing grass in a public park.

Yard waste can also be used as a biomass fuel. These wastes are ground and screened, and wood chips 0.5 in. or larger are sold as a fuel. Pieces smaller than 0.5 in. and green wastes are used for composting. (A case study follows at the end of this chapter.)

Markets for Recyclables

In the recycling business, the suppliers of collected materials and the demands of the manufactures who form the end-use markets are linked. Collectors, processors, brokers, and convertors have functions that include acquiring recyclable materials by either buying or salvaging, separating, and classifying them into some physical form. The material is then sold to the manufacturers.

Many communities do some or all of this processing at the recycling centers or materials recovery facilities (MRFs), which are either publicly or privately owned. There were approximately 900 such U.S. processing facilities in 1992 (10).

As with most commodities, the supply and corresponding demand of recyclable material tend to be cyclical. At times, supply exceeds demand; at other times, demand exceeds supply. The recycling movement has increased in popularity, but it has brought with it need to ensure that once materials are collected, there will be a market for them. Established, secure markets are vital to any successful MRF operation. The market economy is of concern in the political arena, where local governments are committed to recycling major portions of the solid waste stream and need to find markets for the recyclable materials to pay for the recycling programs.

Table 14–4 lists prices for recyclable materials as of November 1997 (8). There is significant volatility in prices for recyclables. At different times or in different locations, they are likely to be very different. For example, the supply of used aluminum

Table 14–4   National Average Prices for Recycled Materials

Material	November 1997 Price
Newspaper #6 ($/ton)	24
Corrugated cardboard ($/ton)	70
High-grade paper ($/ton)	73
Steel cans ($/ton)	1080
Aluminum cans (¢/lb)	56
Clear PETE, flaked (¢/lb)	100
Green PETE, flaked (¢/lb)	100
Natural HDPE ($/ton)	320
Glass ($/ton)	25

Source: Market Page, *Recycling Times,* November 1997; with permission.

cans is typically higher during the summer months compared with other seasons. During the summer, beverage consumption increases; consequently, the supply of cans also increases, which results in lower prices for empty aluminum cans. On the other hand, demand by the can makers for empty aluminum cans is good during the spring months, when additional metal cans are needed for recycling to meet the anticipated summer demands. Consequently, the prices rise.

During the last few years, industry has blamed successful residential curbside collection programs for causing a glut of waste paper—and the recession—in the waste paper market. The waste paper market has experienced dramatic downturns in the past, however. The industry experienced ups and downs long before residential collection programs were ever instituted, yet some industry managers today blame local governments for the potential problems of the waste paper business. Most local governments have had to develop aggressive recycling programs, as means of reducing operational costs, extending landfill life, and reducing the environmental hazards of landfilling.

Newsprint manufacturers are under pressure from the general public to use more recycled newsprint. Many offices are also being asked to use recycled paper. Manufacturers have both environmental and economic incentives for using recyclable materials instead of virgin resources. Environmental benefits such as reduced energy and water use are attained when aluminum, steel, paper, and glass products are manufactured using recycled materials (Table 14–5). Thus, manufacturers can realize economic benefits through reduced energy, decreased raw material costs, and lower pollution-control costs. There is a conflict, however. Many manufacturers have major investments in facilities and equipment that use virgin materials and either own the source of these raw materials or have long-term contracts for an adequate supply at a reasonable price.

Environmental Impacts

Table 14–5 illustrates the reduction in both pollution and energy use, from recycling, thus resulting in environmental benefit (11). When secondary materials are used in manufacturing, virgin resources are conserved. For example, recycled paper takes the place of paper made from virgin fiber obtained from trees, and when scrap iron is used in steel manufacturing, iron ore is saved. There are also limits and problems, however, associated with recycling and resource recovery.

Table 14–5   Environmental Benefits Derived from Substituting Secondary Materials for Virgin Resources

Environmental Benefit Reduction of:	% Reduction			
	Aluminum	Steel	Paper	Glass
Energy use	90–97	47–74	23–74	4–32
Air pollution	95	85	74	20
Water pollution	97	76	35	—
Mining wastes	—	97	—	80
Water use	—	40	58	50

Source: Pollock, C. *Mining Urban Wastes: The Potential for Recycling.* Worldwatch Paper 76. Washington, D.C.: Worldwatch Institute, 1987; with permission.

Limitations to the recoverability of materials result from physical and economic constraints. For example, paper fibers become shorter and weaker as they are reused, thus limiting the number of times they can be recycled. A certain amount of heat energy is lost when fuel is used, and that energy cannot be recovered. When metals are reprocessed, some portion is lost through oxidation.

An economic analysis should include the potential revenues and benefits of recycling *and* any possible economic limitations. If a reliable and satisfactory supply of secondary materials is not available at a price competitive with primary (i.e., virgin) materials, manufacturers will not use the former source. The most obvious source of revenues is the sale of recovered and/or processed materials, but if the processor of the secondary material cannot make profit, the recycling program will not work. Communities that are not required to do so by law may not want to implement a recycling program if the cost of collecting and processing the recyclables exceeds those of collection and disposal.

The effects of recycling are not always positive. Recycling involves reprocessing or remanufacturing materials that have negative environmental impacts. One example of how recycling carries potentially negative environmental impacts is the de-inking of waste paper. Colored inks used in magazines and newspapers may contain hazardous heavy metals (e.g., lead, cadmium). After the de-inking process, these constituents may be found in the wastewater treatment sludge, and if improperly disposed of, these metals could eventually leach from the sludge into the groundwater. On a positive note, many de-inking facilities are reducing or eliminating the use of hazardous cleaning agents. Collection of recyclable material also usually involves additional collection vehicles on the road, which could potentially cause increased traffic and affect air quality.

ECONOMIC ANALYSIS OF MRFs

The environmental and resource benefits of recycling are clear, but the economic value under the current market and regulatory conditions, public attitudes, and labor conditions are not. To make an economic assessment, consider the cost of collection, the cost of processing, the market value of the recycled materials, and the avoided landfill costs (i.e., tipping fees) from reducing the amount of waste sent for disposal. Table 14–6 lists the estimates of collection costs based on those for 10 actual MRFs (12).

Composition of Recyclables

To perform a cost analysis for an MRF, the composition of the recyclable materials is assumed. It is also assumed that commingled paper will arrive in the facility separated from commingled containers (e.g., aluminum, steel, plastic, glass).

Capital Costs

Facility Construction.   Estimated capital costs have been developed for facility construction (13). The difference between low and high cost ranges include project-specific elements such as subsurface conditions, local topography, structural construction materials (e.g., steel, concrete), and local building code requirements. Table 14–7 presents total and unit construction costs for a facility with a capacity of 100 tons per day. Economics of scale are achieved with larger facilities.

Equipment.   Estimated capital costs have also been developed for equipment (13). Table 14–8 presents these typical equipment costs for a throughput capacity of 100 tons per day. In this category, sorting systems include conveyors, trommel screens, and magnets. Processing systems include balers, HDPE granulators, and glass crushers.

Total Capital.   Table 14–9 presents estimated total capital costs for a throughput capacity of 100 tons per day. The information in this table is divided into facility construction

Table 14–6   Estimated Collection Costs for a Typical Curbside Recycling Program

Crew Size (n)	Set-Out Rate ($/ton)[a]		
1	175	134	123
2	164	126	115

[a] In 1997 dollars.

Source: Miller C. "The Cost of Recycling at the Curb," *Waste Age,* Vol. 24, No. 10, October 1993, pp. 46–54; with permission.

Table 14–7   Estimated Construction Cost Range for a Throughput Capacity of 100 Tons/Day[a]

	Cost		
	Low	*High*	*Average*
Total cost ($)	1,206,219	3,568,744	2,262,700
Unit cost ($/ton/day)	12,062	35,687	22,627

[a] In 1997 dollars.

Source: U.S. Environmental Protection Agency. *Material Recovery Facilities for Municipal Solid Waste.* Washington, D.C.: U.S. Govt. Printing Office, 1991; with permission.

Total 14–8   Estimated Equipment Cost for a Throughput Capacity of 100 Tons/Day[a]

Equipment	Low	High	Average
Sorting system	1,318,900	2,335,300	1,827,100
Processing system	516,543	731,281	623,912
Rolling stock	242,000	302,500	272,250
Installation	183,545	245,326	214,435
Contingency	226,098	361,440	293,770
Total equipment cost	2,487,086	3,975,847	3,231,467
Unit cost ($/ton/day)	24,871	39,758	32,315

[a] In 1997 dollars.

Source: U.S. Environmental Protection Agency. *Material Recovery Facilities for Municipal Solid Waste.* Washington, D.C.: U.S. Govt. Printing Office, 1991; with permission.

costs, equipment costs, and engineering fees. The ranges of total capital cost presented are at the upper end of the cost range for existing facilities, because most current facilities have inadequate floor area, buildings housing these facilities do not meet the current stringent building codes, and most do not have a sorting area and equipment for commingled mixed paper. Engineering costs are approximately 10 percent of the project costs.

Operating Costs

Labor.   A range of labor requirements based on facility throughput capacity is presented in Table 14–10. In general, labor requirements for sorting per ton of material decrease with increased capacity, because of the increased need for mechanical separation equipment (e.g., classifiers, separators). In addition, an MRF that receives separated material categories (e.g., clear glass vs. color-mixed glass) will require significantly fewer sorters than an MRF that receives commingled wastes.

Table 14–9   Estimated Total Capital Cost Range for a Throughput Capacity of 100 Tons/Day[a]

Item	Cost ($)		
	Low	High	Average
Construction	1,206,219	3,568,744	2,262,700
Equipment	2,487,086	3,975,847	3,231,467
Engineering	369,300	754,500	549,400
Total	4,062,605	8,299,091	6,043,600
Unit cost ($/ton/day)	40,626	82,990	60,436

[a] In 1997 dollars.

Source: U.S. Environmental Protection Agency. *Materials Recovery Facilities for MSW.* Washington D.C.: U.S. Govt. Printing Office, 1991; with permission.

Table 14–10   Estimated Labor Cost Range for a Throughput Capacity of 100 Tons/Day[a]

	Cost ($)		
Labor	Low	High	Average
Sorters	194,688	374,400	284,544
Other	239,520	359,424	299,520
Total	434,208	733,824	584,064
Annual cost ($/ton/day)	16.7	28.22	22.46

[a] In 1997 dollars.

Source: U.S. Environmental Protection Agency. *Materials Recovery Facilities for MSW.* Washington, D.C.: U.S. Govt. Printing Office, 1991; with permission.

Operations and Maintenance.   Table 14–11 presents estimated operations and maintenance costs. Debt service has been included based on an interest rate of 10% and amortized over 20 years for facilities and 7 years for equipment. Taxes and depreciation have not been included, however, because of their dependence on plant location and the tax structure of each particular business and financial arrangement.

Based on a 75-percent set-out rate and a two-man collection crew, the cost per ton for collection and processing of recyclables at an MRF averages $204 per ton ($115 for collection and $89 for processing). The economic question is how does the cost of collecting and processing recyclables, which seems to be approximately $204 per ton, compare with the revenues received when recovered materials are sold and avoided costs of disposal are taken into account? This comparison is not easy for a general case. Prices of recyclables vary significantly, both within a given year and from year to year. For example, clear PETE prices dropped an average of 13 percent from December 1996 to June 1997 (14). During the same period, clear glass dropped from $41.20 per ton to $34.40 per ton. Prices for recycled aluminum cans tend to rise in the spring because of the demand for beverages during the summer months. In the fall, prices drop as the

Table 14–11   Estimated Annual Operations and Maintenance Cost Range for a Throughput Capacity of 100 Tons/Day[a]

	Cost ($)		
Item	Low	High	Average
Operations and Maintenance	877,130	1,441,065	1,159,098
Residue disposal	78,650	314,600	196,625
Debt service	677,912	1,292,673	969,397
Total annual cost	1,633,692	3,048,338	2,325,120
Annual Cost ($/ton/day)[b]	62.83	117.24	89.43

[a] In 1997 dollars.

[b] Based on 260 working days per year.

Source: U.S. Environmental Protection Agency. *Materials Recovery Facilities for MSW.* Washington, D.C.: U.S. Govt. Printing Office, 1991; with permission.

supply of returned cans increases and the demand drops. Recently, newsprint prices have also dropped, and paper processors often end up paying mills to take it off their hands. In 1994, however, it was not uncommon for mills to pay close to $100 per ton.

Table 14–4 shows the prices paid for various processed, recycled materials in November 1997 (8). Prices such as these, along with estimates of the amount of materials recovered, provide a basis for evaluating the potential economic value of any recycling program. Table 14–12 provides an example of such estimates. The quantities given in this table are representative of resource-recovery rates based on national averages that are adjusted to 100 tons of recovered materials. Note that with only 3 percent of mass, aluminum provides 37 percent of revenues, whereas corrugated cardboard accounts for 48 percent of mass but only 38 percent of revenues.

The data in Table 14–12 indicates the revenues generated by the recycling program are approximately $88 per ton. Using an estimate of approximately $204 per ton to collect and process recyclables and of avoided disposal costs ranging from $22 to $46 per ton, depending on the region, the average net cost of recycling ranges from $70 to $94 per ton (15).

It is calculated as follows:

$$\text{net cost} = \text{gross cost} - (\text{income} + \text{avoided cost} + \text{other benefits})$$

$$= \$204 - (\$88 + \$22 \text{ to } \$46)$$

$$= \$70 \text{ to } \$94$$

Is this cost of recycling high and unnecessary? Should recycling programs be cut, because they are uneconomical? If the alternative is simply to send the MSW to a landfill, then the tipping fee and cost of long-term environmental damage (e.g., groundwater contamination) must be considered. So must the other benefits of recycling, such as increasing employment that, in turn, thus increases the community's tax base.

A general statement cannot be made, because there are many uncertainties and variations. Recycling has many benefits, but it is difficult to assign a dollar value to them. If the market forces are favorable (i.e., the price of materials is high enough), recycling programs can make money. If the collection costs can be reduced, recycling

Table 14–12   Estimates of Recycling Rates and Revenues for 100 Tons of Recovered Materials

Materials	Mass (tons)	Price ($/ton)	Revenue ($)	% of Revenue
Newspaper #6	23	24	552	6
Corrugated cardboard	48	70	3,360	38
High-grade paper	8	73	584	7
Steel cans	5	56	280	3
Aluminum cans	3	1080	3240	37
Clear PETE	1	100	100	1
Green PETE	1	100	100	1
Natural HDPE	1	320	320	4
Glass	10	25	250	3
Total	100	1848	8,786	100

programs can be of economic value as well. Note that the collection cost is significantly higher than the processing cost ($115 vs. $89). Each community has different circumstances. In some, recycling will make money; in others, it will not.

CASE STUDY

Marion County Recycling Center

The Marion County Recycling Center is located in San Rafel, California. It provides curbside collection of recyclable materials and has a dump-site separation program. The facility was started by Joseph J. Garbarino out of concern for the loss of valuable, recyclable materials and the decreasing capacity of the local landfill.

The waste processing facility was financed by West American Bank and built in 1986. The purpose was to recover the valuable resources and, at the same time, extend the life of the landfill. At the time, participation in the recycling service was on a volunteer basis. Approximately 50 percent of the county's population participated in the beginning, but this gradually increased to more than 75 percent. Initially, 30 percent of the recyclable materials were recycled, but now the aim is to recycle 100 percent.

The recycling center's main building is a steel-frame structure that is 294 ft wide, 461 ft long, and 54 ft wide and houses the recycling facility. Its floor is built to resist wear and tear caused by heavy machinery and abrasive material brought into the facility. Figure 14–6 shows recyclable materials on the facility's floor, which consists of a dump area with 17 belt conveyors. The first system of conveyors is used for handsorting recyclables such as paper and glass. The second system is used for wood grading, and the third is used as waste (i.e., nonrecyclable) conveyors. Waste that cannot be recycled is collected in bins and sent to

Figure 14–6   Recyclable material on the recycling center's floor.
(Courtesy of Marion County Recycling Facility.)

a transfer station on the way to the landfill for disposal. Figure 14–7 shows a close-up view of the handsorting.

Figure 14–8 shows a pile of the waste material before sorting, and Figure 14–9 shows a sorting line for paper, cardboard, and wood. After sorting, the paper and cardboard are baled by an automatic baler. The bales are then sold, mostly overseas, for use in making paper products.

Wood scrap is sent up a rubber-belted conveyor that is lined with magnets. Figure 14–10 shows wood or yard waste being loaded. The magnets remove any type of metal pieces (e.g., nails) in the wood, which is then ground and sent to a shaker table. Figure 14–11 shows wood being fed into a grinder. Any dust or fines are collected and sold as potting soil or soil conditioner. The rest of the wood is ground into chips, which are sold to paper mills. Figure 14–12 shows a nail-removing device and wood-chip conveyor; Figure 14–13 shows chips coming out of a grinder.

The top soil and chips made from waste wood are marketed. Figure 14–14 shows the two bins. The chips are sold to paper mills for use as fuel to produce steam for driving the generators that produce electricity and the top soil is used for landscaping.

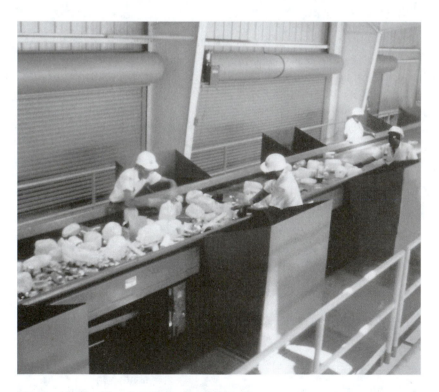

Figure 14–7   Handsorting of solid waste.
(Courtesy of Marion County Recycling Facility.)

Figure 14–8   Solid waste pile before sorting.
(Courtesy of Marion County Recycling Facility.)

Figure 14–9   Solid waste sorting line.
(Courtesy of Marion County Recycling Facility.)

Figure 14–10   Loading yard waste and wood waste.
(Courtesy of Marion County Recycling Facility.)

Figure 14–11   Wood being fed into a grinder.
(Courtesy of Marion County Recycling Facility.)

Figure 14–12   Nail-removal mechanism and conveyor.
(Courtesy of Marion County Recycling Facility.)

Figure 14–13   End product from a grinder.
(Courtesy of Marion County Recycling Facility.)

Figure 14–14   Top soil and wood chips processed from waste wood.
(Courtesy of Marion County Recycling Facility.)

Different types of metals are sorted into separate bins and sold as scrap metal. Glass is also separated by hand into different bins according to color. The glass can then be recycled to make new products.

The Marion County recycling program is a success story. It makes money by selling the products processed from the recyclable materials and by saving the cost of landfilling these materials. Thus it is able to pay $130,000 per month mortgage on the facility's building loan, pay salary to employees, and still make a profit.

REVIEW QUESTIONS

1. a. What is the difference between reuse and recycling?
   b. Give examples of recycling and reuse.
2. a. List three toxic materials in MSW.
   b. Suggest ways to reduce the toxic materials in MSW.
3. Assume that you plan to start a recycling center:
   a. How would you organize an effective recycling program?
   b. List the items likely to generate money for the program.
4. a. List some common types of plastics and their characteristics.
   b. What problems are associated with recycling plastics?

5. a. Why do landfills refuse to take tires or charge more for their disposal?
   b. List three uses for old tires in engineering works.
6. a. How does the market economy affect the recycling efforts of a city?
   b. List the cost factors involved in a recycling program.
7. Discuss the environmental impacts of a recycling program.
8. Check the quantity and price of newspaper available for recycling in your town, and compute the cost of collection and processing for recycling with the help of the cost data given in this chapter.

REFERENCES

1. U.S. Environmental Protection Agency. *Agenda for Action.* Washington, D.C.: U.S. Govt. Printing Office, 1994.

2. Keep America Beautiful, Inc. *The Role of Recycling in Integrated Solid Waste Management to the Year 2000.* Prairie Village, KS: Franklin Associates, 1994.

3. U.S. Environmental Protection Agency. *Characterization of Municipal Solid Waste in the U.S.: 1990 Update.* Washington, D.C.: U.S. Govt. Printing Office, 1990.

4. Johnson, R., and C. Hirth. "Collecting Household Batteries," *Waste Age,* Vol. 21, No. 6, June 1990, pp. 48–49.

5. U.S. Environmental Protection Agency. *Generation and Recovery of Materials in the U.S.A.* Washington, D.C.: U.S. Govt. Printing Office, 1994.

6. Pfeffer, J. *Solid Waste Management Engineering.* Upper Saddle River, N.J.: Prentice-Hall, 1994.

7. U.S. Environmental Protection Agency. *Household Hazardous Waste Management: A Manual for 1-day Community Collection Programs.* EPA/530-R-93-026. Washington, D.C.: U.S. Govt. Printing Office, 1993.

8. Market Page, *Recycling Times,* November 1997.

9. Pillsbury, H. "Markets for Scrap Tires on EPA Assessment," *Resource Recycling,* Vol. X, No. 6, June 1991.

10. Steuteville, R., and N. Goldstein. "The State of Garbage in America, Part I," *Biocycle* May 1993, pp. 42–50.

11. Pollock, C. *Mining Urban Wastes: The Potential for Recycling.* Worldwatch Paper 76. Washington, D.C.: Worldwatch Institute, 1987

12. Miller, C. "The Cost of Recycling at the Curb," *Waste Age,* Vol. 24, No. 10, October 1993, pp. 46–54.

13. U.S. Environmental Protection Agency. *Material Recovery Facilities for Municipal Solid Waste.* Washington, D.C.: U.S. Govt. Printing Office, 1991.

14. U.S. Environmental Protection Agency. Editor's Page, *Waste Age,* Vol. 28, No. 9, September 1997.

15. U.S. Environmental Protection Agency. *Economic Analysis of Materials Separation Requirement, Emissions Standards Division.* EPA-45013-1991. Washington, D.C.: U.S. Govt. Printing Office.

Chapter 15 Disposal of Solid Waste and Residual Matter in Landfills

This chapter considers the site selection, layout, design, construction, and operation of landfills. A landfill is defined as a system that is designed and constructed to dispose of discarded waste by burial in land to minimize the release of contaminants to the environment. Landfills are currently a significant part of municipal solid waste (MSW) management and hazardous waste management practice, and they are likely to remain so in the future.

Since early in the twentieth century, landfills in one form or another have been the most economical and acceptable method for solid waste disposal in the United States and throughout the world. Many landfills constructed in the past have been associated with dumps, because waste was deposited with little consideration for public health and the environment. Unfortunately, that impression of a "dump," which attracts flies, rats, gulls, and produces odors and smoke as well as noise and airborne paper and garbage, is deeply rooted in the minds of many people. Thus, the not-in-my-backyard attitude toward a new site comes as no surprise, even though modern landfills are engineered facilities.

Landfills have been the most widely used method of waste management in the United States; approximately 80 percent of the nation's MSW is landfilled (1). Many communities are finding it difficult to site new landfills, however. This difficulty has resulted from increased concern among citizens and government regarding the adverse environmental impact associated with improperly located, designed, and operated landfills. The "3 R's"—reduction, reuse, and recycling—are beginning to have some effect on public opinion and on the desirability of landfilling all the waste.

Today, landfill management incorporates the planning, design, operation, environmental monitoring, closure, and postclosure control of landfills. Modern solid waste landfills are now subject to increased regulatory scrutiny, and many sites (mostly small facilities unable to meet the new Resource Conservation and Recovery Act (RCRA), Subtitle D landfill regulations) have been closed. Therefore, the number of MSW landfill sites has declined from 8000 in 1988 to 3558 in 1994 (2).

LANDFILL CLASSIFICATION

A number of landfill classification systems have been proposed. The following is used in this book:

Class	Designed to Handle
I (secure landfills)	Hazardous waste
II (monofills)	Designated waste
III (sanitary landfills)	MSW

Some aspects of hazardous waste and MSW landfills are common, but hazardous waste landfills are discussed in Chapter 19. This chapter deals only with Class III, MSW landfills.

Most U.S. landfills are designed for commingled MSW. These Class III landfills may include nonhazardous industrial wastes and sludge from water and wastewater treatment plants. These sludges are accepted for landfill after they have been dewatered, however, to increase the solid content to 51 percent or more and to remove all free-flowing liquid. Federal regulations no longer allow liquid wastes into MSW landfills.

SITING CONSIDERATIONS

Proper siting of sanitary landfills is crucial to providing economical disposal while protecting human health and the environment. The process of selecting a landfill site is complex, and it involves four major issues: data collection, location constraints, assessment of public reaction, and area requirements.

Data Collection

Many maps and other information (e.g., solid waste volume, landfill volume) must be studied to obtain data within the search area (Table 15–1). The search area is usually indicated by a circle, which is the maximum distance the waste can be hauled economically. Hauling MSW is one of the costlier items in landfill operation. Wenger and Thyner (3) suggest that once a landfill site is identified, the waste generation points within the search area that can be served by the landfill at a reasonable cost can be identified. Both the landfill cost and the hauling cost are considered in creating an optimal service region.

Location Constraints

A search for a suitable landfill site typically begins by eliminating environmentally unsuitable locations. In general, certain types of land are environmentally unsuitable. These include flood plains, wetlands, land near airports, geological fault zones, seismic impact zones, or other unstable areas. State and local regulatory agencies are also known to specify additional restrictions, such as maintaining specified distances from highways, ponds, lakes, parks, or water supplies because of the threat of contaminating ground or surface water or of endangering plant or animal species. Table 15–2 lists various siting limitations.

Table 15–1   Information Needed in Landfill Site Selection

Items	Uses/Needs	Possible Sources
Topographic maps	Show the high and low areas, natural drainage, streams, and wetlands. Can help locate a site that is not a natural drainage or wetland.	U.S. Geological Survey State Geological Survey
Soil maps	Show the type of surface soil. Availability of soil is important.	Local County Extension Agent U.S. Soil and Conservation Service U.S. Department of Agriculture
Transportation maps	Show the locations of various roads, railways, airports. Used to determine the transportation needs in developing a site (e.g., hauling distance). All-weather-access roads for transportation of solid waste are important.	State Highway Department U.S. Department of Transportation County and City Officials
Food plain maps	Delineate areas within a 100-year or a 500-year flood plain.	U.S. Corps of Engineers U.S. Geological Survey County Extension Agent
Geologic maps	Indicate geologic features from which general ideas about soil type can be developed. Hydrogeologic conditions should also be noted to protect against groundwater contamination.	U.S. Geological Survey Well logs of existing wells U.S. Soil and Conservation Service State Geological Surveys

Plans

Land use plans	For delineating areas with zoning or other restrictions. Used to locate sites that are not restricted areas and to meet the zoning criteria within the search area.	Comprehensive plans from local planning or zoning commissions
Water use plans	Indicate public and private wells, other water-supply sources and facilities, and their capacities. Needed to satisfy regulatory requirements of minimum distance from the proposed landfill.	Usually not readily available City Utilities or Water Departments
Aerial maps (photographs)	Show surface features such as small springs, small lakes, intermittent stream beds, and so on that may not appear on other maps.	May not be available for the areas of interest U.S. Geological Survey.

Waste

Waste type and volume	Waste type should be known and characterized. Waste volumes for industrial waste can be estimated from production and disposal records. For new plants, study of similar, existing plants can be helpful. Possible future expansions should be noted.	Long-term study of waste generation (characteristics, volumes). Industries or Public Works Departments
Landfill volume	Used for estimating daily landfill cover.	
Recycling and incineration options	Technical and economic feasibility and regulatory requirements should be considered. Recycling is mandated in most cases.	Federal and state regulatory requirements

Existing disposal	Available landfill volume nearby should be considered. The cost of disposing in the existing landfill could be less than planning, designing, developing, and operating a new landfill.	State and local solid waste management authorities
Financing		
Financing	Should be discussed with persons responsible for budgeting.	Consulting engineers in the solid waste disposal field. Finance or Budget Departments

Assessment of Public Reaction

The public should be informed regarding the possibility of a landfill in their area as soon as a list of potential sites is developed. The public is less suspicious and more open to discussion if they are informed by the landfill owner or municipal officials rather than by other sources. In many situations, public hearings are used to air concerns about a landfill site. The most commonly expressed concerns relate to odors, traffic, noise, health hazards, property values, birds, dust, debris, and leachate. The time and place for such hearings should be publicized, and the best approach is to respond to each concern in the siting study.

Area Requirements

In selecting potential sites, it is necessary to ensure that sufficient land area is available. There are no fixed rules concerning the area required, but it is desirable to include an adequate buffer zone for a designated time period (typically 10 years or longer). Smaller sites rarely justify the investment in development and operating costs. Site size is based on the quantity of solid waste to be deposited in the proposed landfill during its lifetime. This quantity depends on the waste generation rates of the population and the institutional, industrial, and commercial units in the service region and on the recycling, reuse, and recovery activities. In addition, the potential for future expansion should be considered.

Additional land is required for a buffer zone, access roads, office and service buildings, and utilities. Typically, this allowance is made by using factors ranging from approximately 1.25 for the area method and 2.0 for the trench method. A reasonable "rule of thumb" estimate for the annual volume requirement of an MSW landfill is approximately 1.4 acre-ft per 1000 persons (4).

Example 15–1

Estimate the volume of a landfill and the accompanying space needed to receive MSW from a community of 30,000 persons. Assume the following conditions:

$$\text{per-capita solid waste generation} = 6 \text{ lbs/day}$$

$$\text{compacted specific weight of solid waste in landfill} = 1000 \text{ lbs/yd}^3$$

$$\text{average depth of compacted solid waste} = 16 \text{ ft}$$

Table 15–2   Siting Limitations Contained in Subtitle D of the RCRA and Other Limitations

Location	Siting Limitations
Airports	Landfills must be located at least 10,000 ft from an airport used by turbojets and 5,000 ft from an airport used by piston-type aircraft. Closer landfills must demonstrate they do not pose a bird hazard to aircraft.
Flood plains	Landfills located within a 100-year flood plain must be designed not to restrict flood flow, reduce the temporary water storage capacity of the flood plain, or result in the washout of solid waste, which would pose a hazard to human health and the environment.
Lakes or ponds	No landfills should be constructed within 1,000 ft of a lake or pond.
Wetlands	Landfills are prohibited within wetlands.
Rivers	No landfill should be constructed within 300 ft of a river or stream. Exceptions may sometimes be made for nonmeandering rivers, but a minimum of 100 ft should be maintained.
Water-supply wells	No landfill should be constructed within 1,200 ft of any water-supply well.
Highways	No landfill should be constructed within 1,000 ft of the right-of-way of any state or federal highway unless trees or berms are used to screen the site.
Public parks	No landfill should be constructed within 1,000 ft of a public park unless some kind of screening is used.
Critical habitat area	No landfill should be constructed within critical habitat areas (i.e., the area in which one or more endangered species live).
Fault areas	New landfill units cannot be constructed within 200 ft of a fault line that has had a displacement during the past 10,000 years.
Seismic impact zone	New landfills within a seismic impact zone must demonstrate that all containment structures (e.g., liners, leachate collection systems) can resist the maximum horizontal acceleration.
Unstable areas	Landfills in unstable areas must demonstrate stability of structural components. Unstable areas include landslide prone, possibility of sinkhole formation, and undermined by subsurface mines. Existing facilities that cannot demonstrate stability must close within 5 years of the regulation's effective date.

	Degree of Limitations		
	Severe	Moderate	Minimal
Land slope	>15%	3–15%	<3%
Surface	Clean sand/gravel	Sand/gravel	Silty
Deposits	Heavy organic clay	Silt	Clay
Bedrock depth	<10 ft	10–25 ft	>25 ft
Bedrock type	Fractured limestone	Sandstone	
Groundwater depth	<10 ft	10–25 ft	>25 ft

Source: Data from *Waste Age*, Vol. 17, No. 6, June 1986, p. 88.

Solution

First, determine the daily solid waste generation rate in tons per day:

$$\frac{30,000 \ people \times 6.0 \ lbs/person/day}{2000 \ lbs/ton} = 90 \ tons/day$$

Thus, the volume required per day is

$$\frac{90 \ tons/day \times 2000 \ lbs/ton}{1000 \ lbs/yd^3} = 180 \ yd^3/day$$

and the area required per year is

$$\frac{180 \ yd^3/day \times 365 \ day/yr \times 27 \ ft^3/yd^3}{(16 \ ft) \ (43,560 \ ft^2/acre)} = 2.55 \ acres$$

Note: The actual site requirement will be greater than the value computed because additional land is required for a buffer zone, access roads, office and service buildings, utilities, and so on. As mentioned, this allowance typically is made by using factors ranging from approximately 1.25 for the area method to 2.0 for the trench method. Therefore, the land required for the area method is $2.55 \times 1.25 = 3.18$ acres/year, and for the trench method is $2.55 \times 2.0 = 5.1$ acres/year.

SITE SUITABILITY

After estimating the area for the proposed landfill, the search for a suitable site begins. This search often starts by ruling out environmentally unacceptable areas. Subsurface conditions such as types of soil, underlying rock strata, and groundwater conditions are important factors for determining if an environmentally safe landfill can be built economically at a specific site.

Siting criteria can be divided into two groups: exclusive, and nonexclusive criteria. Exclusive criteria include federal, state, and local restrictions and physical restrictions. If a site fails to meet any exclusive criterion, it is excluded from consideration. For nonexclusive criteria, a graphical approach for selecting potential landfill sites within a large geographical area is to prepare a series of overlay maps (Figure 15–1). On transparent sheets, a separate map is prepared for each major evaluation factor. Those areas having limitations for landfilling are darkened. These transparent overlays with darkened, restricted areas can then be placed over the base map, such as a U.S. Geological Survey quadrangle map. The clear areas that show through have the lowest restrictions and can be considered as potential landfill sites (5).

After the exclusive criteria are applied, a limited number of potential landfill sites are selected from those areas that remain. The final selection process uses nonexclusive criteria, such as hydrogeological conditions, hauling distance, site accessibility, and land use.

Soil Properties

The types and quantities of soil available are significant factors in the cost of operating a landfill. Soils are needed as a cover and as a moisture barrier in the landfill bottom. The soil should also be able to support the equipment used to transport and place the solid waste, and on completion of a landfill, a soil capable of supporting a good vegetative cover is provided. Chapter 1 briefly discussed soils, but the soil properties

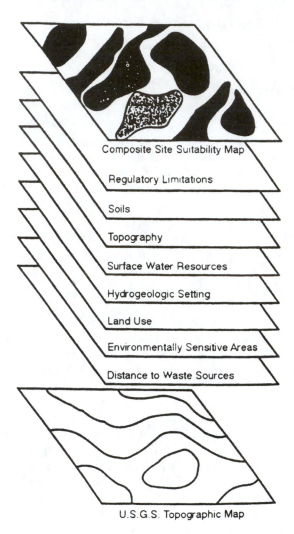

Composite Site Suitability Map

Regulatory Limitations

Soils

Topography

Surface Water Resources

Hydrogeologic Setting

Land Use

Environmentally Sensitive Areas

Distance to Waste Sources

U.S.G.S. Topographic Map

Figure 15–1   Overlay maps for various site criteria in the screening of potential landfill sites.

(From Siddiqui, M. Municipal Solid Waste Landfill Site Selection using Geographical Information Systems. Master's Thesis, University of Oklahoma, 1994; with permission.)

that are important in a sanitary landfill relate to those activities associated with operating the landfill. Such properties include permeability, swelling and cracking, and support of vegetation.

Permeability.   Soil permeability is a function of the particle size and distribution. A soil with low permeability will prevent the passage of water into the landfill and the loss of leachate from it. A tight clay is effective for this purpose; silty clay is still satisfactory but less so than the tight clay. Porous soils with large, mostly uniform, particle sizes are very suitable for the control of landfill gas. Such media permit the gas to flow to the desired location by providing a path of least resistance. These soils can also serve as drainage layers to direct the flow of liquids within the landfill and cover.

Swelling and Cracking.   Certain soils are prone to swelling when wet, and when these soils dry, they crack. Examples include silty clays and clay. Clays with a high organic content have a plastic characteristic and very compressible under load. When wet, such soils are unsuitable for supporting operational vehicles.

Support of Vegetation.   When a landfill is completed, a final cover of soil is placed. This soil must support good vegetation to protect against erosion and to dissipate the water that may infiltrate into the top layers of the cover. These soils absorb and retain appreciable quantities of moisture and plant nutrients needed for growth of the vegetative cover. Suitable soils include silt, sandy silt, and clay-sandy silt. Thus, a landfill cover may consist of several layers of soil, each with a separate purpose.

Hydrogeological Properties

To determine the best possible location for a solid waste facility, a potential site must be evaluated for hydrogeological conditions, technical and engineering features, and site-specific characteristics. To predict the fate of leachate that leaves a landfill site and its potential for contaminating groundwater, understanding how water flows through subsurface materials is necessary. Characterization of the contamination leaching from a waste disposal site involves groundwater movement of all types of substances dissolved in groundwater.

The objective is to reduce the possibility of groundwater contamination by leachate from the fill. This can be achieved by maximizing the distance between the landfill base to the groundwater table and by having the maximum possible depth of impervious material between the landfill and the water table. Table 15–2 indicates the desirability of such conditions, and Figure 15–2 illustrates groundwater flow beneath a landfill. The deep aquifer is protected by impervious strata and is at a considerable distance below the surface; therefore, the travel time is very long. The water level will mostly remain constant. Contamination of this aquifer is unlikely because of the reasons stated earlier. A shallow aquifer, however, is more vulnerable, because there is no protective layer and the distance between it and the landfill is much less than that between a deep aquifer and the landfill. A shallow aquifer has a high probability of contamination if the landfill is not designed and constructed to minimize production of leachate and to contain all that may be produced.

As discussed, it is important to select a landfill site relative to the groundwater table to prevent contamination. A shallow aquifer is replenished by local precipitation. Consequently, a shallow aquifer's level may vary depending on the local precipitation. A landfill built on a high ground (i.e., the distance to the aquifer is considerable) is less likely to contaminate the groundwater. As illustrated in Figure 15–3, however, if the landfill site is closer to a stream, it is possible for the groundwater to find its way

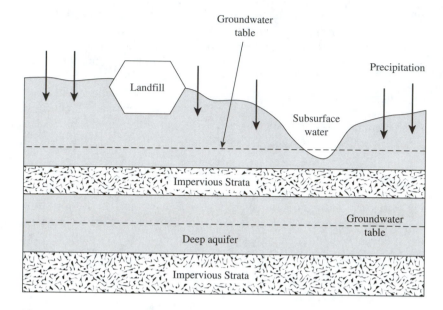

Figure 15–2   Hydrogeological considerations in landfill siting.

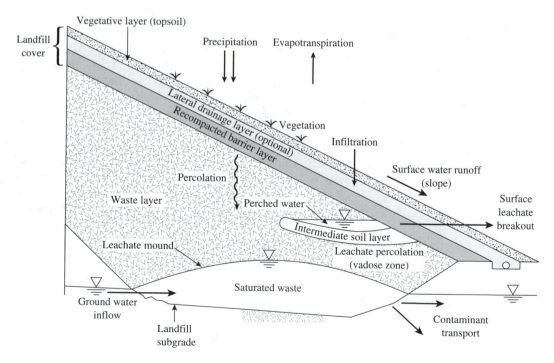

Figure 15–3   Typical mechanisms of leachate flow.

into the landfill during periods of precipitation. Rupture of the lining may also occur, allowing the landfill to become saturated with groundwater.

Figure 15–3 also illustrates another local groundwater problem. An impervious layer of soil (e.g., as clay lens) above the main body of groundwater may be a few feet below the surface and cover a significant area. Downward-percolating water is trapped above this layer, thus creating a perched water table. The level of this water may vary significantly depending on the precipitation. It may come very close to the surface during periods of heavy precipitation, and groundwater may penetrate the landfill. When the water level drops, the contaminated water then drains from the fill.

The goal of hydrogeological investigation is to ensure a site is suitable to retain the solid waste. In most cases, a liner and or recovery system is employed. As part of this investigation, soils that may be suitable for use as a liner are sought. Potential cover soils are investigated as well. Obviously, a complex geological formation will need more extensive investigation than a uniform formation.

Table 15–1 indicates that much of the information on hydrogeological conditions within a given area is available from the the state geological survey or the U.S. Soil Conservation Service. The drilling logs from water, oil, and gas wells are also sources of information and data. This valuable information is used to map the site and to guide the initial stages of site selection. This initial study will eliminate sites that are unsuitable because of water or soil problems; those remaining then become potential sites. On-site drilling is necessary for these potential sites. The soil-boring, test-pit logs and the collection of soil samples are used to identify the subsurface geology and the physical characteristics of the soil. The soil should also be tested for hydraulic conductivity gradation, moisture–density relationship, and Atterberg limits (i.e., the effect of moisture on soil properties). Each state permitting agency has a required number of spacing for test drillings. Soil borings on a 100- to 200-ft grid may be necessary to characterize the subsoil sufficiently. The borings should extend 20 ft or more below the intended bottom of the site. Borings located on the periphery of the site can be converted into monitoring wells after the site is in use. Groundwater monitoring wells are used to identify groundwater elevations and direction of flow and to collect water samples for testing.

In addition to horizontal movement, groundwater and potential contaminants from the landfill may move vertically within the subsurface. This movement can be detected by use of groundwater monitoring wells. Precipitation over the landfill recharges the subsurface flow, which then moves almost horizontally and then may move vertically along with the contaminants to a discharge area. Figure 15–3 illustrates this subsurface movement.

If the landfill is in a recharge area, percolation of water into the soil will cause movement of any contamination released from the landfill. Thus, surface water infiltration into the landfill must be prevented or at least minimized, and a well-designed containment and removal system should be installed. In addition, all the surface water from the surrounding area should be drained away from the landfill.

A landfill in a discharge area may result in the flow of groundwater into the landfill. An underdrain system or a well field surrounding the area should be installed to remove the groundwater inflow before it reaches the landfill. If the groundwater is not controlled, it will find its way into the landfill, and the leachate may flow from the site to contaminate the surface water.

LANDFILLING TECHNIQUES

After the site-selection process and consideration of environmental factors such as leachate and gas controls, exactly how the landfill will be built is determined. Various titles are used to describe landfilling, but only two basic techniques are involved: the area method, and the trench method. At many sites, both methods are used either simultaneously or sequentially.

Area Method

In the area method, solid waste is deposited on the surface, compacted, and then covered with a layer of compacted soil at the end of the working day. This method is suitable for most terrain. Site preparation includes installation of a liner and leachate control system; therefore, the subsoil conditions must be completely known. Figure 15–4 illustrates the area method. An embankment with a 3:1 slope is built from excavated

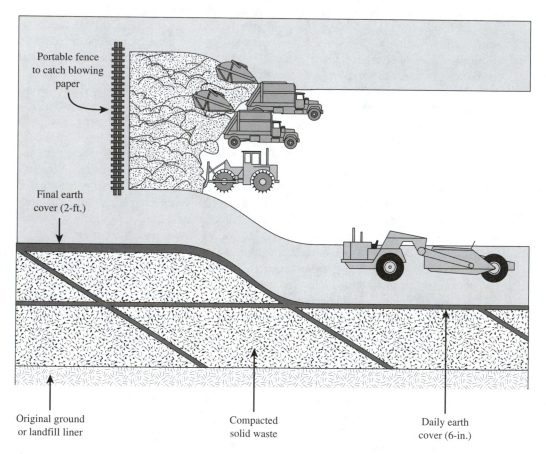

Figure 15–4   Area method of sanitary landfilling.

(Adapted trom Brunner, D.R., and J.J. Keller. Sanitary Landfill Design and Operation. *Publication SW-65. Washingtin, D.C.: U.S. Govt. Printing Office, 1972; with permission.)*

soil and becomes the starting point for MSW placement. Waste is deposited at the toe of the working face and is then spread by compactors in layers of approximately 1 to 2 ft. Compaction is done to achieve maximum MSW density in the landfill. The process of layering is continued, and the vertical lift of the waste may range from 10 to 14 ft. The waste is then covered by spreading a 6- to 12-in. layer of soil over the working face. The top of the intermediate lift is covered with at least 12 in. of soil. This cover becomes a base for the trucks and equipment for the next lift. After the last lift, a final cover, which is designed to prevent moisture infiltration of the waste, is placed. The soil-cover thickness may be 2 ft or more.

A variety of earth-moving equipment is used in area landfills. Scrapers are used to excavate the area, stockpile the soils, and to spread the cover soil on complete areas. Tractors and compactors are used to spread and compact the waste and the cover material. Road graders are used to maintain the access road. Figure 15–5 illustrates some typical equipment; Table 15–3 provides their performance characteristics.

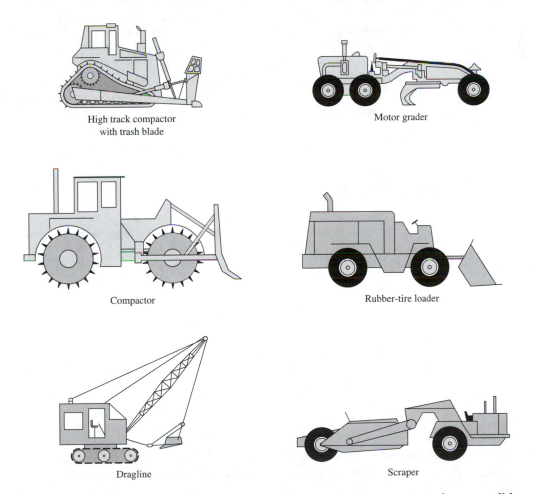

High track compactor
with trash blade

Motor grader

Compactor

Rubber-tire loader

Dragline

Scraper

Figure 15–5   Typical equipment used at landfills to place, compact, and cover solid waste.

Table 15–3   Performance Characteristics of Landfill Equipment

| Equipment | Solid Waste | | Cover Materials | | | |
	Spreading	Compacting	Excavating	Spreading	Compacting	Hauling
Wheeled compactor	Excellent	Excellent	Poor	Fair	Excellent	—
Rubber-tired loader	Good	Good	Fair	Good	Fair	—
Scraper	—	—	Good	Excellent	—	Excellent

Trench Method

In the trench method, solid waste is spread and then compacted in an excavated trench that may be 10 to 15 ft deep. A tractor/compactor spreads the solid waste by pushing it up the slope while simultaneously compacting it. The equipment makes several passes to achieve maximum density. The cover material is excavated by extending the trench, which creates space for the solid waste. Figure 15–6 illustrates the trench method of construction. The cover material is spread and compacted over the waste to form the basic cell structure. Spoil material not needed for a daily cover, however, is usually stockpiled for later use. The excavated soil must be of the quality needed for a cover. A clay or clay silt that can be used for a layer of relatively impervious soil and a silt that can be placed atop the clay to support a vegetative cover should be satisfactory.

The bottom of the trench should be slightly sloped for drainage, and provision should be made for surface water to run off at the low end of the trench. Excavated soil can be used to form a temporary berm on the sides to divert surface water. The trench can be as deep as the soil and groundwater conditions permit.

Combination of Methods

Both the area and trench methods have advantages depending on the location and amount and type of solid waste to be handled. For example, the area method can be

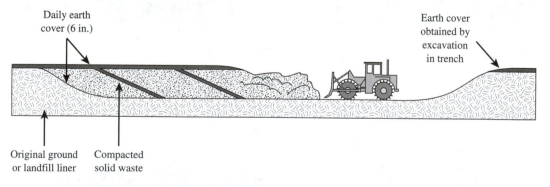

Daily earth cover (6 in.)

Earth cover obtained by excavation in trench

Original ground or landfill liner

Compacted solid waste

Figure 15–6   Trench method of sanitary landfilling.

(Adapted from Brunner, D.R., and J.J. Keller. Sanitary Landfill Design and Operation. *Publication SW-65. Washington, D.C.: U.S. Govt. Printing Office, 1972; with permission.)*

used for large-volume operations and where no excavation below the original grade is feasible. The trench method permits variation in terrain conditions, use of on-site material for a cover, and can be adapted for a wide variation in the size of operation. The two methods have also been used at the same site.

Commonalities

All landfills have certain commonalities. Buffer zones and screens are necessary to isolate them from people living close by, and planting proper trees and shrubs sometimes provides such a buffer. An all-weather access road from prime roadways is mandatory as well. This road should be maintained so that it can be used for heavy vehicles even during the inclement weather. Figure 15–7 shows a modern landfill system.

The building block common to all methods of landfill construction is the cell. All the solid waste received is spread and compacted in layers within a confined area, and it is covered at the end of each working day. Figure 15–8 shows a construction of a typical landfill cell. (Landfill components are discussed in Chapter 16 as well.)

The cell dimensions are determined by the volume of the compacted waste, which in turn depends on the density of the in-place solid waste. The field density of the

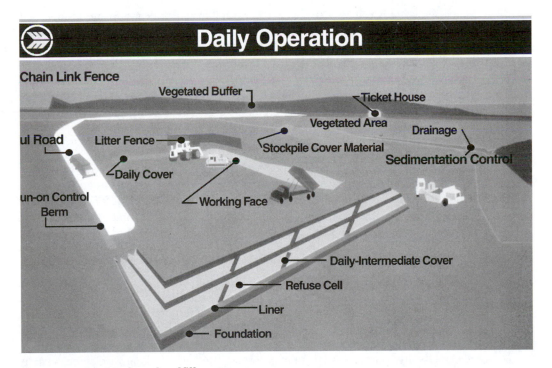

Figure 15–7 Modern landfill system.

(Courtesy of the Waste Management Company.)

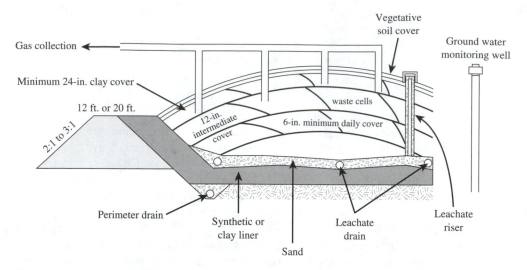

Figure 15–8   Detail of a typical landfill cell.

compacted waste is usually between 800 and 1200 lbs/yd^3. Because materials normally tend to rebound when compacting load is released, it is desirable to spread in layers of less than 2 ft in depth.

No definite rules or guidelines exist regarding the proper height of a cell. In general, this height ranges from 8 to 14 ft, but heights of 30 ft or more are common in large operations. Volume requirements for the cover material depend on the surface area of the waste being covered. Therefore, the cell surface areas to be covered should be kept to a minimum.

Cover Material Requirements

Cover material is incorporated into a landfill at each stage of its construction. A cover of 6 in. to 1 ft of soil is applied to the working faces of the landfill at the end of each working day. The cover prevents insects and rats from getting into landfill, and it stops material from blowing away. Final covers are usually 3 to 6 ft in thickness and include a layer of compacted clay as well as other layers for drainage and support of surface vegetation.

The amount of cover material required for a landfill is a function of the size of the cell. The ratio of cover material to solid waste in the cell decreases as the cell size increases. The quantity of cover material is an important factor in determining the capacity of a landfill site. Typically, daily and interim cover needs are expressed as a waste-to-soil ratio (i.e., volume of waste deposited per unit volume of cover provided). Usually, waste:soil ratios range from 4:1 to 6:1. Working face slopes usually range from 2:1 to 3:1.

Example 15–2

A solid waste stream of 100 tons per day is to be placed in 10-ft lifts with a cell width of 14 ft. The slope of the working face is 3:1. Assume that the waste is initially compacted to a specific weight of 800, 1000, and 1200 lbs/yd^3 and that the daily cover thickness is 6 in. Determine the ratio of waste to cover soil.

Solution

First, calculate the daily volume of the deposited solid waste. For 800 lbs/yd^3,

$$V = 100 \text{ tons/day} \times 2000 \text{ lbs/ton} \times \frac{\text{yd}^3}{800 \text{ lbs}} = 250 \text{ yd}^3$$

For 1000 lbs/yd^3,

$$V = 100 \text{ tons/day} \times 2000 \text{ lbs/ton} \times \frac{\text{yd}^3}{1000 \text{ lbs}} = 200 \text{ yd}^3$$

For 1200 lbs/yd^3,

$$V = 100 \text{ tons/day} \times 2000 \text{ lbs/ton} \times \frac{\text{yd}^3}{1200 \text{ lbs}} = 167 \text{ yd}^3$$

Second, calculate the length of each daily cell. For 800 lbs/yd^3,

$$L = \frac{250 \text{ yd}^3 \times 27 \text{ ft}^3/\text{yd}^3}{10 \text{ ft} \times 14 \text{ ft}} = 48.21 \text{ ft}$$

For 1000 lbs/yd^3,

$$L = \frac{200 \text{ yd}^3 \times 27 \text{ ft}^3/\text{yd}^3}{10 \text{ ft} \times 14 \text{ ft}} = 38.57 \text{ ft}$$

For 1200 lbs/yd^3,

$$L = \frac{167 \text{ yd}^3 \times 27 \text{ ft}^3/\text{yd}^3}{10 \text{ ft} \times 14 \text{ ft}} = 32.20 \text{ ft}$$

Third, calculate the cell surface areas. For the top of the cell,

$$A_{T800} = 48.21 \text{ ft} \times 14 \text{ ft} = 675 \text{ ft}^2$$

$$A_{T1000} = 38.57 \text{ ft} \times 14 \text{ ft} = 540 \text{ ft}^2$$

$$A_{T1200} = 32.20 \text{ ft} \times 14 \text{ ft} = 451 \text{ ft}^2$$

For the side of the cell surface area,

$$A_s = 14 \text{ ft} \times \sqrt{(10 \text{ ft})^2 + (3 \times 10 \text{ ft})^2} = 443 \text{ ft}^2$$

and for the face of the cell,

$$A_{F800} = 48.21 \times \sqrt{(10\,ft)^2 + (3 \times 10\,ft)^2} = 1524\,ft^2$$
$$A_{F1000} = 38.57 \times \sqrt{(10\,ft)^2 + (3 \times 10\,ft)^2} = 1219\,ft^2$$
$$A_{F1200} = 32.20 \times \sqrt{(10\,ft)^2 + (3 \times 10\,ft)^2} = 1018\,ft^2$$

Fourth, calculate the volume of soil for the daily cover using the following equation:

$$V_S = 6\,in. \times \frac{1}{12\,in.}\,(A_T + A_F + A_S)$$

Thus,

$$V_{S800} = 6\,in. \times \frac{1}{12\,in.} \times (675\,ft^2 + 1525\,ft^2 + 443\,ft^2) = 1321\,ft^3$$

$$V_{S1000} = 6\,in. \times \frac{1}{12\,in.} \times (540\,ft^2 + 1220\,ft^2 + 443\,ft^2) = 1101\,ft^3$$

$$V_{S1200} = 6\,in. \times \frac{1}{12\,in.} \times (451\,ft^2 + 1018\,ft^2 + 443\,ft^2) = 956\,ft^3$$

Now, calculate the ratio of waste to soil cover:

$$r_{w.s_{800}} = \frac{250\,yd^3 \times 27\,ft^3/yd^3}{1321\,ft^3} = 5.10:1$$

$$r_{w.s_{1000}} = \frac{200\,yd^3 \times 27\,ft^3/yd^3}{1101\,ft^3} = 4.90:1$$

$$r_{w.s_{1200}} = \frac{167\,yd^3 \times 27\,ft^3/yd^3}{956\,ft^3} = 4.71:1$$

(*Note*: As the compacted specific weight of the waste deposited in the landfill increases, the ratio of the waste to cover material decreases.)

LANDFILL COVER DESIGN

Water infiltrating the landfill cover picks up soluble contaminants during its passage through the solid waste. This liquid (i.e., leachate) is a potential groundwater contaminant. Therefore, one important design consideration is to minimize leachate production during the operation of the landfill and after its closure.

Lowering leachate quantities via surface cover design reduces the expenses involved in handling and treatment systems. It is still desirable, however, to evaluate the cost-effectiveness. As with all designs, there is a trade-off. In this case, it is between the effectiveness of the moisture barrier and the amount of leachate to be collected and treated. Figure 15–9 shows a final cap.

To minimize leachate production, the design elements should include an increase in the surface slope to increase the surface runoff and provision of a relatively impermeable cover that will minimize the costs of the removal, treatment, and disposal of

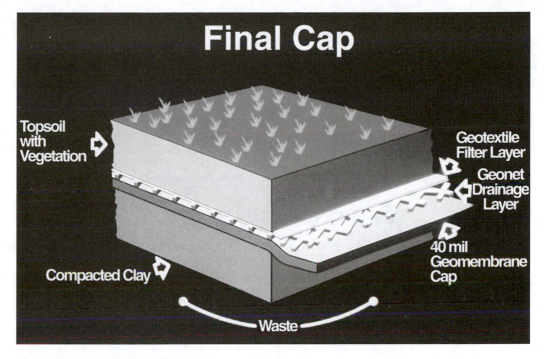

Figure 15–9   Final landfill cap.
(Courtesy of the Waste Management Company.)

leachate. In normal practice, the design of a cover includes several components. The role of each individual component is different, but their total effect is to minimize water infiltration into the landfill. Designers have used different combinations of components based on their experience, economics, and the site. A multilayer configuration is generally used, in which each layer has a function to perform.

A modern cover design may contain layers of different soil types. These may range from tight, impervious soils such as clay to coarse sand and gravel. Figure 15–10 shows the layers that can be used in a landfill cover, and Table 15–4 lists the layers and their roles. Certain tasks are also achieved with synthetic materials such as geomembranes, which are moisture barriers, and with geotextiles, which are used for filtration and drainage.

Surface Vegetative Layer

Completed landfill sites are now being developed as parks, golf courses, and bicycle paths. As a result, effective vegetative growth must be established and maintained on the surface layer. This growth helps to maintain the water balance and to protect the surface cover. The vegetation also helps to prevent erosion and encourages evapotranspiration. Selection of vegetation is an important consideration in designing a landfill

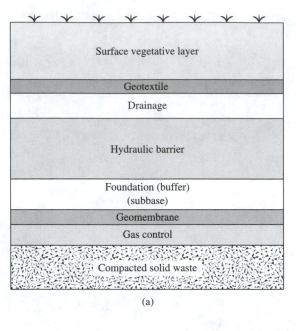

(a)

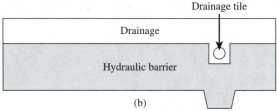

(b)

Figure 15–10   (a) Soil landfill cover system. (b) Drainage layer in a landfill cover.

cover. The vegetation should be able to flourish without its roots penetrating beyond the vegetative layer, and it should be indigenous to area and hardy.

The topsoil forming the protective cover must be selected and constructed to support the vegetation by allowing enough water to infiltrate the topsoil and by retaining enough moisture for plant growth during drought. Suitable soils are silt, silt-loam, loam, and sand-loam soils (see Figure 1–6). The thickness of the cover soil needed to support the vegetation will depend on the type of vegetation and the evapotranspiration in the area. A general guideline is to provide soil cover of approximately two times the depth of the vegetative root zone. Thus, the required soil thickness is 24 to 30 in. for grasses and 36 to 42 in. for shrubs.

Filter Layer

The filter layer protects the overlying cover soil. This soil is selected for its particle size gradation, and geotextile fiber may also be used. The intent is to prevent down-

Table 15–4   Primary Role of Various Components in a Landfill Cover

Component	Primary Role
Vegetative soil layer	Reduces infiltration and wind erosion. Provides root zone for plant growth, and retains moisture.
Filter layer	Prevents sifting of overlying cover soil into the drainage layer.
Drainage layer	Removes water that infiltrates the top layer of the cover.
Hydraulic layer	Minimizes infiltration from reaching the solid waste.
Foundation layer	Separates the geomembrane from the solid waste, and protects it from damage.

ward movement (i.e., piping) of the soil particles from the vegetative layer into the drainage layer but, at the same time, allow passage of the infiltrating water. The movement of soil particles may cause plugging of the drainage layer and/or gas collection in the gravel layer.

Drainage Layer

The drainage layer removes water that infiltrates the top layer of the cover. This function takes on added importance in areas of high precipitation and low evapotranspiration. It is a permeable drainage layer, sloped to a drain line to remove the liquid. A coarse, relatively uniform sand or gravel is suitable for this layer, which lessens the contact time of the leachate with the waste by conducting the percolation away from the waste.

For proper functioning, correct porosity and slope of the drainage layer (as well as of the collection pipes) are essential. As the water from the layers above percolates down through the soil, particles of soil may be transported into the lower, coarse soil and clog the pores. This movement of fine particles is termed as *piping,* and installation of a "filter layer" is suggested to remedy this problem. Layering filter material with an increasing particle size from top to bottom or placing geotextiles is advisable.

Landfills undergo subsidence or settlement over time. Settlement may occur because of one or more of the following factors:

1. Compression of loose materials because of self-weight and the weight of overlying materials.

2. Movement of fine particles into larger voids.

3. Settlement of underlying soil materials beneath the landfill.

4. Settlement because of volume changes from biological decomposition and chemical reaction (i.e., as the waste solids are converted to gases and soluble compounds).

Because of the random characteristics of the solid waste in each cell, settlement is not uniform. Without an adequate initial slope, excessive settlement of waste can produce localized depressions that allow surface water to pond. In turn, the drainage system is likely to fail if the differential settling creates depressions in either the drainage layer or the drain pipes. The ponded water in the depressions also causes the hydraulic gradient to increase the rate at which water penetrates the soil hydraulic barrier. Ponding can be eliminated (or at least reduced) by filling the depression and restoring the original grade.

Hydraulic Layer

The hydraulic barrier minimizes the infiltration that reaches the solid waste. The preferred material for this task is tight clay, synthetic clay liner, or synthetic membrane. Bagchi (6) recommends that a low-permeability layer (i.e., a clay or synthetic clay layer) be used below a synthetic membrane in the barrier layer. Usually, a 2-ft-thick clay layer or a layer of synthetic clay is used for the barrier.

Geomembranes can also be used as the hydraulic barrier when suitable soils are not available (7). Geomembranes can be easily damaged, however, by differential settling or construction equipment.

Foundation Layer

The foundation layer separates the geomembrane (i.e., hydraulic barrier) from the solid waste and protects the geomembrane from damage. The foundation layer can be built from local soils without stones or objects that may damage the membrane.

LANDFILL LINERS

The landfill liner requires most of the same qualities as the cover. The liner is a barrier to intercept leachate and direct it to a leachate collection system. Liners may be constructed of a single material or as a composite of materials, and they are frequently constructed with natural materials serving as the primary barrier.

To construct a liner, the bottom of the landfill is first excavated to the desired depth. If the *in situ* soil is a tight clay, it can serve as a bottom liner. If the soil is unsuitable for liner construction, however, soil may need to be transported from other places or materials used to amend the existing soil to reduce its permeability. The soil should have a permeability of 10^{-7} cm/s or less when compacted to its optimum density. Bentonite is frequently used for this purpose, and fly ash and cementing materials have also been used. (See Figure 16–1 for the components of a liner constructed from soil).

The base for the liner must be prepared to receive the weight of the solid waste. The specific weight of solid waste under overburden pressure varies from 1750 to 2150 lbs/yd^3 (6). If the soil is unstable and the load is not uniform, the liner may not function as intended.

The finished grade of the barrier is sloped (2% minimum) to provide drainage to the leachate collection pipes. These pipes are either plastic or vitrified clay drain tiles, and they are covered with 2 ft of porous sand. The solid waste is then placed on the sand and is compacted with heavy equipment. Care should be exercised during this process, because it is possible to damage the drain pipes. These pipes are then connected to a manifold that discharge the leachate to a wetwell for pumping to treatment or disposal. A minimum slope of 0.05 percent is typically employed, as is a minimum diameter of 6 in. to minimize the possibility of clogging.

When suitable soil is not available, a synthetic membrane can be used as the moisture barrier. Synthetic materials have gained acceptance as landfill barriers because they exhibit permeabilities of less than 10^{-11} cm/s (6). In many cases, they can be installed at a lower cost than clay liners as well. Figure 15–11 shows a truck depositing solid waste in a cell lined with a synthetic liner.

Figure 15–11   Prepared, lined cell ready for solid waste.

Loose soil pockets that may not be able to carry the superimposed load must be removed from the soil liner. The base soil is graded and compacted with a 12-in. layer of sand to support the synthetic membrane. An underdrain system is installed in landfills where there is a possibility of the groundwater table rising to the fill bottom. The function of the underdrain system is to remove such water.

Geomembranes are available in a series of different compositions. Each material has different properties. Selection of a liner for a given application will depend on many factors, including the site requirements, length of storage, and the waste to be contained. Table 15–5 lists the American Society of Testing Materials (ASTM) tests used for the physical properties of synthetic membrane. A 30-mil (0.75-mm) or thicker membrane is generally used. Many types of polymers are available, including polyvinyl

Table 15–5   Standard Tests for the Physical Properties of Synthetic Membranes

Test Number	Physical Properties	Standard Test Method
1	Tensile strength	ASTM D638
2	Tear resistance	ASTM D1004
3	Puncture resistance	ASTM D4833
4	Low-temperature brittleness	ASTM D746
5	Environmental stress crack resistance	ASTM D1693
6	Permeability	ASTM E96
7	Density	ASTM D1505

chloride, and fabrics such as high-density polyethylene and Hypalon are being used for these applications as well.

After the membrane has been placed, its seams are sealed. Field seaming is a critical factor for minimizing leakage, because most geomembrane leaks occur at the seams. The membrane is covered with a 6- to 12-in. layer of sand to serve as a protective layer for the membrane. This is followed by a 12- to 18-in. layer of soil with low permeability. This latter layer is compacted to serve as a moisture barrier and to conduct the leachate to the drainage pipes.

MOISTURE IN LANDFILLS

General

Leachate is generated by the percolation of water or some other liquid through any waste and the squeezing of that waste by self-weight. The quantity of leachate generated during the active life of a landfill and after its closure is important in managing a landfill. Understanding the hydrological cycle as it relates to landfills is also essential in managing water at a particular site. Because of climate and the associated hydrological cycle, there is a significant variation in leachate production from one area to another.

Hydrological Cycle

Figure 15–12 illustrates what happens to precipitation falling on a landfill site. Precipitation can become a surface runoff, and the water retained on the surface can evaporate, can infiltrate into the cover (where it is extracted by plant transpiration), or can infiltrate into the waste (where it may eventually become leachate). The surface evaporation and plant transpiration (i.e., removal of moisture by plant roots) are usually combined and called evapotranspiration (discussed later).

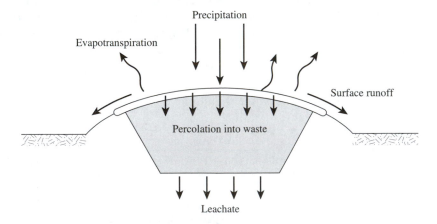

Figure 15–12   Leachate generation.

Infiltration.   When there is free-standing water, such as in ponds and depressions, this water is likely to penetrate (i.e., infiltrate) the ground surface. Many factors influence the infiltration rate, including the condition of the soil surface, the soil porosity, the current moisture content of the soil, and hydraulic conductivity. Water flow is defined by the Darcy Equation:

$$Q = PIA \qquad\qquad (15\text{-}1)$$

where: Q is the flow rate, P is the permeability of the soil, I is the hydraulic gradient, and A is the surface area.

Permeability (i.e., hydraulic conductivity) is defined by laboratory measurements of the flow rate of water through a column of soil at 60°F and hydraulic gradient of 1 ft/ft. Table 15–6 lists the permeability of several soils.

Because permeability is the direct indication of a material's tendency to allow the percolation of water, proper selection of the cover soil can significantly affect the amount of water that can infiltrate the landfill. More than one type of soil will be required, however, with lower layers impeding percolation and upper layers supporting vegetation and protecting against erosion.

The importance of soil permeability is evident from the amount of water that infiltrates, which is given by the Q values in Table 15–6. A good clay cover or liner can significantly reduce the rate of infiltration into a landfill. For example, a P of 1×10^{-9} means that the infiltrating water flows downward at the rate of 1×10^{-9} cm/s with a hydraulic gradient of 1 ft/ft. Through a 2-ft-deep clay layer, the time required to penetrate this layer is approximately 1934 years, as calculated here:

$$P = \cfrac{\cfrac{(1 \times 10^{-9}\,cm) \times 1\,in \times ft}{2.54\,cm \times 12\,in.}}{\dfrac{sec}{60\,sec} \times \dfrac{1\,min}{60\,min} \times \dfrac{1\,hour}{24\,hour} \times \dfrac{day}{365\,days} \times years} = \frac{3.28 \times 10^{-11}\,ft}{31.70 \times 10^{-9}\,year}$$

That is, the moisture will flow downward at the rate of 1.034×10^{-3} ft/year, and at this rate, it will take approximately 1934 years for this moisture to penetrate the

Table 15–6   Typical Values of Permeability Coefficients

Soil Type	P (cm/s)	Q (gal/day/acre)
Uniform, coarse sand	0.47	4×10^8
Uniform, fine sand	4.7×10^{-3}	4.4×10^6
Silty sand	1.1×10^{-4}	95,700
Sandy clay	5.6×10^{-6}	5,220
Silty clay	1×10^{-6}	957
Clay	1×10^{-7}	96

Source: Soil Survey of Hardin County Ohio, 1992; with permission.

2-ft clay layer. This flow will increase, however, if the surface grades are not maintained. Water will stand, thus resulting in a static head on the cover. A 7-ft static head of water on the top of the 2-ft clay cover will increase the hydraulic gradient to $\left(\dfrac{7+2}{2}\right)$ ft = 4.5 ft. Under this condition moisture will travel through a 2-ft barrier in $\left(\dfrac{1934 \text{ years}}{4.5 \text{ ft}} \times 1 \text{ ft}\right)$ = 430 years.

This example illustrates that if carefully selected and built, covers and liners in a landfill can exclude surface water or retain moisture that enters the landfill as long as the integrity of such barriers is maintained. Any subsidence because of differential settlement may break the moisture barrier, thus allowing water to pass through it.

Evapotranspiration.   The term *evapotranspiration* combines *evaporation* and *transpiration*. Evaporation is the loss of water from the soil surface. Transpiration is the loss of water from uptake by plants and later, partial release to the atmosphere. Because of the difficulties in measuring these two items separately, they are measured as one: evapotranspiration. The goal of a water budget is to "predict" the future leachate generation rate, so potential rather than actual evapotranspiration is of interest.

During precipitation, a certain amount of water wets the cover surface and vegetation. This moisture evaporates soon after, but the soil cover captures any water that infiltrates the soil. This moisture then moves down through the soil, and if the soil cover is deep enough, it can hold a significant amount of moisture and be the source of water for the vegetation on the landfill cover. The water is taken up from the soil by the roots and is returned to the atmosphere by transpiration. Thus, the amount of moisture reaching the solid waste is very little or even none.

Obviously, for the evapotranspiration to occur, moisture must be present in the cover material. The quantity of water removed by transpiration depends on the plants and climatic conditions. Only plants with moderate moisture needs and a root system that penetrates the depth of the soil cover are suitable for a landfill. Deep-rooted plants may not work because of the acidic nature and high salt concentration of the leachate. In turn, plants with shallow roots cannot remove moisture from the entire depth of the cover. During warm and sunny weather, plants are most active; evapotranspiration decreases as the weather cools. Table 15–7 gives typical transporation losses for selected vegetation.

Surface Runoff.   Surface runoff is the water that drains away from the landfill. Quantities of runoff are a function of the surface slope, antecedent moisture conditions, vegetative cover, surface, and so on. For example, precipitation on a clay surface will produce more surface runoff than precipitation on a sandy-soil surface. The opportunity to infiltrate depends on the time that precipitation remains on the surface of the landfill. A good surface drainage system removes precipitation from the landfill area rapidly, thus reducing the time for infiltration to occur. If the surface is relatively impervious and is adequately sloped for drainage, more precipitation will end up as surface runoff rather than as infiltration.

Table 15–7 Transpiration Losses from Vegetation

Vegetation	Transpiration (in./year)
Meadow grass	22–60
Rye grass	18–24
Clover	18–24

Source: Soil Survey of Hardin County Ohio, 1992; with permission.

Surface runoff can be estimated by the rational formula:

$$Q = CIA \qquad (15\text{-}2)$$

where Q is the rate of runoff, C is the coefficient of runoff, I is the rainfall intensity, (in./hour), and A is the surface area (acres). Table 15–8 provides the coefficient of runoff values.

Surface runoff can be discharged into a holding pond constructed near the landfill with a relatively impermeable bottom or allowed to flow into a stream.

An investment in a good surface drainage system will pay off, because as the surface runoff from the landfill increases, the infiltration decreases.

Table 15–8 Runoff Coefficients for the Rational Formula

Topography and Vegetation	Values of C^a		
	Open Sandy Loam	Clay and Silt Load	Tight Clay
Woodland			
Flat (0–5% slope)	0.10	0.30	0.40
Rolling (5–10% slope)	0.25	0.35	0.50
Hilly (0–30% slope)	0.30	0.50	0.60
Pasture			
Flat	0.10	0.30	0.40
Rolling	0.16	0.36	0.55
Hilly	0.22	0.42	0.60
Cultivated			
Flat	0.30	0.50	0.60
Rolling	0.40	0.60	0.70
Hilly	0.52	0.72	0.82

[a] In $Q = CIA$.

Source: U.S. Department of Interior. Washington, D.C.: U.S. Govt. Printing Office, 1982; with permission.

Moisture Retention.   The moisture retention capacity of soil used in a cover is important, and two important characteristics are field capacity and permanent wilting percentage (PWP). During precipitation, soil pores fill with water, and even after precipitation ceases, water continues to be pulled down into the soil by gravity until the soil reaches a certain moisture level (i.e., the field capacity).

If the area has vegetation and roots grow into the moisture area of the soil, the moisture will be used by the plants for transpiration. If moisture decreases to a level at which plants can no longer extract water, however, the PWP occurs. The difference between field capacity and PWP represents the available soil moisture. If the moisture content is less than the field capacity, any increase in moisture because of infiltration will be used for the transpiration needs of the plants. The moisture content will not drop below the PWP unless conditions are extremely dry. If the moisture content exceeds the field capacity, water will move deeper into the cover and, eventually, into the landfill.

Soils and MSW have different moisture-retention capacities depending on the characteristics of the materials and on the researchers' methods. Table 15–9 presents data on representative soil and MSW field capacities.

Soils with a high field capacity are suitable for moisture barriers. Soils with a low field capacity (e.g., sand, especially with uniform particle sizes) are suitable as drainage layers, because they allow water to move through (i.e., they are unable to hold water). The soil moisture reservoir capacity is determined by the difference between the two moisture levels.

Example 15–3

Determine the maximum potential evapotranspiration of available moisture from a 4-ft-deep clay loam cover vegetated with alfalfa. The root zone depth of this vegetation is 3 ft.

Table 15–9   Representative Soil Field and MSW Capacities

	Field Capacity (in./ft)
Clay	4.5
Clay loam	3.6
Silt	4.1
Coarse sand	0.5
MSW	3.9–4.5

Source: Sigh, R.J. *Boone County Field Study, Interim Report.* Washington, D.C.: U.S. Govt. Printing Office, 1979; with permission.

Solution

As Table 15–9 shows,

$$\text{field capacity} = 3.6 \text{ in./ft}$$

$$\text{wilting point (PWP)} = 1.6 \text{ in./ft}$$

The wilting point is the moisture limit below which a particular type of vegetation can no longer extract moisture. The wilting point differs according to the characteristics of the vegetation.

Choosing the smaller of the depth of soil or root zone, the available moisture is

$$(3.6 - 1.6)\, 3.0 = 6 \text{ in. of water}$$

This is the maximum possible moisture available for evapotranspiration.

ESTIMATION OF LEACHATE GENERATION RATES

The design of leachate collection, treatment, and disposal systems should consider the estimated leachate generation. The estimated quantity of leachate resulting from infiltration/vertical percolation through the waste can be estimated by using several alternate methods. Two methods are discussed here: the HELP computer model, and the water balance method.

HELP Computer Model

The model HELP computer was developed in 1984 to estimate water movement across, into, through, and out of landfills for the U.S. Environmental Protection Agency. The model accepts climatological, soil, and landfill design data, and it uses a solution technique that accounts for the effects of surface storage, runoff, infiltration, percolation, evapotranspsiration, soil moisture storage, and lateral drainage. Figure 15–13 illustrates the landfill cover, and various combinations of vegetation, cover soil, drainage layers, synthetic membrane covers, and liners can be modeled by HELP. The program was developed for rapid estimation of runoff, drainage, and leachate that can be expected from the operation of a given design. Because of its nonproprietary nature and flexibility in representing a wide array of landfill configurations, the model provides a calculation sequence on a daily basis. Substantial data are required; however, the model does provide default parameters that may be used. Table 15–10 provides a summary of the imput data.

Water Balance Method

The water balance method was developed by Thornthwaite and Mather (8) for quantifying evapotranspiration and was adapted by Fenn *et al.* (9) for landfill conditions. The calculations are simple and can be performed by using a calculator or computer spreadsheet. As an example, the sequence of calculations for the water balance method is outlined in Table 15–11, the provisional water-holding capacities for different

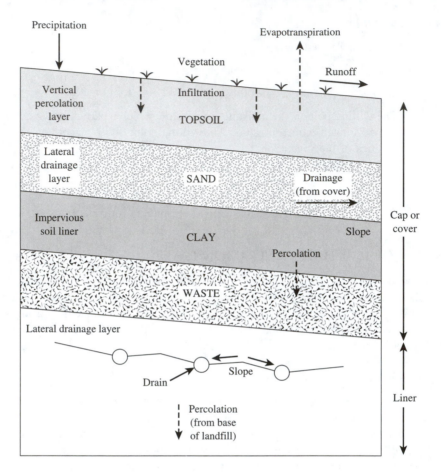

Figure 15–13   Detail of a soil landfill cover system.

combinations of soil and vegetation in Table 15–12, and an application of the model in Table 15–13.

LANDFILL OPERATION

A good landfill operating schedule, plan, and methodology result in public health and safety, minimized operating costs, and minimized leachate production.

Landfill disposal has become a complex process, however, because of public opposition to such facilities and regulatory restrictions for their siting, construction, operation, closure, and postclosure. Therefore, the overall costs of the facility have increased considerably.

Table 15–10 Input Data for the HELP Model[a]

Data	Description
Climate	Daily precipitation:
	1. Input precipitation data
	2. Generate data using a precipitation generator
	3. Use default precipitation data (based on 5 years of historical data)
Vegetation	Crop and crop cover
	Winter cover factor
	Evaporative zone depth
Soil	Porosity
	Saturated hydraulic conductivity
	Evaporation coefficient
	Field capacity
	Wilting point
	Minimum infiltration rate
Design	Number of layers
	Layer thickness
	Layer slope
	Lateral flow distance
	Runoff percentage
	Leakage fraction of liners

[a] Limitations of the model include:

1. It does not allow for failure of the moisture barrier cover, which can crack and introduce significant quantities of water into solid waste, thus resulting in production of leachate.

2. It does not take into account aging of liner.

3. It underpredicts the surface runoff coefficient, because the precipitation rate from a brief but intense rainfall is averaged out over the daily time increment. Thus, the intensity is much lower.

Projections of MSW Quantity

Reliable projections regarding the quantity characteristics, and composition of MSW during the life of the landfill are needed to calculate the landfill site capacity. Quantity projections should be based on population, per-capita generation, consumption trends, characteristics of the waste (both current and future), recycling, the regional and national economy, and regulations.

The type, size, design, location, equipment, and personnel needs for the facility should be based on reliable projections. Projections, however, are just that. For example, recycling, recovery, and reuse are all based mainly on the market economy, which is impossible to predict.

Table 15–11   Sequence of Calculations for Water Balance Model

PET	Potential evapotranspiration can be estimated by using the Thornthwaite and Mather method (8) (in. of water)
P	Average monthly precipitation (in. of water)
C_{RO}	Applicable runoff coefficient to calculate runoff for each month (see Table 15–8)
RO	Multiply the monthly precipitation by the monthly runoff coefficient to calculate the runoff (in. of water)
I	Subtract the monthly runoff from the monthly precipitation to obtain the monthly infiltration (in. of water)
I − PET	Subtract the monthly potential evapotranspiration from the monthly infiltration to obtain the water available (in. of water)
	Add the negative $(I - PET)$ values on a cumulative basis to obtain the cumulative water loss (in. of water)
S	Determine the monthly soil moisture storage (in. of water) as follows:

1. Determine the initial soil moisture storage for the soil depth and type (see Table 15–12)

2. Assign this value to the last month with $(I - PET) > 0$

3. Determine *S* for each subsequent month with $(I - PET) < 0$ (see Table in Appendix D)

4. Add the $(I - PET)$ value for months with $(I - PET) \geq 0$ to the preceding month's storage. Enter the field capacity if the sum exceeds this maximum

Δ *S*	Calculate the change in soil moisture for each month by subtracting *S* from that for preceding month (in. of water)
AET	Calculate the actual evapotranspiration (in. of water) as follows:

$$AET = PET \text{ for } (I - PET) \geq 0$$

$$AET = PET + (I - PET - \Delta S) \text{ for } (I - PET) < 0$$

When $(I - PET) < 0$, the evapotranspired amount is the potential evapotranspiration plus that available from excess infiltration that would otherwise add to the soil moisture storage plus that available from previously stored soil moisture

PC   Compute the percolation (in. of water) as follows:

$$(I - PET) < 0, PC = 0$$

$$(I - PET) \geq 0, PC = (I - PET - \Delta S)$$

$$PC = P - AET - \Delta S - RO$$

Weather Considerations

Weather has a lot to do with the operation of a landfill. Day-to-day operations are hindered by bad weather such as blowing wind, snow, and rain. For example, paper and other material can be blown away during high winds, and the daily cover material is difficult to use during wet weather. Modified operations should be considered at these times. For example, work at the lower end of the cell and use portable

Table 15–12 Provisional Water-Holding Capacities for Different Combinations of Soil and Vegetation

	Available Water		Root Zone		Applicable Soil Moisture Retention Table	
	mm/m	in./ft	in.	ft	mm	in.
Shallow-Rooted Crops (spinach, peas, beans, beets, carrots)						
Find sand	100	1.2	.50	1.67	50	2.0
Fine sandy loam	150	1.8	.50	1.67	75	3.0
Silt loam	200	2.4	.62	2.08	125	5.0
Clay loam	250	3.0	.40	1.33	100	4.0
Clay	300	3.6	.25	.83	75	3.0
Moderately Deep-Rooted Crops (corn, cotton, tobacco, cereal grains)						
Fine sand	100	1.2	.75	2.50	75	3.0
Fine sandy loam	150	1.8	1.00	3.33	150	6.0
Silt loam	200	2.4	1.00	3.33	200	8.0
Clay loam	250	3.0	.80	2.67	200	8.0
Clay	300	3.6	.50	1.67	50	6.0
Deep-Rooted Crops (alfalfa, pastures, shrubs)						
Fine sand	100	1.2	1.00	3.33	100	4.0
Fine sandy loam	150	1.8	1.00	3.33	150	6.0
Silt loam	200	2.4	1.25	4.17	250	10.0
Clay loam	250	3.0	1.00	3.33	250	10.0
Clay	300	3.6	.67	2.22	200	8.0
Orchards						
Find sand	100	1.2	1.50	5.00	150	6.0
Fine sandy loam	150	1.8	1.67	5.55	250	10.0
Silt loam	200	2.4	1.50	5.00	300	12.0

screens during high winds. Access roads to the site should be designed and built so that equipment and trucks can get to the site during wet weather. In northern climates, frost can be a problem when excavating cover material. For this reason, excavation should be attempted from the side walls of borrow pits, because the frost depth may be less there. In addition, cover material can be stored in advance for use during these conditions.

Compaction

A variety of components form MSW. A few items do not compact well, and some do not compact at all. For example, rubber tires act as a cushion and bounce back during compaction. Most MSW landfills either do not accept or shred rubber tires before disposal.

Table 15–13 Application of the Water Balance Model

Water Balance Components	Jan	Feb	Mar	Apr	May	Jun	Jul	Aug	Sep	Oct	Nov	Dec	Totals[a]
$(T°F)$	19.60	23.70	32.00	46.50	61.70	73.60	78.80	84.20	64.30	52.10	37.20	25.10	
P (in.)[a]	3.24	2.62	3.78	3.70	3.74	4.05	4.49	3.34	3.17	2.19	2.96	2.34	39.62
C_{RO}[b]	0.18	0.18	0.18	0.18	0.18	0.18	0.18	0.18	0.18	0.18	0.18	0.18	
RO	0.58	0.47	0.68	0.67	0.67	0.73	0.81	0.61	0.57	0.39	0.53	0.42	7.13
I	2.66	2.15	3.10	3.03	3.07	3.32	3.68	2.74	2.60	1.80	2.43	1.92	32.50
PET[a]	0.67	0.92	1.80	3.21	4.56	4.89	5.25	4.61	3.18	2.22	1.00	0.79	33.10
$(I - PET)$	1.99	1.23	1.30	-0.18	-1.49	-1.57	-1.57	-1.87	-0.58	-0.42	1.43	1.13	
$(\Sigma I - PET)$			0.00	-0.18	-1.67	-3.24	-4.81	-6.68	-7.26	-7.68			
S	4.00	4.00	4.00	3.82	2.60	1.73	1.17	0.72	0.62	0.56	1.99	3.12	
ΔS	0.00	0.00	0.00	-0.18	-1.22	-0.87	-0.56	-0.45	-0.10	-0.06	1.43	1.13	-0.88
AET	0.67	0.92	1.80	3.21	4.29	4.28	4.14	3.19	2.70	1.86	1.00	0.79	28.85
$PERC$	1.99	1.23	1.30	0.00	0.00	0.00	0.00	0.00	0.00	0.00	0.00	0.00	4.52

[a] P and PET data are for the Coshocton Agricultural Experiment Station.

[b] $C_{RO} = 0.18$ (coefficient recommended by Fenn et al. [9] for heavy soil at a 2–7% ground slope).

[c] Final cover has 4.52 in. of available water.

If the waste is properly spread in layers of approximately 2 ft and then compacted, a compacted density of 800 to 1000 lbs/yd^3 can be easily achieved by making three to six passes. The slope of the working face is usually 3:1 to 5:1 depending on the machinery and the landfill. Steeper slopes require more energy and effort for the heavy equipment.

Good compaction has several benefits:

1. Reduces waste volume, which translates into more waste per unit volume of landfill. In other words, the landfill can last longer.
2. Reduces the needed volume of daily cover.
3. Reduces the waste debris from blowing in the wind.
4. Discourages rodents.
5. Decreases future settlement of the solid waste and, thus, the ponding and consequent infiltration of precipitation in the surface.
6. Reduces the possibility that waste will wash away during heavy rainfall.

There is a downside to excessive compaction. The biodegradation of buried waste is slowed, because bacterial activity is inhibited. More energy is also spent in additional compaction.

Special Wastes

Asbestos.   The U.S. Environmental Protection Agency regulates the disposal of asbestos under the RCRA. Required disposal procedures include:

1. Asbestos should be put in closed containers or bags.
2. The material must be dampened and covered before shipment.
3. The landfill owner/operator must be notified in advance about the waste being shipped for disposal.
4. Wind velocities should be less than 10 mph during disposal of the asbestos load.
5. Employees should wear protective clothing and masks.

Automobile Tires.   The earlier discussion indicated that tires are generally not accepted in landfills. If they are, they are placed at the bottom of the working face, and other solid waste is placed on top. Tires can also be shredded for easy disposal.

White Goods (Appliances), Furniture, and Tree Stumps.   Disposal of these wastes is usually done in one or more of the following steps:

1. Crushing to compact the objects.
2. Salvaging for scrap.
3. Disposing in a different section of the landfill.

Construction and Demolition of Waste.   Most of this material can be disposed of with domestic wastes, and some can be used as a sub-base after shredding.

Biomedical Waste.   These wastes originate from hospitals, nursing homes, veterinary facilities, and laboratories. The material is considered to be infectious and requires a permit for disposal. The material to be disposed should be covered, and most waste of this nature is incinerated.

Other Wastes.   These wastes are of unknown characteristics and, sometimes, of unknown origin, such as sludges and ashes. Any chemicals from businesses or industrial processes may be hazardous wastes, and MSW landfills are not allowed to accept such waste without prior approval from the appropriate regulatory agency. Hazardous waste in an MSW facility can cause serious environmental and regulatory problems and, consequently, significant additional costs.

SITE OPERATIONS

Almost all site operations are designed and scheduled to promote the health and safety of both workers and the public. There are several typical problems.

Litter

Litter can be a problem during high winds and a source of complaints from residents close to the landfill. Open loads, high winds, and dumping all contribute to the litter problem. Ways to control litter include:

- Stop or reduce the level of operations during windy conditions.
- Cover open loads.
- Restrict dumping to a minimal surface area.
- Use movable fences or windbreaks.
- Unload waste at the bottom of the slope, and move the material up the slope.

Odor

As with litter, odors can be a source of complaints from nearby residents. Odors result from organic wastes, landfill gases, and leachate. Possible remedies include:

- Covering the solid waste.
- Proper venting and flaring of landfill gases.
- Collection and treatment of leachate.

Noise

Noise can cause psychological problems and loss of hearing in workers. Nearby residents frequently complain about noise as well. Noise can be controlled by the following methods:

- Create a buffer zone or barrier between the source of the noise and residents.
- Plant trees and/or build berms.
- Fix the noisy equipment, and repair the mufflers of trucks.

Dust

Dust can cause allergic reactions and nuisance conditions for both workers and nearby residents. Dust is caused by trucks carrying wastes, dumping, earth-moving equipment, and wind. Dust problems can be reduced by:

- Building concrete or asphalt roads instead of dirt roads.
- Spraying water or applying calcium chloride.
- Using vegetation and windbreaks to reduce windspeed.
- Reducing the dust-transport capacity of wind is also done by erecting a fence or planting shrubs around the work area, which will cut off the wind's capacity to blow dust.

Insects and Rodents

Mosquitoes and flies are a nuisance and have the potential for spreading disease, but they can be controlled by covering the solid waste. A good, compacted cover effectively eliminates food, shelter, and breeding areas.

Rodents can also spread many dangerous diseases (e.g., rabies). They are attracted by food and shelter in the landfill cells. Waste should be covered with a compacted layer of clay so that rodents cannot get in easily.

ON-SITE OPERATION FACILITIES

Fences and Signs

Fences should be built around the landfill to limit access by people and animals. Signs and notices should be posted at various locations for safety and better site operation. Easy-to-read signs include:

- Directional signs to the facility.
- No-entry signs for unauthorized people.
- Speed-limit signs at site access roads.
- Caution signs at manholes and gas-valve chambers.
- Notification and stop signs for public drop-off areas.
- Instructions that traffic must stop at the cabin for directions regarding the unloading of waste.
- Signs at the main entrance indicating the operating days and hours.

Public Drop-Off Area

Some landfill sites may have a public drop-off area for recyclable items. This area may also be combined with a transfer station. The drop-off facility should be so designed and located so that items can be dropped off or transferred from small vehicles to large vehicles.

Example 15–4

A solid waste manager estimates 45,500 tons of solid waste per year at the gate of a proposed, modern landfill facility. After detailed investigations, a suitable site has been selected with the following details:

$$\text{Landfill footprint} = 720 \times 1050 \text{ ft}^2 \times \frac{\text{acre}}{43,560 \text{ ft}^2} = 17.4 \text{ acres}$$

Bedrock at an elevation of 670 ft; natural grade at an elevation of 710 ft.

Base of excavation at an elevation of 685 ft; average excavation for bottom liner at 25 ft below existing grade.

Base of waste placement at an elevation of 690 ft; average depth of waste at 20 ft below existing grade.

60,000 tons of waste per year at gate.

10% of landfill airspace occupied by intermediate, daily cover soils and intercell berms.

1,250 lbs of waste per cubic yard.

3 ft of recompacted clay required for bottom liner.

3:1 waste slopes below grade; 4:1 waste slopes above grade.

Costs: assume continued off-site hauling of leachate at the existing rate of $0.065 per gallon.

No postclosure costs for operation of the active gas removal system included.

Solution

If the landfill footprint is 17.4 acres, then the total volume is

$$0.5(A_t + A_g) \times \text{height} + 0.5(A_g + A_b) \times \text{depth}$$

where the height is 77 ft above ground level, the waste depth is 20 ft below ground level with an excavation depth of 25 ft, A_t is the area of the top of the waste at an elevation of 787 ft, A_g is the area of waste at an elevation of 710 ft (at ground level), and A_b is the area of waste at an elevation of 690 ft (Figure 15–14). Thus, the total volume is $\approx 44,000,000 \text{ ft}^3$ or $1,630,000 \text{ yd}^3$.

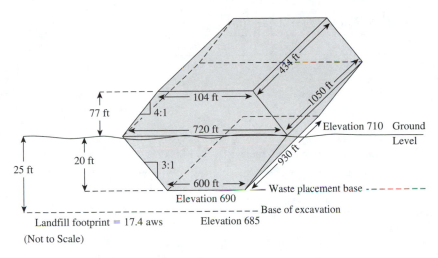

Figure 15–14   Data for Example 15–4.

Assuming that 10 percent of the area will be occupied by soil from interme-diate and daily cover, 163,000 yd³ will be required for such cover. The remaining volume (1,467,000 yd³) will be available for waste. Assuming that 1,250 lbs of waste can be compacted to 1 yd³ in the landfill, then approximately 917,000 tons can be placed. Other design details include:

45,500 tons of solid waste per year received at the gate.

Life of the landfill is 20 years.

3 percent additional material for cap.

17.9 acres of clay are required for cap.

Assuming 2 percent additional materials for liner, 17.7 acres and 3 ft of clay are required for liner.

Table 15–14 provides a cost estimate for such a facility.

Table 15–14   Cost Estimate for a Landfill[a]

	Unit	Quantity (n)	Unit Price ($)	Total Amount ($)
Permit to install Hydrogeo investigation	Lump	1	300,000	300,000
site preparation (fencing, cleaning and grubbing, field office)	Lump	1	150,000	150,000
Earthwork (liner)	Cubic yard of clay	85,680	5	428,400
Earthwork (excavation)	Cubic yard removed	294,000	4	1,176,000
Geomembrane/Leachate Control System drainage system/pipe	Acre-footprint	17.4	86,000	1,496,400
Leachate removal system (piping and storage)	Acre-footprint	17.4	30,000	522,000
Roads (paved)	Linear foot	8,100	42	340,200
Surface water management system (ditches/grading)	Acre-footprint	17.4	7,800	135,720
Cover system (4.5-ft thick)	Acre-cover	17.9	66,000	1,181,400
Sedimentation basin (excavation and piping)	Lump	1	200,000	200,000
Support facility (scales, buildings)	Lump	1	450,000	450,000
Gas-monitoring probes	Each	20	2,000	40,000
Active gas removal and treatment system	Lump	1	500,000	500,000
Monitoring wells	Each	6	6,500	39,000
Subtotal				6,959,120
Construction quality assurance	Percent	15	6,959,120	1,043,868
Additional land for borrow pit/sediment basin	Acre	0	2,500	0
Wetlands restoration	Acre	0	10,000	0
Land acquisition	Acre	0	2,500	0
Subtotal				8,002,988
Contingency	Percent	15	8,002,988	1,200,448
Total construction				9,203,436

Operating Costs (Including Existing Site)

	Unit	Quantity (n)	Unit Price ($)	Total Amount ($)
Personnel and expenses	Annual	1	300,000	300,000
Equipment and expenses	Annual	1	225,000	225,000
Supplies and services	Annual	1	65,000	65,000
Maintenance of cover/roads/ buildings and systems	Annual	1	100,000	100,000
Leachate management	Gallons	1,600,000	0.065	104,000
Maintenance of active gas system	Annual	1	50,000	50,000
Engineering/legal services	Annual	1	150,000	150,000
Groundwater monitoring	Annual	1	50,000	50,000

Passive landfill gas monitoring	Annual	1	18,000	18,000
Subtotal ($ per year)	Annual			$1,062,000
Projected life	years	15.3		$16,248,600
Contingency	percent	15	16,248,600	$2,437,290
Costs during operation				$18,685,890

30-Year Postclosure Costs (Not Including Existing Site)

Groundwater monitoring (6 monitoring wells twice/year)	Each	6	4,000	24,000
Landfill gas passive monitoring (quarterly)	Event	4	1,500	6,000
Maintenance of cover/roads/ditches and systems	Acre	17.9	300	5,370
Leachate management	Gallon	76,000	0.065	4,940
Annual reporting/administration	Lump	1	20,000	20,000
Subtotal	$/Year			60,310
Projected life	Years	30	60,310	1,809,300
Closure report	Lump	1	100,000	100,000
Subtotal				1,909,300
Contingency	Percent	15	1,909,300	$286,395
Costs during postclosure				$2,195,695

[a] In 1996 dollars for a footprint of 17.4 acres and a lifetime of 20 years.

REVIEW PROBLEMS

1. Assume you are the consulting engineer hired by a city to recommend a suitable construction site of a landfill.
   a. What investigations would you make?
   b. How would you investigate the suitability of a site?

2. Describe the different methods of landfill construction.

3. What are the commonalities of the various types of landfills?

4. a. What is the purpose of a daily cover?
   b. A solid waste stream of 250 tons per day is to be placed in 12-ft lifts with a cell width of 15 ft. The slope of the working face is 3:1. Assume that the waste is initially compacted to a specific weight of 100 lbs/yd^3 and that the daily cover thickness is 6 in. Determine the waste-to-cover-soil ratio.

5. a. List the characteristics of a good liner.
   b. What type of soil is considered to be satisfactory for a liner?

6. a. List the problems associated with operation of a landfill.
   b. Suggest remedies to avoid or minimize these problems.

REFERENCES

1. U.S. Environmental Protection Agency. *The Solid Waste Dilemma: An Agenda for Action.* Background Document, Office of Solid Waste. Washington, D.C.: U.S. Govt. Printing Office, 1988.
2. U.S. Environmental Protection Agency. *Municipal Waste Characteristics.* Washington, D.C.: U.S. Govt. Printing Office, 1994.
3. Wenger, R.B., and C.R. Thyner. "Optional Service Regions of Solid Waste Facilities," *Waste Management,* Vol. 2, No. 1, 1984, pp. 1–15.
4. Robinson, W. *Solid Waste Handbook.* New York: Wiley, 1986.
5. Siddiqui, M. *Municipal Solid Waste Landfill Site Selection using Geographical Information Systems.* Master's Thesis, University of Oklahoma, 1994.
6. Bagchi, A. *Design, Construction, and Monitoring of Landfills.* New York: Wiley, 1994.
7. Pfeffer, J. *Solid Waste Management Engineering.* Upper Saddle River, NJ: Prentice-Hall, 1992.
8. Thornthwaite, C.W., and J.R. Mather. *Instructions and Tables for Computing Potential Evapotranspiration and the Water Balance.* Centerton, NJ: Drexel Institute of Technology, Laboratory of Climatology, 1957.
9. Fenn, D., K.J. Hanley and T.V. DeGeare. *Use of the Water Balance Method for Predicting Leachate Generation from Solid Waste Disposal Sites.* U.S. EPA-530-SW-169. Washington, D.C.: U.S. Govt. Printing Office, 1975.

Chapter 16   Management and Control of Landfill Leachate and Gases

Chapter 15 focused on the causes of leachate generation and the procedures for estimating its quantity. This chapter discusses the management and control of leachate and gas. One main environmental concern when constructing and operating a landfill is the formation of highly contaminated water that may leak through the landfill. This leachate is formed as the water passes through the landfill, and it may flow into either ground or surface water, thus resulting in their contamination.

The source of water for leachate formation is primarily precipitation and the resulting infiltration of the landfill and the groundwater that may flow laterally from below the landfill. In a few cases, moisture within the solid waste deposited in the landfill and surface runoff from the surrounding areas also contribute to leachate.

As the percolating water moves through the landfill, it reacts chemically and biologically with the solid waste. Biological and chemical reactions occur continuously in the landfill. Depending on the stage of decomposition and the availability of oxygen, biological reactions are either aerobic or anaerobic. These biological and chemical reactions result in the formation of leachate, which moves downward through the landfill.

Once the leachate reaches the bottom of or an impermeable layer in the landfill, it travels laterally to the ground's surface or through the landfill bottom (if not impermeable). The path the leachate follows will depend on the hydraulic conductivities of the materials encountered.

If the landfill has a leachate collection system at its base, the leachate flows into this system. In the absence of an adequate collection system, the leachate moves below the fill and may enter into the groundwater, thus contaminating the aquifer.

LEACHATE CONTROL

Leachate can be controlled by preventing precipitation from infiltrating the solid waste through providing an appropriate cover or by removing and treating the leachate generated. Landfill cover designs and liners were discussed in Chapter 15. Even with a cover, however, leachate formation should be minimized to avoid any contamination of ground or surface water.

LEACHATE REMOVAL SYSTEM

Leachate is removed from a landfill to prevent surface or groundwater contamination. The purposes of the leachate collection system are to collect leachate for treatment or alternative disposal and to reduce the depth of leachate build up or saturation level over

329

the low-permeability liner. Under the 1991 Subtitle D rules promulgated by the U.S. Environmental Protection Agency (EPA), the leachate collection system must prevent the depth of leachate above the liner from exceeding 1 ft. The collection system is designed by sloping the floor of the landfill to a grid of underdrain pipes. A 1- to 2-ft-deep layer of granular material (e.g., sand) is placed over the geomembrane to conduct the leachate to the underdrains. The configuration in the immediate vicinity of the drain is illustrated in Figure 16–1. A tile drain and french drain are used (the french drain in the event of pipe failure or clogging).

The underdrain system is constructed before landfilling and consists of a drainage system that removes leachate from the base of the fill. A collector line located away from the active fill area connects the individual drain lines. All drain lines are designed for gravity flow, and Figures 16–2 and 16–3 illustrate leachate collection pipe layouts.

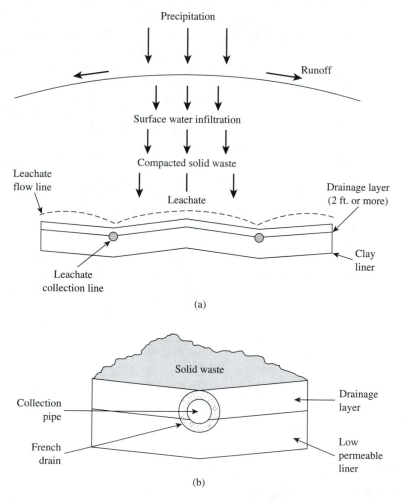

Figure 16–1   (a) Soil liner for a sanitary landfill. (b) Collection tile.

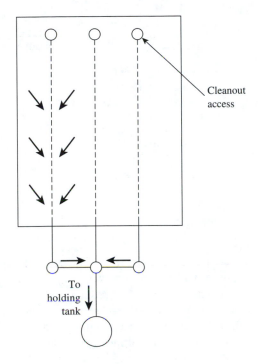

Figure 16–2 Pipe layout for leachate collection.

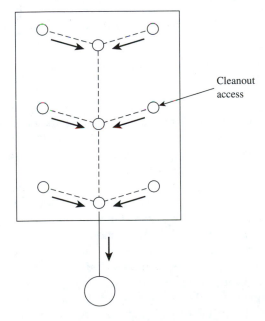

Figure 16–3 Pipe layout for leachate collection.

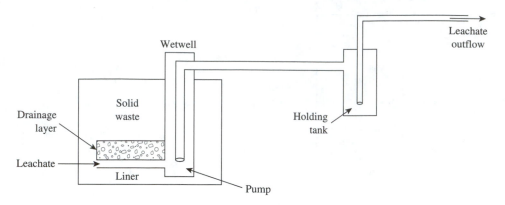

Figure 16–4 Wetwell method for pumping leachate.

Collection of the leachate that accumulates in the bottom of a landfill is usually accomplished by providing terraces, usually at a 1- to 5-percent slope, and a system of collection pipes at a 0.5- to 1.0-percent slope. Leachate accumulates on the surface of the terraces that will drain to collection channels. Perforated pipe in each leachate collection channel is used to convey the collected leachate to a central location, from which it is then removed for treatment or spraying on the surface of the landfill.

Figure 16–4 illustrates a leachate removal method. The leachate is transported by pipe to a wetwell, from which it is then pumped to a leachate holding tank. From this tank, the leachate is taken for treatment and disposal or for spraying on the landfill. The size of the wetwell should be sufficient to keep the pump from running too often.

When the projected volume of the leachate is low or a modern wastewater treatment plant is within reasonable distance from the landfill, the leachate can be hauled to the treatment plant. Alternatively, it can be piped to the plant.

LEACHATE CHARACTERISTICS

Characterization of the leachate is necessary to treat the waste. Numerous complex biological and chemical reactions occur as water infiltrate percolates through the landfill, and organic and inorganic compounds appear as leachate and gas (i.e., products of complex reactions). Figure 16–5 illustrates this process. Contaminants in leachate may include organic acids that exert a biochemical oxygen demand (BOD), volatile organic compounds, toxic organic compounds (e.g., halogens), high total dissolved solids, soluble metal (e.g., lead, mercury, zinc, iron), and many other chemicals. Leachate from municipal waste landfills can have quite high concentrations of BOD and of metals, and trace organics. Leachate from relatively newer sites is more contaminated than that from mature landfills. With time, the pH changes from slightly acidic to neutral, BOD and chemical oxygen demand (COD) decrease, and so on.

The ranges from concentrations of different parameters in the leachate of various nonhazardous wastes are included in Table 16–1 to provide some idea about leachate quality. Leachate from each landfill is unique, so the constituents and concentrations

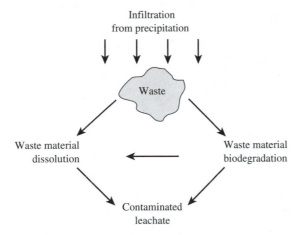

Infiltration
from precipitation

Waste

Waste material
dissolution

Waste material
biodegradation

Contaminated
leachate

Figure 16–5   Leachate formation.

provided in the table form a database. Design of treatment systems for landfill leachate must consider the variability in flow rates and the complex and varying compositions. Therefore, treatment options as illustrated in Figure 16–6 are considered.

The factors affecting the leachate composition include solid waste composition, landfill age, moisture, depth of waste, and waste temperature. Household and commercial wastes contain a variety of components, but they are similar in overall composition (1). The biological decomposition of the organic material, however, will significantly change the chemical environment in the landfill. Organic material decomposes to form gases (mainly CH_4 and CO_2) and water, and many solids may become soluble.

Table 16–1   Phases of the Landfill Stabilization Process and Approximate Ranges of Leachate Constituent Concentrations

	Concentration (mg/L)			
Leachate Constitutent	Transition Phase (Aerobic–Anaerobic, Years 0–5)	Acid-Formation Phase (Anaerobic, Years 5–10)	Methane Fermentation Phase (Anaerobic, Years 10–20)	Final Maturation Phase (Aerobic, >20 years)
BOD (5-day)	100–11,000	1,000–57,000	100–3,500	4–120
COD	500–22,000	1,500–71,000	150–10,000	30–900
Total organic carbon	100–3,000	500–28,000	50–2,200	70–260
NH_3-N	0–190	2–1,000	6–430	6–430
NO_3-N	0.1–500	0.1–20	0.1–1.5	0.5–0.6
Total dissolved solid	2,500–14,000	4,000–55,000	1,100–6,400	1,460–4,640

Source: Data from Chian, E.S.K, and F.B. DeWalle. "Sanitary Landfill Leachate," *Journal of the Environmental Engineering Division, ASCE,* Vol. 102, 1976, pp. 411–413; and Pohland, F.G., and S.R. Harper. *Critical Reivew and Summary of Leachate and Gas Production from Landfills.* EPA-600/2-86-073. Washington, D.C.: U.S. Govt. Printing Office, 1986.

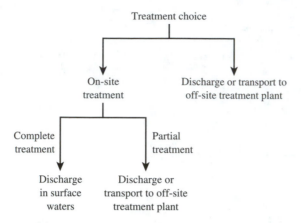

Figure 16–6 Leachate treatment options.

A landfill can be considered to be a biochemical reactor with solid waste and water as the main inputs and leachate and landfill gases as the main outputs. Solid waste input includes organic and inorganic materials. During the early stages of decomposition, the more readily biodegradable materials (e.g., sugars, starches, proteins, fats) are degraded. The remaining organic matter (e.g., cellulose materials) is slow to decompose. A few insecticides and pesticides may also be present.

The age of the landfill affects the leachate characteristics. For example, the BOD concentration from a landfill that has been in operation for 5 years may be approximately 5000 mg/L and for 10 years approximately 1000 mg/L. Clearly, the longer the period of biological stabilization, the more organics become stabilized, thus resulting in a lower BOD. Table 16.1 presents the phases of the landfill stabilization process (2, 3), but it should be noted that leachate characteristics and the stabilization period for waste organics differ from landfill to landfill.

The moisture in a landfill influences the quality of leachate. Low moisture encourages anaerobic microbial activity, but high moisture may flush soluble organics and microbes out from the solid waste. The deeper the solid waste depth, the more time the water takes to percolate through. Thus, the liquid reaches its solubility limit as a result of increased contact time with the waste and increases the leachate strength.

Waste temperature affects waste decomposition because of the increased bacterial activity and growth. Optimum bacterial activity occurs in the mesophyllic range, which is between 30°C to 40°C. Higher temperatures increase the solubility of most salts and the rate of chemical reactions. A study by Kmet and McGinley (4) of 16 operating landfills in Wisconsin concluded that leachate compositions can vary significantly between landfills—and also at individual landfills over time. Table 16–2 summarizes data on the concentrations of trace compounds in landfill leachate and gas samples.

Table 16–2 Organic Compound Analysis Summary for Brown and Ferris Industrial and MSW Landfills

Organic Compound Constituent	Leachate Analyses Compound Detections	Gas Analyses Compound Detections
1,1,1-Trichloroethane	✓	✓
1,1,2,2-Tetrachloroethane	✓	✓
1,1,2-Trichloroethane	✓	—
1,1-Dichlorethylene	✓	✓
1,2-Dichloroethane	✓	✓
1,2,3-Trichloropropane		
1,2-Dibromoethane		✓
1,2-Dichloroethane	✓	✓
1,2-Dichloropropane	✓	
2-Hexanone	✓	
4-Methyl-2-pentanone (MIK)	✓	✓
Acetone	✓	✓
Acrylonitrile		✓
Benzene	✓	✓
Bromodichloromethane		
Carbon Disulfide	✓	✓
Carbon Tetrachloride		✓
Chlorobenzene	✓	✓
Chloroethane	✓	
Chloroform	✓	✓
cis-1,2-Dichlorethylene	✓	✓
cis-1,2-Dichloropropane		
Dibromochloromethane	✓	
Ethylbenzene	✓	✓
Methyl bromide	✓	
Methyl chloride	✓	
Methyl ethyl ketone (MEK)	✓	✓
Methyl iodide		
Methylene bromide		
Methylene chloride	✓	✓
Styrene	✓	
Tetrachloroethylene	✓	✓
Toluene	✓	✓
trans-1,2-Dichlorethylene	✓	✓
trans-1,3-Dichloropropene		
Trichloroethylene	✓	✓
Trichlorofluoromethane	✓	
Vinyl acetate		
Vinyl chloride	✓	✓
Xylenes	✓	✓
1,2-Dichlorobenzene	✓	✓
1,4-Dichlorobenzene	✓	✓

Source: Courtesy of BFI.

LEACHATE TREATMENT

For large municipal waste landfills and most hazardous waste landfills, landfill leachate generally is treated in an on-site plant. Treatment of leachate from nonhazardous waste municipal landfills is also done in off-site wastewater treatment systems. In some cases, pretreatment of leachate is done, and the effluent is discharged in an existing wastewater treatment facility. Leachate may be transported to the treatment facility by sewers or tanker trucks.

Many techniques used for wastewater treatment are also used for the treatment of landfill leachate. These include biological treatment (i.e., aerobic and anaerobic biological stabilization) and physical–chemical treatment (i.e., precipitation, adsorption, coagulation, air stripping, chemical oxidation and reverse osmosis). Table 16–3 lists wastewater treatment processes along with their applications and comments. Leachate can also be recycled through the landfill as a partial treatment and sprayed onto land or the landfill. Spraying of leachate on a landfill permits further removal or stabilization of contaminants and supplies moisture to the vegetation there.

Table 16–3   Leachate Treatment Processes

Likely Treatment Process	Application	Comments
Biological Processes		
Aerobic Systems	Removal of organics	Refractory or slowly degrading compounds are not removed. Process cannot tolerate influent toxics. Biological sludge is produced.
Activated sludge		Needs separate clarifier.
Aerated stabilization ponds (lagoons)		Requires large land area.
Fixed-film processes (trickling filters, rotating biological contractors)		Temperature-sensitive in cold weather. Cover may be needed.
Anaerobic Systems (anaerobic contractors and lagoons)	Removal of organics	Low operating costs and sludge production. Requires heating. Long detection times are required for high removal levels. Typically cannot tolerate influent toxics or high concentrations of some inorganics.
Nitrification and denitrification	Removal of nitrogen	Nitrification/denitrification can be accomplished along with removal of organics.

Sedimentation	Removal of suspended material	Can remove only heavier material by gravity. Usually used in conjunction with other treatment processes.
Dissolved air flotation	Removal of suspended solids, oil, and grease	Dissolved contaminates are not removed, thus requiring additional treatment or potential release of volatile organic compounds into the air.
Filtration	Removal of suspended matter	Is combined with other technologies. Useful as a polishing process.
Ultrafiltration	Removal of bacteria and high-molecular-weight organics	Possibility of fouling. Limited applicability to leachate.
Air stripping	Removal of volatile organics or ammonia	Not effective for removal of nonvolatile organics. Off-gases may require air pollution—control equipment. Possible problems with media scaling and/or plugging
Steam stripping	Removal of volatile organics or ammonia	Effective removals of volatile organics. Expensive process.
Absorption	Removal of organics	Costs depend on the leachate quality.
Reverse osmosis	Dilute inorganic solutions	Good pretreatment needed.
Ion exchange	Removal of dissolved organic and inorganics	Ions of alkaline earth metals are difficult to remove.
Neutralization	pH adjustment	Frequently used in conjunction with other treatment.
Chemical precipitation	Removal of some metals, specifically cadmium, chromium, copper, lead, nickel	Will not remove most organics. A wet sludge is produced that may require disposal as a hazardous waste.
Oxidation	Removal of organics. Detoxification of some inorganics	Incomplete oxidation can produce compounds that may be more undesirable than the parent compound. Complete oxidation is usually impossible. Works well on dilute waste streams. High oxidant demand by leachate.
Evaporation	Where leachate discharge is not permitted.	Process sludge may be hazardous.

The aqueous waste streams that result from landfill leachate and the cleanup of hazardous waste sites vary widely in volume, level, and type of contaminants and solids. The main sources of such waste include:

- Leachate that has been collected via surface drains.
- Leachate plumes that have been pumped to the surface or collected via sub-surface drains.
- Contaminated water from dredging operations.
- Contaminated runoff from rainfall.
- Contaminated water generated from equipment cleanup.
- Aqueous waste generated from sediment or sludge dewatering.

Treatment of liquid waste at municipal solid waste or hazardous waste sites can be accomplished by using one of the following general approaches:

- On-site treatment using mobile treatment systems.
- On-site construction and operation of treatment systems.
- Pretreatment followed by discharge.
- Hauling of waste to an off-site treatment facility.

Both mobile and on-site treatment systems have broad applicability. Discharged wastewaters often require extensive pretreatment to meet National Pollutant Discharge Elimination System permit conditions.

Because these aqueous waste streams are so diverse in volume, type, and concentration of contaminants, a variety of treatment processes are applicable. Rarely will any one process be sufficient for aqueous waste treatment. Most methods used for treating water and wastewater can also be used for treating aqueous waste from solid and hazardous waste sites. These include primary treatment such as sedimentation followed by secondary treatment such as biological treatment. In addition to or in place of biological, physical and/or chemical treatment can also be provided.

Biological Treatment

The objective of biological treatment is to change the organic constituents of leachate into harmless compounds. Biological treatment reduces the biodegradable organics as defined by the BOD, which is the amount of oxygen needed to stabilize organics as measured over a 5-day period at 20°C. Biological treatment also removes ammonia nitrogen and organic nitrogen through bio-uptake. Nitrification occurs under aerobic conditions. Metals are removed through biosorption and precipitation as oxides and carbonates.

As the pH and valence increase, many metals change from a lower oxidation state and higher solubility to a higher oxidation state and lower solubility. For example, Fe^{2+}, which is more soluble, oxidizes to Fe^{3+}, which is less soluble. In aerobic treatment, metals oxidize and precipitate as metal hydroxides and carbonate. In anaerobic treatment, metals precipite as sulfides.

Aerobic Systems   As discussed in Chapter 6, aerobic processes require oxygen to complete the reaction. The objective of aerobic systems is to grow microbes that will convert a variety of organic compounds to cell material (for removal by sedimentation) and nonhazardous products such as carbon dioxide and water.

Some organics may not degrade and others only partially degrade to some secondary compounds depending on the environment. This is true for both hazardous and nonhazardous organics. Fatty acids are a product of the anaerobic decomposition of organic materials within the landfill, and they are present in leachate. If toxic chemicals are not present, these fatty acids are easily biodegradable using aerobic processes.

In general, aerobic processes can be represented by the following equation:

$$\text{organic} + O_2 \rightarrow CO_2 + H_2O + \text{other products} + \text{energy} \qquad \textbf{(16-1)}$$

The other products depend on the type of organics and nutrients. The dissolved molecular oxygen in the equation must be maintained at a concentration of approximately 1.5 to 2 lbs per pound of solids destroyed. Lower concentrations may cause parts of the aerobic system to become anaerobic, thus adversely affecting biological destruction.

Aerobic processes include activated sludge, trickling filters, aerobic ponds, and rotating biological contactors. All these processes work on the principle of microorganisms biodegrading organics in the presence of oxygen, the difference being whether the microorganisms are in a fixed or a suspension medium. Figure 16–7 shows common

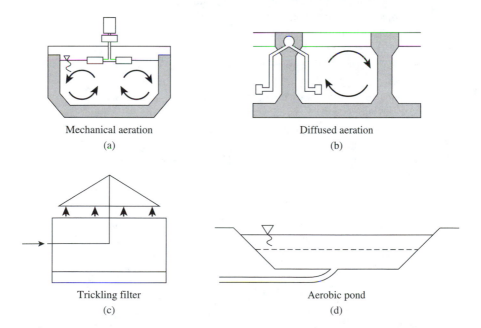

Mechanical aeration
(a)

Diffused aeration
(b)

Trickling filter
(c)

Aerobic pond
(d)

Figure 16–7   Aerobic reactors. (a) Mechanical areation. (b) Diffused aeration. (c) Trickling filter. (d) Aerobic pond.

aerobic reactors. Good estimates of the organic load of leachate are essential to supply sufficient oxygen. For example, leachates from relatively new landfills (i.e., those built during the last 5 years or so) have high concentrations of organics, which include easily degradable matter consisting mainly of volatile fatty acids. Leachates from older landfills, which are more stabilized, have a greater fraction of organics as refractory material and therefore, are difficult to biodegrade.

Activated Sludge Treatment.   Figure 16–8 shows the basic components of an activated sludge treatment system. Leachate or effluent from an anaerobic process flows into a primary tank to permit the settling of solids. From the primary tank, the liquid flows into an aeration tank. Air is supplied to the liquid in this tank to provide oxygen for the microbes to absorb and aerobically decompose the organic solids, and water, carbon dioxide, and other stable compounds are formed. Air is supplied either by mechanical aeration or by pumping through diffusers.

The aerobic microorganisms in the tank grow, thus forming an active suspension of biological solids called activated sludge. After approximately 6 to 8 hours of aeration in the tank, most organics in the liquid are stabilized. The liquid then flows to the secondary clarifier, in which activated sludge solids settle out by gravity. These solids are suspensions of biological solids, and a portion of these solids from the secondary or final clarifier are recycled back to the aeration tank to provide acclimated microbes for biodegradation. The remaining solids are pumped from the tank for further processing in another unit before discharge into the environment. Several modifications of this process are available, but the principle is the same.

Ehrig (5) reported on the study of leachate treatment by activated sludge plants in which the average influent BOD was 5294 mg/L and the average effluent BOD was 254 mg/L. In addition to BOD reduction, nitrification of ammonium is an important aspect of waste treatment. This conversion of ammonia to nitrate requires oxygen, so nitrification exerts its own oxygen demand. The main purpose of secondary treatment is to provide additional removal of BOD and suspended solids. Commonly used approaches take advantage of the ability of microorganisms to convert organic wastes into stabilized compounds by providing good contact with food and a proper environment. One such approach, the activated sludge system, has been discussed; a discussion of other approaches now follows.

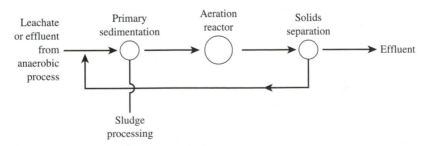

Figure 16–8   Treatment of anaerobic leachate by an aerobic process with sludge recirculation.

Trickling Filters. The effluent from the primary tank is sprayed over a bed of rock, plastic, or other medium coated with biological films. Figure 16–7(c) illustrates the major components of a trickling filter: the filter medium, underdrain system, and rotary distributor. The filter medium provides a surface for biological growth and voids for the passage of liquid and air. The medium is 3 to 5 in. of rocks or some artificial medium. The underdrain system is an open channel that carries away the effluent and permits air to circulate through the bed. A rotary distributor provides a uniform hydraulic load by spraying the leachate (i.e., waste liquid) on top of the filter surface. The finely suspended and dissolved organics are transformed into biological film, which can then be removed from the wastewater.

The biological layer (Figure 16–9) forms on a solid surface and consists of aerobic microbes to a depth of approximately 0.1 to 0.2 mm, and the remaining portion of the layer on the solid surface is anaerobic. Microorganisms in the upper zone (i.e., near the surface of the bed), have a high food concentration (i.e., organics), but the lower zone of the bed is in a starvation stage. Dissolved oxygen in the waste-liquid layer that is used by aerobic bacteria is replenished from the surrounding air. As the leachate (or liquid waste) passes through the trickling filter bed, the microbial film removes organic waste for use as food. Aerobic processes produce oxidized organic and inorganic end products, which are then released into the downward-flowing water film. The microbial film next to the solid surface does not receive dissolved oxygen and sufficient food, because the upper film extracts most of the organic material and oxygen. Consequently, the biological film on the solid surface is anaerobic and in a starvation state.

The film undergoes bio-oxidation and becomes old and heavier. It then breaks away from the filter surface and flows via the underdrain to a settling tank, where it is removed as sludge.

Aerated Lagoons. Aerated lagoons or ponds are created by a series of aerators (mechanical or diffuser). Generally, the depth of such ponds ranges from 8 to 15 ft, and they require a retention time from 1 to 2 months. Ample oxygen should be supplied to maintain a dissolved oxygen concentration of approximately 2 mg/L through

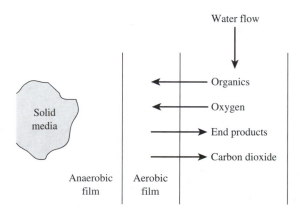

Figure 16–9   Microbial film on a trickling-filter medium.

the entire pond. The diffused air can be used to create mixing as well as to supply oxygen. Figure 16–10 illustrates a typical system.

Aerated ponds are usually rectangular, with a length-to-width ratio of 2:1. A system may have several cells, which may be installed so that parallel or series operation is possible. Dikes are protected from erosion by vegetation on the outside slopes. The lagoon bottom and sides should have a liner and a freeboard of 3 ft.

Aerated lagoons are furnished with concrete pads under each aerator to protect the bottom from erosion. Strong wastes are treated using long retention times and floating or platform-mounted mechanical aeration units. Complete mixing and adequate aeration are essential. Robinson (6) reported good treatment of high ammonia concentrations in leachate in England using a 15-day retention period. Zaf-Gilje (7) reported a 99-percent reduction in BOD using a retention time of 6 days.

Rotating Biological Contactors. Like a trickling filter, a rotating biological contactor (RBC) is a fixed-film biological process in which biomass accumulates on a rotating disk. The RBC is constructed of bundles of plastic packing attached radially to a shaft, thus forming a cylinder of media. The shaft is placed over a contour-bottomed tank so that the media is submerged approximately 40 percent (Figure 16–11). When rotated out of the tank, the liquid trickles out of the voids between the surfaces and is replaced by air. A fixed-film biological growth adheres to the media surfaces. Excess biomass sloughs from the media and is carried out in the process effluent for separation in a settling tank. The dissolved and finely suspended organics are converted into

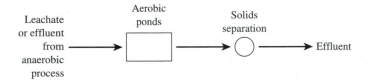

Figure 16–10   Leachate treatment by aerated ponds.

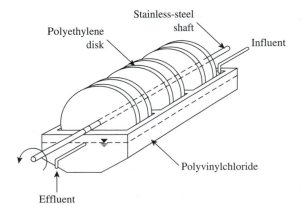

Figure 16–11   Rotating biological contactor.

solids in the form of biological film. A fixed-film biomass system tolerates hydraulic or chemical shock loading, which are common in landfill leachates. These units tend to clog with calcium deposits, however, which prevent the growth of biomass. Hence, many RBC or trickling filter systems raise the pH of the leachate by adding sodium hydroxide to precipitate the metals.

Problems with Aerobic Treatment Processes. Aerobic treatment does not work well in the presence of toxic metals such as copper, zinc, nickel, and so on. These treatment systems are effective for leachates from relatively new landfills, but aerobic biological treatment of leachate is unsatisfactory at high organic loadings and low retention periods.

Anaerobic Systems. Anaerobic biological activity occurs in a landfill after oxygen is no longer available: the organics degrades to liquid and gases. Anaerobic processes are also used to treat leachate.

In an anaerobic treatment system, large and complex organic compounds are biodegraded into organic acids and gaseous byproducts (i.e., CO_2, CH_4, and trace amounts of H_2S). With the BOD concentration in excess of 1000 mg/L, initial treatment is likely to be an anaerobic biological process. The efficiency of heavy metal removal in anaerobic biological processes is generally high.

A problem in anaerobic treatment is that methane-forming bacteria are inhibited by the low pH (i.e., acidic) environment. This acidic environment occurs when the acid bacteria are much more active than the methane bacteria. In addition, higher temperatures (e.g., 15°C to 35°C) are necessary for optimal treatment, with relatively long retention times and only limited NH_3-N reduction. Anaerobic treatment of leachate has several potential advantages over aerobic treatment, however. These include generation of methane gas as byproduct and much lower generation of sludge. There is also a savings in the energy requirement, because there is no need of aeration equipment.

A variety of anaerobic processes are available. These include anaerobic lagoons, anaerobic filters, and anaerobic bioreactors.

Anaerobic filters slowly pass the leachate through a medium. The filter is submerged in the leachate, and a film of anaerobic bacteria builds on the surface of the filter material. Bacteria grow and are retained on this material.

Anaerobic bioreactors are closed, agitated vessels that are used to biodegrade organics and produce methane. Influent containing a high concentration of organics (e.g., leachate) is fed into the bioreactor. The anaerobic digestion is typically carried on at 90°F to 95°F for 15 to 20 days.

Figure 16–12 illustrates typical anaerobic bioreactors, including the packed bed reactor, completely mixed heated reactor, and completely mixed reactor with sludge recycle or sludge and mixed liquor recycle. BOD reduction is greater in the latter case.

Figure 16–12(a) shows a fixed-film reactor, which can be a packed bed or a fluidized bed reactor. The completely stirred tank reactor (CSTR) in Figure 16–12(b) is an enclosed, gas-tight tank with a mixing system to keep the contents from stratifying. This process is normally performed at 35°C because of the low fermentation rate.

Figure 16–12(c) shows a CSTR with sludge recycle. The complete mixing of the tank contents accelerates the biodegradation process, because the microorganisms are mixed with the tank liquor, thus providing an opportunity for good contact with the waste. Sludge recycling has several benefits as well. For example, it provides a seed to the influent to get an early start for the biodegration of waste.

Figure 16–12(d) shows another CSTR with sludge and mixed liquor recycle. A CSTR with recycling of sludge and mixed liquor also accelerates the biodegration process, because the healthy and viable microorganisms in the recycled sludge and mixed liquor biodegrade the incoming organics.

Physical and Chemical Treatment

Physical and chemical methods that have been used to treat leachate include equalization, sedimentation, coagulation, dissolved air flotation filtration, ultrafiltration, air stripping, steam stripping, adsorption, reverse osmosis, ion exchange, neutralization, precipitation, oxidation, and evaporation.

Physical treatment usually transfers constituents from one medium to another, but the basic characteristics of those constituents remain the same. Physical treatment is

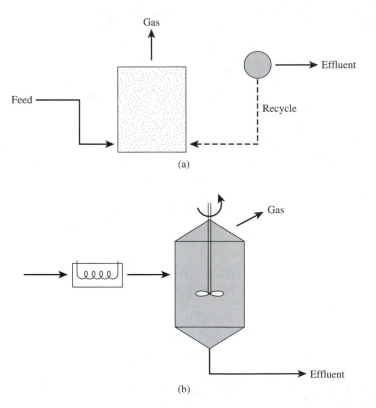

Figure 16–12   (a) Fixed-film reactor. (b) Completely stirred tank reactor. (c) Completely stirred tank reactor with sludge recycling. (d) Completely stirred tank reactor with sludge and mixed liquor recycling.

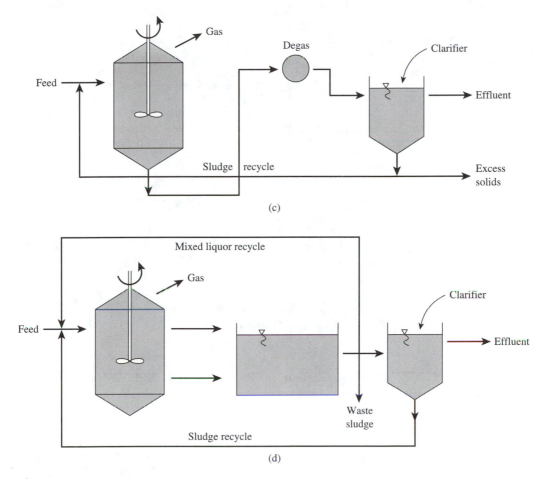

Figure 16–12   (cont.)

generally used in combination with chemical and/or biological treatment. Table 16–3 lists the applications and comments on those processes.

Equalization.   Equalization entails mixing the incoming leachate in a large tank or basin and then discharging it to the treatment plant at a constant rate, which improves both the efficiency and control of downstream processes by providing a more uniform feed. Because leachate is subject to large fluctuations in both volume and strength, operational problems can arise in leachate-treatment plants.

Equalization tanks or basins may be designed as in-line or side-line units (Figure 16–13). With in-line equalization, the entire daily flow passes through the basin, and the leachate is discharged to the treatment plant at a constant rate. With side-line equalization, only the flow in excess of the average daily flow rate is diverted into the basin. When the flow rate is less than below the daily average, leachate from the

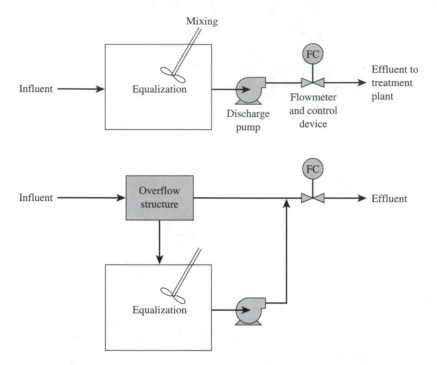

Figure 16–13   Flow equalization. (a) In-line method. (b) Slide-line method.
(From U.S. Environmental Protection Agency. A Handbook on Treatment of Hazardous Waste
Leachate. *Washington, D.C.: U.S. Govt. Printing Office, 1987; with permission.)*

equalization basin is discharged to the treatment plant to increase the flow to the average rate. In-line equalization is the preferred arrangement for leachate treatment applications, because contaminant concentrations as well as flow rates are equalized in
this approach. Equalization tanks or basins can be constructed of steel, concrete, or
lined earthen materials.

When placed ahead of chemical operations in the treatment system, equalization
improves the chemical feed control. When placed ahead of biological operations, equalization minimizes shock loadings, dilutes inhibitory substances, stabilizes the pH, and
improves secondary settling. Disadvantages associated with equalization include possible accumulation on the basin walls of solids, oils, and grease present in the leachate.
Mixing of tank/basin contents may strip highly volatile components from the leachate,
which results in air pollution because volatiles can escape into the air.

Primary considerations in the design of equalization tanks or basins include the
required storage volume, mixing equipment and control devices for pumping, and discharge flow rates. The required storage volume can be determined from the average
daily flow rate and the magnitude of inflow fluctuations. The tank is generally designed
with excess capacity to provide storage volume during periods of maintenance or regeneration of downstream processes. Mixing requirements for leachate containing 200

mg of suspended solids per liter range from 0.02 to 0.04 hp per 1000 gal of storage. Pumping may be required and may precede or follow equalization. Because of the low flow rates and relatively high contaminant concentrations associated with leachate, aboveground tanks or concrete basins with a lined surface are preferred.

Sedimentation.   Sedimentation is the gravitational settling in a large tank or basin of suspended particles that are heavier than water. The basin (i.e., a clarifier) may be rectangular or circular (Figure 16–14). The settled solids are collected mechanically on the bottom of the clarifier and are pumped out as sludge underflow. Clarifiers also have surface-skimming equipment to remove floating scum.

Sedimentation is widely used to remove solids and immiscible liquids, including oil, grease, and some organics. Leachate from landfills typically contains only small amounts of suspended solids. Sedimentation may be included as a pretreatment step, however, because of the sensitivity of many downstream processes to fouling interference from suspended solids. Frequently, sedimentation is included in leachate treatment to separate solids generated by chemical and biological processes.

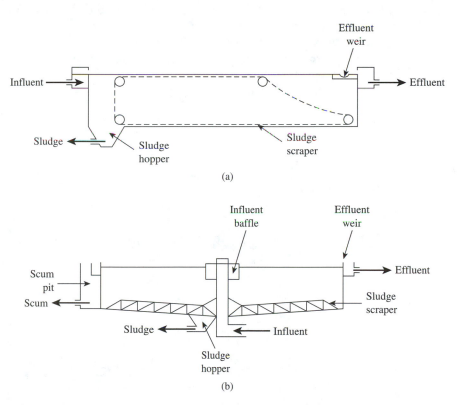

Figure 16–14   (a) Rectangular clarifier. (b) Circular clarifier.

(From U.S. Environmental Protection Agency. A Handbook on Treatment of Hazardous Waste Leachate. *Washington, D.C.: U.S. Govt. Printing Office, 1987; with permission.)*

The sludge underflow produced by sedimentation may be hazardous. It requires further treatment and disposal.

The design of clarifiers is based on the settling rate or the rise rate of the smallest particles to be removed expressed as flow rate per unit area. Most applications fall within the range of 0.2 to 1.0 gal/min/ft^2 (300 to 1500 gal/day/ft^2) (8). Horizontal velocities must be limited to prevent scouring of settled solids from the sludge bed. If designed and operated properly, clarifiers can remove 50 to 65 percent of the total suspended solids.

Granular-Media Filtration.   Filtration is a physical process whereby suspended solids are removed from leachate by forcing the fluid through a porous medium (Figure 16–15). The filter consists of a bed of granular material (typically a coarse layer of sand or anthracite coal). As leachate laden with suspended solids passes through the filter medium, the particles become trapped within the bed. This either reduces the filtration rate at a constant pressure or increases the pressure needed to force the water through the filter. When the maximum pressure drop is reached, the filter is backwashed to dislodge the trapped contaminants. The backwash water, which contains high concentrations of solids, is typically recycled at the head end of the treatment plant. Periodic backwashing of the filter medium is essential. The driving force in the operation of granular-media filtration systems is either gravity or applied pressure.

The most common application of granular-media filtration involves pretreatment of leachate before carbon adsorption. Filtration may also be used as a polishing step after precipitation flocculation or biological processes to remove residual suspended solids in the clarifier effluent.

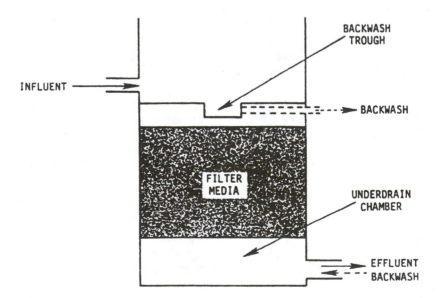

Figure 16–15   Granular-media gravity filter.

(From U.S. Environmental Protection Agency. A Handbook on Treatment of Hazardous Waste Leachate. *Washington, D.C.: U.S. Govt. Printing Office, 1987; with permission.)*

Key variables in the design of such systems include the size of the filter bed porosity, filter bed depth, filtration rate, and influent leachate characteristics. The filtration rate, which ranges from 2 to 15 gal/min/ft^2, affects the size of the filter and the maximum head loss affects the length of the run.

Dissolved Air Flotation.   Dissolved air flotation is achieved by releasing fine air bubbles into the waste liquid that attach to sludge particles. The particles then rise to the surface, where they are removed by skimming. Figure 16–16 illustrates a dissolved air flotation tank.

The solid (i.e., particle) recovery rate of 85 percent is possible. Polymers or other flocculents may be added to increase this rate to 95 percent or greater. Practical loading rates vary depending on the contaminants. Recommendations include a depth-to-width ratio between 0.3 to 0.5, maximum horizontal velocity of 3 ft/min, a maximum horizontal velocity to theoretical rise rate of particle (obtained from bench-scale tests) of approximately 15:1, and an optimum length-to-width ratio of 4:1.

Ultrafiltration.   Ultrafiltration is a membrane process capable of separating solution components on the basis of molecular size, shape, and flexibility. The semipermeable membrane acts as a sieve to retain dissolved and suspended molecules that are too large to pass through its pores, which range in size from 0.001 to 0.02 mm. Pressure is the driving force for the separation. A simplified schematic of an ultrafiltration system would resemble that of the reverse osmosis process (Figure 16–17), except that a low-pressure pump would be used in place of the high-pressure pump.

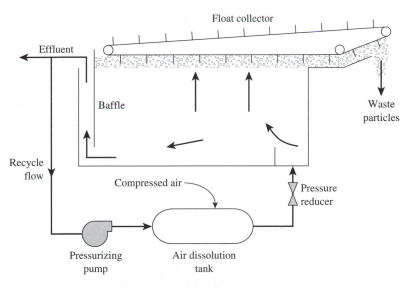

Figure 16–16   Dissolved air flotation.
(From U.S. Environmental Protection Agency. A Handbook on Treatment of Hazardous Waste Leachate. *Washington, D.C.: U.S. Govt. Printing Office, 1987; with permission.)*

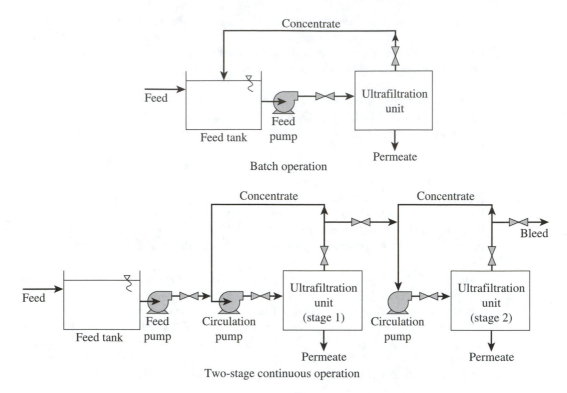

Figure 16–17   Ultrafiltration. (a) Batch operation. (b) Two-stage continuous operation.
(From U.S. Environmental Protection Agency. A Handbook on Treatment of Hazardous Waste Leachate. *Washington, D.C.: U.S. Govt. Printing Office, 1987; with permission.)*

Ultrafiltration has many industrial applications, but the greatest potential for this technology probably involves sites where leachate contains only one primary contaminant. Minimum pretreatment requirements for ultrafiltration include equalization, oil–water separation, and suspended-solid removal through sedimentation and/or filtration.

The number and type of ultrafiltration modules required for a specific application are functions of several system variables, including the desired recovery, separation, and product flow. Recirculation of the feed stream is required, and this can be achieved with either batch or continuous operation. With batch operation, the feed is recirculated to a feed tank, and the batch is processed until the desired reduction in volume is obtained. With continuous operation, the feed is added to and concentrate is bled from a continuous, recirculating loop at equal rates.

Air Stripping.   Air stripping is a mass-transfer process that uses air to remove volatile organic compounds from leachate. Types of air-stripping devices include diffused aerators, mechanical surface aerators, sprays and spray towers, and packed towers. Countercurrent packed towers (Figure 16–18) are best suited for treatment of leachate.

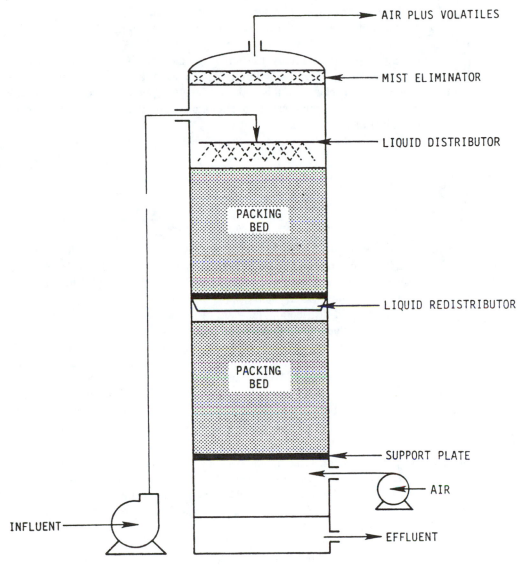

Figure 16–18   Countercurrent packed tower air stripper.

(From U.S. Environmental Protection Agency. A Handbook on Treatment of Hazardous Waste Leachate. Washington, D.C.: U.S. Govt. Printing Office, 1987; with permission.)

A countercurrent packed tower consists of a cylindrical shell containing random packing on a support plate. Leachate is distributed uniformly atop the packing with sprays or distribution trays and flows downward by gravity. Air is blown upward through the packing by forced or induced draft and flows countercurrently to the descending liquid. The volatile organics stripped from the leachate by the rising air are discharged to the atmosphere through the top of the column; effluent is discharged from the bottom of the column.

Air stripping can effectively treat leachate containing volatile organics that are only slightly water soluble. Pretreatment requirements for air stripping include removal of suspended solids and separation of nonaqueous phases.

Because air stripping essentially transfers volatile contaminants from the aqueous leachate to the air stream, air emission limitations for volatile organic compounds typically preclude direct discharge of this vapor to the atmosphere. Carbon adsorption is commonly used to control emissions of volatile organic compounds. The contaminant-laden carbon must then be regenerated.

Four principal factors govern the design of packed column air stripping towers:

1. Type of packing.
2. Tower diameter.
3. Height of packing.
4. Air-to-water ratio.

The common packing shapes are ring and saddle. Packings may be manufactured of ceramics, metals, or plastics and in nominal sizes ranging from 0.5 to 4 in. Packings should be selected for their material characteristics, such as strength, durability, resistance to corrosion, and mass transfer.

To avoid poor liquid distribution, the ratio of tower diameter to nominal packing size should be no less than 8:1. The preferred ratio is 15:1.

The height of packing required for various air-stripping applications depends on the desired degree of contaminant removal. The volume of air required to treat a unit volume of leachate (i.e., the air-to-water ratio) is a function of the volatility of the contaminants. Air-to-water ratios typically range from 10:1 to 300:1 for removal of volatile organics during warm-weather operations. As the temperature drops, however, stripping becomes more difficult, and air requirements increase if the desired performance is to be maintained.

Steam Stripping.   Steam stripping involves injecting steam into liquid waste to volatilize and separate the lighter constituents. Stream stripping removes these volatile contaminants from the liquid waste and makes them part of the vapor and is used to remove volatile organics from leachate. Steam stripping is performed in a simple distillation column (Figure 16–19).

Vapor–liquid contact may occur on discrete plates or trays or over continuous beds of packing (i.e., packed columns). Leachate is introduced at the top of the column, above the plates or packing, and flows downward by gravity. Steam is introduced at the bottom of the column and flows upward. Volatile components of the leachate are vaporized by contact with the rising steam and carried overhead. The vapor (i.e., steam plus volatiles) is then cooled and condensed, and the condensate, which is rich in organics, is treated further before disposal.

Steam stripping is considerably more expensive than air stripping. Pretreatment requirements include removal of suspended solids, oil, and grease to prevent fouling of the steam stripping equipment and separation of nonaqueous phases.

Major design considerations associated with steam stripping include selection of a packed column, column diameter, spacing between plates, height of packing, liquid

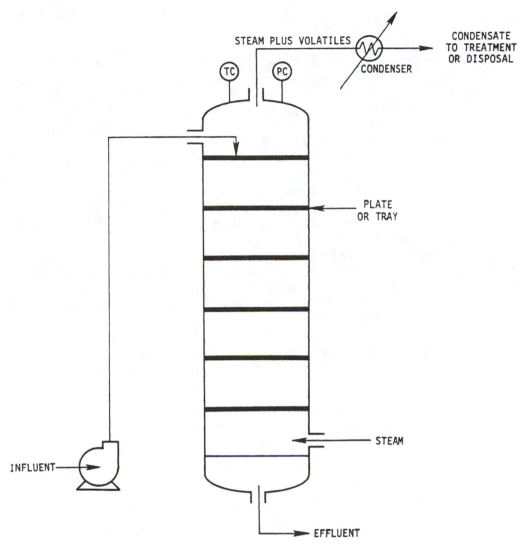

Figure 16–19   Plate-type steam-stripping column.

(From U.S. Environmental Protection Agency. A Handbook on Treatment of Hazardous Waste Leachate. Washington, D.C.: U.S. Govt. Printing Office, 1987; with permission.)

and vapor flow rates, and operating temperature. Packed columns are applicable for small-scale installation (column diameter, < 5 ft), severe corrosion environments, and foaming systems.

Absorption.   Absorption may be physical or involve reaction with chemicals in a liquid solution. An example of absorption in the treatment of hazardous material is the water absorption of fluorine gas. The solute is recovered by stripping or distillation, and the absorbing liquid is recycled. A packed tower of absorbing material is often used for gas absorption.

Reverse Osmosis.   When two solutions of differing concentrations are separated by a semipermeable membrane (i.e., one with a high permeability for water but a low permeability for dissolved solids), pure water from the more dilute solution will pass through the membrane to the more concentrated solution until the liquid head balances the osmotic pressure. This is called osmosis. In reverse osmosis, a differential pressure that exceeds the osmotic pressure is applied to the membrane, thus causing the solvent to flow from the stronger to the weaker solution. Figure 16–20 shows a reverse osmosis unit.

Reverse osmosis can remove dissolved inorganics (e.g., metals, metal–cyanide complexes) and high-molecular-weight organics from leachate. This process is generally applicable for total dissolved solids concentrations of 50,000 mg/L or less. Normally, some combination of filtration, chlorination, carbon adsorption, and pH adjustment is required for pretreatment.

Both the number and type of reverse osmosis modules required for a specific application are functions of several system variables, including the desired recovery and product flow and the required system pressure. Because reverse osmosis membranes are susceptible to fouling by iron-corrosion products, stainless steel and plastic should be used in their construction.

Ion Exchange.   Ion exchange is a process that reversibly exchanges ions in solution with ions of like charge retained on an insoluble, resinous solid (i.e., ion exchange

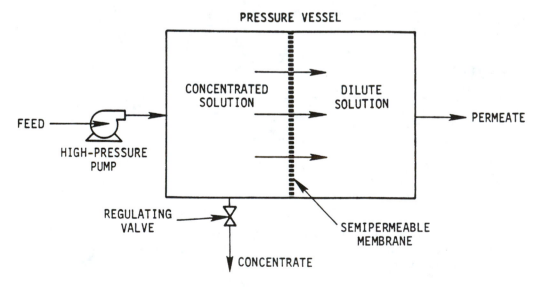

Figure 16–20   Reverse-osmosis unit.

(From U.S. Environmental Protection Agency. A Handbook on Treatment of Hazardous Waste Leachate. *Washington, D.C.: U.S. Govt. Printing Office, 1987; with permission.)*

resin). The ion exchange resin can exchange either positively charged ions (i.e., cation exchange) or negatively charged ions (i.e., anion exchange). For example, the ion exchange resin may be represented as

$$Na_2R + Ca^{+2} \Longleftrightarrow CaR + 2\,Na^+$$

where calcium ions in solution displace sodium ions initially on the resin (R). When the useful capacity of the resin is exhausted, the resin may be regenerated, which essentially reverses the exchange process. The regeneration process may be represented as

$$CaR + 2\,NaCl \Longleftrightarrow Na_2R + CaCl_2$$

where a sodium chloride solution is used to regenerate a sodium from cation exchange resin.

Ion exchange can be used to treat metal electroplating wastes. The applicability of this process to the treatment of hazardous waste leachate, however, is limited to use as a final "polishing" stage, when effluent is discharged to sensitive surface waters. Extensive pretreatment is required before ion exchange in a hazardous waste leachate treatment system. Suspended solids and nonaqueous liquids (e.g., oil, grease) must be removed to prevent fouling and plugging of the resin beds. High concentrations of ionic species should also be removed through less costly processes of precipitation/flocculation/sedimentation.

Selection of an appropriate resin is the primary design consideration for an ion exchange system. The usable exchange capacity of a given unit varies under differing operating conditions; therefore, it must be determined experimentally.

Neutralization.   Neutralization of leachate with an extreme pH involves addition of a base or an acid to the leachate to adjust the pH upward or downward (as required) to an acceptable level (usually between 6 and 9). Bases commonly used for neutralization include lime (CaO), calcium hydroxide $[Ca(OH)_2]$, caustic soda (NaOH), soda ash (Na_2CO_3), and ammonium hydroxide (NH_4OH). Acids commonly used include sulfuric acid (H_2SO_4), hydrochloric acid (HCl), and nitric acid (NHO_3). Salts are formed as reaction byproducts or through neutralization:

$$acid + base \rightarrow salt + water$$

Neutralization of an acidic waste stream can cause precipitation of heavy metals. This precipitate can be removed as sludge by sedimentation or filtration. The hydroxides of heavy metals are usually insoluble, so lime or caustic soda is mostly used to precipitate them. Precipitation depends on pH, temperature, and valence state of metal. For example, copper hydroxide is least soluble at a pH of 8.9. Temperature is an important factor, because it affects solubility, which usually increases with temperature.

Typically, neutralization is performed by complete mixing of the aqueous leachate with the neutralizing agent in a corrosion-resistant tank (Figure 16–21). Neutralization is one of the simplest and most common technologies available for treatment of leachate. In most leachate (especially hazardous waste leachate) treatment applications, neutralization serves as pretreatment to optimize the performance of pH-sensitive processes.

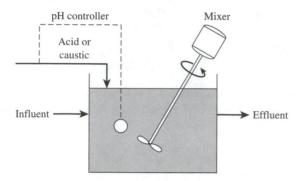

Figure 16–21   Neutralization system.

(From U.S. Environmental Protection Agency. A Handbook on Treatment of Hazardous Waste Leachate. *Washington, D.C.: U.S. Govt. Printing Office, 1987; with permission.)*

Neutralization of pH is common in biological treatment processes and for minimization of corrosion.

Neutralization may also be applied as a posttreatment operation downstream of certain chemical processes that yield acidic or caustic effluents (e.g., oxidation and reduction). Use of posttreatment neutralization to meet final discharge criteria is particularly applicable where treated effluent is discharged to surface or groundwater. Equalization typically precedes neutralization in the treatment process to dampen fluctuating influent conditions, thus improving process control.

Major design considerations for neutralization systems include reagent selection and dosage, mixing, contact time, and process control. Factors that affect reagent selection include availability and cost of chemicals, reaction rate, reaction byproducts, and storage and equipment requirements. For treatment of hazardous waste leachate, selection of vessels, piping, and instrumentation must be given special consideration.

Oxidation and Reduction.   Oxidation and reduction can be used to remove or convert toxic pollutants to harmless or less toxic substances. Oxidation/reduction reactions are those in which the valence state of one reactant is raised and that of another is lowered.

Oxidation/reduction involves the addition of a chemical oxidizing or reducing agent to leachate under controlled pH. Common oxidizing agents include chlorine gas, calcium and sodium hypochlorite, chlorine dioxide, potassium permanganate, hydrogen peroxide, and ozone. Common reducing agents include sulfur dioxide, sodium sulfite salts, and base metals (e.g., iron, aluminum, zinc). Figure 16–22 shows an oxidation/reduction reactor.

Oxidizable leachate constituents include organics (e.g., acids, aldehydes, phenols, mercaptans, polynuclear aromatic hydrocarbons, pesticides, polychlorinated biphenyls (PCBs), ammonia) and some metals. Reducible leachate constituents include a variety of metals, such as chromium, mercury, lead, silver, nickel, copper, and zinc. Metal–cyanide complexes can be treated by first oxidizing the cyanide and then reducing the metal.

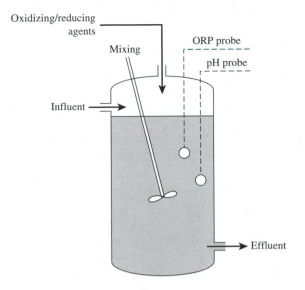

Figure 16–22   Oxidation/reduction reactor.

(From U.S. Environmental Protection Agency. A Handbook on Treatment of Hazardous Waste Leachate. *Washington, D.C.: U.S. Govt. Printing Office, 1987; with permission.)*

Oxidation and reduction are complementary chemical reactions involving the transfer of electrons among ions. The most common applications to hazardous waste leachate include cyanide destruction and reduction of hexavalent chromium to the less hazardous, trivalent form. Both processes are very important.

Oxidation of cyanide wastes is done in the metal-finishing industry. In the following reactions, cyanide is first converted to a less toxic cyanate using chlorination at a pH greater than 10; further chlorination results in oxidation of the cyanate to simple carbon dioxide and nitrogen gases.

$$NaCN + Cl_2 + 2NaOH \rightarrow NaCNO + 2NaCl + H_2O \qquad \text{(16-2)}$$

$$2NaCNO + 3Cl_2 + 4NaOH \rightarrow 2CO_2 + N_2 + 6NaCl + 2H_2O \qquad \text{(16-3)}$$

Reduction of hexavalent chromium (Cr VI) to trivalent chromium (Cr III) is a very important process in electroplating operations. Sulfur dioxide is often used as the reducing agent, as shown by the following reactions:

$$SO_2 + H_2O \rightarrow H_2SO_3 \qquad \text{(16-4)}$$

$$2CrO_3 + 3H_2SO_3 \rightarrow Cr_2(SO_4)_3 + 3H_2O \qquad \text{(16-5)}$$

The trivalent chromium formed in equation (16-5) is much less toxic and more easily precipitated than the hexavalent chromium.

The primary design considerations associated with oxidation/reduction systems include the appropriate type and dosage of chemical reagent, minimum contact time required to assure complete reaction, and optimum pH of the solution. Selection of the appropriate oxidizing/reducing agent is influenced by chemical and equipment costs, ease of handling, and safety.

Oxidation and reduction reactions are typically exothermic, and they can be violent. Chemical oxidizing and reducing agents are nonselective. Consequently, leachate containing multiple oxidizable/reducible constituents will exert higher chemical demand.

Evaporation.   Evaporation of waste liquids is also used for treatment and disposal. For example, lined ponds are used to evaporate liquid. Liquid with a high vapor pressure evaporates readily, whereas liquid with a low vapor pressure evaporates more slowly. Evaporation produces a concentrated liquor that contains contaminants, thus reducing the volume of liquid waste that must be treated. Physical separation techniques are generally used before the evaporation process.

Evaporation depends on the vapor pressure and dissolved salts present in the leachate. Evaporators use steam tubes to heat liquid waste to the boiling point.

To conserve energy and enhance separation of volatile wastes from liquids, a multiple-effect evaporator is suitable. In this device the vapor from the first evaporator goes through the tubes to the second evaporator, and the vapor from the second unit flows to the third unit, thus producing the multiple effect (Figure 16–23).

Precipitation/Flocculation/Sedimentation.   Combined precipitation/flocculation/sedimentation is the most common method of removing soluble metals from leachate. Precipitation involves the addition of chemicals to the leachate to transform dissolved contaminants into soluble precipitates. Flocculation promotes agglomeration of the precipitated particles, which in turn facilitates their subsequent removal from the liquid phase by sedimentation (i.e., gravity settling) and/or filtration.

Metals can be precipitated from leachate as hydroxides, sulfides, or carbonates by adding an appropriate chemical precipitant and then adjusting the pH to favor insolubility. The processes of precipitation, flocculation, and sedimentation can be performed in separate basins (Figure 16–24) or in a single basin. Precipitation requires rapid mixing to affect complete dispersion of the chemical precipitant, whereas flocculation requires slow and gentle mixing to promote particle contact. Frequently, flocculants such as alum, lime, ferric chloride, and polyelectrolytes are added along with the precipitant to reduce the repulsive forces between particles and to bring about agglomeration of particles and settling.

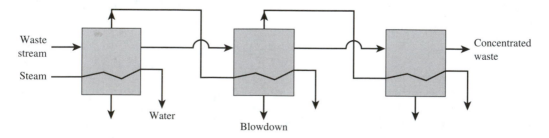

Figure 16–23   Multiple-effect evaporator.

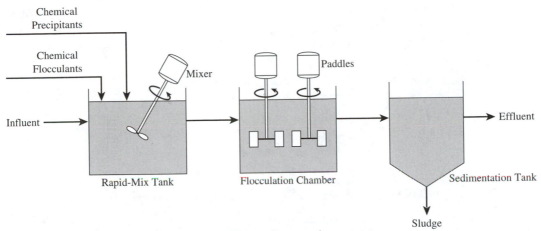

Figure 16–24   Precipitation/flocculation/sedimentation system.

(From U.S. Environmental Protection Agency. A Handbook on Treatment of Hazardous Waste Leachate. *Washington, D.C.: U.S. Govt. Printing Office, 1987; with permission.)*

Precipitation/flocculation/sedimentation is applicable to the removal of most metals (e.g., arsenic, cadmium, trivalent chromium, copper, iron, lead, mercury, nickel, zinc) as well as suspended solids and some anionic species (e.g., phosphates, sulfates, fluorides) from the aqueous phase of leachate. Because precipitation of most metals is conducted at an elevated pH, neutralization of the effluent may be required, particularly if a pH-sensitive biological treatment unit is included downstream. Precipitation/flocculation/sedimentation also generates large amounts of wet sludge, which is generally considered to be hazardous because of its metal content.

Design considerations involve determination of the required chemical dosages, optimum operating pH, degree of precipitation, reaction times, and sludge production and settling rates. These parameters can be determined from bench-scale treatability studies. Because of the corrosive nature of many of the chemicals used for precipitation, special consideration should be given to the materials of which the tanks, pumps, and piping are constructed. Packaged plants are available for low flow rates (10,000 to 2×10^6 gal/day).

Carbon Adsorption.   Carbon adsorption is a technique for removing dissolved organics from leachate. The process involves passing the leachate through beds of granular, activated carbon. Activated carbon is characterized by a large specific surface area. Contaminants are adsorbed from the leachate onto the carbon surface, and they are held there by chemical and physical forces. The carbon surface can be regenerated.

The various means of contacting leachate with granular carbon include fixed-bed, expanded-bed, and moving-bed columns (Figure 16–25). In the fixed-bed adsorber, leachate is distributed at the top of the column, flows downward through the carbon bed (supported by an underdrain system), and is withdrawn at the bottom. When the pressure drop through the column becomes excessive (from the accumulation of suspended solids), the column is taken off-line and is backwashed with the treated

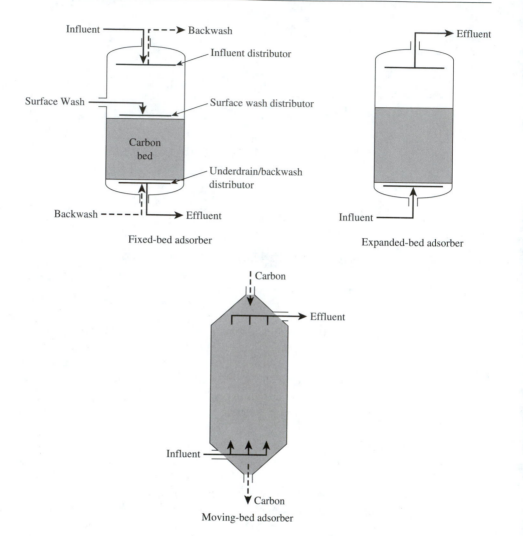

Figure 16–25   Activated carbon adsorbers.

(From U.S. Environmental Protection Agency. A Handbook on Treatment of Hazardous Waste Leachate. *Washington, D.C.: U.S. Govt. Printing Office, 1987; with permission.)*

effluent. The backwash water is then returned to the headworks of the plant treatment. In the expanded-bed adsorber, leachate is introduced at the bottom of the column and flows upward through the bed at a velocity sufficient to suspend the carbon. Backwashing is not needed, because suspended solids pass through the bed with the effluent. In the moving-bed adsorber, leachate is also introduced at the bottom of the column and flows upward through the carbon bed. Spent carbon is withdrawn intermittently from the bottom of the column, however, and is replaced with fresh carbon at the top.

Granular, activated carbon adsorption is a well-developed, standard technology for the treatment of most hazardous waste leachate. It is well suited for removal of mixed

Table 16–4   Operating Parameters for Carbon Adsorption

Parameter	Typical Operating Range
Carbon dosage (1 lb COD/lb carbon)[a]	0.2–0.8
Hydraulic load (gal/min/ft^2)	2–15
Bed depth (ft)	10–30
Contact time (min)	10–50

[a]COD = chemical oxygen demand.

Source: Data from U.S. Environmental Protection Agency. *A Handbook on Treatment of Hazardous Waste Leachate.* Washington, D.C.: U.S. Govt. Printing Office, 1982.

organic contaminants, including volatile organics, phenols, pesticides, PCBs, and foaming agents. Effective pretreatment of leachate is critical to the successful operation of activated carbon adsorption units, however, because if not removed in pretreatment, suspended solids, oil, and grease will accumulate on the surface.

In general, influent concentrations should be limited to 50 mg/L for suspended solids and 10 mg/L for oil and grease. Factors that affect the dynamics of the adsorption process include characteristics of the the adsorbent (e.g., surface area, pore structure size and distribution, particle size), characteristics of the solute (e.g., solubility, molar size ionization), and characteristics of the aqueous system (e.g., temperature, pH, dissolved solids). Pilot tests should be performed to determine the design hydraulic load, bed depth, and contact time; Table 16–4 presents typical ranges for the parameters (9).

LANDFILL CLOSURE AND POSTCLOSURE CONSIDERATIONS

The terms *landfill closure* and *postclosure* are used to describe future plans for a completed landfill. To ensure that completed landfills will be maintained for at least 30 years into the future, many states and the federal government require operators to put aside money so that when the landfill is completed, the facility can be closed, maintained, used, and monitored properly.

In addition to the requirements by federal and state regulating agencies, some plan for productive use of the site must be developed. In general, open-space uses are suitable. Though landfills do not provide firm foundations and sometimes create serious environmental problems, construction is undertaken on many sites (10). It is unsafe to construct dwelling units on decomposing waste, however, because of the explosive hazards associated with landfill gases. Because of the corrosive nature of the fill material, the structural stability and long-range integrity must be considered as well.

The foundation problems of landfills have been discussed in the *Journal of the Sanitary Engineering Division, ASCE* (11). The maximum load the foundation can support is between 500 and 800 lbs/ft^2. Stone columns may be used to provide structural support in landfills. A vibrator is driven into the soil/waste like a pile, and it is used to release stone radially with the subsurface, thus making columns of stone. These columns are designed and constructed for specific uses, including supporting buildings, bridges, overhead tanks, and so on.

Settling of the landfill over time should be carefully studied. Most of the settling will occur during the first 5 to 10 years. Settling also depends on the final use of the landfill. Differential settling can produce stresses on the landfill cover and its components.

The biodegradation products of solid wastes are acidic, thus creating a corrosive environment. High concentrations of CO_2 and organic acids generated by biodegradation of solid waste cause the acidic conditions. Consequently, any material placed in the landfill must be corrosion resistant.

As indicated, the preferred use of a completed landfill site is open space rather than buildings. Open-space uses include parks, parking lots, playgrounds, and hiking areas. In some cases, warehouses are built. The layout, design, and construction of columns for a warehouse are done before erecting the building. Whatever land use is planned after the fill is completed it should be compatible with any applicable zoning regulations. A beneficial and prudent plan for use of the ultimate site—even though far in the future—can generate valuable support among local residents.

One important element in the long-term maintenance of a completed landfill is a plan that clearly states the requirements for closure. A closure plan must include a design for the landfill cover and landscaping of the completed site. It must also include plans for runoff control, erosion control, gas and leachate treatment, and environmental monitoring.

SUBTITLE D REGULATIONS

A number of federal and state regulations have been enacted to improve the performance of sanitary landfills. The principal federal requirements for municipal solid waste landfills (MSWLFs) are contained in Subtitle D of the Resource Conservation and Recovery Act (RCRA).

In October 1991, the U.S. EPA promulgated revisions to the criteria for classification of solid waste disposal facilities and practices as set forth in 40 CRF 257. These revisions were developed in response to the 1984 Hazardous and Solid Waste Amendments to the RCRA. The revision included location restrictions, facility design and operating criteria, groundwater monitoring requirements, corrective action requirements, financial assurance requirements, and closure and postclosure requirements. Differing requirements were established for existing and new units; for example, existing units are not required to remove wastes to install liners.

During the revision, a new part 258 was added, which set forth revised minimum federal criteria for MSWLFs, and was published in the *Federal Register* for October 1991. Part 258 has subparts dealing with the following issues:

Subparts	Issue
A	General
B	Siting limitations
C	Operating criteria
D	Design criteria
E	Groundwater monitoring and corrective action
F	Closure and postclosure
G	Financial assurance

These are discussed here in turn.

Subpart A: General

Subpart A contains the purpose, scope, applicability, and effective date of the revisions.

Subpart B: Siting Limitations

Certain locations have been identified as being incompatible with the operation of a sanitary landfill. These limitations apply to airports, flood plains, unstable areas, wetlands, seismic impact zones, and fault areas (see Table 15–2).

Subpart C: Operating Criteria

Subpart C establishes the operating requirements for new MSWLFs, existing MSWLFs, and lateral expansions. These requirements include hazardous waste exclusions, cover material, control of disease vectors and explosive gases, air criteria, access runoff control systems, surface water requirements, liquid restrictions, and record-keeping requirements. (Cover material, access roads, and runoff control systems were discussed earlier.)

Hazardous Waste Exclusions. Owners or operators of MSWLFs are to implement a program at the facility for detecting and preventing disposal of regulated quantities of hazardous and PCB wastes. This program must include random inspections of incoming loads, records of any inspections, training of facility personnel to recognize regulated hazardous and PCB wastes, and notification to states with authorized Subtitle C programs or to the EPA regional office if hazardous or PCB wastes are discovered at the facility.

Disease Vector Control. Owners or operators of MSWLFs are to prevent or control on-site disease-vector populations by using techniques to protect human health and the environment.

Explosive Gas Control. Owners and operators of MSWLFs are to ensure the concentration of methane generated by the landfill does not exceed 25% of the lower explosive limit in on-site structures such as scale houses or at the facility property boundary. The owner or operator must implement a routine methane-monitoring program, with at least a quarterly monitoring frequency. If the methane concentration limits are exceeded, the owner or operator must notify the director of the state EPA or health department within 7 days that the problem exists and submit—and implement—a remediation plan within 60 days.

Air Criteria. Owners or operators of MSWLFs units are to comply with applicable requirements of state implementation plans pursuant to Section 110 of the U.S. Clean Air Act. Open burning of solid waste is prohibited.

Surface Water Requirements. All MSWLFs must be operated in compliance with the National Pollutant Discharge Elimination System requirements pursuant to the U.S. Clean Water Act. Any discharges of a nonpoint source of pollution from an MSWLF into U.S. waters must be in conformance with the U.S. Clean Water Act.

Liquids Restrictions.   Disposal of bulk or noncontainerized liquid wastes in MSWLFs is prohibited with two exceptions: the waste is household waste (other than septic tank waste), and the waste is a leachate or gas condensate from the MSWLF and the landfill is equipped with a composite liner and leachate collection system.

Record-Keeping Requirements.   Owners and operators of sanitary landfills are required to record and retain environmental monitoring test results or analytical data and inspection records. These records are to be retained at the site and they must be furnished on request by the director of the state EPA or health department.

Subpart D: Design Criteria

Subpart D establishes facility design requirements applicable to new MSWLFs and lateral expansions. The final design criteria provide owners and operators with two basic options: (a) A site-specific design that meets the performance standard and is approved by the director of an approved state EPA or health department or the U.S. EPA, or (b) a composite liner.

Subpart E: Groundwater Monitoring and Corrective Action

Subpart E requires a system of monitoring wells at new MSWLFs, lateral expansions, and existing units. The groundwater-monitoring systems must be in place before any waste is accepted. All existing MSWLFs must have installed a groundwater monitoring system by October 9, 1996. The system must consist of a sufficient number of appropriately located wells to yield representative groundwater samples. The number, spacing, and depths of these wells may be based on site-specific characteristics. The rule under this subpart provides procedures for sampling the monitoring wells and methods for the statistical analysis of groundwater for hazardous constituents released from the MSWLF. Requirements are also included for determination of groundwater elevations and the number of the samples to be collected.

Owners or operators of MSWLFs are required to establish background concentrations and to sample at least semiannually during the active life of the facility, closure, and postclosure periods. If any of the indicator parameters are detected at a statistically significant level above the established background concentrations, the owner or operator must notify the director of the state EPA or health department.

Subpart F: Closure and Postclosure

Subpart F requires owners or operators of new and existing MSWLFs to close each unit in accordance with specified standards and to monitor and maintain the units after closure. In addition, all owners or operators must prepare closure and postclosure plans that comply with the procedural requirements. They must describe what activities will be undertaken to properly close each MSWLF and to maintain them after closure, and they must be placed in the facility operating record. All owners or operators of MSWLFs must install a final cover designed to minimize infiltration and erosion (see Chapter 15). Furthermore, the rule requires all owners or operators to conduct postclosure activities for a period of 30 years after the closure of each MSWLF. During the postclosure care period, all owners or operators must maintain the integrity and effectiveness of the final cover and continue groundwater monitoring, gas monitoring, and leachate management.

Subpart G: Financial Assurance

Subpart G requires all owners or operators new and existing MSWLFs and lateral expansions (except those owned or operated by state or federal government entities) to demonstrate financial responsibility for the costs of closure, postclosure, and corrective action of known releases. The rule requires owners or operators of MSWLFs to demonstrate financial responsibility in an amount equal to the cost of a third party conducting these activities. The cost estimates must be updated annually for inflation or whenever operational or design changes increase the costs at the MSWLF.

CASE STUDY

The Hancock County, Ohio, Sanitary Landfill

The Hancock County Sanitary Landfill is located on Township Road 107 between County Road 140 and Township Road 142 in Allen Township, Section 23. This landfill is owned and operated by the Hancock County Board of Commissioners. Initially the Board purchased 148 acres at $650 per acre. The startup date was September 1, 1969, and the total startup cost was $290,000. This cost included the land, roads, buildings, equipment, and so on. At the time, the tipping cost was set at $3.50 per ton.

The original permit from the Ohio Department of Health allowed approximately 30 acres for landfilling. Subsequent permits were issued by the Ohio EPA for additional landfill area. In 1990, the Ohio EPA notified the Board that the facility must get a permit under new rules based on House Bill 592 (1988) and either conform to these rules or close. A siting permit is now required for lateral expansion or vertical expansion above elevation 901 ft.

In April 1994, the fourth permit was issued by the Ohio EPA to the Board under amended regulations in compliance with the federal Subtitle D regulations. This permit allowed a vertical expansion of 45 acres over the old landfill site, as well as a horizontal expansion of 50 acres to the west of the facility. The permit also addresses the requirements for groundwater and gas monitoring, liner installation, surface water drainage, leachate control, procedures for testing and operation, financial assurance for closure and postclosure care for 30 years, and other regulations. Figure 16–26 shows the layout of the landfill, and Figure 16–27 shows the landfill entrance. In 1996, the permit was modified to allow a "bulge" on east of the existing fill for more waste volume. The landfill receives most of the waste from Hancock County and some from the surrounding

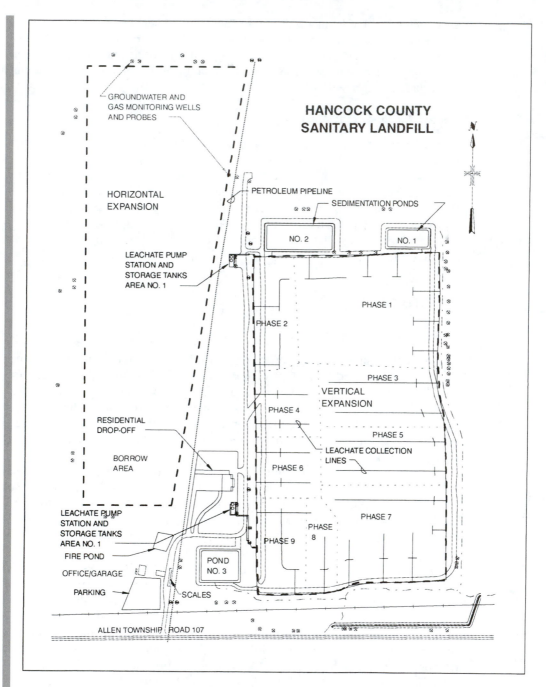

Figure 16–26　Layout of the Hancock County Sanitary Landfill.

Figure 16–27 Entrance to the Hancock County Sanitary Landfill.

counties. Table 16–5 presents the quantities of solid waste received at the landfill from 1991 to 1997.

Currently, the Board has requested the Ohio EPA to allow installation of synthetic liner in the landfill's cap system to replace the present, 2-ft-thick engineered recompacted clay liner system being used. This change in liners will save approximately 1.5 ft of air space for more placement of solid waste in the landfill.

Facility Layout and Technical Description

The Hancock County Sanitary Landfill is permitted as a MSWLF. It can accept municipal and industrial solid waste but no infectious or hazardous waste. The maximum daily waste permitted by the Ohio EPA is 500 tons per day. The average daily tonnage received at the facility was 332 tons per day in 1995, 290 tons per day in 1996, and 314 tons per day in 1997. The expected useful life of the facility is 43 years. More than 76 percent of the waste received is from Hancock County; the remainder is received from the surrounding counties. The landfill is pictured in Figure 16–28, in which the capped cells can be seen in the background. The current vertical expansion will reach a final elevation of approximately 915 ft. The average ground elevation is 795 ft.

Prevailing construction practice is to provide enough capacity for the next year's waste volume. Currently, a 2-ft engineered recompacted clay liner (permeability, 1×10^{-6} cm/s) is used, but plans exist to use high-density polyethylene for both the cap and liner. Figure 16–29 shows a lined cell ready to receive waste, and Figure 16–30 illustrates the waste being received in a prepared cell from trucks. Waste is placed in 2- to 4-ft lifts. It is then spread and compacted to

Table 16–5   Tonnage Totals at the Hancock County Sanitary Landfill

Year	Waste Received (tons)
1985	109,363
1986	98,059
1987	95,719
1988	105,904
1989	70,578
1990	74,240
1991	71,775
1992	75,277
1993	72,271
1994	81,133
1995	94,674
1996	83,085
1997	87,712
1998	—
1999	—
2000	—

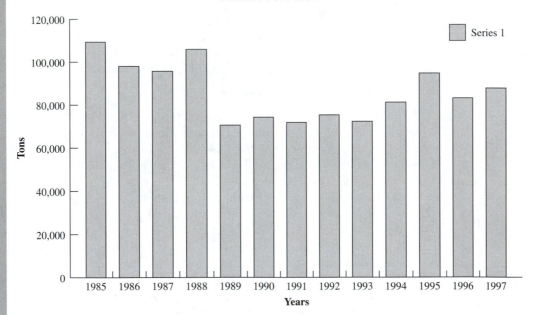

YEARLY TONNAGE

Figure 16–28 Area of the Hancock County Sanitary Landfill. Capped cells can be seen in the background.

Figure 16–29 Lined cell ready to receive waste.

Figure 16–30   Trucks deposting waste in a prepared cell.

1400 lbs/yd^3. Figure 16–31 shows the compactor in operation. To minimize problems created by flying debris, a fence has been erected (Figure 16–32). The landfill is currently being filled in the vertical direction. This expansion is staged into nine phases of operation. As of 1998, waste was being placed in phase 3. An

Figure 16–31   Landfill compactor in operation.

Figure 16–32   Fence to prevent debris from blowing away from the landfill.

interim cover was placed for phases 1 and 2, and a final cover will be placed later. For a daily cover, a spray called Topcoat, which is made from shredded, dense paper, is used. Figure 16–33 illustrates the spraying process.

The landfill has a full-perimeter leachate collection system draining both existing and new waste placement areas. This leachate is hauled off-site, tested, and treated at the Findlay water pollution control facility. On-site, two leachate storage areas are available. Each area is equipped with two 23,500-gal holding tanks with automatic controls, freeze protection, overflow protection, and double

Figure 16–33   Spraying the top coat cover.

containments. Figure 16–34 shows the leachate tanks. The landfill also has two leachate pump stations, each equipped with 100-gal/min pumps. Table 16–6 lists leachate production. Two sedimentation ponds are in operation as well, and these ponds store the surface water drainage. The sediment settles out, and the water is then discharged into the Rocky Ford Creek water shed. Figure 16–35 shows a holding pond for runoff.

The landfill also has dropoff facilities for the recyclables such as paper, plastic, aluminum, and glass (Figure 16–36). In addition, the landfill accepts appliances (Figure 16–37).

The landfill currently has 63 methane gas-monitoring wells. Twenty of these probe wells are 30- to 50 ft in depths, and 43 are punch-bar stations and 5 ft in depth. There are 26 groundwater monitoring wells located around the perimeter of the facility, and these are tested semi-annually. The Ohio EPA wants a gas-collection system for the landfill as well. Therefore, plans are either to flare the gas or to use it as a source of energy for the facility.

Figure 16–34   Leachate tanks and monitoring well.

Table 16–6   Leachate Production from the Hancock County Landfill: Averages for 1994 to 1997

Month	Leachate (gal)
January	68,000
February	120,000
March	140,000
April	244,000
May	308,000
June	92,000
July	232,000
August	376,000
September	412,000
October	116,000
November	100,000
December	280,000
Totals	2,488,000

Ecology and Hydrology

Because of the upper, saturated sandy soil, there may be a possibility of ground-water contamination. Thus, regular monitoring is necessary. The stream that flows near the landfill must be relocated so that the distance between the landfill edge and the stream is at least 200 ft. All drinking-water wells in the area are at least

Figure 16–35   Landfill holding pond for runoff.

Figure 16–36   Residential drop-off center.

Figure 16–37   Appliance drop-off area. (Coolant is removed before disposal.)

1000 ft away from the landfill, however, as mandated by the U.S. EPA. Permits are required for stormwater discharge into the stream. A permit is also required from the U.S. Corps of Engineers for operation near the wetlands.

Equipment

Figure 16–38 shows some of the landfill equipment currently in use. From right to left is a water tank, tractor, Caterpillar (627E), pan, bulldozer, and landfill compactor. Table 16–7 details the various equipment used.

Figure 16–38   Equipment currently in use at the Hancock County Sanitary Landfill.

Table 16–7   Hancock County Landfill Equipment

2 Caterpillar 826C compactor	Ford 445 tractor with loader
Caterpillar 627E pan	Ford WD35 tractor
Dresser TD25G bulldozer	Caterpillar D6 bulldozer
Dresser 250E loader	International 412 waterwagon
Caterpillar D5 bulldozer	White 5000-gal fuel truck
Ford F800 dump truck	2 Miller-Johnson 2200 field disk
Caterpillar padfoot roller	Case 590 backhoe
Dresser SD784 smooth drum roller	Mack rolloff truck
GMC tank truck	4 pick-up trucks
Scale house weigh-in computer	Finn hydroseeding machine
Sweeper	Madvac vacuum

Financial

The financial aspects of the landfill construction and operation are based on the facility's expected life of 43 years. Table 16–8 details costs, revenues, and fees, and Table 16–9 presents revenue for 1997.

Tipping Fee

Details of the current total revenues (i.e., tipping fees less expenses) are given in Table 16–8. The estimated annual cost of operation in 1997 was $1,889,492. For 87,712.32 tons of waste, the tipping fee including district tier and Ohio EPA fees to achieve breakeven cost is

$$\frac{\$1,889,492}{87,712.32 \text{ tons}} = \$21.54 \text{ per ton}$$

Table 16–8   Hancock County Landfill Capital Costs, Expenses, and Revenues

Land

148 acres @ $650.00 per acre in 1969	= $ 96,200.00
In 1997 dollars, assuming 3.5% annual inflation: $96,200 (1 + 0.035)^{28}$	= $252,060.00
Additional Land to be acquired in 1997 or 1998	= $150,000.00
Total	= $402,060.00

Other Starting Costs (roads, buildings, other)

Infrastructure (in 1997 dollars)	= $507,789.00
Contract projects (in 1997 dollars)	= $295,000.00
Total	= $802,789.00

Equipment

In 1997 dollars	= $ 65,000.00

Assumptions

1. The landfill has an expected life of 43 years.
2. Infrastructure will last for 43 years with maintenance.
3. Equipment will have to be replaced at the end of 5 years.

Annual Cost of Capital, (7% interest rate)

present worth (P) × capital recovery factor (CRF)

Land	= $402,060.00 × 0.0744	= $ 29,913.00
Buildings	= $802,789.00 × 0.0744	= $ 59,727.00
Equipment	= $65,000.00 × 0.2489	= $ 15,853.00
Total annual capital cost		= $105,493.00

Expenses in 1997

Capital cost	= $ 105,493
Employee salaries	= $ 380,000
Insurance	= $ 50,000
Workers' compensation	= $ 16,825
Employees retirement	= $ 55,000
Medicare	= $ 6,000
Traveling expenses	= $ 5,000
Office supplies and printing	= $ 8,000
Other expenses	= $ 99,174
Repairs	= $ 40,000
Materials	= $ 374,000
Fees (state and county)	= $ 750,000
Total Annual Cost	= $1,889,492

Current Tipping Fees Per Ton

District (Within County)

Commercial rate	= $21.25
District tier fee	= $ 2.00
Ohio EPA fee	= $ 1.75
Total commercial	= $25.00

Adjacent Counties

Commercial rate	= $23.25
District tier fee	= $ 2.00
Ohio EPA fee	= $ 1.75
Total Commerical	= $27.00

Table 16–9   Hancock County Landfill Revenues for 1997

Month	Total Waste (tons)	Total Revenue ($)	Average Revenue ($/ton)
January	7,545.77	169,363.75	22.44
February	5,982.24	134,779.25	22.53
March	7,400.89	168,145.00	22.72
April	7,963.49	181,252.50	22.76
May	8,043.05	184,705.75	22.96
June	8,849.00	202,267.00	22.86
July	8,551.47	195,310.25	22.84
August	7,520.09	171,517.00	22.81
September	7,507.42	171,182.00	22.80
October	6,783.17	180,832.75	26.66
November	5,162.05	137,418.25	26.62
December	6,403.68	169,615.75	26.49
Total	87,712.32	2,066,389.75	23.56

The district tier and Ohio EPA fees are $2.00 and $1.75, respectively, and they are included in the cost of $21.54 per ton. Thus, the suggested tipping charge should be at least

$$\$21.54 - (\$2.00 - \$1.75) = \$17.79$$

or $18.00.

The Hancock County Landfill competes for waste with another nearby landfill, and it may have to lower its rates to get more business. If it receives approximately 87,712 tons (1997 receipts) of waste or more per year, the rates can be lowered to $18.00 per ton. In fact, the Board is considering lowering the rates. Because of the the ever-increasing costs of meeting regulatory requirements and of operating the facility, however, lowering rates for a significant length of time may not be possible unless a larger tonnage of waste is received for disposal.

Summary and Comments

The Hancock County Landfill was built 28 years ago as little more than a small dump for the area's solid waste disposal. It has gradually evolved into a modern, well-engineered landfill, however, with sufficient capacity to accept waste from adjacent counties. The facility meets the Ohio EPA requirements, as indicated in the permit, and subsequent modifications made under amended regulations in compliance with the federal Subtitle D regulations. The facility has extended both vertically and horizontally to increase the capacity.

The facility has proposed installing an active gas collection system in the near future. It currently has a program to monitor the gas wells and groundwater wells as mandated by the Ohio EPA. The liner and cap systems and the daily cover are technically sound and cost-saving items. Replacement of the 2-ft-thick compacted clay by a synthetic liner will result in a space-savings of approximately 1.5 feet over the entire facility.

The landfill has been able to generate sufficient revenues to meet all its operation and maintenance costs and to save some funds for future needs. Another area landfill is in operation now, however, and is competing for waste, so the Board of the Hancock County Landfill Commissioners may need to lower its rates. This landfill is an example of a free-market economy and good management practices.

LANDFILL GAS GENERATION

The principal landfill gases that are generated are carbon dioxide and methane. Generally, decomposition of waste in a landfill is thought to occur in five or fewer sequential phases (12), as illustrated in Figure 16–39 and described here.

Phase I

In phase I, initial adjustments occur. Organic, biodegradable components begin to undergo bacterial decomposition within a few days after the waste is placed in a landfill. The decomposition is aerobic, because air is present in the solid waste voids. The oxygen is rapidly used up by aerobic microorganisms, however.

Phase II

Phase II is called the transition phase. The oxygen is depleted, and the anaerobic phase begins. Under anaerobic conditions, nitrates and sulfates are reduced to nitrogen gas and hydrogen sulfide through biological reactions. Gradually, the microorganisms begin to convert the organics into organic acids. The pH of the leachate (if formed) begins to drop because of the formation of organic acids. Higher concentrations of carbon dioxide are noted as well.

Phase III

Phase III is called the acid phase. The bacterial metabolism process started in phase II now accelerates, resulting in organic acid production at a fast rate. The microorganisms are called acid-formers or acidogens. More carbon dioxide is produced as well. This phase can last for several years. Leachates produced during this period are characterized by a high BOD, acidic leachate (typically with pH of 5 or less), odors, and high ammonia concentrations (as high as 1000 mg/L). Because of the low pH of the

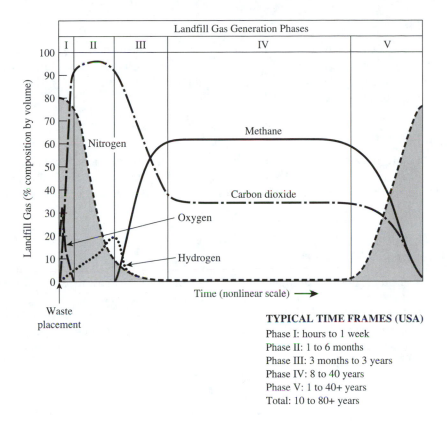

Landfill Gas Generation Phases

TYPICAL TIME FRAMES (USA)
Phase I: hours to 1 week
Phase II: 1 to 6 months
Phase III: 3 months to 3 years
Phase IV: 8 to 40 years
Phase V: 1 to 40+ years
Total: 10 to 80+ years

Figure 16–39   Typical time frames for the phases of landfill gas generation.

(From Furguhar, G.J., and F.A. Rovers. "Gas Production during Refuse Decomposition," Journal of Water, Air and Soil Pollution, *Vol. 2; 1973; with permission.)*

leachate, certain inorganic constituents (mainly heavy metals) will go into the solution. If leachate is not formed, the products will remain within the landfill until leachate is produced.

Phase IV

Phase IV is known as the methane phase, during which a second group of microorganisms (i.e., methanogens or methane-formers) develop. These microorganisms convert the acetic acid and hydrogen gas formed earlier by the acid-formers into methane and carbon dioxide. The bacteria responsible for the formation of methane and carbon dioxide are strict anaerobes. In this phase, both methane and acid fermentation proceed, but the rate of acid fermentation is reduced.

Because the acids produced by acid-formers have been converted to methane and carbon dioxide, the pH within the landfill will rise (usually to close to 7). The concentrations of BOD and the solubility of inorganic matter will decrease as well.

Phase V

Phase V is identified as the maturation phase. It is characterized by conversion of almost all biodegradable organics into methane and carbon dioxide. During the maturation phase, the rate of landfill gas generation decreases significantly, and the leachate will contain higher concentrations of acids.

Duration of Phases

Figure 16–39 illustrates typical time frames for landfill gas-generation phases. The duration of each phase will vary depending on the organic matter, available moisture, and degree of initial compaction. Lack of moisture and increased density will reduce the rate of bioconversion and gas production.

LANDFILL GAS MIGRATION

The production and accumulation of gases increases the gas pressures in the landfill, and the resulting pressure differential is one of the forces driving these gases to travel beyond the confines of the solid waste. These gases can escape into the atmosphere, move into the surrounding soil environment, or leak into the groundwater. Consequently, migration of landfill gases from disposal sites and control of such gases are receiving significant attention.

Landfill gases can result in serious safety and health hazards, because methane gas is explosive. Landfill gases migrate from the waste through soils and into adjacent structures, and there are several recorded instances of explosions in structures built on or near a landfill, resulting in the loss of property and of life. The potential for explosion develops when methane migrates from a landfill and mixes with air in a confined space. In addition, release of hydrogen sulfide (H_2S) can cause odor problems. Both federal and state regulatory agencies have increased the requirements for landfill gas control.

There are two potential sources of gas from landfills: gases produced by the decomposition of organic solids in the solid waste, and volatilization of volatile organic compounds such as organic solvents disposed in the landfill. The latter has been greatly reduced (and will continue to be reduced) because of the ban on the disposal of hazardous wastes in sanitary landfills.

Normally, gas will move through porous media such as soil through two processes: diffusion because of a concentration gradient, and advection because of a pressure gradient. Gas moves preferentially along paths of lowest resistance (i.e., high conductivity) with discharge to the atmosphere. If the soil in the liner is very tight, gas will not move through it easily. If a tight cover is provided that also has resistance to gas diffusion, the internal gas pressure of the landfill may increase enough to rupture the cover. The gas then flows through a free opening driven by the pressure differential. A gas escape or removal plan should be provided to prevent gas migration or rupture of the liner (Figure 16–40).

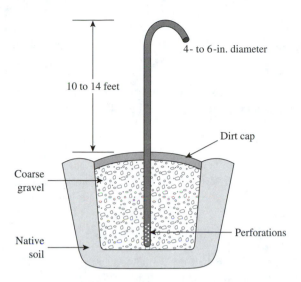

Figure 16–40   Vertical gas vent pipe.

GAS PRODUCTION POTENTIAL

The gas production potential can be determined from the quantity of solid waste in the landfill, and a rough approximation of landfill gas production can be estimated using the amount of waste in place as the only variable. The procedure described here for approximating gas production is derived from the ratio of waste quantity to gas flow observed in the many (and often different) projects in operation at various sites. It reflects an average landfill that has an energy project, and it may not accurately reflect the waste, climate, and other characteristics. Therefore, it should be used primarily as a screening tool.

A simple approximation method only requires knowledge of how much waste is in place at the target landfill. Based on their extensive experience at many landfills, industry experts have developed a rule of thumb: landfill gas generation rates range from 0.05 to more than 0.20 ft^3 of gas per 1 lb of refuse per year, with an average landfill generating 0.10 ft^3 of landfill gas per 1 lb of waste per year (13).

Method I

Using the rule of thumb previously described results in the following equation:

Annual landfill gas generation (ft^3) = 0.10 ft^3/lb

$$\times\ 2000\ \text{lbs/ton} \times \text{waste in place (tons)}$$

Assume a landfill is receiving 100 tons per day for 250 days per year. Thus,

$$\text{waste in place} = 100\ \text{tons/day} \times 250\ \text{days} = 25{,}000\ \text{tons}$$

and

annual landfill gas generation (ft^3) = 0.10 ft^3/lb × 2000 lbs/ton × 25,000 tons

$$= 5 \times 10^6 \text{ ft}^3/\text{year or } 20{,}000 \text{ ft}^3/\text{day}$$

The uncertainty associated with this should be accounted for by adding and subtracting 50 percent, thus yielding a range for the landfill gas flow of from 10,000 to 30,000 ft^3/day.

Method II

The second approach, which is a first-order decay model, can be used to account for changing gas generation rates over the life of a landfill. The basic first-order decay model is as follows:

$$LFG = 2 L_0 R \, (e^{-kc} - e^{-kt})$$

where *LFG* is the total amount of landfill gas generated in the current year (ft^3), L_0 is the total methane generation potential of the waste (ft^3/lb), *R* is the average annual waste acceptance rate during active life (lb), *k* is the rate of methane generation (per year), *t* is the time since landfill opened (years), and *c* is the time since landfill closure (years).

The methane generation potential, L_0, represents the total amount of methane that 1 lb of waste is expected to generate during its lifetime. The decay constant, *k*, represents the rate at which the methane will be released from each pound of waste. If these terms were known with certainty, the first-order decay model would predict methane generation relatively accurately. The values for L_0 and *k*, however, vary widely and are difficult to estimate accurately for any particular landfill.

The values for L_0 and *k* depend in part on the local climatic conditions and waste composition. Therefore, a landfill owner or operator may want to consult others in the local area with similar landfills who have installed gas collection systems to narrow the range of potential values. In March 1996, the U.S. EPA issued final regulations for the control of landfill gas at new and existing MSWLFs with design capacities of 2.5 million metric tons or more (14). Affected landfills model their gas emissions using the first-order decay model. The regulations include the following default values (as well as a nonmethane organic compound default value of 4000 ppm, which can be replaced with a site-specific at a later date):

		Suggested Values		
Variable	Range	Wet Climate	Medium Moisture Climate	Dry Climate
L_0 (ft^3/lb)	0–5	2.25–2.88	2.25–2.88	2.25–2.88
k (per year)	0.003–0.4	0.1–0.35	0.05–0.15	0.02–0.10

New source performance standard (NSPS) values are:

$$L_0 = 2.72 \text{ ft}^3/\text{lb}, \; K = 0.05 \text{ per year}$$

Example 16-1

Estimate the yearly gas generation for a landfill with the following characteristics:

- Open for 20 years.
- Still accepting waste.
- Average annual waste acceptance rate of 25,000 tons.

Solution

The first-order decay model yields a rough estimate by the equation

$$LFG = 2L_0R(e^{-kc} - e^{-kt})$$

Using NSPS k and L_0 values,

$$LFG = 2 \times 2.72 \times 2000 \times 25,000$$
$$- (e^{-0.05(0)} - e^{-0.05(20)}) = 172 \times 10^6 \text{ ft}^3/\text{year}$$

The first-order decay model yields a rough estimate of landfill gas per year of approximately 471,233 ft^3/day (using the NSPS k and L_0 values). The uncertainty associated with this estimate should be accounted for by adding and subtracting 50 percent, thus yielding a range of 235,617 to 706,850 ft^3/day.

GAS RECOVERY AND MIGRATION CONTROL

The U.S. (15) requires that the combustible gas concentration shall not exceed 5 percent at or beyond the property boundary. Several methods are available to control or prevent the migration of gas from landfills.

Pumping of the gas or gas control systems may be used. These can be natural barriers, (e.g., saturated coarse-grained soils) or constructed barriers (e.g., trenches, wells, vents, membranes). Constructed barriers can be categorized into either active- or passive protection systems.

Figure 16-41 illustrates a passive system for control of gas migration. Such systems are installed where gas generation is low and off-site migration is not expected. The system may consist of a series of isolated wells or gas vents to allow the gas to escape. In passive-control systems, the pressure of the gas generated within the landfill serves as the force driving movement of the gas. In addition, a trench filled with granular backfill or with gravel and a membrane intercepts the migrating gas, which is allowed to escape by diffusing upward through the trench to the atmosphere.

Figure 16-42 illustrates an active venting system. This system consists of a series of deep extraction wells that are connected by a header pipe to a blower that either delivers the gas for energy or to an on-site burner or releases it to the atmosphere.

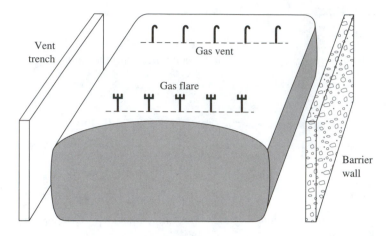

Figure 16–41 Passive methane gas venting systems.

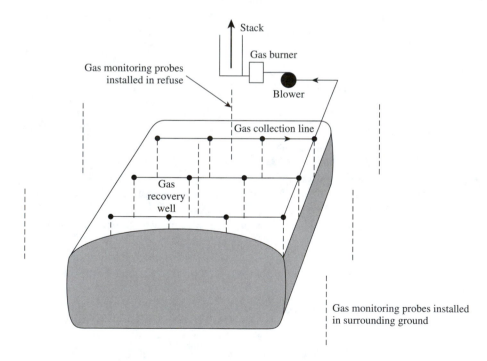

Figure 16–42 Active (mechanical) method of gas withdrawal from the surrounding formation.

For gas recovery, additional wells can be installed in the fill for more efficient recovery. These wells extend deep into the fill.

Typical landfill gas collection systems have three central components: collection wells, a condensate collection and treatment system, and a compressor. In addition, most landfills with energy recovery systems have a flare for the combustion of excess gas and for use during equipment downtime.

Figure 16–43 illustrates a typical landfill gas extraction well. Gas collection typically begins after a portion of a landfill (i.e., a cell) is closed. There are two system configurations: vertical wells, and horizontal trenches. Vertical wells are by far the most

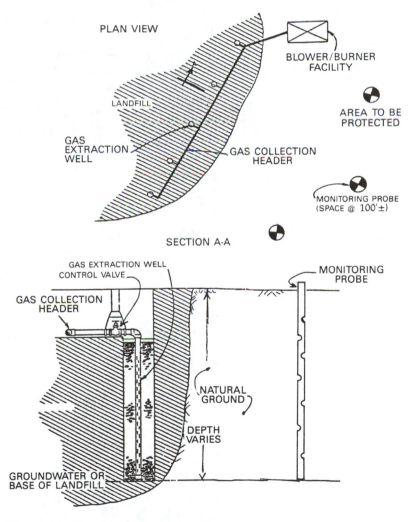

Figure 16–43   Active gas extraction.

(From U.S. Environmental Protection Agency. Remedial Action at Waste Disposal Sites. *Washington, D.C.: U.S. Govt. Printing Office, 1980; with permission.)*

common type used for gas collection. Trenches may be appropriate for deeper land-fills. Regardless of which configuration is used, each wellhead is connected to lateral piping, which transports the gas to a main collector header. Spacing of wells will typ-ically be on a grid at distances ranging from 100 to 200 ft.

The monitoring probe indicates if methane is migrating from the site. The probe can be a simple, perforated polyvinyl chloride pipe installed in the soil adjacent to the landfill and embedded in a bore hole that is backfilled with gravel or coarse sand. The top of the probe is then sealed with concrete or clay. Methane concentrations and in-ternal pressures can be measured with the probe (Figure 16–44), and a complex probe can sample from different levels.

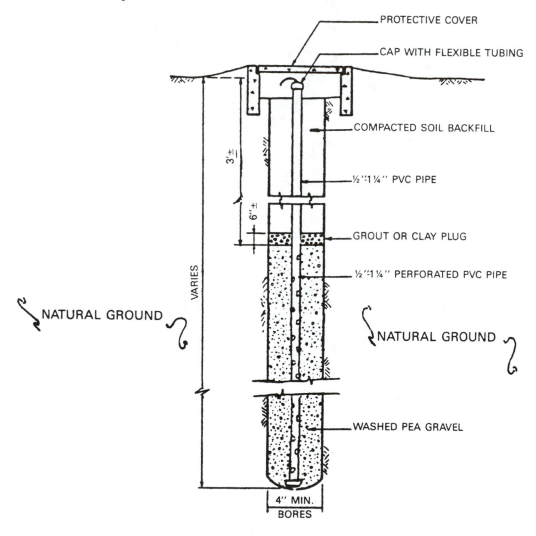

Figure 16–44 Typical gas monitoring probe.

(From U.S. Environmental Protection Agency. Remedial Action at Waste Disposal Sites. *Washington, D.C.: U.S. Govt. Printing Office, 1980; with permission.)*

LANDFILL GAS COMPOSITION AND CHARACTERISTICS

Landfill gas is actually composed of several gases. The principal gases are produced from decomposition of the organic fraction of waste. Although present in small quantities, a few of the trace gases can be toxic and present risks to public health. Table 16–2 summarized data on the concentrations of trace compounds in landfill gas samples from BFI MSW landfills. Many of these compounds are volatile organics, and the concentration of these compounds varies depending on factors such as composition of the waste and age of the landfill.

Principal landfill gases include methane, carbon dioxide, carbon monoxide, ammonia, hydrogen sulfide, hydrogen, nitrogen, and oxygen. Table 16–10 reports the typical percentage distribution of gases in a MSWLF. Methane and carbon dioxide are the principal gases produced from the anaerobic decomposition of biodegradable organics. In the range of 5- to 15-percent concentrations in the air, methane can be explosive, and such mixtures can form if landfill gas migrates off-site and mixes with air.

LANDFILL GAS UTILIZATION

The heat value of raw landfill gas (mostly CH_4 and CO_2) is between 400 and 600 Btu/ft^3. Other gases may also be found in this gas (e.g., H_2S, NH_3, CO) depending on the materials received at the site. Older landfills may have industrial wastes with volatile organic compounds. New landfills will not have this problem, however, because hazardous wastes are prohibited in MSWLFs. Even so, there still may be small amounts of household solvents and similar items going into MSWLFs but that are not detected in the gas.

The goal of a landfill gas-to-energy project is to convert landfill gas into useful energy such as electricity, steam, boiler fuel, vehicle fuel, or pipeline-quality gas. Several factors, including the existence of an available energy market, project costs, potential revenue sources, and many technical considerations, are relevant.

Incineration

Incineration is a control measure intended to limit the migration of gas from a landfill site and to mitigate odors that result from emissions of landfill gas to the atmosphere. If the gas is to be used in a conversion process, it must be treated to remove any

Table 16–10   Typical Constituents of Landfill Gas

Component	Percentage (dry volume basis)
Methane	40–60
Carbon dioxide	30–50
Oxygen	0–2
Nitrogen	0–10
Sulfides, disulfides, and so on	0–1.0
Ammonia	0.1–1.0
Hydrogen	0–0.2
Carbon monoxide	0–0.2

Source: Landfill Gas Seminar Proceedings, Volume II, Baltimore, MD, September 1996; with permission.

condensate, carbon dioxide, and other impurities. Table 16–11 presents various uses for landfill gas.

Gas Treatment Systems

After the landfill gas has been collected but before it can be used in a conversion process, it must be treated to remove any condensate, particulates, and other impurities. Treatment depends on the end-use application. Minimal treatment is required for gas to be used in boilers. For injection into a natural gas pipeline, however, it is necessary to remove the carbon dioxide. Power-production applications typically include a series of filters to remove impurities, and semipermeable membranes have been applied successfully to remove carbon dioxide. Membranes are thin barriers that allow preferential passage of certain molecules while retaining others (e.g., allowing CO_2, H_2S, and H_2O to pass while retaining CH_4. The cost of gas purification with membrane processes can be affected by the design of the process. A single-stage membrane (i.e., a low-cost process) is adequate for some applications. A more expensive design may be needed to achieve a higher product recovery rate.

Separation of carbon dioxide from methane can also be accomplished by physical and chemical absorption. The landfill gas is passed into an absorption column under pressure and is contacted by a solvent. The solvent absorbs the carbon dioxide, and the clean gas is then piped out for use.

Sale of Medium-Btu Gas

The simplest—and often most cost-effective—use of landfill gas is as a medium-Btu fuel for boiler or industrial use. Gas is piped directly to a nearby customer. For example, the Kentucky–Tennessee Clay Company in Aiken County, South Carolina, burns landfill gas in its rotary dryer to dry clay before shipment.

Power Generation

The most prevalent use for landfill gas is as a fuel for power generation, with the electricity then being sold to a utility company. Power generation is advantageous, because it produces a valuable end product: electricity. Facilities that use landfill gas to generate electricity can also qualify as a "small power producer" under the Public Utilities

Table 16–11   Uses for Landfill Gas

Medium-Energy Use[a]	High-Energy Use[b]
Internal-combustion engine	Sale to gas companies
Gas turbine	CO_2 recovery
Power generation	Use in vehicles
Steam production	
Sale to industries	

[a] Gas includes CO_2.

[b] CO_2 removed from gas.

Regulatory Policy Act, which requires electric utilities to purchase the output of such facilities.

Upgrade to High–Btu Gas

Landfill gas can be upgraded through extensive treatment to remove carbon dioxide and other impurities. This high-Btu product is then injected into a natural-gas pipeline.

Alternative Uses

Other options for using landfill gas include on-site use, where is appropriate for small landfills, and for heating greenhouses.

REVIEW QUESTIONS

1. List the problems associated with operation of a landfill.
2. How is landfill leachate formed?
3. a. What is the purpose of leachate removal?
   b. Draw a schematic for a landfill leachate collection system.
4. a. List the parameters used to characterize leachate.
   b. What factors affect leachate composition?
5. a. List the technologies used for treating leachate.
   b. Discuss primary and secondary treatment systems.
6. What are the objectives of the following treatment processes?
   a. Physical.
   b. Chemical.
   c. Biological.
7. Suggest a treatment system for landfill leachate. Explain the function of each unit with the help of schematic.
8. a. Explain the removal of organics from liquid waste by use of an activated sludge treatment system.
   b. What are the main components of such an activated sludge system?

9. a. Compare the aerobic and anaerobic processes.
   b. What are the advantages and disadvantages of each?
10. What physical and biological principles are involved in the trickling filter and aerobic pond treatment systems?
11. a. What reactions occur in a municipal solid waste landfill?
    b. Describe the various phases of decomposition in a landfill.
12. Calculate the expected generation of gas from a landfill that has been operating for 10 years and receives 40,000 tons of waste per year.
13. a. List the problems associated with gas migration.
    b. Suggest methods to control gas migration.
13. Assume you are an engineer with the U.S. EPA in charge of closure and postclosure landfill compliance:
    a. Discuss the current federal and state landfill closure and postclosure requirements.
    b. How would you ensure these requirements are met?

REFERENCES

1. McGinley, P.M., and P. Kmet. *Formation, Characteristics, Treatment and Disposal of Leachate from Municipal Solid Waste Landfills.* Madison, WI: Wisconsin Dept. of Natural Resources, 1984.

2. Chian, E.S.K., and F.B. DeWalle. "Sanitary Landfill Leachate," *Journal of the Environmental Engineering Division, ASCE,* Vol. 102; 1976, pp. 411–431.

3. Pohland, F.G., and S.R. Harper. *Critical Review and Summary of Leachate and Gas Production from Landfills.* EPA-600/2-86-073. Washington, D.C.: U.S. Govt. Printing Office, 1986.

4. Kmet, P., and P.M. McGinley. "Chemical Characteristics of Leachate from Municipal Solid Waste Landfills," Paper Presented at the Fifth Annual Madison Conference of Applied Research and Practice on Municipal and Industrial Waste, University of Wisconsin, September 1982.

5. Ehrig, H.J. "Leachate Treatment Overview," Paper Presented at the Second International Landfill Symposium, Italy, 1989.

6. Robinson, H.S. "On-Site Treatment of Leachates from Landfilled Wastes," *Journal of Institute of Water and Environmental Management,* Vol. 4, No. 1, February 1990, pp. 36–39.

7. Zapf-Gilje, R. "Effects of Temperature on Two-Stage Biostabilization of Landfill Leachate," Master's Thesis in Civil Engineering, University of British Columbia, 1979.

8. Metcalf and Eddy, Inc. *Wastewater Engineering Treatment Disposal.* New York: McGraw-Hill, 1985.

9. U.S. Environmental Protection Agency. *A Handbook on Treatment of Hazardous Waste Leachate.* Washington, D.C.: U.S. Govt. Printing Office, 1982.

10. U.S. Environmental Protection Agency. *Construction of Structures on Landfills.* SW-846. Washington, D.C.: U.S. Govt. Printing Office, 1993.

11. "Foundation Problems in Sanitary Landfills," *Journal of the Sanitary Engineering Division, ASCE,* February 1968.

12. Farquhar, G.J., and F.A. Rovers. "Gas Production during Refuse Decomposition," *Journal of Water, Air, and Soil Pollution,* Vol. 2, 1973.

13. Waste Management of North America, Inc. "Landfill Gas Recovery Projects," Paper Presented at the Solid Waste Association, 15th Annual Landfill Gas Symposium, New York, NY, March 24–26, 1992.

14. 61 *Federal Register,* 9905, March 12, 1996.

15. U.S. Environmental Protection Agency. *RCRA Orientation Manual.* Washington, D.C.: U.S. Govt. Printing Office, 1995.

HAZARDOUS WASTE MANAGEMENT TECHNOLOGY

Chapter 17   Characteristics of Hazardous Waste

This Chapter defines hazardous waste and discusses its characteristics and quantities. Thus, readers can appreciate the problems involved in managing such wastes.

Of all human activities to change or modify nature, none is as alarming as the creation of chemical compounds. Modern chemists develop approximately 1000 new concoctions each year, and 60,000 different commercial chemicals are produced in the United States alone. Many of these chemicals have greatly benefited mankind, but there is a price to pay for an industrial society that relies heavily on chemicals: almost 35,000 of those used in the United States are classified by the U.S. Environmental Protection Agency (EPA) as being hazardous to human health and the environment. Many of these chemicals are discharged in the environment (i.e., land, air, or water) as waste either by design or by accident during storage or transport. Until the late 1970s, in the absence of any meaningful and clear regulations, these hazardous wastes were stored or discharged at various sites without any concern for the public health and the environment. Such sites came to be known as toxic dumps.

No accurate count of these toxic dumps is possible, but in 1980, the EPA estimated there were approximately 50,000 chemical dump sites, of which 2,000 posed serious health hazards. The public got an indication of the seriousness of this problem in 1970 with the revelation of serious problems in New York's Love Canal and other areas. For example, in Love Canal, contamination from a landfill laced with chemicals seeped into the area on the outskirts of Niagara Falls. Some 1200 houses and a school had been built near the site, and many residents suffered serious health problems, which included a high incidence of cancer, respiratory and neurological problems, and birth defects. Concerned by these health problems, the federal government (under pressure from the Carter Administration) finally paid to evacuate the affected families in the Love Canal area. In Times Beach, Missouri, thousands of gallons of dioxin-contaminated oil were sprayed onto the streets, and many people were affected by birth defects, respiratory illness, cancer, and neurological problems. People moved away and boarded up their homes, and suddenly, Times Beach turned into a ghost town.

In response to public demand, the U.S. Congress passed the Comprehensive Environmental Response, Compensation, and Liability Act (CERCLA) in 1980 to deal with contaminated sites. In 1980 and again in 1984, Congress also revised and strengthened the Resource Conservation and Recovery Act (RCRA), which controls new sources of hazardous waste. In other words, CERCLA deals with the problems of the past, whereas RCRA aims to prevent future problems. These two laws have directed almost all efforts at hazardous waste control.

DEFINITION OF HAZARDOUS WASTE

Published definitions and classifications of hazardous waste vary. In the United States, wastes are defined by the U.S. EPA (i.e., RCRA), CERCLA (also known as Superfund), and the U.S. Department of Transport (DOT). The EPA defines a waste as being hazardous if it may cause or significantly contribute to an increase in mortality or serious, irreversible, or incapacitating, reversible illness; pose a substantial present or potential hazard to human health or the environment when improperly treated, stored, transported, disposed of, or otherwise managed; and the characteristics can be measured by a standardized test or reasonably detected by generators of solid waste (1). This may give a broad or general definition, but it is not a workable definition. More specific criteria are needed to identify a waste material as being hazardous or nonhazardous. Here, the distinction must also be made between a hazardous substance and a hazardous waste. A hazardous substance has some commercial value, because it is usable. A hazardous waste, however, is a material that has been used, spilled, or is no longer needed.

A waste (i.e., solid, liquid, semiliquid, or contained gas) can be considered "hazardous" by the EPA under Subtitle C for three reasons:

1. The waste is declared hazardous by the generator.

2. The material exhibits ignitable, corrosive, reactive, or toxic characteristics.

3. It is specifically listed as such by the EPA. (This list includes nonspecific source wastes, specific source wastes, and certain commercial products.)

GENERATION OF HAZARDOUS WASTE

Hazardous waste can originate from a range of industrial, commercial, household, agricultural and institutional activities and from both manufacturing and nonmanufacturing facilities and processes. Nonmanufacturing activities may include service and wholesale trade establishments as well as universities, hospitals, government facilities, and households.

After a waste is generated, the generator can either manage the waste on-site or transport it off-site (usually to a commercial hazardous waste facility) for treatment, disposal, or recycling. Hazardous waste is termed on-site waste or off-site waste depending on whether it is managed either on the site where it is generated or at a site other than where it is generated, respectively. Hazardous waste regulations in the United States place the responsibility for proper disposal on the generator of that waste. The physical removal of hazardous waste from a site constitutes generation. Generators are subject to the regulations contained in the 40 CFR 262 (Generator Standards). The EPA monitors generator activity by assigning an EPA identification number to each generator, each hauler, and each treatment, storage, and disposal facility.

Magnitude of the Problem

An EPA survey conducted in 1985 identified 2959 facilities regulated under the RCRA, which in turn managed a total of 247 million tons of hazardous waste (2). An additional 322 million tons of hazardous waste were handled by units exempt from RCRA reporting requirements. In 1987, there were 17,677 large-quantity generators (>1000 kg/month) in the United States. The reported quantities of hazardous waste generated were in the range of 275 to 300 million tons per year through most of the 1980s. In the 1990s, however, levels are estimated to be below the range of 275 to 300 million tons per year because of the emphasis on pollution prevention, effect of land-disposal restrictions, and escalating land-disposal costs.

Generator Responsibilities

Generators are responsible for determining whether the wastes they generate are hazardous and, if so, for managing them in compliance with 40 CFR 261.20–261.24. The Ohio EPA provides a generator with two options for hazardous waste determination. The generator can make this determination by "applying knowledge of the hazard characteristic of the waste in light of the materials or the process used" or by testing the waste using standard methods specified in the Ohio Administrative Code Chapter 3745-51. Determining if materials are hazardous wastes involves classifying them as waste, ensuring they are not excluded from RCRA regulations, and evaluating them for the hazardous properties outlined in the regulations. Table 17–1 lists wastes that are excluded from RCRA regulations.

The U.S. EPA recognizes that hazardous waste regulations would impose a substantial burden on small generators if they were brought entirely within the Subtitle C regulatory system. Thus, the EPA has exempted small-quantity generators from most hazardous waste requirements. The EPA defines generators as follows:

- *Large-quantity generators* are those facilities producing more than 1000 kg of hazardous waste per calendar month (approximately 2200 lbs) or more than 1 kg of acutely hazardous waste per calendar month (approximately 2.2 lbs).

- *Small-quantity generators* are those facilities producing from 100 kg (approximately 220 lbs) to 1000 kg (approximately 2,200 lbs) of hazardous waste per calendar month and accumulating less than 6,000 kg (approximately 13,200 lbs) of hazardous waste at any time.

- *Conditionally exempt small-quantity generators* are those facilities producing less than 100 kg of hazardous waste or less than 1 kg of acutely hazardous waste per calendar month. The requirements for this category additionally limit the facility's waste accumulation to less than 1000 kg of hazardous waste, 1 kg of acute hazardous waste, or 100 kg of any residue from the cleanup of a spill of acutely hazardous waste at any time.

Table 17–1   Waste Excluded from RCRA Regulations

Domestic sewage and mixtures of domestic sewage and other wastes.

Industrial wastewater point source discharges subject to regulation under the Clean Water Act.

Irrigation return flows.

Special nuclear or byproduct material.

Materials subject to *in-situ* mining techniques.

Pulping liquors that are reclaimed in a pulping furnace.

Spent sulfuric acid used to produce virgin sulfuric acid.

Secondary materials that are reclaimed and returned to the original process.

Spent wood-preserving solutions that are reclaimed and reused, and wood-preserving wastewaters reclaimed and reused to treat wood.

Toxicity characteristic and listed, hazardous coke byproduct residues that are used in producing coke and coal tar.

Household waste.

Agricultural waste returned to the soils as fertilizers.

Mining overburden returned to the mining site.

Ash from combustion of coal or other fossil fuels.

Drilling fluids and other waste associated with production of crude oil, natural gas, or geothermal energy.

Wastes failing the total carbon test for chromium if the chromium is in the trivalent state and generated in the leather tanning and finishing industry.

Waste from the extraction and processing of ores and minerals.

Cement kiln dust.

Arsenical-treated wood or wood products failing the total carbon test for arsenic, used as intended and discarded.

Petroleum-contaminated media and debris failing the total carbon test for D018 through D043 (newly identified organic constituents only) that are subject to corrective action under the Underground Storage Tank program described in 40 CFR 280.

Reclaimed chlorofluorocarbon refrigerants.

Non-teme plated used oil filters that are drained.

Waste generated in a product or raw material storage tank or in a manufacturing process unit is not considered hazardous until the waste exits the unit or until the unit ceases to operate for more than 90 days.

Samples of solid waste or samples of water, soil, or air tested for hazardous characteristics or composition as long as conditions under OAC 3745-51-04(D) are met.

Treatability study samples as described under OAC 3745-51-04(E) and (F).

Residues of hazardous waste in empty containers as defined under OAC 3745-51-07.

Recycled materials that are (1) used or reused as ingredients in an industrial process to make a product, provided the materials are not being reclaimed, or (2) used or reused as effective substitutes for commercial products, or (3) returned to the original process from which they are generated without first being reclaimed. OAC 3745-51-06 provides the following list of recyclable materials not subject to hazardous waste regulations:

a. Industrial ethyl alcohol that is reclaimed unless otherwise specified in an international agreement.

b. Used batteries returned to a battery manufacturer for regeneration.

c. Used oil that exhibits one or more of the characteristics of a hazardous waste that is recycled in some manner other than burned for energy recovery.

d. Scrap metal.

e. Fuels produced from the refining of oil-bearing hazardous wastes.

f. Oil reclaimed from hazardous waste resulting from petroleum refining, production, and transportation practices.

g. Coke and coal tar from the iron and steel industry that contains K087.

h. Hazardous waste fuel produced from oil bearing hazardous wastes from petroleum production and transportation practices.

i. Petroleum coke produced from petroleum refinery hazardous wastes containing oil.

Source: Ohio Administrative Code (OAC), Chapter 3745-51, 1996.

■ All generators are required to meet certain requirements. For example, they must test their waste, store it properly, and dispose of it by approved methods. They are also to maintain records as hazardous waste generators, get an EPA permit number, and use an EPA manifest form.

IDENTIFICATION AND CHARACTERIZATION OF HAZARDOUS WASTE

For a material to be classified as hazardous waste, it must first be considered a waste and also meet at least one of the following criteria:

1. It exhibits any characteristic of a hazardous waste.

2. It has been named a hazardous waste and listed as such in the EPA regulations.

3. It is a mixture containing a listed hazardous waste and a nonhazardous solid waste.

4. It is a mixture containing a listed or characteristic hazardous waste and special nuclear material.

5. It is a waste residue generated from the treatment, storage, or disposal of a listed hazardous waste (i.e., is a "derived-from" waste).

Characteristics

Characteristic wastes are those that exhibit measurable properties indicating that a waste poses enough of a threat to deserve regulation as a hazardous waste. The U.S. EPA decided that the characteristics of hazardous waste should be detectable by using a standardized test method or by applying general knowledge of the waste properties. Given these criteria, the EPA established four hazardous waste characteristics: ignitability, corrosivity, reactivity, and toxicity.

Ignitability.   Ignitible wastes are easily combustible or flammable, and they may cause fires during their transport, storage, or disposal. Examples include paint waste, certain degreasers, or other solvents. Regulations describing the characteristic of ignitability are codified at 40 CFR 261.21. A waste that exhibits any of the following properties is considered to be a hazardous waste identified by the waste code D001:

■ A liquid, except aqueous solutions, containing less than 24-percent alcohol and that has a flash point of less than 60°C (140°F)

■ A nonliquid capable of spontaneous and sustained combustion under normal conditions.

- An ignitable, compressed gas as defined by U.S. DOT regulations.
- An oxidizer as defined by U.S. DOT regulations.

Corrosivity.   Corrosive wastes can react dangerously with other wastes, dissolve metal or other materials, or burn the skin. Examples include waste rust removers, waste acid or alkaline cleaning fluids, and waste battery acid. The regulations describing the corrosivity characteristic are found at 40 CFR 261.22. A waste that exhibits either of the following properties is considered to be a hazardous waste identified by the waste code D002:

- An aqueous material with a pH of less than or equal to 2 or of greater than or equal to 12.5.
- A liquid that corrodes steel at a rate greater than 0.25 in. per year at a temperature of 55°C (130°F).

Reactivity.   Reactive wastes are unstable or undergo a rapid or violent chemical reaction with water or other materials. Examples include cyanide plating wastes, waste bleaches, and other waste oxidizers. Reactivity characteristic is described in the regulations at 40 CFR 261.23. A waste that exhibits any of the following properties is considered to be a hazardous waste identified by the waste code D003:

- Normally unstable and reacts violently without detonating.
- Reacts violently with water (i.e., causes sudden flash of fire accompanied by spattering).
- Forms an explosive mixture with water (i.e., causes an explosion) that may damage the container and anything or anyone near it.
- Generates toxic gases, vapors, or fumes when mixed with water that are not necessarily accompanied by a violent reaction or explosion.
- Contains cyanide or sulfide, and generates toxic gases, vapors, or fumes at a pH of between 2 and 12.5.

Toxicity.   Toxic wastes are considered to be hazardous because of the presence of toxic constituents in the wastes at greater than the established regulatory levels. These constituents (including metals, insecticides, herbicides, and other organics) and their current regulatory levels are included in Table 17–2. To determine if a waste displays a toxicity characteristic, a test method (i.e., the Toxicity Characteristic Leaching Procedure) is performed. If the waste contains any of the toxic constituents listed in Table 17–2 at greater than the regulatory levels, it is required to carry the specific hazardous waste code associated with the constituent(s). The regulations describing the toxicity characteristic are codified at 40 CFR 261.24.

Other Characteristics.   Medical waste that may be infectious includes used bandages and hypodermic needles, human tissues and blood, and biological substances generated by hospitals and research institutions. Because of its infectious nature, this waste is handled and disposed of according to specific EPA guidelines.

Table 17–2   List of Regulated Toxic Characteristic Leaching Procedure Constituents.

U.S. EPA Code	Original EPTC[b] Constituent	Regulatory levels (mg/L)	U.S. EPA Code	Original EPTC Constituent	Regulatory levels (mg/L)
D004	Arsenic	5.0	D020	Chlordane	0.3
D005	Barium	100.0	D021	Chlorobenzene	100.0
D006	Cadmium	1.0	D022	Chloroform	6.0
			D023	o-Cresol	200.0
			D024	m-Cresol	200.0
D007	Chromium	5.0	D025	p-Cresol	200.0
D008	Lead	5.0	D026	Cresol[a]	200.0
D009	Mercury	0.2	D027	1,4-dichlorobenzene	7.5
			D028	1,2-Dichloroethane	0.5
D010	Selenium	1.0	D029	1,1-dichloroethane	0.70
D011	Silver	5.0	D030	2,4-dinitrotoluene	0.13
			D031	Heptachlor (and its hydroxide)	0.008
D012	Endrin	0.02			
D013	Lindane	0.4	D032	Hexachlorobenzene	0.13
			D033	Hexachloro-1,3-butadiene	0.5
D014	Methoxychlor	10.0			
D015	Toxaphene	0.5	D034	Hexachloroethane	3.0
D016	2,4-Dichloro-phenoxy acetic acid (2,4-D)	10.0	D035	Methyl ethyl ketone	200.0
			D036	Nitrobenzene	2.0
			D037	Pentachlorophenol	100.0
D017	2,4,5-Trichloro-phenoxypropionic acid (2,4,5-TP, Silvex)	1.0	D038	Pyridine	5.0
			D039	Tetrachloroethylene	0.7
			D040	Trichloroethylene	0.5
			D041	2,4,5-Trichlorophenol	400.0
D018	Benzene	0.5	D042	2,4,6-Trichlorophenol	2.0
D019	Carbon tetrachloride	0.5	D043	Vinyl chloride	0.2

[a] If o-, m-, and p-Cresol concentrations cannot be differentiated, the total cresol (D026) concentration is used.

[b] EP toxicity contaminants

Source: 40 CFR 261.24.

Testing for Characteristics.   Wastes generally must be tested by a qualified laboratory to determine if they possess any of the hazardous characteristics described here. Tests must be applied to each individual waste; they cannot be used to assess a type of waste. The test must also be performed on representative samples to obtain results that adequately characterize the nature of the waste.

When a generator has sufficient knowledge to determine a waste is likely to possess a characteristic of a hazardous waste, the generator can use that knowledge to declare the waste as being hazardous without actually testing the waste. If, however, a waste determination is based on process knowledge alone, the generator is required to keep any supporting documentation used to make this determination (e.g., material safety datasheets or other information) at the facility.

Listed Wastes

Listed wastes are those from generic industrial processes, from certain sectors of industry, and unused, pure chemical products and formulations. The U.S. EPA has four different criteria for listing a waste as being hazardous:

1. It exhibits one of the four characteristics of hazardous waste (i.e., ignitability, corrosivity, reactivity, and toxicity).

2. It meets the statutory definition of hazardous waste (i.e., it could pose a threat to human health and the environment in the absence of special regulation).

3. It is acutely toxic or hazardous (i.e., even low does are fatal to humans and animals).

4. It is otherwise considered to be toxic by the EPA for some other reason.

Listed wastes are defined and identified by a specific chemical name or production process name. These include:

1. *Nonspecific source wastes:* These are generic wastes and are commonly produced by manufacturing and industrial processes. Examples include spent halogenated and nonhalogenated solvents used in degreasing, wastewater treatment sludge from electroplating processes, electroplating and heat-treating wastes, and dioxin-bearing production wastes. These wastes are also known as "F" wastes (waste codes F001 through F039). Some of these wastes, specifically dioxin wastes, are classified as being inherently wastelike materials because of the danger they present to human health and the environment. The "F" list is codified at 40 CFR 261.31.

2. *Specific source wastes:* These wastes result from specifically identified industries such as wood preserving, petroleum refining, and organic chemical manufacturing or as production wastes from specific sources, including inorganic pigments, organic chemicals, pesticides, explosives, iron and steel, secondary lead, veterinary pharmaceuticals, ink formulation, and coking. These wastes typically include sludges, still bottoms, wastewaters, spent catalysts, and residues (e.g., wastewater treatment sludge from the production of pigments). These wastes are known as "K" wastes (waste codes K001 through K136). The "K" list is codified at 40 CFR 261.32.

4. *Commercial chemical products:* These wastes consist of specific commercial chemical products or manufacturing chemical intermediates. These wastes are known as "P" wastes (waste codes P001 through P123) and "U" wastes (waste codes U001 through U359). These wastes are discarded (or intended to be discarded) toxic commercial chemical products, off-specification species, container residues, and spill residues. "P" wastes are also considered to be acutely hazardous or very dangerous in small amounts. The Ohio EPA has determined that because of their dangerous properties, "P" wastes must be regulated in the same way as large amounts of other hazardous wastes are. Examples of "P" and "U" wastes include commercial chemical products such as chloroform and creosote and acids such as sulfuric acid and hydrochloric acid. To be considered a listed waste, these chemicals must be in a pure, unused form. They are not considered to meet the "P" and "U" listings after they have been blended with other materials to formulate a product or have been used and now must be discarded because of contamination. In these instances, the

waste should be evaluated to determine if it meets another listing criteria (e.g., "F" and "K" wastes) or possesses a hazardous waste characteristic. The "P" and "U" lists are codified at 40 CFR 261.33.

The U.S. EPA recognized, however, that its procedures for listing hazardous wastes might not be applicable in all cases. Therefore, it created a process called "delisting" that allows any person to petition the EPA to exclude a listed waste from regulation. If after further evaluation the waste is not considered to be hazardous because of conditions at the facility, the waste is removed from RCRA regulatory jurisdiction. Delisting is done on a case-by-case basis, so if a waste is delisted at one facility, it is not automatically delisted at others.

A facility that wishes to go through the delisting process must conduct detailed studies and provide a comprehensive, technical demonstration showing that the waste does not pose a significant human health or environmental hazard. Thus, the delisting process itself can be very complex and time-consuming, and it does not waive a company's obligation to manage a waste as being hazardous during the petitioning process.

Waste Mixtures

Any waste mixture that contains a listed hazardous waste and a nonhazardous waste is considered to be a hazardous waste. This "mixture rule" applies regardless of the percentage composition of the listed waste in the waste mixture. For example, a mixture of spent methylene chloride formulation, which is listed as F002 because of its hazardous constituent, and used oil would be defined as a hazardous waste and be designated as F002 whether or not the mixture itself exhibited a hazardous waste characteristic. The regulation was developed to prevent generators from avoiding the hazardous waste requirements simply by commingling listed with nonlisted wastes.

A few exceptions to the mixture rule (Figure 17–1) include:

1. Wastewater (the discharge of which is subject to regulation under 40 CFR or Section 307(b) of the Clean Water Act) that has been mixed with low concentrations of listed waste.

2. Mixtures of nonhazardous wastes and listed wastes that were listed for exhibiting a characteristic, if the mixture no longer exhibits that characteristic.

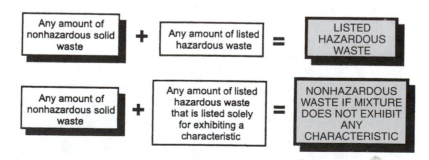

Figure 17–1   The mixture rule.

(From U.S. Environmental Protection Agency. RCRA Orientation Manual. *Washington, D.C.: U.S. Govt. Printing Office, 1998; with permission.)*

3. Mixtures of nonhazardous wastes and characteristic hazardous waste if the mixture no longer exhibits any of the characteristic.

4. Certain concentrations of spent solvents and laboratory wastewater that are discharged in low concentration and pose no threat to human health and the environment.

5. *De minimis* losses of discarded commercial chemical products or intermediaries used as raw materials in manufacturing or produced as byproducts. These include minor losses from spills and transfer of materials, process leaks, and similar incidental discharges.

Purposeful dilution of hazardous waste to render it nonhazardous is considered to be a regulated activity and requires a hazardous waste permit. Illegal treatment of hazardous waste to render it nonhazardous may subject a company to civil penalties.

Any waste mixture that contains a listed or characteristic hazardous waste and special nuclear material as defined by the Atomic Energy Act of 1954, amended by 42 USC 2011, is considered to be a hazardous waste only for those portions of the mixture subject to RCRA regulations. Special nuclear material is not otherwise regulated under the RCRA.

Wastes Derived from Hazardous Waste

The hazardous waste rules state that any residue generated from the treatment, storage, or disposal of a listed hazardous waste remains a hazardous waste unless it can be shown they pose no threat to human health and the environment. The U.S. EPA established this "derived-from" rule, because wastes derived from hazardous waste are generally hazardous. Examples of wastes defined as being hazardous through the derived-from rule include ash resulting from the incineration of off-specification toluene (i.e., code U220) and leachate resulting from disposal of separator sludge from the petroleum-refining industry (i.e., code K051) in a landfill.

PRIORITY TOXIC POLLUTANTS AND HAZARDOUS WASTE

The U.S. EPA has listed 129 specific toxic pollutants for priority action. These have been divided into various categories as shown in Table 17–3.

The major generators of hazardous waste among 15 industries studied by the EPA (3) in million tons per year are:

Primary metals	8.3
Organic chemicals	6.7
Electroplating	5.3
Inorganic chemicals	3.4
Textiles	1.8
Petroleum refining	1.8
Rubber and plastics	0.8
Miscellaneous	0.7

Examples of hazardous waste types generated by businesses and industries are given in Table 17–4.

Table 17–3 Priority Toxic Pollutants

Pollutant	Characteristics	Sources	Remarks
Pesticides (Generally chlorinated hydrocarbons.)	Readily assimilated by aquatic animals, fat soluble, concentrated through the food chain (i.e., biomagnified), and persistent in soil and sediments.	Direct application to farm and forestlands, runoff from lawns and gardens, urban runoff, and discharge in industrial wastewater.	Several chlorinated hydrocarbon pesticides already restricted by EPA; aldrin, dieldrin, DDT, endrin, heptachlor, lindane, and chlordane.
Polychlorinated biphenyls (PCBs) (Used in electrical capacitors and transformers, paints, plastics, insecticides, and other industrial products.)	Readily assimilated by aquatic animals, fat soluble, subject to biomagnification, persistent, and chemically similar to chlorinated hydrocarbons.	Municipal and industrial waste discharges disposed of in dumps and landfills.	Toxic Substance Control Act ban on production after 6/1/79 but will persist in sediments. Restrictions on many freshwater fisheries as a result of PCB pollution (e.g., lower Hudson River, upper Housatonic River, parts of Lake Michigan)
Metals (Antimony, arsenic, beryllium, cadmium, copper, lead, mercury, nickel, selenium, silver, thallium, and zinc)	Nonbiodegradable, persistent in sediments, toxic in solution, and subject to biomagnification.	Industrial discharges, mining activity, urban runoff, erosion of metal-rich soil, and certain agricultural uses (e.g. mercury as a fungicide).	
Asbestos	May cause cancer when inhaled; aquatic toxicity not well understood.	Manufacture and use as a retardant, roofing material, brake lining, and so on; runoff from mining.	
Cyanide	Variably persistent; inhibits oxygen metabolism.	Wide variety of industrial uses.	
Halogenated aliphatics (Used in fire extinguishers, refrigerants, propellants, pesticides, solvents for oils and greases, and in dry cleaning)	Largest single class of "priority toxics." Can cause damage to central nervous system and liver. Not very persistent	Produced by chlorination of water; vaporization during use.	Large-volume industrial chemicals, widely dispersed, but less threat to the environment than persistent chemicals.
Ethers (Used mainly as solvents for polymer plastics.)	Potent carcinogen, aquatic toxicity; fate well understood	Escape during production and use.	Though some are volatile, ethers have been identified in some natural waters.
Phthalate esters (Used chiefly in production of polyvinyl chloride and thermoplastics as plasticizers.)	Common aquatic pollutant. Moderately toxic, but teratogenic and mutagenic properties in low concentrations. Aquatic invertebrates are particularly sensitive to toxic effects. Persistent and can be biomagnified.	Waste disposal vaporization during use (in nonplastics).	

Table 17–3 (cont.)

Pollutant	Characteristics	Sources	Remarks
Monocyclic aromatics (excluding phenols, cresols, and phthalates) (Used in the manufacture of other chemicals, explosives, dyes, and pigments and in solvents, fungicides, and herbicides.)	Central nervous system depressant; can damage liver and kidneys.	Enters environment during production and byproduct production states by direct volatilization; wastewater.	
Phenols (Large-volume industrial compounds used chiefly as chemical intermediates in the production of synthetic polymers, dyestuffs, pigments, pesticides, and herbicides.)	Toxicity increases with degree of chlorination of the phenolic molecule. Very low concentrations can taint fish flesh and impart objectionable odor and taste to drinking water. Difficult to remove from water by conventional treatment. Carcinogenic in mice.	Occur naturally in fossil fuels, wastewater from coking ovens, oil refineries, tar distillation plants, herbicide manufacturing, and plastic manufacturing; can all contain phenolic compounds.	
Polycyclic aromatic hydrocarbons (Used as dyestuffs, chemical intermediates, pesticides, herbicides, motor fuels, and oils.)	Carcinogenic in animals and indirectly linked to cancer in humans. Most work done on air pollution; more is needed on the aquatic toxicity of these compounds. Not persistent and are biodegradable, though bioaccumulation can occur.	Fossil fuels (e.g., use, spills, and production); incomplete combustion of hydrocarbons.	
Nitrosamines (Used in the production of organic chemicals and rubber; patents exist on processes using these compounds.)	Tests on laboratory animals have shown nitrosamines to be some of the most potent carcinogens.	Production and use can occur spontaneously in food-cooking operations.	

Source: Council on Environmental Quality. *Environmental Quality: The Ninth Annual Report on the Council on Environmental Quality.* Washington, D.C.: U.S. Govt. Printing Office, 1978; with permission.)

Table 17–4   Examples of Hazardous Waste Generated by Business and Industry

Waste Generator	Waste Type
Chemical manufacturers	Strong acids and bases, spent solvents, reactive wastes
Vehicle maintenance shops	Heavy metal–paint wastes, ignitable wastes, used lead–acid batteries, spent solvents
Printing industry	Heavy metal solutions, waste inks, spent solvents, spent electroplating wastes, ink sludges containing heavy metals
Leather products manufacturing	Waste toluene and benzene
Paper industry	Paint wastes containing heavy metals, ignitable solvents, strong acids and bases
Construction industry	Ignitable paint wastes, spent solvents, strong acids and bases
Cleaning agents and cosmetics manufacturing	Heavy metal dusts, ignitable wastes, flammable solvents, strong acids and bases
Furniture and wood manufacturing and refinishing	Ignitable wastes, spent solvents
Metal manufacturing	Paint wastes containing heavy metals, strong acids and bases, cyanide wastes, sludges containing heavy metals

Source: U.S. Environmental Protection Agency. *Solving the Hazardous Waste Problem: The EPA and RCRA Program.* EPA/530-SW-86-037. Washington, D.C.: U.S. Govt. Printing Office, 1986; with permission.

Hazardous wastes that are characterized as ignitable, corrosive, explosive, or toxic should be removed from industrial waste or treated before discharge to a municipal sewer. Many toxic wastes upset biological wastewater treatment processes.

Solid and liquid hazardous materials disposed of at hazardous waste sites may also become airborne or evaporate if not controlled. Disintegration of solids as well as absorption and adsorption of liquids both in and on soil may permit contaminated particulates to become airborne. These may be inhaled or end up on land or water, thus entering the food chain.

REVIEW QUESTIONS

1. Discuss at least three environmental disasters caused by hazardous chemicals.
2. a. What is a hazardous waste?
   b. What are the responsibilities of a hazardous waste generator?
3. a. What are the sources of hazardous waste?
   b. List at least three commercial or residential products that contain hazardous chemicals, and then list those chemicals.
4. List the characteristics of various types of hazardous waste.
5. Ask your city administrator if emergency plans are in place to deal with a large chemical spill within the city limits.

REFERENCES

1. 40 CFR 260.10.
2. U.S. Environmental Protection Agency. *Hazardous Waste Treatment, Storage, and Disposal Facility Report for 1985.* Washington, D.C.: U.S. Govt. Printing Office, 1985.
3. U.S. Environmental Protection Agency. *Solid Waste Facts.* SW-694. Washington, D.C.: U.S. Govt. Printing Office, 1978.

Chapter 18   Transportation of Hazardous Waste

This chapter covers important aspects of transporting hazardous wastes. This should familiarize readers with the complexity of the risks and problems involved.

In 1975, the U.S. Congress passed the Hazardous Materials Transportation Act (HMTA). The intent of this law was to improve regulatory and enforcement activities by giving the Secretary of Transportation broad authority to regulate all aspects of hazardous material transport. The modes of such transportation include aircraft, watercraft, motor vehicle, and railway.

The U.S. Federal Aviation Administration enforces regulations pertaining to air transportation. The U.S. Coast Guard enforces regulations pertaining to water transportation. Similarly, the U.S. Federal Highway Administration and Federal Railroad Administration enforce regulations associated with hazardous waste transportation by highways and railroads, respectively.

Table 18–1 lists the classifications of hazardous materials that are subject to federal transportation regulations. The purpose is to regulate materials likely to cause injury to the public and the carrier personnel if they are accidently released during transport. The Hazardous Materials Table in 49 CFR 172.101 further defines the labeling, packaging, and carrier requirements for specific hazardous chemicals.

Transportation of these wastes is regulated by both the HMTA and the Resource Conservation and Recovery Act (RCRA), Subtitle C. The RCRA is administered by the U.S. Environmental Protection Agency (EPA), and it is the primary federal statute governing hazardous waste. It contains certain standards for hazardous material transporters and coordinates the regulatory aspect with the U.S. Department of Transportation (DOT).

As discussed in Chapter 8, a cradle-to-grave chain of hazardous waste management has been instituted under RCRA. This chain is composed of the generator, transporter, treatment systems, storage facilities, and disposal sites. The hazardous waste transporter is an important link between the generator and the ultimate treatment, storage, and disposal (TSD) facility. The generator produces the hazardous waste, and the transporter receives that waste from the generator and then transports it to a treatment plant, storage facility, and so on.

REGULATIONS

Hazardous waste transporters play an integral role in the cradle-to-grave management system by delivering hazardous waste from its point of generation to its ultimate destination. Under Subtitle C, a hazardous waste transporter is any person engaged in the

Table 18–1   Classification of Hazardous Materials for Shipping and Packaging

Hazard	Class	Division
Forbidden materials	—	—
Forbidden explosives	—	—
Explosives (with a mass explosion hazard)	1	1.1
Explosives (with a projection hazard)	1	1.2
Explosives (with predominantly a fire hazard)	1	1.3
Explosives (with no significant blast hazard)	1	1.4
Very insensitive explosives, blasting agents	1	1.5
Extremely insensitive detonating substances	1	1.6
Flammable gas	2	2.1
Nonflammable compressed gas	2	2.2
Poisonous gas	2	2.3
Flammable and combustible liquid	3	—
Flammable solid	4	4.1
Spontaneously combustible material	4	4.2
Dangerous when wet material	4	4.3
Oxidizer	5	5.1
Organic peroxide	5	5.2
Poisonous materials	6	6.1
Infectious substance (etiological agent)	6	6.2
Radioactive material	7	—
Corrosive material	8	—
Miscellaneous hazardous material	9	—
Other regulated material	—	—

Source: 40 CFR 173.

off-site transportation of hazardous waste within the United States. Because these transporters move regulated wastes on public roads, highways, rails, and waterways they are regulated by DOT as well as by RCRA standards.

To ensure that generators and transporters handle hazardous waste properly before transportation, the EPA works closely with the DOT to promulgate transportation regulations to avoid conflict between the two agencies. Generators and transporters are to comply with EPA and DOT regulations covering hazardous waste, but even though these regulations are integrated, they are not located in the same part of the CFR. The DOT regulations (i.e., HTMA) are found in 49 CFR 171–179; the EPA requirements (i.e., RCRA Subtitle C) are found in 40 CFR 263. The regulations cover three main areas:

1. Preparation for transport, including identification, notification, and packaging.

2. Manifest requirements.

3. Record-keeping and cleanup regulations.

A generator, transporter, or TSD facility must meet certain requirements. For example, the generator and the transporter must have an EPA identification number (ID). The generator must not give waste to a transporter who does not have an ID; likewise, the transporter must not accept waste from a generator who does not have the ID. This system enables the EPA to monitor and track all hazardous waste that is produced. The generator must comply with the packaging regulations, which require the generator to identify and quantify the wastes that are put in a drum or container for transport. Figure 18–1 shows examples of DOT labels for such containers. The transporter must also display certain placards, which are symbolic representations of the hazardous cargo, on the ends and sides of the transport vehicles, railcars, and freight containers. Figure 18–2 illustrates several placards, and Figure 18–3 illustrates a truck with a placard. The transporter must also comply with all reporting requirements in case of an accident during transport. The accident must be reported immediately to the EPA, DOT, and local authorities.

Central to the cradle-to-grave chain of hazardous waste management is the hazardous waste manifest. This one-page form must accompany every waste shipment. A manifest is a multicopy shipping document that large- and small-quantity generators must complete and use to accompany hazardous waste shipments. It is an EPA document with DOT-required information. The manifest form (Figure 18–4) is designed so that hazardous waste shipments can be tracked from their point of generation to their final destination. The hazardous waste generator, transporter, and designated management facility must each sign this document and keep a copy. The designated facility operator must also send a copy back to the generator to confirm the shipment arrived at its destination. The generator must keep this copy, which is also signed by the transporter and the designated facility, on file for 3 years. Figure 18–5 illustrates a hazardous waste tracking system.

Manifest requirements are slightly different for water and rail transporters. These transporters must comply with the directions on the manifest, obtain an EPA ID, and be listed on the manifest. The manifest itself, however, is not required to physically accompany the waste shipment at all times; both water and rail transporters can use another shipping document instead of the manifest. The initial water or rail transporter must sign and date the manifest or shipping document and ensure that it reaches the designated facility. The final water or rail transporter must ensure that the owner and operator of the designated facility signs the manifest or shipping paper. Intermediate water and rail transporters are not required to sign the manifest or shipping paper.

EMERGENCY RESPONSE

The transportation of hazardous materials over the vast system of U.S. roads, rails, waterways, and airlines requires emergency response capability at all levels of government. Regulations are designed to ensure that hazardous waste shipments are conducted safely, but the transportation of hazardous waste remains dangerous. There is always the possibility of an accident. Therefore, the regulations require transporters to take immediate action to protect human health and the environment if a release occurs (e.g., notifying local authorities and diking the discharge area). When accidents happen, local police and firefighters are usually the first to respond. Medical personnel are called for any injuries or exposure to hazardous material. When a serious accident or spill

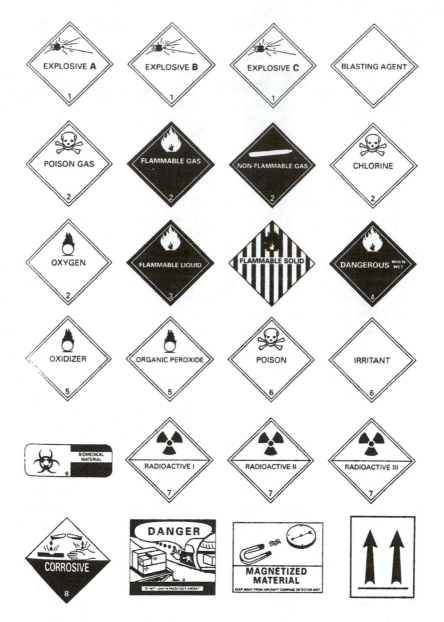

Figure 18–1   U.S. Department of Transportation labels for hazardous material packages.

(From U.S. Environmental Protection Agency. RCRA Manual. *Washington, D.C.: U.S. Govt. Printing Office, 1998; with permission.)*

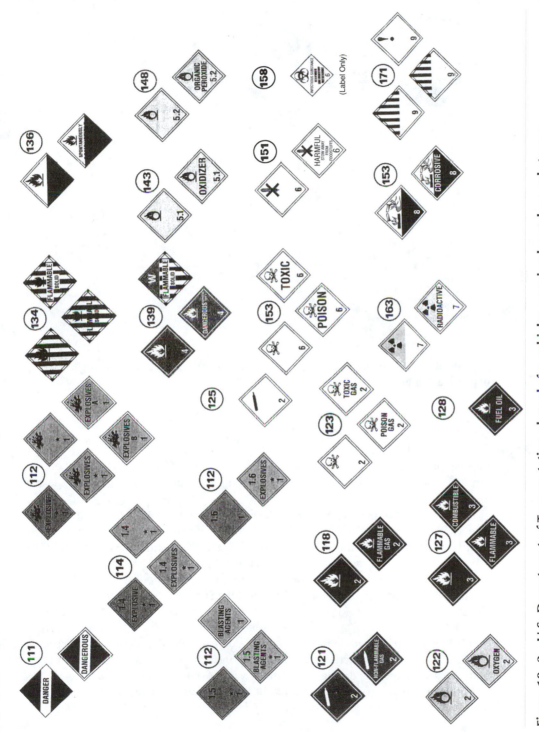

Figure 18–2 U.S. Department of Transportation placards for vehicles carrying hazardous substances.

(From U.S. Environmental Protection Agency. RCRA Manual. Washington, D.C.: U.S. Govt. Printing Office, 1998; with permission.)

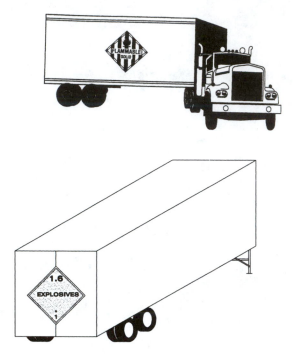

Figure 18–3   (a) Truck carrying a flammable solid. (b) Truck carrying explosive material.

occurs, the transporter must notify the National Response Center by phone. The Centers for Disease Control must also be informed if the spill involves disease-causing agents.

Regulations authorize certain federal, state, or local officials to handle transportation accidents. Specifically, if immediate removal of waste is necessary to protect human health or the environment, one of these officials may authorize a nonmanifested removal of waste by a transporter without an EPA ID number.

The 1996 North American Emergency Response Guidebook (1) was developed jointly by Transport Canada, the U.S. DOT, and the Secretariat of Communications and Transportation of Mexico for use by firefighters, police, and other emergency personnel who may be first on the scene of a transportation accident involving dangerous goods. It is primarily a guide to aid first responders in quickly identifying the specific or generic hazards of material involved in the incident and in protecting themselves and the general public during the initial response phase. This guide incorporates dangerous-goods lists from the most recent United Nations recommendations as well as from other international and national regulations.

<table>
<tr><td colspan="2">UNIFORM HARARDOUS WASTE MANIFEST</td><td>1. Generator's US EPA ID No.</td><td>Manifest Document No.</td><td>2. Page 1 of</td><td>Information in the shaded areas is not requested by Federal law.</td></tr>
</table>

UNIFORM HARARDOUS WASTE MANIFEST

1. Generator's US EPA ID No. | Manifest Document No. | 2. Page 1 of | Information in the shaded areas is not requested by Federal law.

3. Generator's Name and Mailing Address

A. State Manifest Document Number

B. State Generator's ID

4. Generator's Phone (   )

5. Transporter 1 Company Name | 6. | US EPA ID Number

C. State Transporter's ID
D. Transporter's Phone

7. Transporter 2 Company Name | 8. | US EPA ID Number

E. State Transporter's ID
F. Transporter's Phone

9. Designated Facility Name and Site Address | 10. | US EPA ID Number

G. State Facility's ID

H. Facility's Phone

11. US DOT Description (Including Proper Shipping Name, Hazard Class, and ID Number)	12. Containers No.	Type	13 Total Quantity	14. Unit Wt/Vol	I. Waste No.
a.					
b.					
c.					
d.					

J. Additional Descriptions for Materials Listed Above | K. Handling Codes for Wastes Listed Above

15. Special Handling Instructions and Additional Information

16. Generator's Certification: I hereby declare that the contents of this consignment are fully and accurately described above by proper shipping name and are classified, packed, marked, and labeled, and are in all respects in proper condition for transport by highway according to applicable international and national government regulations.

Unless I am a small quantity generator who has been exempted by statute or regulation from the duty to make a waste minimization certification under Section 300(b) of RCRA, I also certify that I have a program in place to reduce the volume and toxicity of waste generated to the degree I have determined to be economically practicable and I have selected the method of treatment, storage, or disposal currently available to me which minimizes the present and future threat to human health and the environment.

Printed/Typed Name | Signature | Month Day Year

17. Transporter 1 Acknowledgement of Receipt of Materials

Printed/Typed Name | Signature | Month Day Year

18. Transporter 2 Acknowledgement of Receipt of Materials

Printed/Typed Name | Signature | Month Day Year

19. Discrepancy Indication Space

20. Facility Owner or Operator: Certification of receipt of hazardous materials covered by this manifest except as noted in Item 19.

Printed/Typed Name | Signature | Month Day Year

EPA Form 8700-22 (Rev. 4-85) Previous edition is obsolete

GENERATOR / TRANSPORTER / FACILITY

Figure 18-4   Hazardous waste manifest.

(From *U.S. Environmental Protection Agency.* Transportation of Hazardous Materials. *Washington, D.C.: U.S. Govt. Printing Office, 1986; with permission.)*

411

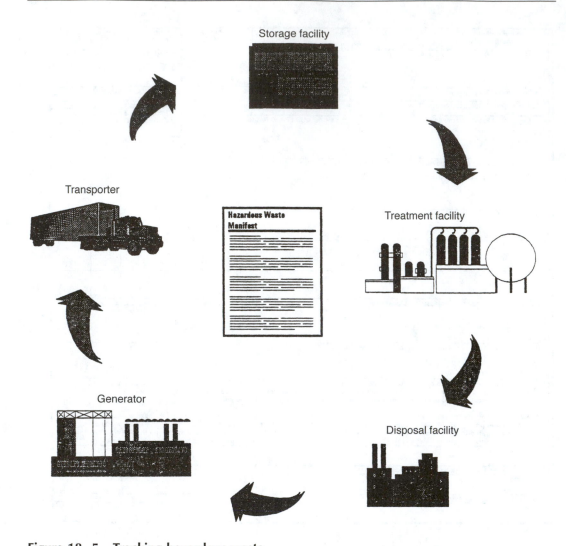

Figure 18–5 Tracking hazardous waste.

(From Ohio Environmental Protection Agency Publication 1996; with permission.)

SAFETY

Pretransport regulations are designed to ensure the safe transportation of hazardous waste from origin to ultimate disposal. In developing these regulations, the EPA adopted the DOT regulations for the transportation of hazardous wastes. These regulations require:

1. Proper packing to prevent leakage of hazardous waste during both normal and potentially dangerous transport situations, such as when a drum falls off the truck.

2. Identification of the characteristics and dangers associated with the waste being transported through labeling, marking, and placarding of the packaged waste.

These pretransport regulations only apply to generators of off-site waste.

In addition to adopting the DOT regulations, the EPA also developed pretransport regulations that cover the accumulation of waste before transport. A large-quantity generator may accumulate hazardous waste on site for 90 days or less. Small-quantity generators may accumulate hazardous waste on site for 180 days, and for 270 days if TSD facilities are more than 200 miles away, with the following requirements:

1. *Proper storage:* The waste is properly stored in containers or tanks marked with the words *Hazardous Waste* and the date on which the accumulation began.

2. *Emergency plan:* There is a contingency plan and emergency procedures in place.

3. *Personnel training:* Facility personnel are trained in the proper handling of hazardous waste.

REVIEW QUESTIONS

1. List the various modes of hazardous waste transport and the agencies responsible for these modes.
2. What requirements must a generator meet?
3. What is the purpose of a manifest form?
4. State the DOT and EPA pretransport regulations.

5. Assume you are a town's Safety Director. Traffic through the town includes a significant number of trucks carrying hazardous waste.
   a. List the problems associated with such traffic.
   b. What steps would you take to minimize the threat to residents?

REFERENCE

1. U.S. Department of Transportation Research Administration. *1996 North American Emergency Response Guidebook.* Washington, D.C.: U.S. Govt. Printing Office, 1996.

Chapter 19   Hazardous Waste Minimization and Environmental Audits

This chapter examines practices that reduce or eliminate hazardous waste at the source. It also presents methods to identify environmental problems at the source.

Until the late 1970s, the general public in the United States—and even the U.S. government—did not know the hazardous waste problem existed. The poor disposal practices of the past, however, have resulted in serious environmental problems. The tragedy of Love Canal, near Niagara Falls, New York, is a well-known horror story of the 1970s (see Chapter 17). Hazardous chemicals were buried until 1952 in part of the canal, and in the early 1970s, liquid and gaseous wastes migrated into house basements, causing severe health problems.

In the United States, burial of the waste in landfills is a fairly old practice that still continues. Regulatory restrictions and better-engineered designs, however, have made current landfills safer than older ones. Even so, because of public pressure and governmental mandate, the current approach is gradually changing from pollution control to pollution prevention.

Pollution is usually considered as being resource gone to waste. The focus in the recent past has been on end-of-pipe treatment methods. Emphasis has now shifted from end-of-pipe pollution control, which began in 1970, to an upfront pollution prevention. End-of-pipe pollution control does not really solve the problem; it simply moves it from one medium to another. A recent Chemical Manufacturers Association survey indicates that approximately 20 percent of new capital expenditures are now for pollution abatement control (1). Obviously, in the interests of preserving the environment and saving the cost of transportation, treatment, and disposal of hazardous waste, generation of waste at the source should be minimized.

The current trend is toward waste minimization. The widespread interest in and implementation of hazardous waste reduction initiatives is a significant change from the previous treat-and-dispose practice. Terms such as *waste minimization, waste reduction,* and *pollution prevention* are frequently used, and all convey the same idea; taking care of the waste at the front end of the pipe. *Pollution prevention* is a broad term for hazardous waste control, whereas *waste minimization* and *waste reduction* narrow the focus mainly to reduction processes. The acronym *P2* has been established for pollution prevention, and the U.S. Environmental Protection Agency (EPA) uses the term *waste minimization* to mean the reduction (to the extent feasible) of solid and hazardous waste. Both programs emphasize source reduction (i.e., reducing waste before it is even generated) and environmentally sound recycling.

POLLUTION PREVENTION

The U.S. Pollution Prevention Act of 1990 established pollution prevention as a "national objective." The Act notes there are significant opportunities for industry to reduce or prevent pollution at the source through cost-effective changes in production, operation, and use of raw materials. Opportunities for source reduction are often not realized, however, because existing regulations and industrial resources required for compliance focus on treatment and disposal rather than on source reduction. Source reduction is fundamentally different from—and is more desirable than—waste management and pollution control. This Act requires industries to provide annual reports on the effectiveness of source reduction for all chemicals they must already report on under the Community Right-to-Know provisions of the Superfund Amendments and Reauthorization Act (SARA). Because these reports become public information, the Act provides another incentive to develop a source reduction program.

The Act also establishes the pollution prevention hierarchy as a national policy, declaring that pollution should be prevented or reduced at the source wherever feasible. In the absence of feasible prevention or recycling opportunities, pollution should be treated in an environmentally safe manner. Disposal or other release into the environment should be used only as a last resort.

Source reduction is defined to mean any practice that reduces the amount of any hazardous substance, pollutant, or contaminant entering any waste stream or otherwise released into the environment (including fugitive emissions) before recycling, treatment, or disposal and that reduces the hazards to public health and the environment associated with the release of such substances, pollutants, or contaminants. The EPA developed a strategy in 1991 to incorporate pollution prevention into their ongoing environmental protection efforts. The EPA also listed 17 high-risk industrial chemicals that present great risk to human health and environment and that can provide opportunities for reduction or prevention. These 17 pollutants are listed in (Table 19–1). The EPA instituted a voluntary 33/50 program, which had an overall national goal of reducing environmental releases of these chemicals by 33 percent by the end of 1992 and at least 50 percent by the end of 1995.

ENVIRONMENTAL AUDITS

The U.S. EPA began proposing environmental audits as an element of enforcement case settlements during the mid-1970s (2). Such settlements are frequent. The EPA defines environmental audits as a systematic, documented, periodic, and objective review by regulated entities of facility operations and practices related to meeting environmental requirements. Audits can be designed to verify compliance with environmental requirements, evaluate the effectiveness of environmental management systems already in place, and assess risks from both regulated and unregulated materials and practices.

Environmental auditing is a management tool that is considered to generally improve the overall environmental performance of manufacturing facilities. These audits usually focus on operating facilities that may involve processing, manufacturing, ma-

Table 19–1   EPA Target Chemicals

Chemical	Amount Released (million pounds/year)
Benzene	33.1
Cadmium	2
Carbon tetrachloride	5
Chloroform	26.9
Chromium	56.9
Cyanide	13.8
Dichloromethane	153.4
Lead	58.7
Mercury	0.3
Methyl ethyl ketone	159.1
Methyl isobutyl ketone	43.7
Nickel	19.4
Tetrachloroethylene	37.5
Toluene	344.6
1,1,1-Trichloroethane	190.5
Trichloroethylene	55.4
Xylene	201.6

Source: Data from *Preventing Pollution in the Chemical Industry.* Washington, D.C.: Chemical Manufacturers Association, 1992.

terial procurement, storage, maintenance, distribution, and waste management. Environmental auditing should detect problems such as noncompliance and improper or inadequate waste management operations and, hopefully, permit timely correction. Managers of hazardous waste facilities, however, are doubtful. They are concerned about the confidentiality of trade secrets and possible discoveries by outsiders who may compete for similar manufacturing or processing business.

The initial step to a waste minimization program is to conduct a waste minimization assessment (i.e., audit). This is a basic review of an industry's potential to reduce or recycle its waste. It also balances the input and output of wastes to study possible processes by which pollution prevention efforts can be established (Figure 19–1). Waste minimization audits involve identifying and characterizing waste streams (i.e., the quantity of wastes generated and the production processes generating such wastes).

The EPA recommends the following steps for a waste management audit:

- Prepare background material for the audit.

- Arrange a preaudit visit to identify waste streams of interest.

- Perform detailed analyses on waste streams of interest.

- Undertake a detailed site visit to collect data on selected waste streams and associated controls and process data.

- Develop a series of potential waste minimization options.

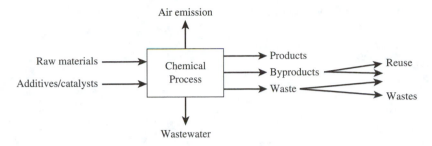

Figure 19–1   Inputs and outputs of a chemical process.

- Evaluate preliminary options.
- Study and rank options based on waste reduction effectiveness, extent of current use in the industry, and potential for future.
- Present results of a preliminary study to plant personnel along with a ranking of options.
- Prepare and submit a final report and recommendations to plant management.
- Develop an implementation plan and schedule.
- Conduct periodic reviews.

Pollution Prevention Hierarchy

The EPA considers source reduction and recycling to be viable pollution prevention techniques. In its "Pollution Prevention Policy Statement" (3), the EPA encouraged organizations, facilities, and individuals to fully use source reduction and recycling practices and procedures.

To encourage hazardous waste minimization nationwide, the EPA has developed the Waste Minimization National Plan. Consequently, minimization and pollution prevention have gradually become industry's approach to achieve less waste and more cost-efficient operations. For example, because of pollution prevention activities, 95-percent reuse of wastewater from coating/painting manufacture saves $375,000 annually (4). (Several other examples of benefits are given later in this Chapter.)

Figures 19–2 and 19–3 indicate the EPA hierarchy of recommended approaches to pollution prevention. As waste at the source is reduced or recycled, the amount of waste for treatment and ultimate disposal is also reduced, as is the release to the environment. The working definition of waste minimization currently used by the EPA consists of two approaches: source reduction, and recycling.

The process for identifying options should follow the hierarchy in which reduction options are explored first and recycling options second. This hierarchy of effort stems from the environmental desirability of source reduction as the preferred means

Source reduction

Recycling

Treatment

Ultimate disposal

Figure 19–2   U.S. Environmental Protection Agency hierarchy of recommended approaches to pollution prevention.

of minimizing waste. Treatment options should be considered only after acceptable waste minimization techniques have been identified. Figure 19–4 presents a logical way to think about waste minimization technique.

Recycling techniques allow hazardous materials to be put to a beneficial use. Source reduction techniques avoid the generation of hazardous wastes, thereby eliminating the problems associated with handling them. Recycling techniques may be performed on site or at an off-site facility designed to recycle the waste.

Benefits of Waste Minimization

In 1992, the United States produced 656 billion pounds of chemicals, including 422 billion pounds of inorganics and 234 billion pounds of organics (5). Many of these are hazardous chemicals, including acids, bases, corrosive and toxic chemicals, and flammable hydrocarbons. Waste reduction can save money and protect both public health and the environment. Money is saved by reducing waste treatment and disposal costs, raw material purchases, and other operating costs. In addition, the waste generator can reduce any potential environmental liabilities.

Many industries have realized the economic and environmental value of pollution prevention measures. These measures are technologies and practices that are designed to replace or improve current industrial practices so that wastefulness and pollution are reduced. It is often the case that a second look at a manufacturing process can identify how waste can be reduced, sometimes at very little cost. For example, one of the major corporations engaging in a pollution prevention scheme was the 3M Company, with a program called 3P (*Pollution Prevention Pays*) (6). The following quantities of waste were eliminated through source reduction, such as process modification, equipment redesign, and recovery of waste material for reuse programs:

Waste Type	Eliminated
Water pollutants (tons)	13,500
Wastewater (10^6 gal)	1,000
Air pollutants (tons)	110,000
Sludge/solid waste (tons)	303,000

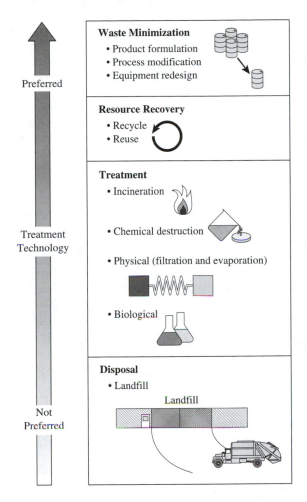

Figure 19–3   U.S. Environmental Protection Agency preferred approach to hazardous waste management.

(From U.S. Environmental Protection Agency. EPA Publication 625. Washington, D.C.: U.S. Govt. Printing Office, 1993; with permission.)

Since this program's inception in 1975, the "process modifications, product changes, and waste recovery for reuse have yielded cumulative first year savings of $350 million" (7). These savings were possible because of savings on the capital costs of unneeded pollution control facilities and sales from products made from waste.

The EPA has developed suggestions in the form of "how-to" publications on waste minimization. Through these publications, the EPA advocates approaches to hazardous waste prevention, reduction, or minimization through source reduction or recycling.

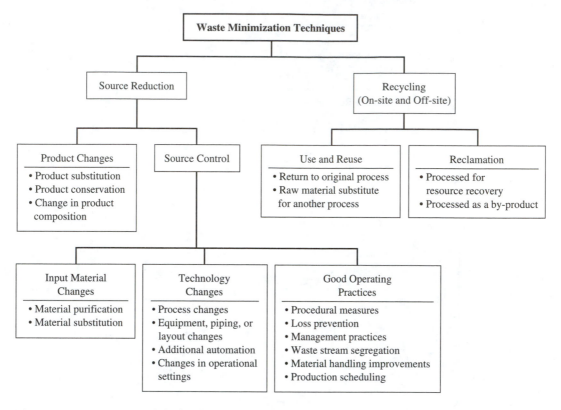

Figure 19–4   Waste minimization techniques.

(From U.S. Environmental Protection Agency. Waste Minimization Opportunity Assessment Manual. *Washington, D.C.: U.S. Govt. Printing Office, 1988; with permission.)*

Source Reduction

Source reduction can be achieved through product changes, good operating practices, technological changes, and input material changes.

Product Changes.   Product changes are performed by the manufacturer to reduce the waste resulting from a product's use. Product changes include product substitution, product conservation, and changes in product composition.

In the paint manufacturing industry, for example, water-based coatings are increasingly used in place of solvent-based paints. These products do not contain toxic or flammable solvents, which make solvent-based paints hazardous during disposal. In addition, cleaning the applicators (e.g., brushes, rollers) with solvents is not necessary.

Good Operating Practices.   Good operating practices are procedural, administrative, or institutional measures to minimize waste. These practices apply to the human aspect of manufacturing operations. Good operating practices include management and personnel practices, material handling and inventory practices, loss prevention, and waste segregation.

Management and personnel practices should focus on employee training and encouragement to reduce waste. Material handling and inventory practices include programs to reduce loss of input materials and to ensure proper storage conditions. Loss prevention minimizes waste by avoiding leaks from equipment and spills. Waste segregation reduces the volume of hazardous waste by preventing hazardous and nonhazardous wastes from mixing.

A large consumer product company in California adopted a corporate policy to minimize generation of hazardous waste. To implement this policy, the company mobilized quality circles made up of employees representing areas within the plant that generated hazardous wastes. The company experienced a 75-percent reduction in the amount of waste generated by instituting proper maintenance procedures as suggested by the quality circles. Because the circle members were also line supervisors and operators, they made sure the procedures were followed.

Technological Changes. Technological changes are oriented toward process and equipment modifications to reduce waste. Such changes can range from minor adjustments in processes to full replacement of processes, involving large capital costs.

One manufacturer of fabricated metal products cleaned nickel-and-titanium wire in an alkaline chemical bath before using the wire in their product. In 1986, however, the company began to experiment with a mechanical abrasive system, which used silk and carbide pads and pressure to brighten the metal. The system worked, but it required passing the wire through the unit twice for complete cleaning. In 1987, the company bought a second abrasive unit and installed it in series with the first unit. This system allowed the company to completely eliminate use of the chemical cleaning bath.

Input Material Changes.   Input material changes minimize waste by reducing or eliminating the hazardous material that enters the production process. Changes in the input materials can also be made to avoid generation of hazardous wastes within the production processes. Input material changes include material purification and material substitution.

An electronic manufacturing facility of one large, diversified corporation originally cleaned printed circuit boards with solvents. The company found that by switching from a solvent-based to an aqueous-based cleaning system, the same operating conditions and workloads could be maintained. The aqueous-based system was also found to clean six times more effectively than the solvent-based system. In turn, this resulted in a lower product rejection rate and eliminated a hazardous waste.

Recycling

Reclamation.   Reclamation is the recovery of valuable material from a hazardous waste. Reclamation techniques differ from use-and-reuse techniques in that the recovered material is not used in the facility; rather, it is sold to another company.

One photoprocessing company uses an electrolytic-deposition cell to recover silver from the rinsewater of film-processing equipment. The silver is then sold to a small recycler. By removing the silver, this wastewater can be discharged to the sewer without additional pretreatment by the company. This unit paid for itself in less than 2 years by the value of silver recovered.

The company also collects used film and sells it to the same recycler. The recycler then burns the film and collects silver from the residual ash. By removing the silver, this ash becomes nonhazardous as well.

Use and Reuse.   Recycling via use and/or reuse involves the return of a waste material to the originating process (or to another process) as an input material. A printer of newspaper advertising in California, for example, purchased an ink-recycling unit to produce black newspaper ink from its various waste inks. The unit blends the different colors of waste ink together with fresh black ink and black toner to create the recycled black ink, which is then filtered to remove flakes of dried ink. This ink is used in place of fresh black ink, and it eliminates the need for the company to ship waste ink for off-site disposal. The price of the recycling unit was paid off in 18 months based only on the savings in purchases of fresh black ink. The payback improved to 9 months when the costs for disposal of ink as a hazardous waste were included.

CASE STUDY

Crown Equipment Corporation

Crown Equipment Corporation of Bremen, Ohio, manufactures electric lift trucks and antennae rotators. In 1988, Crown used 208,000 pounds of trichloroethane (TCA) in cold-cleaning degreasing and operation of two vapor degreasers. The substitution project began in 1990 with bench-testing of the various parts. Metallurgical testing was also performed before deciding on the best equipment to purchase. Crown first substituted a water-based cleaning compound in cold-cleaning (i.e., dip) operations. Two large vapor degreasers were removed and replaced with a series of agitating wash tanks as well.

This central cleaning system employs a wash, rinse, rust-inhibiting, and drying stage to clean parts. No change in the production rate was required for the new process. New waste streams, however, include 900 gal of oily wastewater, which are treated on site every 2 weeks. Energy usage is expected to rise as a result of the additional heating requirements of the water-based wash tanks.

Following implementation of this project, use of TCA was eliminated. The agitating technology used for this project was commercially available, and the central cleaning system was purchased "off the shelf." The chemistry, however, was customized to meet the cleaning requirements and to provide adequate rust protection. The biggest obstacle Crown faced was providing the capital outlay required to purchase the equipment.

Operation and maintenance costs remain the same for the new equipment compared with those for vapor degreasing. The payback period for the entire project is 10 months. This figure is based on the purchase of 17,175 gal of TCA in 1989 at a cost of $103,000. The central washing equipment cost $78,000, with an additional $3,000 for chemical costs.

Benefits of this project included cost savings, improved public relations, and improved employee safety. It has reduced regulatory requirements through a reduction in hazardous waste materials sent for disposal. Nonhazardous waste streams that had previously been contaminated by solvents are no longer rendered hazardous by such contamination. This project has allowed Crown to drop TCA from its SARA reporting requirements, and it has reduced the liability for Crown in sending listed wastes to disposal facilities (8).

REVIEW QUESTIONS

1. Discuss the Pollution Prevention Act of 1990.
2. Define *source reduction.*
3. a. What is an environmental audit?
   b. Can environmental auditing be considered a management tool? As an auditor, how would you audit an industry?
4. Describe the EPA hierarchy of recommended approaches to pollution control.
5. List the benefits of waste minimization.
6. Contact the Department of Pollution or Environment Control of an industry, and find out their current approaches to source reduction or recycling of hazardous waste.
7. Give examples (other than those given in the book) of the following ways to reduce the hazardous waste:
   a. Product changes.
   b. Input material changes.
   c. Good housekeeping practices.

REFERENCES

1. *Preventing Pollution in the Chemical Industry.* Washington, D.C.: Chemical Manufacturers Association, 1992.
2. 59 *Federal Register* 38455.
3. U.S. Environmental Protection Agency. *Waste Minimization Opportunity Assessment Manual.* Washington, D.C.: U.S. Govt. Printing Office, 1988.
4. Quinn, B. "The Surface Coating Industries Try on New Coats," *Pollution Engineering,* Vol. 27, No. 2, 1995, p. 67–70.
5. U.S. Environmental Protection Agency. *Pollution Prevention Case Studies Compendium.* Washington, D.C.: U.S. Govt. Printing Office, 1992.
6. Baringer, R.P., and D.M. Benforado. "Pollution Prevention as Corporate Policy: A Look at the 3M Experience," *The Environmental Professional,* Vol. 11, 1989.
7. *Ohio Environmental Protection Agency Newsletter,* October 1994.
8. Ohio Environmental Protection Agency. *Fact Sheet,* No. 10. Columbus, OH: Ohio EPA, 1993.

Chapter 20 Hazardous Waste Treatment, Storage, and Disposal

This chapter discusses current methods of hazardous waste treatment, storage, and disposal, which are very important aspects of waste management. Some types of hazardous wastes can be made less hazardous or even be detoxified by using chemical, biological, or physical treatment methods. Treatment of hazardous waste may be costly, but the material can then be prepared for recycling or for an ultimate disposal that is safer than disposal without treatment. The hazardous waste industry is constantly developing innovative treatment methods that are more economical and effective.

Even with a vigorous, ongoing hazardous waste reduction program as required by the U.S. Resource Conservation and Recovery Act (RCRA), significant quantities of hazardous waste will still require treatment and disposal. In the past, there was usually little treatment before disposal on land. Both the Superfund legislation and the 1984 Hazardous and Solid Waste Amendments of the RCRA emphasize the development and use of technologies that reduce the toxicity, volume, and mobility of hazardous waste—or that even destroy such waste. Land disposal is discouraged and restricted under the 1984 RCRA amendments.

TREATMENT TECHNOLOGIES

The current treatment technologies for hazardous waste are physical, chemical, biological, thermal, or stabilization/fixation. Chemical, biological, and physical treatment processes are the common methods of treating aqueous hazardous waste. Chemical treatment transforms waste into less hazardous materials by using neutralization, precipitation, and oxidation or reduction. Biological treatment uses microorganisms to degrade organic compounds in the waste. Physical treatment includes gravity separation, filteration, carbon adsorption, and both air and steam stripping of volatile organics from aqueous waste. Thermal treatment includes incineration and pyrolysis. Latter is the chemical decomposition of waste by heating it in the absence of oxygen. Waste stabilization and solidification techniques involve removing excess water from waste and then solidifying the waste either by mixing it with a stabilizing agent (e.g., Portland cement) or by creating a glassy material by (i.e., vitrification).

The chemical, biological, physical, and thermal processes have been discussed in Chapter 13. The methods and design of such processes have been discussed in Chapter 16. Many treatment systems combine two or three of these processes to provide good, economical treatment of industrial wastewaters. There are many treatment technologies for each of these processes. Table 20–1 presents typical technologies for the

Table 20–1 Typical Technologies for Treatment of Industrial Wastewater

Treatment Technology	Textile Mills	Organic Chemicals	Pulp and Paper	Petroleum Refining	Paint and Ink formation	Plastic & Synthetic Materials	Inorganic Chemicals Manufacturing	Electroplating	Iron and Steel Manufacturing	Nonferrous Metals Manufacturing	Battery Manufacturing	Leather Tanning and Finishing	Pesticides Manufacturing	Photographic Processes
Clarification			x		x	x			x	x				
Flotation								x						
Oil–Water separation						x								
Coagulation/ precipitation	x	x	x		x		x	x	x	x	x	x	x	x
Neutralization									x	x			x	x
Biological treatment	x	x	x	x	x	x			x			x	x	

Industry (column group header)

treatment of industrial wastewater, but these represent only a fraction of available technologies. A description of the waste stabilization and solidification techniques follows.

WASTE STABILIZATION AND SOLIDIFICATION TECHNIQUES

Stabilization and solidification processes have different goals. Stabilization attempts to reduce the solubility or chemical reactivity of waste either by changing its chemical state or by physical entrapment (i.e., microencapsulation). Solidification attempts to convert waste into an easily handled solid with reduced hazards from volatilization, leaching, and spillage. Stabilization and solidification share a common purpose of improving the containment of potential pollutants in treated wastes. Combined processes are often termed *waste fixation* or *encapsulation*.

Waste stabilization/solidification systems that are potentially useful in remedial action activities include sorption, lime–fly ash pozzolan processes, pozzolan–portland cement systems, thermoplastic microencapsulation, macroencapsulation or jacketing systems, or vitrification.

Sorption

Sorption involves adding a solid to soak up any liquid in the waste, and it may produce a soil-like material. The major use of sorption is to eliminate all free liquid. Nonreactive, nonbiodegradable materials are most suitable for sorption. The most common sorbents used include soil, fly ash, or kiln dust from cement and lime manufacturers. When fly ash or kiln dust is used to sorb an oil sludge (50% oil, 20% water), soil, fly ash, or kiln dust ratios (absorbent-to-sludge) from 1:1 to 2.5:1 typically would be satisfactory. Sorption has been widely used to eliminate free water and to improve waste handling. Some sorbents have also been used to limit the escape of volatile organic compounds.

Lime–Fly Ash Pozzolan Processes

Pozzolanic materials dry and set to a solid mass when mixed with hydrated lime. Natural pozzolanic materials consist of either volcanic lava masses or deposits of hydrated silicic acid (i.e., diatomaceous earth). These are natural cements. During solidification or stabilization of waste, lime and pozzolanic material should be mixed with a selected, reactive fly ash to a pasty consistency.

Lime–fly ash treatment is relatively inexpensive, and with careful selection of materials, an excellent solid product can be prepared. The cementing system is strongly alkaline, however, and can react with certain wastes to release undesired materials (e.g., gas).

Pozzolan–Portland Cement Systems

Waste-solidifying formulations based on a Portland and pozzolan–Portland system vary widely. Many materials have been added to change performance characteristics; these include bentonite clay, silicate, and so on. Cement-based solidification and stabilization systems have proved to be versatile and adaptable methods in which the waste is incorporated into the rigid matrix of the hardened concrete. This method either physically or chemically solidifies the waste, depending on its characteristics.

Thermoplastic Microencapsulation

Thermoplastic microencapsulation has been successfully used in nuclear waste disposal. The technique involves drying and dispersing the waste through a heated, plastic matrix. The most common material used for waste incorporation is asphalt, but other materials (e.g., polyethylene or wax) can be employed for specific wastes.

The major advantage of thermoplastic (i.e., asphalt) encapsulation is the ability to solidify very soluble, toxic materials. Compatibility of the waste and the matrix, however, becomes a major consideration when using thermoplastic microencapsulation. Thermoplastic encapsulation requires complex mixing equipment. It has been widely used in nuclear waste disposal, and for industrial waste disposal as well.

Macroencapsulation or Jacketing Systems

Macroencapsulation systems contain potential pollutants either by bonding an inert coating or jacket around a mass of cemented waste or by sealing them in polyethylene-lined drums or containers. Macroencapsulation can be used to contain very soluble, toxic wastes; leaching of the waste can be eliminated for the life of the jacketing material. This process has been used at remedial sites as drum overpacks to contain weak or leaking drums and containers. Macroencapsulation requires special molds and heating equipment to fuse the waste and form the jacket. Molding equipment is custom fabricated for waste handling.

Vitrification

Vitrification involves combining waste with molten glass at a temperature of 1350°C or greater. This process is costly, so it has been restricted to radioactive or very highly toxic wastes.

Future Innovations

There are hundreds of treatment technologies. Industry practice, however, has mainly concentrated on four systems. Most metal-bearing wastes are treated by precipitation, and organic wastes are treated by incineration and solvent extraction. Activated carbon is used for a variety of wastes in both liquid and gaseous phases. The U.S. Environmental Protection Agency (EPA) frequently publishes documents on innovative technologies, such as *Innovative Hazardous Waste Treatment Technologies; A Developer's Guide to Support Services* (1). This booklet is intended as a point of departure for technology developers seeking assistance. It lists available technical assistance, test and evaluation facilities, and university-affiliated research groups that can provide a range of developmental aid.

STORAGE TECHNIQUES

Proper storage of hazardous waste is essential, because accidental discharge of such waste can cause both environmental damage and harm to public health. Figure 20–1

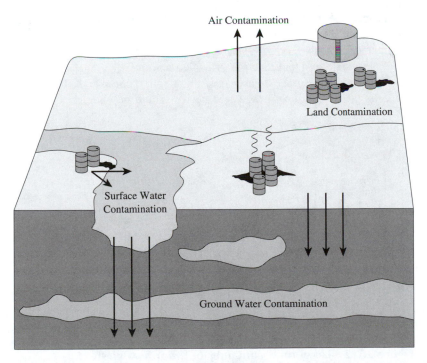

Figure 20–1   Environmental contamination caused by leaks from stored drums.

shows how leaks from stored drums of hazardous chemicals can contaminate surface water, groundwater, and air. Storage facilities represent a temporary management technique, a place where the waste is kept before reaching its final destination. These facilities hold waste before transport for treatment such as incineration, physical or chemical processes, and reuse or recycling.

Large-quantity generators are allowed as many as 90 days and small-quantity generators as many as 180 days to store hazardous waste before disposal under the RCRA (as discussed in Chapter 18). Many generators store hazardous waste on site for varying periods of time. Some hazardous wastes, both solid and liquid, may be temporarily stored in pits, ponds, lagoons, tanks, drums, piles, and vaults. Relatively large quantities may be stored in aboveground basins (e.g., ponds, lagoons), which may be constructed of clay, asphalt, concrete, or soil-cement and lined with bentonite soil or polymeric membranes. Membrane linings are made of polyvinyl chloride, polyethylene, and special rubber (2). Unless diluted with water, solvents (e.g., strong acids and caustics) could damage clay- or soil-based linings.

Because of the many limitations and variations, wastes and lining materials should be tested for effectiveness and compatability before use. The design and layout of a hazardous waste storage facility is generally governed by the waste characteristics, volumes, and safety measures in accordance with regulatory standards (3).

Relatively small amounts of hazardous waste that are generated on an intermittent basis may be placed in plastic, fiber glass, or steel drums for temporary storage and transportation. Noncorrosive liquid waste may be stored in metal drums; corrosive materials are stored in fiber-lined drums.

Proper labeling of containers or drums of hazardous waste before storage and transport is essential (Figure 20–2). For example, labels must identify the contents as toxic, flammable, explosive, or corrosive material. Containers of different chemicals should be kept separate to avoid reaction or contamination.

The design of storage areas should allow for potential leaks to be easily inspected. For example, in a drum storage area, aisles should be wide enough to permit access

Figure 20–2   Label on a container of hazardous waste.

for routine inspection. Some hazardous wastes are stored in covered facilities that are far from populated areas, but proper protection and monitoring must still be provided for leaking drums that might threaten groundwater supplies. Drums that remain on concrete pads often corrode due to "concrete sweating".

Underground Storage Tanks

Underground storage tanks (USTs) have been used for bulk storage of liquid chemicals and petroleum products to reduce the potential for explosion and fire. Environmental damage such as groundwater contamination caused by leaking USTs has been common. Most cases have involved leaks from old USTs containing gasoline or oil, but many USTs containing hazardous wastes have also leaked.

USTs are a problem, because leaks cannot be easily detected. Such leaks pose a threat to the underground water supply. The U.S. EPA finalized regulations for USTs in 1988. These stringent regulations, which are codified in 40 CFR 280–281, have caused many owners of hazardous chemical storage facilities to consider using aboveground tanks, because spills and leaks from these tanks are much easier to detect. The Clean Air Act Amendments of 1990, however, now require complex regulations for aboveground storage emissions.

The 1988 UST regulations are intended to prevent leaks and spills as well as to detect them if and when they occur. EPA standards to control UST systems also include tanks with 10 percent or more of their volume below the ground and containing either petroleum products or hazardous chemicals and any underground piping and pumps connected to the tanks.

Regulations governing the design, installation, and operation of aboveground and underground tanks are complex. New UST systems must be made of fiberglass or cathodically protected steel to protect the tank from corrosion. A backup containment system in the form of double-walled tanks and installation of a leak detection system and alarm are also required. The double-walled tanks prevent leaks in case the main wall fails, and the leak detection and alarm system warn of any problems. Many facilities also have double-walled piping. Owners or operators are required to inspect their tanks monthly and to remove sludge periodically from them.

All existing storage tanks without corrosion or leakage protection are to be taken out of service. (As of this writing, the deadline has not yet been set.) The tanks must be emptied of all liquids, vapors, and sludge and then either be removed or filled with sand and gravel.

Surface Impoundments

Before implementation of the RCRA regulations, liquid hazardous wastes were deposited in ponds, lagoons, and pits, which were collectively called surface impoundments. Because there were no requirements or regulations, most of these excavations were constructed as unlined impoundments. No consideration was given to possible leakage and groundwater contamination. They provided sedimentation, surface aeration, and evaporation of volatile organic chemicals, but this limited treatment was offset by the potential for air and groundwater pollution.

Because of poor siting and construction as well as management problems, surface impoundments are the principal source for contamination of air and groundwater. Under the RCRA, all surface impoundments of hazardous waste must now meet required design criteria. Impoundments must have at least two liners, a lechate collection system, and a groundwater monitoring system.

LAND DISPOSAL TECHNIQUES

The main land disposal techniques include landfills, surface impoundments, and injection wells. Because of the environmental dangers involved, land disposal of hazardous waste is not a desirable option. With proper site selection, engineering design, and safe operation, however, land disposal is generally the least expensive alternative for many types of hazardous waste.

Landfills

Certain hazardous wastes may be disposed of on land. Site selection, design, construction, and operation of municipal solid waste landfills have been discussed in Chapters 15 and 16. Most of the site selection and construction techniques for hazardous waste landfills are similar.

New and more stringent RCRA requirements for the design and operation of hazardous waste landfills, however, have placed limitations on their use. Certain hazardous wastes are now banned from land disposal; these include PCBs, cyanides, halogenated organic compounds, and acids with a pH of less than 2.

A secure landfill is the repository for all hazardous waste residuals, and protection of groundwater is the major concern in designing a secure landfill. The general practice in the United States is to design secure landfills for total containment.

A secure landfill must have a minimum of 10 ft separating the base of the landfill from the bedrock or a groundwater aquifer; this is twice the minimum separation needed for a municipal solid waste landfill. Beneath the hazardous waste, there must also be a double-liner system to stop the flow of leachate from contaminating the soil and groundwater beneath the site. Figure 20–3 illustrates a typical double-liner system with leachate collection above the top liner and leakage detection (or secondary leachate collection) between the liners.

The upper liner must be a flexible-membrane lining (FML). It is usually made from high-density, plastic polymer resins or rubber. The lower liner is usually an FML as well, but recompacted clay with a thickness of at least 3 ft is also considered to be acceptable.

Leachate above each liner is collected in a series of perforated drainage pipes and then pumped to the surface for treatment (see Chapter 16). To minimize the amount of leachate formed by precipitation seeping into the landfill, a low-permeability cap is placed over the completed cells. On closure of the landfill, a final landfill cover cap, which may consist of an FML along with a layer of compacted clay, is placed atop the landfill with sufficient slope to assure drainage.

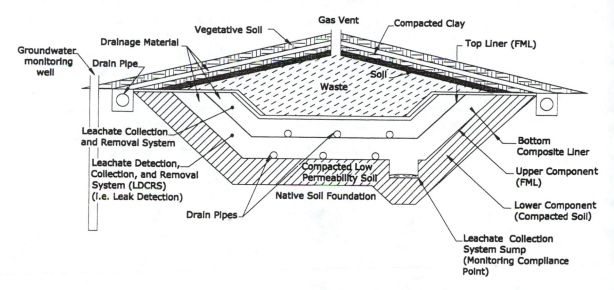

Figure 20–3   Typical secure hazardous waste landfill.

(From U.S. Environmental Protection Agency. How to Meet Requirements for Hazardous Waste Landfill Design, Construction, and Closure. *Washington, D.C.: U.S. Govt. Printing Office, 1984; with permission.)*

The landfill must have a network of monitoring wells both upstream and downstream of the site, and groundwater flowing beneath the landfill should be tested for contamination. In addition, the soil beneath the site should be tested with a lysimeter.

Surface Water Impoundments

Surface water impoundments are excavated or diked areas that are used to contain and store liquid wastes (as discussed previously). In many instances, these impoundments, which should be used for temporary storage, have been closed as a landfill would be. Such facilities have been a significant source of contamination in many Superfund sites.

Deep Well Injection

The U.S. EPA recognizes five classes of injection wells. These classes are defined, in part, by the well's relationship to an underground source of drinking water (USDW). The U.S. Safe Drinking Water Act of 1974 designated as a USDW any aquifer whose water contains a concentration of less than 10,000 mg/L of total dissolved solids. The five classes of injection well are:

1. *Class I wells:* Used for injection of industrial or municipal waste fluids beneath the lowermost formation containing a USDW.

2. *Class II wells:* Used for injection of brines resulting from oil and gas production or of fluids used for enhanced recovery of oil or natural gas.

3. *Class III wells:* Used for injection of fluids for the extraction of soluble minerals, such as salt-solution mining in northeastern Ohio.

4. *Class IV wells:* Used for injection of hazardous or radioactive wastes into or above a USDW. These wells have been banned in the United States since 1984.

5. *Class V wells:* Used for the disposal of nonhazardous fluids. These wells not covered by classes I through IV include stormwater drainage wells, heat-pump and air-conditioning return wells, cesspools, septic systems, floor drains, and sumps.

Because of their use for injection of industrial or municipal waste fluids, only Class I wells are discussed here. The general construction of a typical Class I injection well is illustrated in Figures 20–4 and 20–5. The main design goal for such

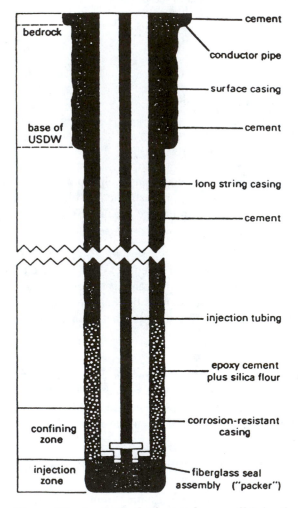

Figure 20–4   Typical Class I deep well injection.

(Courtesy of B.P. Chemicals.)

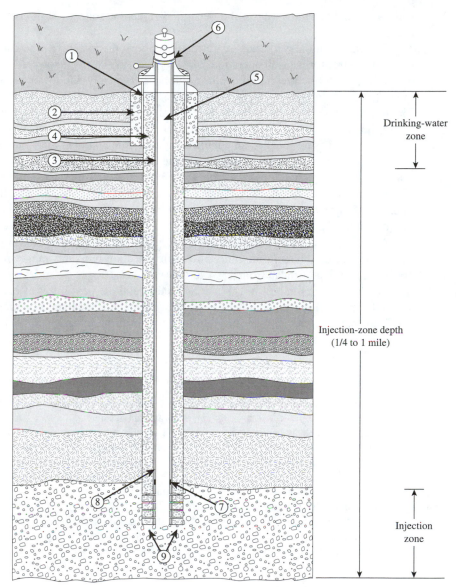

1. Steel surface casing.
2. Cemented outside of casing into the hole.
3. Another protective steel casing.
4. Cemented outside another casing.
5. Injection tubing.
6. Inside tubing sealed at the top.
7. Inside tubing sealed at the bottom.
8. Space between the tubing and casing called annulus filled with liquid.
9. Injection zone.

Figure 20–5 Deep well disposal.

(Courtesy of B.P. Chemicals.)

wells is to deliver the waste to the permitted injection zone without contaminating the USDW.

The casing of the well seals off formations above the injection zone and provides pressure control for the well operation. The "long string" casing is lowered through the surface casing to the prescribed depth in the hole. Typically, the lower portion of the long string casing is constructed of fiberglass, fibercast, or corrosion-resistant steel. The injection tubing is placed inside the long string and then sealed by a packer at the top of the injection zone. The space between the injection tubing and the inner wall of the long string casing is called an annulus, and it is filled with an inert fluid (e.g., water, sodium chloride) and then pressurized. The operator is required to constantly monitor the annulus pressure. Leakage from the tubing to the casing or from the casing to the surrounding rocks will cause either a pressure increase or decrease. A fluctuation in either of these monitored pressures will automatically trigger alarms and shut down the injection pumps.

A maximum allowable injection pressure is set for each well. This limit must be below the fracture pressure. In Ohio, allowable injection pressures range from 150 to 1690 psi, and depths range from 1294 to 6000 ft (4).

In addition to failure of a well's construction and the possibility of encountering an unplugged well bore, the potential for upward migration along a naturally occurring fault plane is probably the most serious threat to the integrity of waste confinement at a Class I site.

REVIEW QUESTIONS

1. a. List the technologies available for hazardous waste treatment.
   b. Discuss each technology and its application in the treatment of hazardous waste.
2. What are the objectives of waste stabilization and solidification?
3. Describe the pozzolan–Portland cement system.
4. Assume you are in charge of storage for a hazardous waste facility. What precautions would you take in the design and operation of this facility to prevent accidental discharge of waste?
5. How have EPA regulations made underground storage tanks safer?
6. a. List the advantages and disadvantages of landfill disposal for hazardous waste.
   b. As an engineer working for the EPA to minimize land disposal, what steps would you take to discourage it?
7. a. Sketch a deep well injection system, and show the various components.
   b. Which aspects of the system present a hazard to groundwater?

REFERENCES

1. U.S. Environmental Protection Agency. *Innovative Hazardous Waste Treatment Technologies: A Developer's Guide to Support Services.* Washington, D.C.: U.S. Govt. Printing Office, 1994.
2. U.S. Environmental Protection Agency. *Draft Environmental Impact Statement on the Proposed Guidelines for the Landfill Disposal of Solid Wastes.* Washington, D.C.: U.S. Govt. Printing Office, 1979.
3. U.S. Department of Defense. *Hazardous Materials Storage and Handling Handbook.* DLAH 4145.6. Alexandria, VA: U.S. DOD, 1987.
4. Ohio Geological Survey. *Industrial Waste Disposal Well in Ohio.* Columbus, OH: Ohio Geological Survey, 1998.

Chapter 21   Site Remediation

This chapter discusses the Superfund program to restore contaminated sites, sources of contamination, and technologies used to remediate such sites. Before legislation that placed restrictions on the disposal of hazardous waste, many industrial and manufacturing establishments stored and disposed of such waste in landfills, ponds, lagoons (mostly unlined), and unlined waste piles. The cradle-to-grave tracking provision under the RCRA was primarily responsible for eliminating unstructured management of hazardous wastes.

During the 1970s, Americans learned the dangers of dumping hazardous waste on land (see Chapter 17). It was discovered that hazardous wastes buried over a period of more than 30 years had contaminated New York's Love Canal, thus endangering both the environment and the health of the residents. Oil laced with toxic dioxin also contaminated the land and water in the small Missouri community of Times Beach. These are just two examples of many such problems discovered between 1970 and 1980.

THE SUPERFUND PROGRAM

At both Love Canal and Times Beach, public health and the environment were adversely affected, lives disrupted, and property values depreciated. The magnitude of this problem moved the U.S. Congress to pass the Comprehensive Environmental Response, Compensation, and Liability Act (CERCLA) in 1980 and then amended by the Superfund Amendments and Reauthorization Act in 1984. CERCLA authorized the federal government to take action on releases (or potential releases) of hazardous substances that might endanger the environment or public health. Legal action could now be taken to force polluters to clean up contaminated sites or to reimburse the cost of cleanup to the Superfund. If the U.S. Environmental Protection Agency (EPA) cannot find those responsible or if polluters were unable or unwilling to clean a site, the EPA could clean the site using the Superfund, which is the trust fund that finances these activities. The initial funding was $1.6 billion, which came from money generated by taxes on oil and chemicals.

Congress directed the EPA to set priorities for federal action under the Superfund and hoped that $1.6 billion would be sufficient to clean all the sites on the priority list. The problem turned out to be much bigger and more complex than anyone imagined, however, and $1.6 billion hardly made a dent in the number of contaminated sites. Thousands of these sites existed, and the funding was not enough to clean them all.

As more and more contaminated sites were discovered, the cost estimates skyrocketed. Obviously, the $1.6 billion originally approved fell short of the amount needed to clean even the nation's most serious hazardous waste sites. Mismanagement of the money is frequently associated when a large sum is involved, and the Superfund was no exception. Because of the nature of the problem, Congress reauthorized the Superfund for another 5 years and provided an additional $8.5 billion. As mentioned, money for the Superfund came from excise taxes on petroleum and certain chemicals, including imported chemical derivatives, an environmental tax on corporations, and monies collected from the parties responsible for site contamination.

The EPA identified 1236 hazardous waste sites between 1980 and 1990. These sites are listed on the National Priorities List (NPL) of targets for cleanup under the Superfund. At the beginning of 1990, only 29 of the 1236 sites had been cleaned and removed from the NPL. The EPA estimates, however, that an average of 100 sites per year will need to be added to the list. Congress is reluctant to let the federal government be responsible for addressing the environmental and public health problems associated with previous hazardous waste disposal practices. Consequently, the EPA was directed to set priorities and to create a list of the worst sites to target.

The work of the Superfund got off to a slow start in the 1980s mainly because of political and legal problems. It was expected that most of the contaminated sites would be not only cleaned but returned to almost pristine conditions. Unfortunately, and in spite of all the efforts and funds expended, the outcome has not been satisfactory. This dissatisfaction has been expressed by all parties concerning the methods of site selection and determining the cleanup standard, criteria, and liability.

SITE CONTAMINATION

A contaminated site is usually characterized by one primary contaminant or group of contaminants associated with the activities conducted at that location. The contaminant profile is established by sampling and analysis, and then detailed studies focus on the contaminant compound (or compounds) identified. At a chemical-manufacturing site, for example, various organic compounds may be present, whereas at a battery-manufacturing site, the main contaminants may be heavy metals (e.g., mercury, cadmium). A metal-products industry or electronic industry site will most likely be characterized by organic solvents. A fertilizer factory would be characterized by ammonia and nitrogen-related compounds. A site devoted to the wood-preserving process, in contrast, would be characterized by phenols and creosote-related compounds. A municipal waste disposal landfill would generally have organic compounds and some heavy metals, and at a commercial hazardous waste disposal facility, one would expect a variety of inorganic and organic contaminants. The EPA publishes several manuals that enumerate industry-specific contaminants.

In general, environmental contaminants can be divided into three categories: organic, inorganic, and biological contaminants. Table 21–1 lists the relevant organic and inorganic compounds.

Organic Contaminants

The U.S. EPA divides organic contaminants on the basis of certain chemical characteristics. These characteristics are volatile organics, semivolatile organics, and pesticides.

Volatile organics compounds (VOCs) in the environment result from wide use of petroleum products, gasoline and fuel oil, and volatile organic solvents in degreasing

Table 21–1   Target Compound/Analyte List

Volatiles		
Acetone	1,2-Dibromo-3-chloropropane	2-Hexanone
Benzene	1,2-Dibromoethane	Methylene chloride
Bromochloromethane	1,2-Dichlorobenzene	4-Methyl-2-pentanone/MIBK
Bromodichloromethane	1,3-Dichlorobenzene	Styrene
Bromoform	1,4-Dichloroethane	1,1,2,2-Tetrachloroethane
Bomomethane/methyl bromide	1,1-Dichloroethane	Tetrachloroethene
2-Butanone/MEK	1,2-Dichloroethane	Toluene
Carbon disulfide	1,1-Dichloroethene	1,1,1-Trichloroethane
Carbon tetrachloride	cis-1,2-Dichloroethene	1,1,2-Trichloroethane
Chlorobenzene	trans-1,2-Dichloroethene	Terichloroethane
Chloroethane	1,2-Dichloropropane	Vinyl chloride
Chloroform	cis-1,3-Dichloropropene	Xylenes (total)
Chloromethane/methyl chloride	trans-1,3-Dichloropropene	
Dibromochloromethane	Ethylbenzene	

Semivolatiles		
Acenaphthene	p-Cresol/4-methylphenol	Indeno[1,2,3-cd]pyrene
Acenaphthylene	Di-n-butyl phthalate	Isophorone
Anthracene	Dibenz[a,h]anthracene	2-Methylnaphthalene
Benzo[a]anthracene	Dibenzofuran	Naphthalene
Benzo[b]fluoranthene	3,3'-Dichlorobenzidine	o-Nitroaniline
Benzo[k]fluoranthene	2,4-Dichlorophenol	m-Nitroaniline
Benzo[ghi]perylene	Diethyl phthalate	p-Nitroaniline
Benzo[a]pyrene	2,4-Dimethylphenol	Nitrobenzene
bis(2-Chloroethox)methane	Dimethyl phthalate	o-Nitrophenol
bis(2-Chloroethyl)ether	4,5-Dinitro-o-cresol	p-Nitrophenol
bis(2-Ethylhexyl)phthalate	2,4-Dinitrophenol	n-Nitrosodiphenylamine
4-Bromophenyl phenyl ether	2,4-Dinitrotoluene	n-Nitrosodi-n-propylamine
Butyl benzyl phthalate	2,6-Dinitrotoluene	2,2'-Oxbil(1-chloropropane)
p-Chloroaniline	Di-n-octyl phthalate	Pentachlorophenol
p-Chloro-m-cresol	Fluoranthene	Phenanthrene
2-Chloronaphthalene	Fluorene	Phenol
2-Chlorophenol	Hexachlorobenzene	Pyrene
4-Chlorophenyl phenyl ether	Hexachlorocyclopentadiene	1,2,4-Trichlorobenzene
Chrysene	Hexachloroethane	2,4,5-Trichlorophenol
o-Cresol/2-methylphenol	Hexachlorobutadiene	2,4,6-Trichlorophenol

Pesticides/PCBs		
Aldrin	Dieldrin	Aroclor 1016
α-BHC	Endosulfan I	Aroclor 1221
β-BHC	Endosulfan II	Aroclor 1232
γ-BHC (Lindane)	Endosulfan sulfate	Aroclor 1242
δ-BHC	Endrin	Aroclor 1248
α-Chlordane	Endrin aldehyde	Aroclor 1254
γ-Chlordane	Endrin ketone	Aroclor 1260
4,4'-DDT	Heptachlor	Toxaphene
4,4'-DDE	Heptachlor epoxide	
4,4'-DDED	Methoxychlor	

Table 21–1   (*cont.*)

Inorganics		
Aluminum	Cobalt	Nickel
Antimony	Copper	Potassium
Arsenic	Cyanide, total	Selenium
Barium	Iron	Silver
Beryllium	Lead	Sodium
Cadmium	Magnesium	Thallium
Calcium	Manganese	Vanadium
Chromium	Mercury	Zinc

Source: U.S. Environmental Protection Agency. Washington, D.C.: *U.S. Govt. Printing Office,* 1982; with permission.

operations. VOCs have a high vapor pressure. They volatilize to the gaseous state at standard temperature and pressure, and they have a low water solubility. In addition, VOCs have a specific gravity lighter than that of water without chlorine atoms and heavier than that of water with chlorinated compounds.

Chlorinated aliphatic compounds are the most common contaminants. The compounds of tetrachloroethane are used as dry-cleaning fluid and for metal degreasing. Methylene chloride has a wide range of industrial uses, including wood-stripping operations.

Aromatic compounds in the environment are commonly those present in petroleum and those used in solvents. Common examples include benzene, ethylbenzene, toluene, and xylene. These are unchlorinated compounds; therefore, they are lighter than water.

Semivolatile organic compounds are byproducts of various industrial and nonindustrial processes. A few of the more common include phenol (used in wood preservatives) and phthalates (used as plasticizers). The semi-VOCs have lower pressures than VOCs, and because of their low volatility, they normally exist in the solid or liquid state at standard temperature and pressure. Many semi-VOCs are heavier than water and have a low solubility.

Pesticides are semivolatile compounds, and these chemicals find their way into the environment through both agricultural and residential use for pest and weed control. Pesticides may be in the form of a powder or be dissolved in solvents. PCBs are also semivolatile compounds. They were generally dissolved in oil for application, but their manufacture was halted in the 1970s because they were considered to be environmentally unsafe.

Dioxin and its related compounds (i.e., furans) are considered to be very dangerous to the environment. Dioxin is an incidental byproduct from the manufacture of other products or from processes such as the bleaching of paper. Dioxin can also form as a combustion byproduct of some organic materials.

Inorganic Contaminants

Inorganic contaminants are naturally occurring substances that can become concentrated in the environment because of human activities. The main problem contaminants in this category are heavy metals, which have a low vapor pressure and are nonvolatile at standard pressure and temperature. Mercury and lead, however, are volatile at high temperature, and this fact is significant when waste contaminated with such metals is thermally treated. Lead, mercury, and a few other heavy metals are very toxic, whereas iron, manganese, and potassium, are less toxic.

Biological Contaminants

Biological contaminants include human and animal wastes, blood and flesh from hospitals and laboratories, contaminated needles, and so on. The principal concerns in this category are bacteria and pathogens.

Contaminated Media

Media that can be contaminated by chemical compounds or biological contaminants include soil, groundwater, surface water, and air. These media can be contaminated directly by spillage and/or disposal of liquids or indirectly by infiltration of contaminated water. Such contamination of ground or surface waters can occur directly or by non-aqueous-phase liquids and/or contaminants in solution or suspension in air, which may transport contaminated dust or contaminants in gas.

Waste Materials

Many types of waste can cause environmental contamination. These include solid waste, liquid chemicals, municipal solid waste, and soil.

Solid Waste.   Solid waste consists mostly of residuals generated by a manufacturing or a treatment process. For example, these residuals may include ashes from the combustion of waste or sludges from refineries and electroplating processes. Sludges are mostly contained in drums or in lagoons as precipitates. Solids may also include mining spoils and industrial powders and salts.

Liquid Chemicals.   Raw liquid chemicals or waste liquid may be spilled or leak when placed in containers or disposed of in bulk on the ground. Such liquid may also spill during transportation. Spillage or improper disposal can contaminate soil, groundwater, surface water, and air. For example, old transformers filled with PCBs have been reported to have leaked.

Municipal Solid Waste.   Older landfills that were built without proper lining or drainage and that now require remediation typically contain household solid waste. This waste is usually mixed with some waste generated from industrial facilities. Leachate from such landfills usually contains low concentrations of hazardous substances, which originate from the disposal of household products (e.g., pesticide containers with some residuals) and from industrial disposal.

Soil.   As discussed earlier, soil can be contaminated in various ways. The degree to which the soil at a site is contaminated, however, depends on the nature of the disposal activities and the quantity and characteristics of the material disposed. Contaminated soil must be remediated immediately after the discovery of contamination to minimize or even prevent adverse effects on the public health and damage to the environment. There is always a serious threat to the groundwater underlying the site and/or the surface water at or near it. Contaminants in the soil often provide a nearly continuous source of contamination to underlying ground and/or surface water, especially after a rainfall.

REMEDIAL INVESTIGATION/FEASIBILITY STUDY

The U.S. CERCLA requires that the EPA establish procedures to ensure the Hazardous Substance Response Trust Fund (i.e., Superfund) be used effectively in responding to releases of hazardous substances into the environment. Therefore, the EPA has established a procedure for discovering releases, evaluating remedies, determining the appropriate extent of response, and ensuring that cost-effective remedies are selected. This procedure is commonly referred to as the Remedial Investigation/Feasibility Study (RI/FS) process and is outlined in the revised National Contingency Plan (40 CFR 300).

The National Contingency Plan requires that a detailed RI/FS be conducted for every site targeted for remedial response action under CERCLA. The RI emphasizes data collection and site characterization to define the nature and extent of contamination at a site to the extent necessary to evaluate, select, and design a cost-effective remedial action. The FS emphasizes data analysis and decision making by using the data from the RI to develop objectives and alternative remedial responses. These alternatives are then evaluated in terms of engineering feasibility, public health protection, environmental impacts, and costs.

The RI and FS are interdependent processes that are generally performed concurrently rather than sequentially. Figure 21–1 presents a flow chart of the RI/FS process, illustrating the major tasks. The interactive nature of the RI/FS process results from the integration of data collected during the RI and the data analyses performed during the FS (1).

The first step in the FS is to identify existing site problems using preliminary RI data (e.g., site background data, previous studies, initial RI activities) and to determine the remedial technologies (e.g., surface water controls, air pollution controls) that are most applicable. Table 21–2 presents remedial technology categories for specific site problems; more than one category may be applicable at any given site.

The next step in the FS is to identify and screen potentially applicable remedial technologies to eliminate those that may be difficult to implement, that rely on unproven technologies, or that may not achieve the remedial objectives within a reasonable period of time. The screening process focuses on eliminating those technologies with severe limitations regarding site-specific conditions. Site, waste, and technology characteristics are used to screen for inapplicable remedial technologies.

Screening of Technologies

Before undertaking remedial action at a site, several questions must be answered. What is the size of the contaminated site? What is the source of the contaminants? What are the contaminants? What is the best estimate of the quantity or extent of contamination?

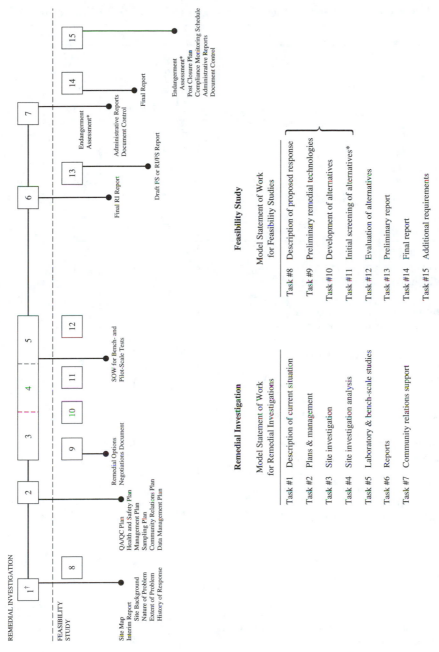

RI/FS PROCESS

REMEDIAL INVESTIGATION

FEASIBILITY
STUDY

Site Map
Interim Report
Site Background
Nature of Problem
Extent of Problem
History of Response

QA/QC Plan
Health and Safety Plan
Management Plan
Sampling Plan
Community Relations Plan
Data Management Plan

Remedial Options
Negotiations Document

SOW for Bench- and
Pilot-Scale Tests

Final RI Report

Draft FS or RI/FS Report

Endangerment
Assessment*

Administrative Reports
Document Control

Final Report

Endangerment
  Assessment*
Post Closure Plan
Compliance Monitoring Schedule
Administrative Reports
Document Control

Remedial Investigation

Model Statement of Work
for Remedial Investigations

Task #1 Description of current situation

Task #2 Plans & management

Task #3 Site investigation

Task #4 Site investigation analysis

Task #5 Laboratory & bench-scale studies

Task #6 Reports

Task #7 Community relations support

Feasibility Study

Model Statement of Work
for Feasibility Studies

Task #8 Description of proposed response

Task #9 Preliminary remedial technologies

Task #10 Development of alternatives

Task #11 Initial screening of alternatives*

Task #12 Evaluation of alternatives

Task #13 Preliminary report

Task #14 Final report

Task #15 Additional requirements

†Numbers in the boxes refer to tasks described in the Model Statement of work for RI/FS under CERCLA Guidance issued, February, 1985
*Endangerment assessment may be prepared at any point in the RI/FS Process in support of enforcement actions

Figure 21–1 Remedial investigation/feasibility study.

(From U.S. Environmental Protection Agency: Handbook on Remedial Action at Waste Disposal Sites. Washington, D.C.: U.S. Govt. Printing Office, 1985; with permission.)

Table 21–2 General Remedial Technology Categories for Specific Site Problems

Site Problem	Surface Water Controls	Air Pollution Controls	Leachate and Groundwater Controls	Gas Migration Control	Waste and Soil Excavation and Removal and Land Disposal	Contaminated Sediments Removal and Containment	In Situ Treatment	Direct Waste Treatment	Contaminated Water Supply and Sewer Line Controls
Volatilization of chemicals into air		•							
Hazardous particulates released to atmosphere		•							
Dust generation by heavy construction or other site activities		•							
Contaminated site runoff	•								
Erosion of surface due to wind or water	•								
Surface seepage of leachate	•								
Flood hazard or contact of surface water body with wastes	•								
Leachate migrating vertically or horizontally			•				•		
High water table, which may result in groundwater contamination or interfere with other remedial technologies			•						
Precipitation infiltrating into site to form leachate	•		•						
Evidence of methane or toxic gases migrating laterally underground				•					
On-site waste materials in non disposed form (drums, lagooned waste, watepiles)					•		•	•	
Contaminated surface water, groundwater, or other aqueous or liquid waste					•		•	•	
Contaminated soils					•		•	•	
Toxic and/or hazardous gases that have been collected		•						•	
Contaminated stream banks and sediments					•	•		•	
Drinking water distribution system contamination									•
Contaminated sewer lines									•

Source: U.S. Environmental Protection Agency. *Handbook on Remedial Action at Waste Disposal Sites.* Washington, D.C.: U.S. Govt. Printing Office, 1985; with permission.

What is the size of the contamination plume in the groundwater, its location, and direction of movement? What are the feasible alternatives to clean up the site? Which alternative is best? What will the duration of clean up be? What end product may be expected? And, finally, what will the project cost, and what level of cleanup should be achieved?

All these questions (except the last one) require engineering and scientific knowledge and extensive data collection at the contaminated area. The last question is subjective, however, and it requires political and social considerations.

The cleanup process and the technologies involved are complex. Restoring sites to background (i.e., precontamination) levels is rarely practical or economical, based on benefit/cost analysis.

Site Characteristics.   Site data should be reviewed to identify conditions that may limit or aid use of certain remedial technologies. Such information is generally gathered during the RI process. Table 21–3 lists the site characterization data necessary to evaluate remedial technologies. Those technologies precluded by the site characteristics should be eliminated from further consideration. For example, very-low-permeability soil would preclude use of *in situ* methods, because it would be impossible to ensure complete mixing of treatment chemicals with waste components. Similarly, contaminated sediments below 65 ft would preclude the use of certain types of dredges.

Waste Characteristics.   Identification of waste characteristics that limit the feasibility of remedial technologies is another important part of the screening process. Table 21–4 presents waste characteristics that may influence the feasibility of remedial technologies.

Technological Limitations.   The level, performance record, and inherent problems for each technology being considered should be identified. Technologies that are unreliable or not fully demonstrated may be eliminated.

Development of Remedial Action Alternatives

Technologies that pass the initial screening process comprise the overall remedial action alternatives to address the problems at the site. Each alternative may consist of an individual technology or some combination of technologies.

Once a list of remedial action alternatives has been developed, additional screening is done to eliminate those that are more costly than other alternatives, provide inadequate public health protection, or have adverse environmental impacts. The EPA then selects the lowest-cost alternative that both mitigates and minimizes damages and adequately protects public health and the environment. It is very important that these monies are spent effectively. Unfortunately, however, this has not always been the case, because in many instances, a significant amount of money is spent on legal and political wrangling.

To meet this goal, each alternative is analyzed in terms of the following factors:

- Technical considerations.
- Environmental concerns.

Table 21–3   Identification or Remedial Investigation Site Characterization Data Needed to Evaluate Remedial Technologies

Required Data	Gas Migration Controls			Waste Excavation and Removal	Contaminated Sediment Removal and Containment		In Situ Treatment
	Capping	Collection (with or without Recovery)	Containment		Removal	Containment and Turbidity Control	
General Site Conditions							
Accessibility	•		•	•	•	•	•
Topography	•	•	•	•			•
Native vegetation	•			•			•
Waste Characteristics							
Physical state	•	•	•	•	•	•	
Chemical composition	•	•	•	•	•	•	•
Disposal/burial practices	•	•	•	•	•	•	
Physical chemical properties	•	•	•		•	•	•
Site Geology							
Seismic history	•		•				
Depth to bedrock	•		•	•			•
Bedrock type							•
Bedrock profile							•
Structural configuration							•
Structural strength	•		•				
Bedrock permeability/porosity							•
Soil Characteristics							
Profiles to bedrock	•	•	•	•		•	•
Type/texture	•	•	•	•	•	•	•
Permeability/porosity	•	•	•	•	•	•	•
Engineering properties	•			•	•		•
Soil chemistry	•	•	•	•	•	•	•
Erosion potential	•			•		•	
Contaminant profiles	•	•	•	•	•	•	•
Moisture content	•	•		•			•

- Public health concerns.
- Institutional concerns.
- Costs.

Technical Considerations.   Technical considerations involves evaluating an alternative in terms of performance, reliability, implementability, time, and safety. Performance is measured based on effectiveness and useful life. (Effectiveness is evaluated

Table 21–3 (*cont.*)

Required Data	Gas Migration Controls			Waste Excavation and Removal	Contaminated Sediment Removal and Containment		In Situ Treatment
	Capping	Collection (with or without Recovery)	Containment		Removal	Containment and Turbidity Control	
Groundwater characteristics							
Natural groundwater chemistry							•
Seasonal potentiometric surfaces	•	•	•	•			•
Aquifer profile							•
Aquifer characteristics							•
Groundwater velocity and direction of flow							•
Groundwater recharge and discharge areas				•			•
Contaminant profiles				•			•
Supply well characteristics							•
Surface Water							
Proximity of nearest surface waters				•			•
Presence of leachate seeps	•	•	•	•	•	•	•
Floodplain or coastal storm surge boundaries	•	•	•	•	•	•	•
Stream profiles					•	•	
Surface water use					•	•	•
Drainage area/runoff	•	•	•	•	•	•	•
Local surface water quality					•	•	•
Stream flow characteristics					•	•	
Climatology							
Evapotranspiration parameters	•	•	•				•
Wind speed and direction	•	•	•	•			•
Temperature parameters	•	•	•	•	•	•	•
Precipitation	•	•	•	•			•
Local air quality		•		•	•		•
Regional air quality							

Source: U.S. Environmental Protection Agency. *Handbook on Remedial Action at Waste Disposal Sites.* Washington, D.C.: U.S. Govt. Printing Office, 1985; with permission.

Table 21–4   Waste Characteristics that May Affect Selection of Remedial Technologies

Quantity/concentration	Infectiousness
Chemical composition	Solubility
Acute toxicity	Volatility
Persistence	Density
Biodegradability	Partition coefficient
Radioactivity	Compatibility with other chemicals
Ignitability	Treatability
Reactivity/corrosivity	

Source: U.S. Environmental Protection Agency. *Handbook on Remedial Action at Waste Disposal Sites.* Washington, D.C.: U.S. Govt. Printing Office, 1985; with permission.

in terms of containment, diversion, removal, destruction, or treatment at the site and useful life in terms of the length of time the level of effectiveness can be maintained.) Reliability is based on technologies not requiring frequent or complex operation and maintenance activities. Implementability can be described as the relative ease of constructability and the time needed to actually see beneficial results. Safety includes consideration of fire, explosion, and exposure to hazardous substances.

Environmental Concerns.   Proposed alternatives should be evaluated to determine any adverse environmental impacts, methods for mitigating these impacts, and the cost of mitigation. The environmental assessment should also analyze environmental impacts in the absence of remedial action (see Chapter 2).

Public Health Concerns.   Remedial sites should also undergo a public health evaluation. All sites must undergo a baseline evaluation (i.e. current site conditions), and an analysis should be performed regarding the extent and duration of human exposure to site contaminants in the absence of remedial action.

Institutional Concerns.   The effects of federal, state, and local standards and of other institutional considerations on the implementation and operational timing of each alternative should be determined.

Costs.   Even though the cost factor has been listed last, it is a very important consideration. A detailed analysis of costs should include the estimated capital and the operation and maintenance costs. It should also include a present-worth analysis. (Economic analysis was discussed in Chapter 2.)

Excavation and Removal at Hazardous Waste Sites

Excavation and removal followed by land disposal or treatment are performed in the remediation of hazardous waste sites. There are no absolute limitations on the types of waste that can be excavated and removed, but worker health and safety are prime considerations in the decision to excavate explosive, reactive, or highly toxic waste material. Other factors that should be considered include the mobility of the waste, the feasibility of on-site containment or of *in situ* treatment, and the cost for disposing of the waste or rendering it nonhazardous once it has been excavated. A frequent practice at hazardous waste sites is to excavate and remove contaminant "hot spots" and to use other remedial measures for less-contaminated areas. Excavation and removal are applicable to almost all site conditions but may become cost prohibitive at great depths or in complex hydrogeologic environments (2).

Excavation Equipment. The basic types of excavation equipment for site remediation include backhoes, cranes and attachments (e.g., draglines, clamshells), dozers, and loaders. Hauling equipment includes scrapers, haulers, dredges, dozers, and loaders. This equipment is used for both on-site and off-site transport of waste.

Pumps are also important machinery during these tasks. Pumping is required to remove liquids and sludges from ponds, waste lagoons, and surface impoundments. Liquid waste pumped from these sites is then managed to prevent degradation of the surrounding environment. The liquid waste may be pumped to a treatment system or transported off-site to a commercially operated treatment, storage, and disposal facility.

Planning Excavation and Removal Activities. Many activities are performed as part of the excavation and removal of contaminants at a site, including design and construction of site operating areas, implementation of controls to minimize environmental releases and to protect workers, and equipment selection and mobilization. Proper layout of the work area is critical to safe and cost-effective excavation and removal. "Hot," "transition," and "clean" zones should be established and posted using air-monitoring data and available information on waste locations, and the location of these zones should govern where the activities are performed. For example, any staging or on-site treatment of waste would be conducted in the contaminated zone, personnel decontamination in the transition zone, and administrative and/or emergency medical care in the clean zone.

Distinct operating areas should be provided for staging, treating, storing, and transporting waste and for equipment decontamination. Each area should be designed to provide adequate room to maneuver equipment and for emergency evacuation. Within each operating area, reactive, corrosive, explosive, flammable, and incompatible wastes must be made to segregate. Explosive or radioactive waste should be staged or stored in isolated areas until arrangements can be made for its safe detonation or off-site disposal. For each operating area, necessary measures must be taken to minimize environmental releases, to prevent incompatible waste reactions, and to contain any contaminants that are released.

Environmental Controls. Both the nature and extent of the preventive and mitigative measures required to control environmental releases during excavation and removal are site specific, but several general procedures apply to all sites. To prevent puddling, operating areas for staging and treating drummed wastes and contaminated soils should be graded, lined with polyethylene or clay, and bermed or diked. Where temporary impoundments must be used to store liquids, it may be acceptable to provide a thin clay liner and to excavate contaminated soils after use of the impoundment is finished. Long storage periods or poor site conditions (e.g., wastes in the water table or permeable, unsaturated zone) may necessitate use of a synthetic liner system. The equipment decontamination area should be a hard-surfaced area that will retain wash water by perimeter curbing and will collect the liquid through a central trough and perimeter sump.

In addition to the mentioned preventive measures, releases can be mitigated and minimized in several ways, including:

- Covering contaminated soils to prevent leaching and fugitive dusts.
- Using sorbents and pumps to clean up spills promptly.
- Maintaining drums or other containers in work areas and on access roads for use in cleaning spills.
- Controlling runon and runoff water with diversions.
- Containing contaminated runoff downslope with a holding pond.
- Avoiding mixing of incompatible wastes by promptly overpacking or transferring leaking drums.
- Suppressing small fires before they spread.
- Avoiding storage of explosives or reactive wastes near buildings.
- Covering water-reactive wastes.

Health and Safety of Field Personnel.   The U.S. Occupational Safety and Health Administration and the EPA have published guidelines on health and safety procedures applicable to the cleanup of uncontrolled hazardous waste sites (3). These guidelines should be considered at all remedial action sites.

Excavation and Removal Procedures.   Table 21–5 details the choices of equipment for excavation and removal. Regardless of the type of equipment used, certain standard operating procedures and safety practices should be followed.

As soils are excavated, air monitoring should be conducted to warn of unsafe levels of various hazardous constituents. The portable direct-reading instruments currently available include:

- Combustible gas detectors for measuring the lower explosive limit.
- Meters for measuring the oxygen level.
- Detectors for measuring levels of gases and vapors.
- Meters for detecting radiation.

Drums and other containers should be inspected visually for any leaks or deformations. Drum identification and inventory should be done before excavation. Location, date of removal, drum identification number, overpack status, and any other relevant information should be recorded. If the drums contain explosive or shock-sensitive materials, they should be handled remotely or, at a minimum, by using vehicles equipped

Table 21–5   Capabilities and Limitations of Excavation/Removal Equipment

Equipment Type	Excavation of Cover	Excavation of Depths	Removal of Liquids	Hauling	Comments
Backhoes	x	x	x	x	Maximum depth of 30 ft. Maximum reach of 100 ft.
Bulldozer and loaders	x	x	x	x	Limited in mobility in marshy and swampy areas. Used with other excavation equipment.
Cranes and attachments (clamshells, draglines)	x	x			Limited mobility.
Scrapers	x			x	Excellent loading action.
Haulers				x	Used only for transport.
Pumping system			x		Not well suited for removal of liquids with high solids.

Source: U.S. Environmental Protection Agency. *Handbook on Remedial Action at Waste Disposal Sites.* Washington, D.C.: U.S. Govt. Printing Office, 1985; with permission.

with safety shields. Likewise, if a drum is overpressurized, it should be isolated with a barricade or steel demolition net until the pressure can be relieved remotely.

Soils and drums containing radioactive material should be immediately drummed and/or overpacked using remotely operated equipment and then moved to a separate staging area. Special shielding devices may be needed to protect the personnel involved. Any drum that is leaking, badly corroded, or deformed should be overpacked or have its contents transferred to a new or reconditioned drum.

If gas cylinders are encountered, they should be moved promptly to an area where the temperature can be controlled. These cylinders should not be rolled, dragged, or slid—even for short distances. Care should be taken not to drop the cylinders or to allow them to strike another cylinder.

As contaminated soils are excavated from the disposal area, they should be transferred to trucks or a temporary storage area, preferably a dike or bermed area lined with plastic or low-permeability clay. Gas analyzers are used to determine the approximate level of soil contamination, and the soil can then be segregated based on containment levels. Pools of liquid waste should be promptly removed using pumps.

Off-Site Disposal.   Determining the feasibility of off-site disposal requires a good knowledge of the RCRA and the various state regulations. The RCRA manifest requirements must be complied with for all waste that is shipped off the site. The waste generator or other responsible party must follow all the RCRA requirements, including those regarding characterization of hazardous waste.

In addition, the generator should ensure that the facility receiving the waste complies with all applicable federal, state, and local public health statutes. Storage and disposal facilities are required to notify the generator in writing that they are capable of managing the waste. The generator must keep a copy of this written notification on file as part of the operating record.

The transportation of hazardous wastes is regulated by the U.S. Department of Transportation (DOT), the EPA, the states, and in some instances, by local ordinances and codes. In addition, more stringent federal regulations govern the transportation and disposal of highly toxic and hazardous materials (e.g., PCBs, radioactive wastes). The EPA regulations under the RCRA adopt the DOT regulations pertaining to labeling, placarding, packaging, and spill reporting. These regulations also impose requirements for compliance with the manifest system and record-keeping. (Transportation of hazardous waste was discussed in Chapter 18).

Vehicles for off-site transport of hazardous waste must be approved by the DOT and display the proper DOT placard. Liquid waste must be hauled in tanker trucks that meet DOT requirements for this type of waste. Contaminated soil in drums is hauled in box trailers or flatbed trucks, and these trucks should be lined with plastic and/or absorbent materials.

Bulk liquid containers and trailers should be checked for proper placarding, cleanliness, tractor-to-trailer hitching, and excess levels of waste. Bulk liquid containers should also be checked for proper venting, closed valve positions, and secured hatches. Trailers should be checked to ensure correct liner installation, secured cover tarpaulin, and locked gates.

Selection/Evaluation Considerations.   Excavation and removal can eliminate contamination at a site and the need for long-term monitoring. Once excavation begins, the time to achieve beneficial results can be short relative to alternatives such as *in situ* treatment, subsurface drains, and in some instances, even pumping. Excavation and removal can be used in combination with almost any other remedial technology.

The major limitations of excavation, removal, and off-site disposal are worker safety, short- and long-term impacts, cost, and institutional aspects. Where highly hazardous or toxic materials are present, excavation can pose a substantial risk to worker safety. Short-term impacts such as fugitive dust emissions, toxic gases, and contaminated runoff are major concerns as well. Costs associated with off-site disposal are high and frequently result in exclusion of complete excavation and removal as a cost-effective alternative; the location of an RCRA-approved landfill or incinerator also has a substantial impact on costs.

On-Site Land Disposal.   On-site disposal of waste by landfilling involves the design and construction of new landfills to comply with the RCRA the standards, which are concerned with the proper location, design, construction, operation, and maintenance of hazardous waste management facilities. These requirements preclude landfilling in areas of seismic instability, flood plains, and where the integrity of the liner system could be adversely affected. They also preclude landfilling of liquids and of highly mobile or highly toxic wastes. Evaluation of an on-site landfill program must address any potential risks posed by the depth to groundwater at the site and the degree of naturally available groundwater protection if the liner system should fail. Other factors in this evaluation include costs for monitoring the groundwater, collecting any accumulated leachate, and implementing further corrective action if the groundwater becomes contaminated. It may be technically unfeasible to develop a groundwater-monitoring program at sites where the groundwater has already been contaminated.

Air Pollution Controls.   Air pollution at hazardous waste sites results from gaseous emissions and fugitive dusts. Gases may be emitted by the vaporization of liquids, the venting of entrained gases, or through chemical and biological reactions with solid and liquid waste. VOCs may be released slowly but continuously from surface impoundments or landfills. Fugitive dusts are particulates that are lifted up from the ground by wind erosion of exposed waste material or cover soil, particulate matter from vehicular traffic on haul roads and exposed surfaces, and excavation of waste material during remedial action.

Methods for controlling gaseous emissions to the atmosphere include covers to block volatile emissions from impoundments and active gas-collection systems. Covers involve placement of a barrier at the water–air interface to reduce gaseous emissions. Lagoon covers or floating and immiscible liquids can be used for this purpose. Floating lagoon covers consist of a synthetic lining placed in one piece over an impoundment, with proper anchoring at the edges and floats to prevent the lining from submerging. Floating, immiscible liquids include a variety of water-insoluble substances, particularly aliphatic alcohols, that form a monolayer on the surface of an impoundment and thereby inhibit the volatilization of water-soluble compounds. Another type of cover material involves the use of floating polypropylene spheres, which arrange themselves into a closely packed configuration and thereby reduce emissions of VOCs by reducing the exposed surface area.

GAS COLLECTION AND RECOVERY SYSTEMS

An active, interior gas-collection system alters the pressure gradients and pathways of gas migration. Such a system typically consists of four components: gas extraction wells, gas collection headers, vacuum blowers or compressors, and a treatment system. The centrifugal blowers create a vacuum through the collection headers and wells to the wastes and the ground surrounding the wells. A pressure gradient is thereby established, which induces the flow from the site (normally under positive pressure) to the blower that in turn creates a negative (or vacuum) pressure.

Gas collection and recovery systems are used to collect gases from beneath a landfill (i.e., site) surface. They can be installed at almost any site where it is possible to drill or excavate through the waste to the required depth. Limiting factors, however, can include the presence of free-standing leachate or impenetrable materials within the landfill.

Fugitive Dusts

Measures for controlling fugitive dusts from waste piles and active cleanup sites include use of chemical dust suppressants, wind screens, water spraying, and other methods. Dust suppressants include a wide range of natural and synthetic waste materials that strengthen the bonds between soil particles. Various resins, bituminous materials, and polymers are marketed as dust suppressants. Table 21–6 lists a few of these commercially available formulations and their application rates.

Use of dust suppressants provides effective, temporary control of fugitive dust emissions. There remains, however, the potential for secondary impacts (i.e., soil and groundwater contamination) from the use of certain chemical dust suppressants that contain toxic substances.

The most commonly used method of controlling dust emissions is to spray water on exposed surface areas. This method is mainly used to reduce fugitive dusts along active travel paths, excavation areas, and from truck boxes loaded with soils.

Gas Controls

The disposal of solid and hazardous waste (see Chapters 15 and 16) by landfilling creates conditions in which gases are produced and vented to the atmosphere. These gases also migrate laterally through the soil. Accidents resulting in injury, loss of life, and extensive property damage can occur where site conditions favor gas migration. Degradation of air quality has resulted from venting or escaping gases to the atmosphere (2).

As discussed in Chapter 13, organic matter in the waste is transformed into various simpler organic materials and byproduct gases through microbial biodegradation. The major components of landfill gas are carbon dioxide and methane; lesser amounts of hydrogen sulfide, oxygen, and nitrogen are frequently produced as well.

The rate of gas production depends on the oxygen level, moisture content, pH, temperature, and waste composition. Gas may be formed through microbiological degradation of organic matter and through volatilization of organic liquids.

The landfill gas moves with relative ease through highly permeable sand or gravel and with the least ease through clay soils. Migrating gas may result in damage to vegetation, odors (from volatile organic gases and hydrogen sulfide), and acute health effects from exposure to toxic gases. Landfill gases vent by vertical or lateral migration. Most of the gas produced normally vents through the cover material. If the vertical path is sealed by frost, rain-saturated cover soil, pavement, or capping with a liner, however, there is a greater tendency toward lateral migration.

Because gas migration and venting can cause hazards, control systems have been developed, and these may be classified into high- and low-permeability control systems. First, gas-monitoring probes are installed (Figure 21–2) to detect movement of gases. High-permeability control systems include highly permeable trenches or wells

Table 21–6   Commercially Available Dust Suppressants

Formulation	Concentration (%)	Rate of Application (gal/yd^2)
Vinyl acetate resin	10.00	0.2
Synthetic resin	3.00	0.3
Latex	7.00	0.5
Petroleum resin	25.00	0.5
Vegetable gum	0.30	1.4

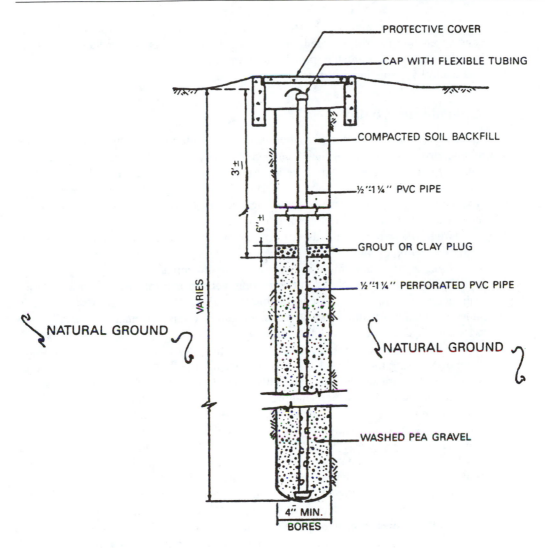

Figure 21–2 Gas-monitoring probe.

(From U.S. Environmental Protection Agency. Handbook on Remedial Action at Waste Disposal Sites. *Washington, D.C.: U.S. Govt. Printing Office, 1985; with permission.)*

between the landfill and the area to be protected. Because the permeable material is more conducive to gas flow than the surrounding soil, paths of flow to points of controlled release are established. Low-permeability systems block the flow of gas into areas of concern by use of barriers between the landfill and the area to be protected. Thus, gases are not collected and therefore cannot be conveyed to a point of controlled release and treatment. High- and low-permeability control systems are often combined to provide controlled venting of gases and to block available paths for gas migration.

Gas control systems can be used at almost any site where it is possible to excavate a trench or to drill the depth of the waste. Passive vents are generally less effective in areas of high rainfall or prolonged, freezing temperatures. (Chapter 16 discusses gas control systems in more detail.)

Groundwater Controls

Control of groundwater contamination involves a containment plume, removal of the plume, diversion to prevent flowthrough contamination of a drinking-water source, or prevention of leachate formation by lowering the water table. Remedial technologies for groundwater involve pumping, subsurface drains, low-permeability barriers, and *in situ* biological or chemical treatment to attenuate subsurface contaminants. These technologies can be used either alone or in combination.

Groundwater pumping is used to contain or remove a plume or to adjust groundwater levels to prevent formation of a plume. Pumping is most effective at sites where underlying aquifers have high interangular hydraulic conductivity.

Figure 21–3 depicts the use of extraction wells to halt the advance of a contaminant plume, thus preventing contamination of a drinking well. Use of extraction wells alone is suited to situations in which contaminants are miscible and move readily with water, the hydraulic conductivity is high, and the hydraulic gradient is steep. Extraction wells are frequently used in combination with slurry walls to prevent groundwater from overflowing the wall and to minimize contact of leachate with the wall, thus preventing wall degradation. Slurry walls also reduce the amount of contaminated water that requires removal, thus reducing both costs and pumping time.

A combination of extraction and injection wells is frequently used where the hydraulic gradient is relatively flat and the hydraulic conductivities are moderate. The function of an injection well is to direct contaminants to the extraction well. This method has been used with some success for plumes that are not miscible with water. Extraction or injection wells can also be used to adjust the groundwater level, but this application is not widely employed. Groundwater barriers can be created by using injection wells to change both the direction of a plume and the speed of the plume's migration. Figure 21–4 shows an example of plume diversion using a line of injection wells to protect domestic water sources. By creating an area with a higher hydraulic head, the plume can be forced to change direction.

Figure 21–5 illustrates the flow of water toward a pumping well. When pumping begins, the water level in the vicinity is lowered. The greatest amount of lowering, or drawdown, is at the well. The velocity of the groundwater increases as it approaches the well. As a result, the lowered water surface develops a continually steeper slope toward the well, thus creating a cone of depression. The distance from the center of the well to the limit of the cone of depression is termed the *radius of influence,* and it increases with a greater hydraulic conductivity. The shape and size of the cone of depression depends on the pumping rate, pumping period, slope of the original water table, hydraulic barriers, aquifer characteristics, and recharge rates.

Determining proper spacing of the wells to capture a groundwater plume completely is an important aspect of design. Adjacent cones of depression or radii of in-

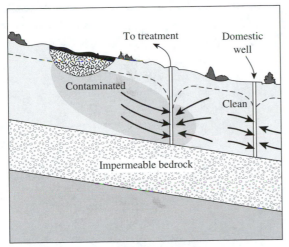

Cross-Sectional View

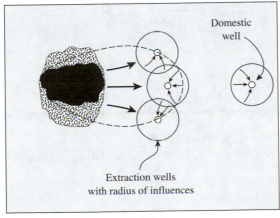

Plan View

Figure 21–3   Containment by using extraction wells.

(From U.S. Environmental Protection Agency. Handbook on Remedial Action at Waste Disposal Sites. *Washington, D.C.: U.S. Govt. Printing Office, 1985; with permission.)*

fluence should overlap for aquifers with low natural flow velocities. For those with high natural flow velocities, velocity distribution plots should be developed to determine the proper well spacing and to ensure capture of the plume. Use of velocity distribution plots is based on determination of the stagnation point, which is the distant downward gradient from the well at which the pull of water toward the well by pumping is exactly countered by the natural flow velocity of water away from the well.

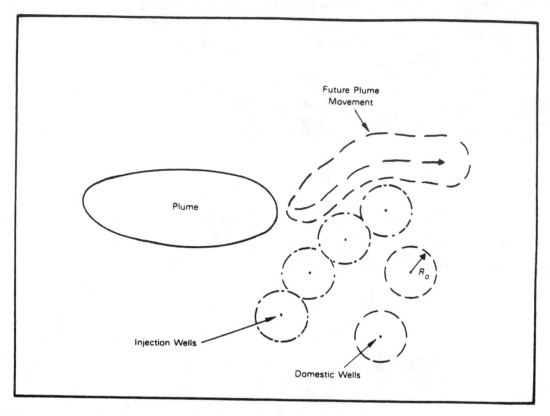

Figure 21–4   Plume diversion by using injection wells.
(*From U.S. Environmental Protection Agency.* Handbook on Remedial Action at Waste Disposal Sites. *Washington, D.C.: U.S. Govt. Printing Office, 1985; with permission.)*

Surface Water Controls

Surface water controls include a range of containment, diversion, and collection methods designed to minimize contamination of surface waters, prevent surface water infiltration, and stop off-site transport of surface waters that have been contaminated. Surface water control technologies include:

- Prevention of runon/interception of runoff by dikes, diversion channels, flood walls, terraces, grading, and vegetation.
- Prevention of infiltration by capping or grading.
- Control of erosion by reducing slope length and steepness and by improving soil management.
- Collection and transfer of water to storage or treatment.
- Storage and discharge of water from seepage basins and ditches, sedimentation basins, and storage ponds.
- Protection from flooding by perimeter structures.

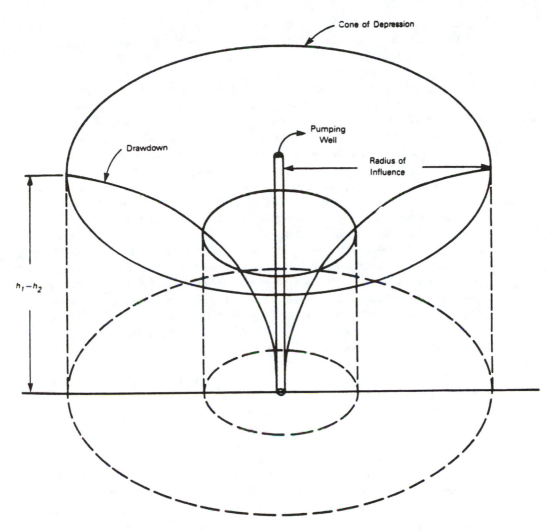

Figure 21–5 Formation of the cone of depression for a pumped well.

The most effective strategy for managing surface flow frequently includes a combination of these technologies.

Capping. Capping is a process used to cover buried waste material to prevent contact with the land surface and groundwater. Capping is necessary whenever contaminated material is to be buried or left in place at the site. It is also used when extensive subsurface contamination precludes excavation and removal of waste because of potential hazards or high costs.

The design of the caps should conform to the RCRA standards. These standards include minimum liquid migration through the waste, low cover maintenance, efficient site drainage, resistance to damage by settling or subsidence, and permeability lower than or equal to the underlying liner system or natural soils. The selection of capping materials and the cap design are both influenced by local availability and costs of cover materials, nature of the waste being covered, local climate and hydrogeology, and projected future use of the site.

Capping is often used with groundwater extraction or containment technologies to prevent or reduce further plume development, thus reducing the time needed to complete groundwater cleanup operations. In addition, groundwater-monitoring wells are often used to detect any unexpected migration of capped waste. When the waste may generate gases, a gas-collection system should be incorporated into the cap as well.

The main disadvantages of capping are the need for long-term maintenance and an uncertain design life. Caps need to be periodically inspected for settlement, ponding of liquids, erosion, and invasion by deep-rooted vegetation. In addition, the groundwater-monitoring wells need to be periodically sampled and maintained.

When a synthetic liner is the only barrier to liquid caps generally have a minimum design life of 20 years. This period may be extended to more than 100 years, however, when a synthetic liner is supported by a low-permeability base, the underlying wastes are unsaturated, there is a great distance between the waste and the groundwater table, and proper maintenance procedures are observed. Rigid barriers (e.g., concrete and bituminous membranes) are vulnerable to cracking and to chemical deterioration, but these cracks can be exposed, cleaned, and repaired (i.e., sealed with tar) with relative ease.

Capping is a high-cost item, but these high costs seldom result in selection of excavation and removal instead. The most probable reason for not selecting use of a cap at a site with extensive subsurface contamination would be unacceptable risk to a source of drinking water from a leachable and highly toxic contaminant.

The primary purpose of a cap is to minimize contact between infiltrating rainwater and waste. The cap functions by diverting infiltrating liquids from the vegetative layer through the drainage layer and away from the underlying waste material.

The low-permeability layer can be composed of a synthetic liner overlying at least 2 ft of low-permeability, natural soil. Standard design practice specifies a permeability of less than 1×10^{-7} cm/s for the soil liner. Bentonite, which is a natural clay with high water-swelling properties, is often mixed with on-site soil and water to produce the low-permeability layer.

Chemical stabilizers and cements can be added to relatively small amounts of on-site soils to create even stronger, less-permeable surface sealants. Portland cement or bitumen (i.e., emulsified asphalt or tar) is suitable for mixing with sandy soils to stabilize and waterproof those soils. Site-specific mixing, spreading, and compacting procedures are required, however. For soil cement, approximately 8 percent (by weight) dry cement is blended into the soil with a rotary hoe or tiller as water is added. Intermittent sprinkling over several days may be required before compaction and solidification are achieved (4).

Flexible synthetic membranes are made of polyvinyl chloride, chlorinated polyethylene, ethylene propylene rubber, and synthetic rubbers. Synthetic liners are generally more expensive and involve labor-intensive sealing materials that require special field-installation methods. The slope of the low-permeability layer should be from 3 to 5 percent to prevent erosion and pooling of rainwater. The underlying base of this layer should consist of fine- to medium-grade fill that will support the weight of the entire cap.

The drainage layer of the multilayered cap is placed directly above the low-permeability layer. The permeability of the drainage layer should be sufficiently high to minimize contact between the infiltrating rainwater and the low-permeability layer.

The vegetative layer of the multilayer cap is placed above the drainage layer, usually with a layer of filter fabric in between to prevent piping. This layer is usually greater than 2-ft thick to accommodate any expected root penetration. Its thickness should also be greater than the deepest penetration by frost. The layer should be spread evenly and not compacted. Frost must not be allowed to reach the low-permeability layer, because cycles of freezing and thawing could greatly increase the permeability.

During construction, the first layer of a multilayer cap is the foundation, which should be composed of soil materials capable of supporting the weight of the cap. The foundation material should be spread over waste in 6-in. lifts and compacted to its maximum achievable density. The shape of this layer should be the same as that of the cap. The low-permeability layer should also be placed in 6-in. lifts and compacted with heavy equipment. The low-permeability zone should be 2-ft thick (or more) if settling is expected in the underlying waste.

The synthetic liner should be placed and seamed. The drainage layer should have a permeability of 1×10^{-3} cm/s or greater. This layer should be put in 6-in. lifts and be at least 1-ft thick. The drainage-layer material must be free of sharp objects that might puncture the liner.

Filter fabric should be placed above the drainage layer to prevent clogging from the overlying vegetative layer. The pore size of this layer should be large enough to allow proper drainage but small enough to prevent soil from migrating into the drainage layer.

Diversion and Collection.

Dikes and berms are compacted, earthen ridges or contours that provide short-term protection of critical areas by intercepting storm runoff and diverting its flow to the drainage system. These structures prevent excessive erosion of newly constructed slopes, and they are widely used for temporary isolation of waste until it can be removed or effectively contained. They are particularly used during excavation and removal operations to isolate drums or contaminated soils that have been temporarily staged on the site. Dikes not only prevent runoff but can also prevent mixing of incompatible wastes. Diversion of storm runoff also decreases the amount of water available to infiltrate the soil cover, thereby reducing the amount of leachate production.

Runoff control dikes, which are generally classified as interceptor dikes (Figure 21–6[a]), are built to reduce the slope length. Diversion dikes (Figure 21–6[b]) are designed to intercept and divert surface flow as well as to reduce the slope length.

Ideally, dikes and berms are constructed of erosion-resistant, low-permeability clay soils. Seeding and mulching or chemical stabilization and application of gravel or stone riprap on dikes and berms extends their life expectancy.

Dikes and berms are constructed by using well-established techniques, and they often require excavation and grading equipment, which is likely to be available at the

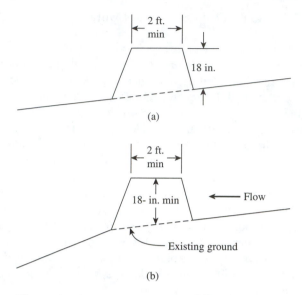

Figure 21–6   (a) Interceptor dike. (b) Diversion dike.

(From U.S. Environmental Protection Agency, Handbook on Remedial Action at Waste Disposal Sites. *Washington, D.C.: U.S. Govt. Printing Office, 1985; with permission.)*

disposal site. The required earthen fill is often available on the site as well. Dikes should be constructed in several lifts, with each lift being tested for adequate compaction. Periodic inspection and maintenance of dikes is also required.

TREATMENT TECHNOLOGIES FOR REMEDIATION OF CONTAMINATED SITES

Some of the typical treatment technologies for aqueous waste were discussed in Chapter 16, and they included leachate treatment. Further discussion on treatment of aqueous waste from the cleanup of hazardous waste sites is presented here.

Dewatering

Dewatering is a physical unit operation to dewater slurries or sludges and thus facilitate handling and prepare the materials for final treatment or disposal. Selection of the technology to dewater slurries or sludges depends on the volume, solids content, land availability, and degree of dewatering required. Dewatering methods remove most of the water, but solids may not be sufficiently dry to meet the requirements for final disposal and thus may require further treatment to stabilize them. The contaminated water generated during dewatering usually contains hazardous substances.

Gravity Thickening.   Gravity thickening of a sludge or slurry is generally done in a circular tank similar in design to a conventional clarifier. The slurry enters the thickener through a center feedwell designed to dissipate the velocity and stabilize the density of the incoming stream.

Centrifugal. Centrifugal dewatering is a process using the force developed through fast rotation of a cylindrical drum or bowl to separate solids and liquids by density differences. Types of centrifuges include the solid bowl and the basket.

The solid-bowl centrifuge consists of a long bowl that is normally mounted horizontally and is tapered at one end. Sludge is continuously introduced into the unit, and solids concentrate on the periphery. A helical scroll within the bowl, which is spinning at a slightly different speed, moves the accumulated sludge toward the tapered end, where the solids are concentrated further before discharge.

In the basket centrifuge, the flow enters the machine at the bottom and is directed toward the outer wall of the basket. Cake continually builds up within the basket until the centrate, which overflows a weir at the top of this unit, begins to increase in solids. At that point, the feed to the unit is shut off, the machine decelerates, and a skimmer enters the bowl to remove the liquid layer. A knife is then moved into the bowl to cut out the cake, which falls through the open bottom of the machine.

Centrifuges can be used to concentrate or dewater soils and sediments ranging in size from fine gravel down to silt. The effectiveness of centrifugation depends on the particle size and shape and the solids concentration (among other factors). According to data from the dewatering of municipal sludges (2), solids concentrations from approximately 15 to 40 percent are achievable with the solid-bowl centrifuge. For the basket centrifuges, the cake solids concentration typically ranges from approximately 9 to 25 percent. Capture of solids typically ranges from approximately 85 to 97 percent with chemical conditioning for both the solid-bowl centrifuge and the basket centrifuge.

Centrifuges can remove particles as small as 1 μm in diameter. Removal efficiency is reduced, however, for particle sizes smaller than 10 μm.

Filtration. Belt filter presses employ moving belts to dewater sludges continuously. The belt press filtration uses three stages: chemical conditioning of the feed, gravity drainage to a nonfluid consistency, and dewatering (Figure 21–7). A flocculent is added before feeding the slurry to the belt press. In the next step, free water drains from the conditioned sludge, which then enters a two-belt contact zone, where a second, upper belt is gently set on the forming sludge cake. The cake between the belts passes through rollers of generally decreasing spacing, thereby subjecting the sludge to continuously increasing pressure and shear force. Progressively more and more water is expelled throughout the roller section, and a scraper blade is often employed for each belt at the discharge point to remove the cake from the belts.

A vacuum filter consists of a horizontal, perforated cylindrical drum (Figure 21–8). It is partially submerged in a vat of sludge and is covered with a continuous belt of fabric or wire mesh. A vacuum is applied to the inside of the drum to pull the liquid in the vat through the filter medium, thereby leaving wet solids adhering to the outer surface. As the drum continues to rotate, a sludge cake forms and enters a drying zone. Finally, the cake enters a discharge zone, where it is removed.

A filter press (Figure 21–9) consists of vertical plates that are held in a frame and pressed together. The liquid to be filtered enters the cavity formed by the frame. Pressed against this hollow frame are perforated metal plates, which are covered with fabric-filter medium. The plate operates on a cycle of filling, pressing, cake removing, media

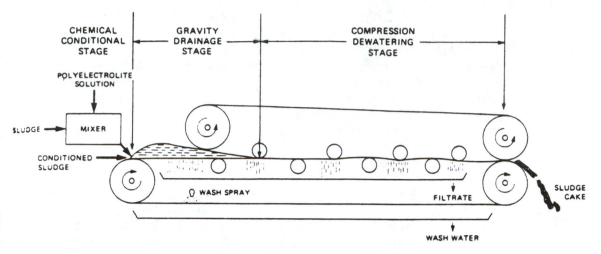

Figure 21–7　Belt press.

(From U.S. Environmental Protection Agency. Handbook on Remedial Action at Waste Disposal Sites. *Washington, D.C.: U.S. Govt. Printing Office, 1985; with permission.)*

washing, and press closing. As the liquid flows through the filter medium, solids are trapped in the cavity. Once the cavity is full, the press is opened, the dewatered slurry removed, and the plates cleaned. Then, the cycle is repeated. In certain applications, the filter medium is precoated with diatomaceous earth or other filter aids to improve performance.

The process of filtration can be used to dewater solids over a wide range of particle sizes and concentrations. Effectiveness for a particular application depends on the type of filter, particle size distribution, and solids concentrations. For dewatering of sludges, typical ranges for filter-cake solids content and solids removal or capture are:

	Solids content (%)	Solids capture (%)
Belt press	15–45	85–95
Vacuum filter	15–35	88–95
Filter press	30–50	98

The filter press achieves a dry filter cake and has the greatest capacity for capturing solids, but several other factors affect the decision to use a particular method of filtration. Filter presses have the highest capital and operating costs, and they also require the largest amount of space. In addition, replacement of the filter medium is both expensive and time-consuming. Vacuum filtration is the most energy intensive of the three methods—and the least effective. The incoming feed must have a solids content of at least 3 percent to achieve adequate cake formation. The belt press filter is the least energy intensive, but the major limitation of this process is sensitivity to incoming feed characteristics and chemical conditioning.

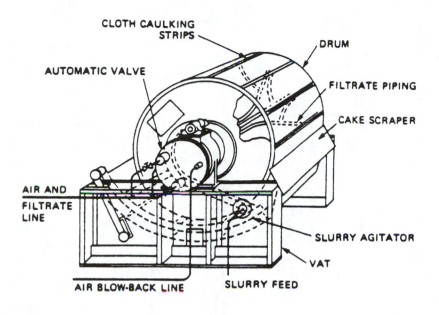

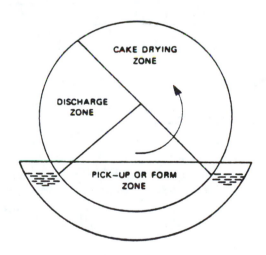

Figure 21–8   Rotary vacuum filter.

(From U.S. Environmental Protection Agency. Remediation of Waste Disposal Sites. Washington, D.C.: U.S. Govt. Printing Office, 1979; with permission.)

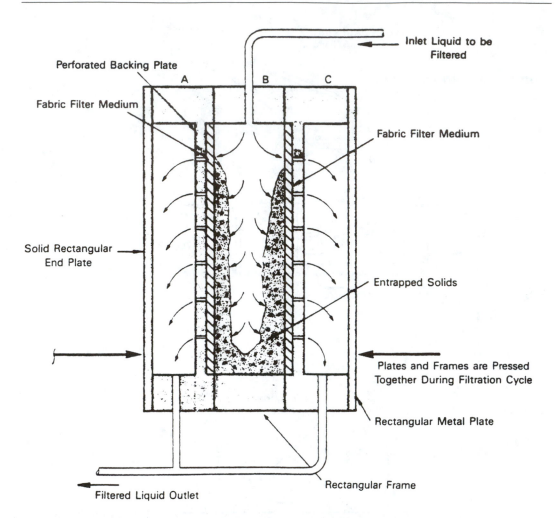

Figure 21–9   Filter press. When the cavity formed by plates *A* and *C* is filled with solids, the plates are separated. The solids are removed, and the medium is washed clean. The plates are then pressed together, and filtration is resumed.

(From U.S. Environmental Protection Agency. Handbook on Remedial Action at Waste Disposal Sites. *Washington, D.C.: U.S. Govt. Printing Office, 1985; with permission.)*

Provided that slurries have been properly prescreened and conditioned, filtration offers an effective method of dewatering. Both the filterate and dewatered sludge require further treatment before disposal. The water generated from washing the filter medium also requires treatment.

Thermal Destruction

Thermal destruction is a process of high-temperature oxidation under controlled conditions to degrade a substance into products that generally include carbon dioxide, water vapor, sulfur dioxide, nitrogen oxides, hydrogen chloride gases, and ash. Air pollution–control equipment is needed to prevent the release into the atmosphere of hazardous products such as particulates, sulfur dioxide, nitrogen oxides, hydrogen chloride, and products of incomplete combustion. (Air pollution–control requirements and systems were discussed in Chapter 13.) Thermal destruction can be used to destroy organic contaminants in liquid, gaseous, and solid waste streams. The most common incineration technologies applicable to hazardous waste sites are the rotary kiln, the fluidized bed (also discussed in Chapter 13), and liquid injection. Mobile systems have been used successfully to clean many hazardous waste sites.

Rotary kilns can burn waste in any physical form. They can incinerate solids and liquids independently or in combination, and they can accept waste feed without any preparation of the feed. Because of their ability to handle waste in any physical form and their high incineration efficiency, rotary kilns are the preferred method of treating mixed hazardous solid residues.

Rotary kiln incinerators are cylindrical, refectory-lined shells (Figure 21–10). They are fueled by natural gas, oil, or pulverized coal. Wastes are injected into the kiln at the higher end and are passed through the combustion zone as the kiln rotates. The rotation creates turbulence and improves combustion.

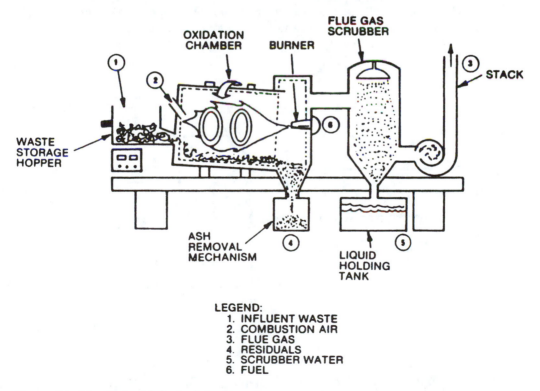

Figure 21–10   Rotary kiln incinerator.

(From U.S. Environmental Protection Agency. Handbook on Remedial Action at Waste Disposal Sites. *Washington, D.C.: U.S. Govt. Printing Office, 1985; with permission.)*

Residence times within the kiln can range from a few seconds to an hour or more for bulk solids. Combustion temperatures range from 1500°F to 3000°F.

Liquid injection can be used to destroy almost any pumpable waste or gas. Unlikely candidates for destruction include heavy metal wastes and other wastes with a high inorganic content. A continuous ash removal system is not required.

Liquid incinerators have no moving parts, and they require very little maintenance. The major limitations of liquid injection are its ability to incinerate only wastes that can be atomized in the burner nozzle and the burner's susceptibility to clogging. Supplemental fuel is required as well. Liquid-injection incinerators are highly sensitive to waste composition and to changes in flow. Therefore, storage and mixing tanks are necessary to ensure a reasonably steady and homogenous flow of waste. Figure 21–11 depicts a liquid injection–incineration system.

Mobile incineration has been widely used at sites because of bans on landfill disposal of certain wastes and the lower resistance from both government and the public to temporary operation permits. Existing mobile systems include liquid injection and rotary kiln incinerators equipped with secondary combustion chambers and environmental controls. These incinerators can handle a variety of wastes, including PCBs, carbon tetrachloride, other liquid hazardous wastes, and contaminated soils. The primary advantage of a mobile incinerator is the ability to treat on the site and thus eliminate the need for off-site transport of waste. Mobile incinerators must meet all applicable state requirements, however, including air emission standards.

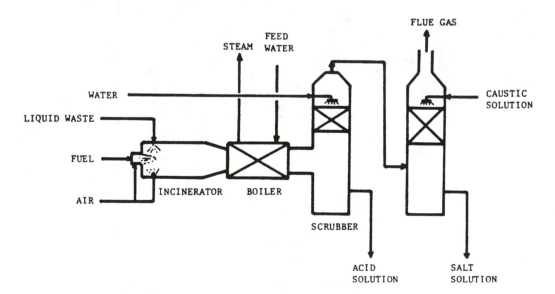

Figure 21–11   Liquid injection–incineration system.

(From U.S. Environmental Protection Agency. Handbook on Remedial Action at Waste Disposal Sites. *Washington, D.C.: U.S. Govt. Printing Office, 1985; with permission.)*

In Situ Treatment

One alternative to waste excavation, removal, and conventional pump-and-treat methods is to treat waste *in situ*. These treatment processes are divided into biological, chemical, and physical. *In situ* biodegradation, which is commonly referred to as bioreclamation, is based on developing and stimulating microbes to decompose contaminants. Chemical treatment involves the injection of chemicals into the subsurface to degrade contaminants. Physical methods involve physical manipulation of the soil to immobilize or separate contaminants. Generally, a combination of *in situ* and aboveground treatment is the most cost-effective at an uncontrolled waste site. The applicability of *in situ* methods is determined on a site-specific basis by using laboratory and pilot-scale testing.

Bioreclamation.   Bioreclamation is a technique for treating zones of contamination through microbial degradation. Suitable environmental conditions are created to enhance the microbial degradation of organic contaminants, thus resulting in their breakdown and detoxification. Bioremediation appears to be a very promising technique and encompasses both *in situ* treatment approaches as well as withdrawal and treatment of groundwater in biological reactors on the surface.

Extensive research has confirmed that microorganisms can break down many organic compounds considered to be environmental and public health hazards at spill sites and uncontrolled hazardous waste sites. This capability of microorganisms can be used *in situ* to reclaim contaminated soils and groundwater.

The most feasible bioreclamation method for *in situ* treatment relies on aerobic microbial processes (see Chapter 13). This method involves optimizing the environmental conditions by providing an oxygen source and nutrients, which are delivered to the subsurface through an injection well or infiltration system to enhance microbial activity. Given proper nutrients and sufficient oxygen, indigenous microorganisms can generally be relied on to degrade a wide range of compounds.

Anaerobic microorganisms can also degrade certain organic contaminants (see Chapters 1 and 13). The potential for anaerobic degradation has been demonstrated in laboratory studies and in industrial waste treatment processes, but it has not been widely demonstrated in the field as an *in situ* reclamation approach.

The feasibility of bioreclamation is dictated by both waste and site characteristics, such as biodegradability of the organic contaminants, environmental factors that affect microbial activity, and site hydrogeology. A compound's relative aerobic biodegradability can be estimated by using the 5- and 21-day biological oxygen demand (BOD_5 and BOD_{21}) and the chemical oxygen demand (COD). Table 21–7 presents the relative biodegradabilities of various substances in terms of a BOD_5/COD ratio. A higher BOD_5/COD ratio represents a higher relative biodegradability. For most compounds, the most rapid and complete degradation occurs aerobically.

In general, aerobic bioreclamation is suitable for the degradation of petroleum hydrocarbons, aromatics, halogenated aromatics, polyaromatic hydrocarbons, phenols, biphenyls, organophosphates, and most pesticides and herbicides. For the degradation of halogenated, lower-molecular-weight hydrocarbons such as unsaturated alkyl halides

Table 21–7 Ratios of 5-Day Biological Oxygen Demand to Chemical Oxygen Demand for Various Organic Compounds

Compound	Ratio	Compound	Ratio
Relatively Undegradable		*Relatively Degradable*	
Butane	~0	Urea	0.11
Butylene	~0	Toluene	<0.12
Carbon tetrachloride	~0	Potassium cyanide	0.12
Chloroform	~0	Isopropyl acetate	≤0.13
1,4-Dioxane	~0	Chlorobenzene	0.15
Ethane	~0	Kerosene	~0.15
Heptane	~0	Glycerine	≤0.16
Hexane	~0	Vinyl acetate	<0.20
Isobutane	~0	Naphthalene (molten)	≤0.20
Isobutylene	~0	Soybean oil	~0.20
Liquefied petroleum gas	~0	Methyl methacrylate	<0.24
Methane	~0	Acrylic acid	0.26
Methyl bromide	~0	Sodium alkyl sulfates	~0.30
Methyl chloride	~0	Acetic acid	0.31–0.37
Monochlorodifluoromethane	~0	Acetic anhydride	≥0.32
Nitrobenzene	~0	Ethyl acetate	≤0.36
Propane	~0	Octanol	0.37
Propylene	~0	Benzene	<0.39
Propylene oxide	~0	Monoethanolamine	0.46
Tetrachloroethylene	~0	Corn syrup	~0.50
Tetrahydronaphthalene	~0	Propionic acid	0.52
1-Pentene	<0.002	Acetone	0.55
Ethylene dichloride	0.002	Isopropyl alcohol	0.56
1-Octene	>0.003	Phthalic anhydride	0.58
Ethylenediaminetetracetic acid	0.005	Benzaldehyde	0.62
Triethanolamine	≤0.006	Isobutyl alcohol	0.63
Xylene	<0.008	2,4-Dichlorophenol	0.78
Ethylbenzene	<0.009	Phenol	0.81
Moderately Degradable		Carbolic acid	0.84
		Methyl ethyl ketone	0.88
Ethyl ether	0.012	Hydrazine	1.0
Sodium alkylbenzenesulfonates	~0.017	Oxalic acid	1.1
Gas oil (cracked)	~0.02		
Gasolines (various)	~0.02		
Acrylonitrile	0.031		
1-Hexene	<0.044		
Methyl isobutyl ketone	≤0.044		
Styrene	>0.06		
Methyl alcohol	0.07–0.73		
Acetonitrile	0.079		
Ethylene glycol	0.081		
Sodium cyanide	≤0.09		
Linear alcohols (12–15 carbons)	>0.09		

Source: Environmental Protection Agency. *Handbook on Remedial Action at Waste Disposal Sites.* Washington, D.C.: U.S. Govt. Printing Office, 1985; with permission.

(e.g., perchloroethylene [PCE], trichloroethylene [TCE]) and saturated alkyl halides (e.g., 1,1,1-trichloroethane, trihaloethane), anaerobic degradation appears to be most feasible under some conditions.

The availability of the compound to the organism will dictate that compound's biodegradability. Compounds with greater aqueous solubilities are generally more available to degradative enzymes. Environmental factors that affect microbial activity and population size will determine the rate and extent of biodegradation. These factors include:

- Nutrient levels.

- Oxygen concentration.

- pH.

- Moisture content.

- Hydraulic conductivity of the soil.

- Temperature.

- Presence of toxins and growth inhibitors.

- Presence of trace metals.

- Types and concentrations of contaminants.

The optimal temperature for growth of organisms during aerobic biological treatment ranges from 20°C to 37°C (68°F to 99°F). Microbial populations in colder waters are adapted to lower temperatures, but biodegradation rates are much slower in these conditions than at higher temperatures.

Concentrations of inorganic and organic contaminants could be so high as to be toxic to the microbial populations. Table 21–8 lists the concentrations at which certain compounds become toxic in industrial waste treatment. Microorganisms in the

Table 21–8   Concentrations Toxic to Microorganisms for Selected Chemicals

Chemical	Toxic Concentration (mg/L)
Butane	>1000
Formaldehyde	50–100
Acetone	>1000
Methyl isobutyol ketone	>1000
Diethylamine	300–1000
Ethylenediamine	100–300
Phenol	300–600
Sodium acrylate	>500
Dextrose	>1000
Ethylene glycol	>900
Kerosene	>500

subsurface may be more tolerant to high concentrations of these compounds. Conversely, contaminant concentration may be so low that the assimilative processes of the microorganisms are sometimes not stimulated. Thus, adaption to a particular substrate will not occur, and that substrate will not be degraded. Even if the contaminant is present in acceptable concentrations or there is another, "preferred" carbon source available, the microorganisms may catabolize it preferentially.

It may be feasible to manipulate some of these factors *in situ* to optimize environmental conditions (Table 21–9). Nutrients and oxygen or nitrates can be added to the subsurface, and it may be feasible to enhance reducing conditions. The pH can be adjusted as well through addition of dilute acids or bases. Water can be pumped into an arid zone. Bioreclamation could also be preceded by other treatments, which could reduce toxic concentrations to a tolerable level. Even raising the temperature of a contaminated zone by pumping in heated water or recirculating groundwater through a surface heating unit may be feasible at sites with a low groundwater level.

Some factors cannot be corrected, and these include competition between microbial populations and salinity of the groundwater. This points to one advantage of relying on indigenous microorganisms rather than on added microorganisms to degrade waste. The added, specialized microorganisms may have a superior degradation capability as developed in the laboratory or enriched in a surface biological reactor, but they may not be able to survive the actual subsurface conditions. Through generations of evolution, however, natural populations have developed that are ideally suited for survival and proliferation in that environment. This is particularly true of uncontrolled hazardous waste sites, where microorganisms have been exposed to the wastes for years or even decades. Specialized microorganisms can be expected to have the greatest application at spill sites where the exposure time has not been long enough for a substantial, adapted, indigenous population to evolve.

Even if all other conditions are favorable, bioreclamation will not be feasible if the hydrogeology of the site is not suitable. The hydraulic conductivity must be great enough and the residence time short enough so that added substances, oxygen, and nutrients reach the entire treatment zone. Sandy and other highly permeable sites are far easier to treat than sites containing clayey soils.

Added substances may react with soil components as well. For example, oxidation may result in precipitation of iron and manganese oxides and hydroxides that could clog the delivery system and even the aquifer. Phosphates may cause precipitation of calcium and iron phosphates. If calcium concentrations are high, the added phosphate could also be tied up by the calcium and thus not be available to the microorganisms.

Aerobic Bioreclamation. The first approach to bioreclamation, which was tried in the early 1970s, involved stimulating indigenous microorganisms. Since then, many different approaches have been used successfully to enhance biodegradation in contaminated zones. Indigenous microorganisms have been used to degrade wastes, and specialized microorganisms (either adapted strains or genetically altered strains) have also been used. Air was used in earlier site remediations to provide oxygen, but today, use of hydrogen peroxide or ozone are feasible alternatives to air or pure oxygen.

In situ aeration wells (Figure 21–12) are suitable for injection of air into contaminated leachate plumes. A bank of aeration wells can be installed to provide a zone of

Table 21–9 Site Reclamations Using Biodegradation with Indigenous Microorganisms

Contaminant	Waste Site Characteristics	Treatment	Comments
Gasoline spill	Soil and groundwater contamination in a dolomite aquifer	Physical recovery followed by *in situ* treatment. Nutrient solution delivered; air sparged through wells. Producing wells controlled groundwater flow.	No gasoline left in aquifer 10 months after treatment
Gasoline spill	Soil and groundwater contamination in a sandy aquifer	Physical recovery followed by *in situ* treatment. Nutrient solution delivered; air sparged through wells. Producing wells controlled groundwater flow.	No free product detected at end of program
Gasoline leak	Soil and groundwater contamination in a shallow, highly permeable aquifer	Physical recovery followed by *in situ* treatment. Nutrient addition; groundwater recycled through site. Air supplied at bottom of injection trench with diffusers.	System operated for 1 year, after which no free product was detected
Methylene chloride, butanol, dimethyl aniline, and acetone	Soil and groundwater contamination in a layer of glacial till	Physical recovery followed by surface biological treatment with indigenous microorganism: reinjection and *in situ* treatment using aeration wells. Producing wells controlled groundwater flow.	Concentrations reduced from 700 to 0.1 mg/L or less in groundwater after a 2-year program
Petroleum products and cyanide	Soil and groundwater contamination in a sandy/gravelly aquifer	Groundwater pumped to surface, treated with ozone, and then reinfiltrated.	Drinking water–quality produced
Aromatic hydrocarbons	Soil and groundwater contamination in a sandy, high permeability aquifer	Physical recovery followed by *in situ* treatment; addition of nitrate; recirculated flushing water stripped and filtered before reinjection. Water temperature increased 10°C.	Aromatics not detected after 3 months
Jet fuel hydrocarbons, halogenated alkanes and aromatics, and heavy metals	Soil and groundwater contamination in clayey, low permeability aquifer	*In situ* treatment; nutrients and hydrogen peroxide will be added to recycled groundwater	—

Source: U.S. Environment Protection Agency. *Handbook on Remedial Action at Waste Disposal Sites.* Washington, D.C.: U.S. Govt. Printing Office, 1985; with permission.

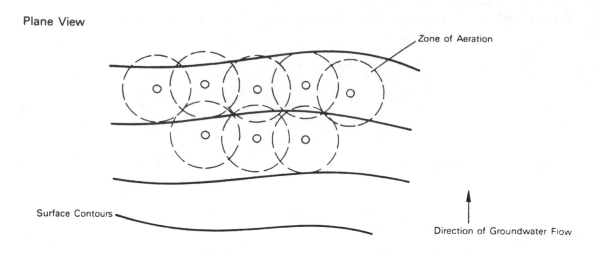

Zone of Aeration

Surface Contours

Direction of Groundwater Flow

Cross-sectional
View

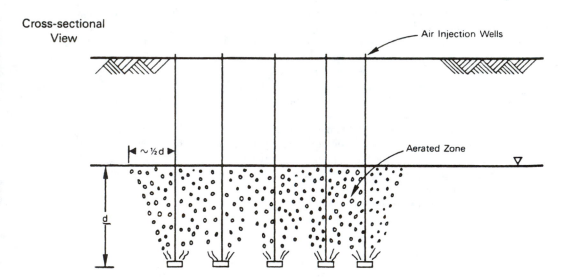

Air Injection Wells

Aerated Zone

~ ½d

d

Figure 21–12   Possible configurations of an *in situ* aeration well bank.

(From U.S. Environmental Protection Agency. Handbook on Remedial Action of Waste Disposal Sites. *Washington, D.C.: U.S. Govt. Printing Office, 1985; with permission.)*

continuous aeration through which contaminated groundwater flows. Conditions of oxygen saturation can be maintained to degrade organics during the residence time of groundwater flowing through the aerated zone.

The zone of an *in situ* aeration well must be wide enough to allow the total plume to pass. The flow of air must be sufficient to provide a large enough radius of aeration while also being small enough to not cause an air barrier to the flow of groundwater. Various methods are available to inject air or pure oxygen. Air has been sparged into wells using diffusers attached to paint sprayer–type compressors, and a blower can also be used.

Oxygenation systems, either in-line or *in situ,* have higher oxygen concentrations and provide efficient oxygen transfer to microorganisms. The solubility of pure oxygen in various liquids is four to five times higher than that with conventional aeration.

Hydrogen peroxide has been successfully used as an oxygen source at several spill sites. The advantages of systems using hydrogen peroxide include:

- Greater oxygen concentrations can be delivered to the subsurface.
- Hydrogen peroxide can be added in-line with the nutrient solution.
- Hydrogen peroxide keeps the well free of heavy biogrowth.

Ozone is used for disinfection and chemical oxidation of organics in both water and wastewater treatment. In commercially available, ozone-from-air generators, ozone is produced at a concentration of 1 to 2 percent in air. During bioreclamation, this mixture could be placed in contact with pumped leachate by using in-line injection and static mixing or by using a bubble contact tank. A dosage of 1 to 3 mg of ozone per liter of air can be used to attain chemical oxidation.

A petroleum-product spill in Germany was cleaned *in situ* by using ozone as an oxygen source for biodegradation (5). The groundwater was pumped out, treated with ozone, and recirculated. Approximately 1 g of ozone per gram of dissolved organic carbon was added to the groundwater and allowed a contact time of 4 minutes in the aboveground reactor. This increased the oxygen concentration to 9 mg/L, with a residual concentration of 0.1 to 0.2 g of ozone per cubic meter of treated water.

Nutrients. Nitrogen and phosphate are the nutrients most frequently present in limiting concentrations in soil. Other nutrients required for microbial metabolism include potassium, magnesium calcium, sulfur, sodium, manganese, iron, and trace metals. Many of these nutrients may already be present in the aquifer in sufficient quantities.

The optimum nutrient mix can be determined from laboratory-growth studies and from geochemical evaluations of the site. Care must be exercised when evaluating microbial needs based on soil and groundwater chemical analysis, because chemical analyses do not always indicate what is truly available to microorganisms.

The form of nutrients may be critical in terms of microbial requirements. For example, an ammonia–nitrogen source is preferable to a nitrate–nitrogen source, because microorganisms more easily assimilate ammonia–nitrogen. The site geochemistry may be a critical factor in determining the form and concentration of nutrients. For example, use of diammonium phosphate could result in excessive precipitation, and a nutrient solution containing sodium could cause dispersion of the clays, thereby reducing permeability. A high calcium level will likely lead to precipitation of added phosphate, thus rendering it unavailable to microbial metabolism. If problems with precipitation are likely, iron and manganese addition may not be desirable, and if the total dissolved solids content in the water is extremely high, addition of salts may be undesirable.

For successful bioreclamation, adequate contact between additives and contaminants to minimize migration and recovery of spent treatment solutions and contaminants are critical. Figure 21–13 illustrates a configuration in which groundwater is extracted in the downgradient of the zone of contamination and then reinjected in the upgradient. *In situ* aeration supplies oxygen directly to the contaminated plume, whereas nutrients and oxygen are added in-line by way of mixing tanks. Treated water flows through the contaminated soil to flush out the contaminants. Extraction and injection wells can be used to treat contaminants to almost any depth, but they become cost-prohibitive in low-permeability soils. Subsurface gravity-delivery systems (e.g., spray irrigation, flooding, ditches) are effective for treating shallow contamination in the unsaturated zone. They can also be used to treat contaminants in the saturated zone if the soil above the saturated zone is sufficiently permeable to allow percolation of treatment solutions to the groundwater and if the groundwater flow rates are sufficient to ensure complete distribution of the treatment solutions. The feasibility and effectiveness of these methods are both affected by topography and climate.

The aquifer flow rate should be sufficiently high to ensure that the aquifer is flushed several times during the period of operation. Thus, if the cleanup occurs over a 3-year period, flow rates between injection and extraction wells should be such that a residence time of 0.5 years or less occurs between the well pairs, which corresponds to six or more flushes. Flushing of soils containing organics prevents clogging by microorganism buildup because of increased flow rate, ensures even distribution of nutrients and organic concentration within the plume, and results in better-controlled degradation. Excessive pumping costs, loss of hydraulic containment efficiency, corrosion, excessive manganese deposition, flooding, or well blowout may occur if the flow rate is too high. The operating period depends on the biodegradation rate of the contaminants in the plume and on the amount of recycling. If the period of operation is long (e.g., >5 years), the operating costs of bioreclamation may outweigh the capital costs of another remedial alternative.

Anaerobic Bioreclamation. Anaerobic treatment is generally not as promising as aerobic treatment for site remediation. Anaerobic processes are slower, fewer compounds can be degraded, and the logistics of rendering a site anaerobic are difficult.

Operation and Maintenance. Operation and maintenance of a bioreclamation process involve the hydraulics system and the biological system. Groundwater monitoring can be performed at the injection and extraction wells as well as at the monitoring wells. Monitoring wells should be placed on the site to measure process performance and off the site to measure pollutant migration and to provide background information on changes in subsurface conditions resulting from seasonal fluctuation. In a biological system, the pH should be maintained between 6 and 8, and concentrations of nutrients and other chemicals should be kept as uniform as possible to protect against shockloading. The level of dissolved oxygen should be maintained above the critical concentration for promotion of aerobic activity, which ranges from 0.2 to 2.0 mg/L and most commonly is from 0.5 to 1.0 mg/L.

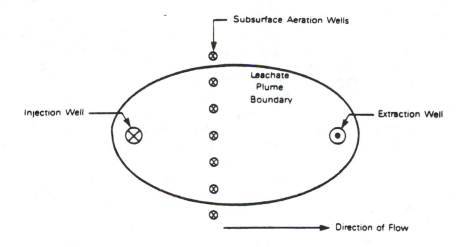

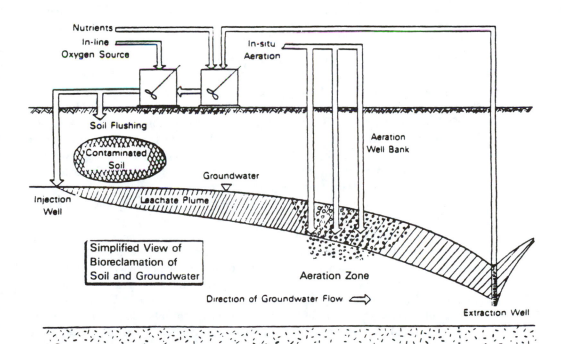

Figure 21–13   Simplified view of groundwater reclamation.

(From U.S. Environmental Protection Agency. Handbook on Remedial Action at Waste Disposal Sites. *Washington, D.C.: U.S. Govt. Printing Office, 1985; with permission.)*

Clogging of the aquifer, the injection wells or trenches, or the extraction wells by microbiological sludge is likely, but hydrogen peroxide is effective in keeping wells free of heavy biological growth. Permeability of the aquifer could be reduced because of precipitation or dispersion of clays. When calcium concentrations are high in the soil, phosphates can be rapidly attenuated through precipitation with calcium, thereby becoming unavailable for microbial metabolism.

Maintenance of the bacterial population at its optimal level is also important. A continuous incubation facility operating at higher temperatures and under more controlled conditions could be used to maintain the microbial population by continuously injecting the subsurface with microorganisms.

Aeration wells may be susceptible to operational problems if injected air fluidizes the material around it and causes a well blowout, thus allowing free passage of air to the surface. The cone of influence in a blown-out well is greatly reduced, therefore requiring installation of a new well. The best way to prevent blowouts is to keep the gas velocities below those necessary to cause fluidization.

Aerobic bioreclamation has effectively degraded organics at many sites. Provided the organics are amenable to aerobic degradation and the hydraulic conductivity of the aquifer is high, it can be expected to be effective at most sites.

Bioreclamation takes longer than excavation and removal of contaminated soils, and depending on the site, it could also take longer than a conventional pump-and-treat approach. The advantage of *in situ* bioreclamation over a pump-and-treat approach is that *in situ* biodegradation treats contaminated subsurface soils, thereby removing the source of groundwater contamination.

Chemical Treatment.   Technologies for immobilizing contaminants include precipitation, chelation, and polymerization. Methods for mobilizing contaminants for extraction are termed *soil flushing* and use surfactants, dilute acids and bases, and water. Detoxification techniques include oxidation/reduction, neutralization, and hydrolysis.

Most of the treatment approaches discussed in this section involve delivery of a fluid to the subsurface. Therefore, the same factors limiting use of injection/extraction wells, drains, or surface-gravity application systems such as flooding and spray irrigation for bioreclamation also limit applicability of most *in situ* chemical treatments. Minimal permeability requirements must be met if the treatment solution is to be delivered successfully to the contaminated zone. Sandy soils are far more amenable than clayey soils to *in situ* treatment. Furthermore, the contaminated groundwater must be contained within the treatment zone. Measures must be taken to ensure that treatment reagents do not migrate and become contaminants themselves. Care must also be taken during the extraction process not to increase the burden of contaminated water by drawing uncontaminated water into the treatment zone from the aquifer or from hydraulically connected surface waters.

Potential chemical reactions of treatment reagents with soil and waste must be considered as well. Most hazardous waste disposal sites contain a mix of contaminants, and a treatment that neutralizes one contaminant may render another more toxic or mobile. For example, chemical oxidation will destroy or reduce the toxicity of many organics, but if present, Cr^{3+} will oxidize to the more toxic and mobile Cr^{+6}. Soil permeability may also be reduced by the treatment. For example, treatment of soils high in iron and manganese may result in precipitation of iron and manganese oxides and hydroxides, which could clog the delivery system and the aquifer.

Soil Flushing. Organic and inorganic contaminants can be washed from contaminated soils by using an extraction process termed *soil flushing*. Water or other solutions are injected into the area of contamination, and the contaminated elutriate is then pumped to the surface for removal, recirculation, or on-site treatment and reinjection. During elutriation, sorbed contaminants are mobilized into solution because of solubility, formation of an emulsion, or chemical reaction with the flushing solution. Substances with the greatest potential for use in soil flushing are water, acids and bases, chelating agents, surfactants, and certain reducing agents.

Water can be used to flush water-soluble or water-mobile organics and inorganics. Organics that are amenable to water flushing can be identified according to their soil/water partition coefficient or estimated using the octanol/water coefficient. Organics considered to be soluble have a partition coefficient (K) of less than 1000 (log $K = 3$). High-solubility organics (e.g., lower-molecular-weight alcohols, phenols, carboxylic acids) and other organics with a partition coefficient of less than 10 (log $K \leq 1$) may have already been flushed from the site through natural flushing processes. Medium-solubility organics ($K = 10$–1000) that may be effectively removed from soils by water flushing include low- to medium-molecular-weight ketones, aldehydes, and aromatics and lower-molecular-weight, halogenated hydrocarbons (e.g., TCE, PCE). Inorganics that can be flushed from soil with water are soluble salts such as the carbonates of nickel, zinc, and copper. Adjusting the pH with dilute solutions of acids or bases will enhance inorganic solubilization and removal.

Solutions of sulfuric, hydrochloric, nitric, phosphoric, and carbonic acid are used in industrial applications to dissolve basic metal salts (e.g., hydroxides, oxides, carbonates). Because of the toxicity of many acids, however, it is desirable to use weak acids for *in situ* treatment. Sodium dihydrogen phosphate and acetic acid have low toxicity and are reasonably stable. A stronger dilute acid (e.g., sulfuric acid) would be used if the soil contained sufficient alkalinity to neutralize that acid. Acidic solutions may also flush some basic organics (e.g., amines, ethers, anilines).

Chelating agents used for *in situ* treatment must result in a stable metal–chelate complex which is resistant to decomposition and degradation. Another possibility for mobilizing metals that are strongly adsorbed to manganese and iron oxides in soils is to reduce the metal oxides, which results in release of the heavy metals into solution. Chelating agents or acids can then be used to keep the metals in solution.

Surfactants can be used to improve the solvent property of the recharge water, to emulsify nonsoluble organics, and to enhance the removal of hydrophobic organics sorbed onto soil particles. Surfactants improve the effectiveness of contaminant removal by improving both the detergency of aqueous solutions and the efficiency by which organics may be transported by aqueous solutions.

If the waste is amenable to this technique and the distribution, collection, and treatment costs are relatively low, soil flushing can be an economical alternative to excavation and treatment.

Immobilization. Immobilization methods are designed to render contaminants insoluble and to prevent leaching of contaminants from the soil matrix and movement from the area of contamination. Immobilization methods currently being investigated include precipitation, chelation, and polymerization.

Precipitation is the most promising method for immobilizing dissolved metals such as lead, cadmium, zinc, and iron. Some forms of arsenic, chromium, mercury, and some organic fatty acids can also be treated by using precipitation. All divalent metal cations can be precipitated by using sulfide, phosphate, hydroxide, or carbonate; however, the solubility product and stability of the metal complexes vary. Because of the low-solubility product of sulfides and the stability of the metal sulfide over a broad range of pH, sulfide precipitation is most promising. The remaining anions decrease in effectiveness in the following order:

$$phosphate \rightarrow hydroxide \rightarrow carbonate$$

Metal carbonates and hydroxides are stable only over a narrow range of pH, and the optimum range varies for different metals. Precipitation of metal as the metal phosphate may require a very high concentration of orthophosphate, because calcium and other naturally occurring soil cations that are present in high concentrations will precipitate first.

Sodium sulfate is used in conjunction with sodium hydroxide to precipitate metals. Such precipitation occurs at a neutral or slightly alkaline pH. Addition of sodium hydroxide minimizes the formation of hydrogen sulfide gas by ensuring an alkaline pH.

Precipitation is most applicable to sites with sand or coarse silt strata. Disadvantages include injection of a potential groundwater pollutant, potential for toxic gases to form, potential for clogging the soil pore spaces, and the possibility of precipitate resolubilization.

Another method for immobilizing metals applies specifically to chromium and selenium. These metals can be present in the highly mobile, hexavalent state, but Cr^{6+} and Se^{6+} can be reduced to the less mobile Cr^{3+} and Se^{4+} by the addition of ferrous sulfate. Arsenic exists in soils as either arsenate (As^{5+}) or arsenite (As^{3+}), which is the more toxic and soluble form. Arsenic can be effectively immobilized by oxidizing As^{3+} to As^{4+} and by treating the As^{4+} with ferrous sulfate to form the highly insoluble $FeAsO_4$.

In situ treatment of a leachate plume by using precipitation or polymerization techniques has the following problems:

- Several closely spaced injection wells are needed, because precipitation will lower hydraulic conductivities near the injection wells.

- Contaminants are not removed from the aquifer, or some chemical reactions can be reversed, thus allowing contaminants to again migrate with the groundwater flow.

- A potential groundwater pollutant is injected, and toxic byproducts may form.

Therefore, before application of an *in situ* precipitation or polymerization technique at a hazardous waste site, thorough laboratory and pilot-scale testing should be conducted to determine any deleterious effects and to ensure complete precipitation or polymerization of the chemical compounds.

Toxicity Control *In situ* treatment techniques that destroy, degrade, or otherwise reduce the toxicity of contaminants include neutralization, hydrolysis, oxidation/ reduction, enzymatic degradation, and permeable treatment beds.

Neutralization may involve injecting dilute acids or bases into the groundwater to adjust the pH. This can serve as a pretreatment before *in situ* biodegradation, oxidation, or reduction to optimize the pH range. It can be used during oxidation, reduction, or precipitation to prevent formation of toxic gases, including hydrogen sulfide and hydrogen cyanide. Spreading crushed limestone or lime on acid spills will neutralize the acid.

The pH adjustment can also be used to increase the hydrolysis rate of certain organics. Hydrolysis involves displacement of a carboxylic group or hydrogen on an organic structure with a hydroxyl group from water. Of the parameters affecting the rate of hydrolysis (e.g., temperature, solvent composition, catalysis, pH), pH adjustment has the greatest effect.

Oxidation and reduction reactions alter the oxidation state of a compound through the loss or gain of electrons, respectively. Such reactions can detoxify, precipitate, or solubilize metals, and they can decompose, detoxify, or solubilize organics. Oxidation may render organics more amenable to biological degradation. As with many chemical treatment technologies, oxidation/reduction techniques are standard approaches to wastewater treatment. *In situ* oxidation of arsenic compounds with potassium permanganate, for example, has been used to successfully reduce the arsenic concentrations in groundwater. Hydrogen peroxide, ozone, and hypochlorites can react with organics and oxidize several different organic contaminants in a hazardous waste site. Selection of the appropriate oxidizing agent depends, in part, on the substance or substances to be detoxified and also on the feasibility of delivery and environmental safety.

Chemical reduction appears promising as a treatment for chromium and selenium in soils. *In situ* reduction of hexavalent to trivalent chromium has been accomplished in well water and groundwater by using a reducing agent.

Several disadvantages of oxidizing and reducing agents limit their use at hazardous waste sites. For example, the treatment compounds are nonspecific, and this may result in degradation of nontargeted compounds. There is also the potential for formation of more toxic or more mobile degradation products, and the introduction of these chemicals into the groundwater system may create a pollution problem in itself. In addition, as with soil flushing, uncertainty exists with respect to obtaining adequate contact with the contaminants in the plume.

Physical Methods.   Several methods have been developed that involve physical manipulation of the subsurface to immobilize or detoxify waste constituents. These technologies include *in situ* heating, *in situ* vitrification, and soil vapor extraction (SVE).

In situ heating destroys or removes organic contaminants in the subsurface through thermal decomposition, vaporization, and distillation with steam injection or radiofrequency heating. *In situ* vitrification stabilizes radioactive wastes and other hazardous wastes. The principle of operation here is Joule heating, which occurs when an electrical current is passed through a molten mass. Contaminated soil is thus converted into

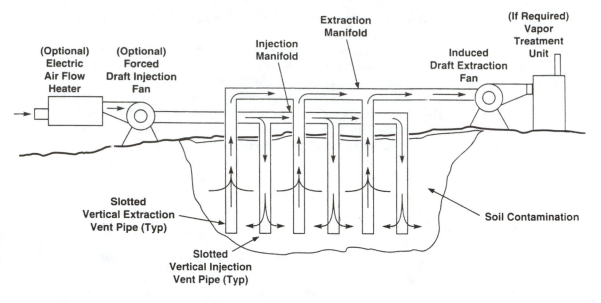

Figure 21–14   Soil vapor extraction system.

(From U.S. Environmental protection Agency. Handbook on Remedial Action of Waste Disposal Sites. *Washington, D.C.: U.S. Govt. Printing Office, 1985; with permission.)*

durable glass, and wastes are pyrolyzed or crystallized. Off-gases released during the melting process are trapped in a hood.

A relatively new remedial option, SVE is used to remove VOCs from soils in the vadose zone (i.e., the unsaturated zone above the groundwater table) or from stockpiled, excavated soils. The SVE process consists of injecting an air stream through the soil, thereby transferring the contaminants from the soil (or the soil/water) to the air stream. SVE systems are put together by installing vapor extraction wells or perforated piping in the contamination zone and then applying a vacuum to induce movement of soil gases. Figure 21–14 shows a SVE system.

CASE STUDIES

Remediation at the Seymour Superfund Site

Seymour, Indiana, is a small rustic town, approximately 70 miles southeast of Indianapolis, with a population of about 13,500. A neat stone wall graces its Freeman Field Industrial Park, which was highly contaminated with hazardous chemicals before 1980. Red and green paint sludges pooled on the porous ground, and chemicals such as arsenic, benzene, toluene, trichloroethylene, and naphthalene oozed from rusty barrels. The trees died, and an acrid scent brought tears to visitors' eyes (6).

Over a period of approximately 12 years before 1980, some sixty-thousand 55-gal drums of waste were heaped on this site by Seymour Recycling

Corporation. In March, hydrogen gas began rising from a shed on the property, where 25 badly corroded drums of chlorosilane had been stored next to barrels of flammable solvents. Rain soaking the chlorosilane created a smoky chemical reaction. "We had a 13-acre keg of dynamite," said one EPA official, who feared extensive contamination of groundwater beneath the sandy soil would show up in wells.

In 1980, after two explosions and complaints from residents of odor, and after state and local officials failed to get results, the U.S. EPA began an emergency response cleanup of Seymour Recycling Corporation's old chemical waste storage and treatment facility. The preliminary cleanup was completed in 1984 and entailed removal of all drums and 100 large tanks containing chemicals, removal of 1 ft of contaminated soil from approximately 75 percent of the 15-acre site, and replacement of the excavated soil with a silty-clay cap to limit infiltration. In 1986, investigation revealed that the remaining soil contained approximately 500,000 lbs (225,000 kg) of semi-VOCs and VOCs. Organics and chlorinated organics were also detected in the groundwater. After a feasibility study, a remedial action plan was created, and work began in December 1988.

The Seymour remediation project employed several methods, including an innovative vapor extraction system. Cleanup proceeded well ahead of schedule, in part because of cooperation by the responsible parties, regulating agencies, citizens, and contractors.

Three plume stabilization wells were installed to prevent more contamination from entering the local groundwater. Pumping was restricted from nearby wells to prevent any interference with remedial pumping. VOCs had decreased by 90 percent in the soil by the end of 1989; this may have resulted from anaerobic biodegradation. The anaerobic conditions may have developed after the installation of the initial clay cap in 1980. An increase in the soil concentration of vinyl chloride, however, was also discovered. Vinyl chloride is the anaerobic biodegradation product of trichloroethylene, which was one of the original contaminants. Because of the high water table, which was approximately 6 ft below grade, vertical walls were unsuitable for vapor extraction. Instead, a network of horizontal pipes was provided, to which vacuum was applied, with alternating pipes being left open to the atmosphere to facilitate air circulation. The vapor extraction system strips VOCs from and adds oxygen to the soil, thus stimulating aerobic biodegradation. Production of vinyl chloride is inhibited during aerobic degradation, because it forms anaerobically, as stated previously.

Recovery, Treatment, and Disposal of Contaminated Soil and Surface Water

On a cold day in January 1980, a railroad tanker car carrying approximately 20,000 gal (76,000 L) of a toxic liquid chemical (mostly chlorophenol and phenol) derailed near the small town of Sturgeon, Missouri. The tanker was punctured, and the liquid spilled at the site. Soon after the spill, noxious vapors from the spilled chemicals prompted the authorities to evacuate the area, which was inhabited by approximately 4000 people. The author worked as a consultant on the cleanup of this spill.

Cleanup Operations

A contractor was employed by the railroad to clean up the spill, and U.S. EPA officials visited the site. Efforts were directed at removing the contaminated soils and railroad bed from track areas by excavation. This material was stockpiled and later transported in dump trucks for disposal at an approved landfill. The contractor then treated the site with lime. The cleanup operation continued for the next 2 months.

After the cleanup, a problem surfaced in late March. Spring thaw and rain caused contaminated runoff from the spill area to collect in the adjacent farm area. A fish kill and chlorophenol concentration of greater than 50 ppm were reported in the freshwater ponds by the Missouri Department of Natural Resources on March 23–24, 1980. Immediate steps were taken to treat the contaminated water from the ponds. Equipment, a mobile laboratory, and workers were moved to the site (Figure 21–15). The EPA established allowable concentrations of no more than 200 ppm for soil and 60 ppb (parts per billion) for water.

During the following weeks, heavy spring rainfall increased the level of contaminated runoff. Most of this runoff found its way into the existing North Pond, which was used for farm cattle. Subsequent investigations, which included soil borings, revealed a few highly contaminated areas. Chlorophenol concentration in excess of 75,000 ppm were obtained in the vicinity of the track areas.

Severe odor emanated from the contaminated area during the spring months. Surface soil was treated with hydrogen peroxide to reduce the odor, because excavation of the contaminated area was ruled out for three reasons:

Figure 21–15   Toxic chemical spill site near Sturgeon, Missouri.

1. Increase in the severity of odor because of atmospheric exposure of the chemical that had seeped into the subsurface.

2. Excavation of the large area would have been uneconomical.

3. Dust resulting from excavation would have caused a health hazard to people living close to the site.

To reduce the contamination, flushing systems were installed in the tract area, and underdrains were also installed to prevent migration of contaminated flush water, which was contained and diverted to North Pond. Water from this pond was then pumped through an activated carbon and filtration unit into a pool, where it was seeded with microbes before being discharged into a holding pond for aeration and storage. Water from this pond was used to clean the major part of the spill area by using an injection system and surface flushing techniques (Figure 21–16).

Heavy rains in the area caused a significant amount of contaminated runoff. One of the main sources was a perforated underdrain that paralleled the main track and carried runoff from the previously excavated areas to ditches that drained into a creek. This contamination caused a second fish kill. Therefore, runoff from these ditches was diverted for treatment by pumping it through a carbon filter unit before discharge into the creek (Figure 21–17).

Continuous flushing, treatment of flushed water by carbon filters, recycling, and monitoring of the spill area for an extended period resulted in a steady reduction of soil contamination. During dry periods, make-up water for flushing was purchased from the city of Sturgeon. Commercially available bacteria (Phenobac) was used for biotreatment in soil and water. Use of Phenobac prolonged the life of carbon in the filters and reduced the phenol concentration to insignificant levels.

Figure 21–16   Site plan of Sturgeon, Missouri, toxic chemical spill cleanup operations.

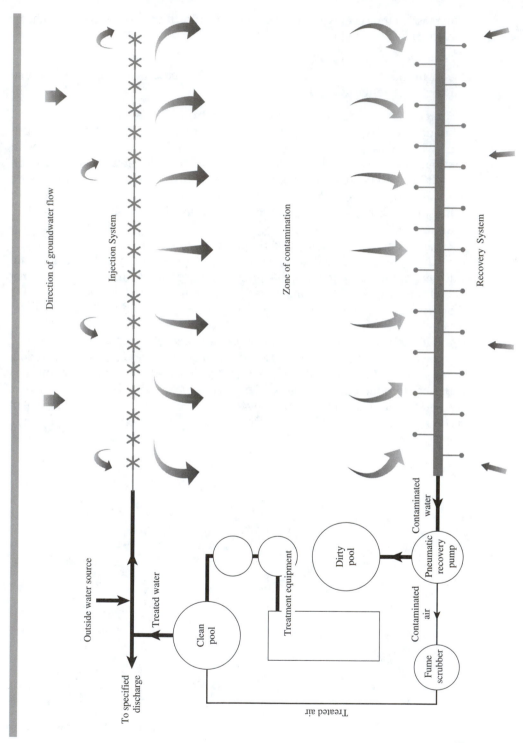

Figure 21–17 Carbon filter unit used during cleanup of Sturgeon, Missouri, toxic chemical spill.

In August 1980 and throughout the following months, daily water samples indicated periphery drainage concentrations below allowable chlorophenol limits of 60 ppb. As a result of flushing, many soil samples also showed a decrease in chemical concentrations. For example, at several locations, the contamination was reduced by approximately 90 percent at a depth of 1.5 ft.

The 90-percent removal of chlorophenol and chemicals from the soil was significant, but it was not enough to reduce the overall concentration to 200 ppm, as required by the E.P.A. For example, several soil samples had concentrations close to 1000 ppm. Excessive lime used earlier at the spill site also proved to be a hindrance. It inhibited the biodegradation of chemicals, and the reduction rate of soil contamination was very slow. Further treatment of these areas included additional well points, intensified flushing, and inoculation with Phenobac. Water samples from the North Pond taken after flushing showed higher concentrations than usual because of chemicals being flushed from soil.

Final Cleanup

Monitoring of air, water, and soil in the area continued throughout the winter months. Chemical concentrations in the runoff samples from the track area continued to be greater than 60 ppb, the limit set by the EPA. To remedy this problem, in the fall of 1980, the highly contaminated soil at the spill site was again excavated (depth, 2–6 ft) and removed, but this failed to eliminate all of the contamination. To oxidize the contaminant, the area was treated with hydrogen peroxide. The excavated area was then filled with carbon and clay, and it was sealed with an impermeable clay.

Results

After 2 years, concentrations in the runoff samples taken after the first significant rainfall were within the acceptable limits set by the EPA. These water samples were taken from the south ditch, north ditch, and the North Pond. Three of the samples were below the lower detectable limit of chlorophenol (i.e., <15 ppb), and the two others were measured at 19 and 23 ppb. Phenol was not detected. Hundreds of local residents complained of the taste and odor problems, headaches, upset stomach, dizziness, loss of love life, miscarriage, and many other health-related problems after the spill. More than $900 million worth of lawsuits were filed. Several cases were settled out of court.

Observations

Cleaning a spill of this type is a very complex operation. Several government agencies, citizen groups, chemical and product manufacturers, and cleanup contractors are involved, and each has a different viewpoint and interest. Many difficult decisions that require technical, managerial, communications, and legal skills must be made. Cleaning a spill of this nature is often done by trial and error, especially in the absence of reliable data and knowledge of the characteristics of the chemical, topography, soil and subsurface conditions, and the possible effects on the environment.

REVIEW QUESTIONS

1. a. Describe the Superfund Program. How did it come about?
   b. List the benefits and problems of the program.
2. Assume you are the mayor of Good Town, USA, which has an old hazardous waste site. What steps would you take to have the site placed on the NPL (for the Superfund)?
3. Contact your district EPA office, and find out the Superfund sites in your district and their current status.
4. Citizens of Good Town, USA, complain to the state EPA that the site on which an abandoned factory is located has a very strong odor and that the taste of well water near the site has changed. Assume you are an engineer in the district EPA office in charge of site investigation. How would you investigate the complaint? List the reasons for each step in the investigation.
5. a. What are advantages and disadvantages of on-site and off-site hazardous waste disposal?
   b. State the causes of air pollution at a hazardous waste cleanup site. How is air pollution controlled?
6. a. Why is gas movement from a hazardous waste landfill controlled?
   b. How does the permeability of soil affect gas movement?
7. a. What techniques are used to contain a plume in the groundwater?
   b. What factors control the containment and removal of a plume?
8. Sketch a capping system, show the various components, and discuss the significance of each.
9. Compare the following approaches to remediation of hazardous waste sites:
   a. Physical removal and disposal.
   b. Physical removal and physical/chemical treatment and disposal.
   c. Biodegradation and disposal.
   d. Fixation/immobilization.

REFERENCES

1. U.S. Environmental Protection Agency. *Revised Hazard Ranking System.* Publication 93207–02 FS. Washington, D.C.: U.S. Govt. Printing Office, 1990.
2. U.S. Environmental Protection Agency. *Remedial Action at Waste Disposal Sites.* EPA/625/6-85/006. Washington, D.C.: U.S. Govt. Printing Office, 1985.
3. U.S. Environmental Protection Agency. *Risk Assessment Guidance for Superfund Environmental Evaluation Manual.* Washington, D.C.: U.S. Govt. Printing Office, 1989.
4. Lutton, R.G., G. Regan, and L. Jones. *Design and Construction of Covers for Solid Waste Landfills.* EPA-600/12-79-165. Washington, D.C.: U.S. Govt. Printing Office, 1979.
5. Lee, M.D., and C.H. Ward. "Reclamation of Contaminated Aquifers: Biological Techniques," Paper Presented at the Hazardous Materials Spills Conference, Rockville, MD, 1984.
6. Watts, R.J. *Hazardous Wastes: Sources, Pathways, and Receptors.* New York: John Wiley & Sons, 1997.

Appendix A   Interest Tables

Discrete Cash Flow with 6.00% Discrete Compound Interest Factors

	Single Payments			Uniform Series Payments			
n	Compound Amount (F/P)	Present Worth (P/F)	Sinking Fund (A/F)	Compound Amount (F/A)	Capital Recovery (A/P)	Present Worth (P/A)	n
1	1.0600	0.9434	1.00000	1.000	1.06000	0.9434	1
2	1.1236	0.8900	0.48544	2.060	0.54544	1.8334	2
3	1.1910	0.8396	0.31411	3.184	0.37411	2.6730	3
4	1.2625	0.7921	0.22859	4.375	0.28859	3.4651	4
5	1.3382	0.7473	0.17740	5.637	0.23740	4.2124	5
6	1.4185	0.7050	0.14336	6.975	0.20336	4.9173	6
7	1.5036	0.6651	0.11914	8.394	0.17914	5.5824	7
8	1.5938	0.6274	0.10104	9.897	0.16104	6.2098	8
9	1.6895	0.5919	0.08702	11.491	0.14702	6.8017	9
10	1.7908	0.5534	0.07587	13.181	0.13587	7.3601	10
11	1.8983	0.5268	0.06679	14.972	0.12679	7.8869	11
12	2.0122	0.4970	0.05928	16.870	0.11928	8.3838	12
13	2.1329	0.4688	0.05296	18.882	0.11296	8.8527	13
14	2.2609	0.4423	0.04758	21.015	0.10758	9.2950	14
15	2.3966	0.4173	0.04296	23.276	0.10296	9.7122	15
16	2.5404	0.3936	0.03895	25.673	0.09895	10.1059	16
17	2.6928	0.3714	0.03544	28.213	0.09544	10.4773	17
18	2.8543	0.3503	0.03236	30.906	0.09236	10.8276	18
19	3.0256	0.3305	0.02962	33.760	0.08962	11.1581	19
20	3.2071	0.3118	0.02718	36.786	0.08718	11.4699	20
22	3.6035	0.2775	0.02305	43.392	0.08305	12.0416	22
24	4.0489	0.2470	0.01968	50.816	0.07968	12.5504	24
25	4.2919	0.2330	0.01823	54.865	0.07823	12.7834	25
26	4.5494	0.2198	0.01690	59.156	0.07690	13.0032	26
28	5.1117	0.1956	0.01459	68.528	0.07459	13.4062	28
30	5.7435	0.1741	0.01265	79.058	0.07265	13.7648	30
32	6.4534	0.1550	0.01100	90.890	0.07100	14.0840	32
34	7.2510	0.1379	0.00960	104.184	0.06960	14.3681	34
35	7.6861	0.1301	0.00897	111.435	0.06897	14.4982	35
36	8.1473	0.1227	0.00839	119.121	0.06839	14.6210	36
38	9.1543	0.1092	0.00736	135.904	0.06736	14.8460	38
40	10.2857	0.0972	0.00646	154.762	0.06646	15.0463	40
45	13.7646	0.0727	0.00470	212.744	0.06470	15.4558	45
50	18.4202	0.0543	0.00344	290.336	0.06344	15.7619	50
55	24.6503	0.0406	0.00254	394.172	0.06254	15.9905	55
60	32.9877	0.0303	0.00188	533.128	0.06188	16.1614	60
65	44.1450	0.0227	0.00139	719.083	0.06139	16.2891	65
70	59.0759	0.0169	0.00103	967.932	0.06103	16.3845	70
75	79.0569	0.0126	0.00077	1300.949	0.06077	16.4558	75
80	105.796	0.0095	0.00057	1746.600	0.06057	16.5091	80
85	141.579	0.0071	0.00043	2342.982	0.06043	16.5489	85
90	189.465	0.0053	0.00032	3141.075	0.06032	16.5787	90
95	253.546	0.0039	0.00024	4209.104	0.06024	16.6009	95
100	339.302	0.0029	0.00018	5638.368	0.06018	16.6175	100

Discrete Cash Flow with 7.00% Discrete Compound Interest Factors

	Single Payments		Uniform Series Payments				
n	Compound Amount (F/P)	Present Worth (P/F)	Sinking Fund (A/F)	Compound Amount (F/A)	Capital Recovery (A/P)	Present Worth (P/A)	n
1	1.0700	0.9346	1.00000	1.000	1.07000	0.9346	1
2	1.1449	0.8734	0.48309	2.070	0.55309	1.8080	2
3	1.2250	0.8163	0.31105	3.215	0.38105	2.6243	3
4	1.3108	0.7629	0.22523	4.440	0.29523	3.3872	4
5	1.4026	0.7130	0.17389	5.751	0.24389	4.1002	5
6	1.5007	0.6663	0.13980	7.153	0.20980	4.7665	6
7	1.6058	0.6227	0.11555	8.654	0.18555	5.3893	7
8	1.7182	0.5820	0.09747	10.260	0.16747	5.9713	8
9	1.8385	0.5439	0.08349	11.978	0.15349	6.5152	9
10	1.9672	0.5083	0.07238	13.816	0.14238	7.0236	10
11	2.1049	0.4751	0.06336	15.784	0.13336	7.4987	11
12	2.2522	0.4440	0.05590	17.888	0.12590	7.9427	12
13	2.4098	0.4150	0.04965	20.141	0.11965	8.3577	13
14	2.5785	0.3878	0.04434	22.550	0.11434	8.7455	14
15	2.7590	0.3624	0.03979	25.129	0.10979	9.1079	15
16	2.9522	0.3387	0.03586	27.888	0.10586	9.4466	16
17	3.1588	0.3166	0.03243	30.840	0.10243	9.7632	17
18	3.3799	0.2959	0.02941	33.999	0.09941	10.0591	18
19	3.6165	0.2765	0.02675	37.379	0.09675	10.3356	19
20	3.8697	0.2584	0.02439	40.895	0.09439	10.5940	20
22	4.4304	0.2257	0.02041	49.006	0.09041	11.0612	22
24	5.0724	0.1971	0.01719	58.177	0.08719	11.4693	24
25	5.4274	0.1842	0.01581	63.249	0.08581	11.6536	25
26	5.8074	0.1722	0.01456	68.676	0.08456	11.8258	26
28	6.6488	0.1504	0.01239	80.698	0.08239	12.1371	28
30	7.6123	0.1314	0.01058	94.461	0.08059	12.4090	30
32	8.7153	0.1147	0.00907	110.218	0.07907	12.6466	32
34	9.9781	0.1002	0.00780	128.259	0.07780	12.8540	34
35	10.6766	0.0937	0.00723	138.237	0.07723	12.9477	35
36	11.4239	0.0875	0.00672	148.913	0.07672	13.0352	36
38	13.0793	0.0765	0.00580	172.561	0.07580	13.1935	38
40	14.9745	0.0668	0.00501	199.635	0.07501	13.3317	40
45	21.0025	0.0476	0.00350	285.749	0.07350	13.6055	45
50	29.4570	0.0339	0.00246	406.529	0.07246	13.8007	50
55	41.3150	0.0242	0.00174	575.929	0.07174	13.9399	55
60	57.9464	0.0173	0.00123	813.520	0.07123	14.0392	60
65	81.2729	0.0123	0.00087	1146.755	0.07087	14.1099	65
70	113.989	0.0088	0.00062	1614.134	0.07062	14.1604	70
75	159.876	0.0063	0.00044	2269.657	0.07044	14.1964	75
80	224.234	0.0045	0.00031	3189.063	0.07031	14.2220	80
85	314.500	0.0032	0.00022	4478.576	0.07022	14.2403	85
90	441.103	0.0023	0.00016	6287.185	0.07016	14.2533	90
95	618.670	0.0016	0.00011	8823.854	0.07011	14.2626	95
100	867.716	0.0012	0.00008	12381.662	0.07008	14.2693	100

Appendix B   Useful Conversion Factors

LENGTH

1 inch	= 2.540 cm
1 foot	= 0.3048 m
1 yard	= 0.9144 m
1 mile	= 1.6093 km
1 meter	= 3.2808 ft
	= 39.37 in.
1 kilometer	= 0.6214 mi

AREA

1 square inch	= 6.452 cm^2
	= 0.0006452 m^2
1 square foot	= 0.0929 m^2
1 acre	= 43,560 ft^2
	= 0.0015625 mi^2
	= 4046.85 m^2
	= 0.404685 ha
1 square mile	= 640 acre
	= 2.604 km^2
	= 259 ha
1 square meter	= 10.764 ft^2
1 hectare	= 2.471 acre
	= 0.00386 mi^2
	= 10,000 m^2

VOLUME

1 cubic foot	= 0.03704 yd^3
	= 7.4805 gal (U.S.)
	= 0.02832 m^3
	= 28.32 L
1 acre foot	= 43,560 ft^3
	= 1233.49 m^3
	= 325,851 gal (U.S.)
1 gallon (U.S.)	= 0.134 ft^3
	= 0.003785 m^3
	= 3.785 L
1 cubic meter	= 8.11 $\times$ 10^{-4} acre-ft
	= 35.3147 ft^3
	= 264.172 gal (U.S.)
	= 1000 L
	= 10^6 cm^3

LINEAR VELOCITY

1 foot per second	= 0.6818 mph
	= 0.3048 m/s
1 mile per hour	= 1.467 ft/s
	= 0.4470 m/s
	= 1.609 km/hour
1 meter per second	= 3.280 ft/s
	= 2.237 mph

MASS

1 pound (avdp)	= 0.453592 kg
1 kilogram	= 2.205 lbs
	= 35.27396 oz
1 ton (short)	= 2000 lbs
	= 907.2 kg
	= 0.9072 ton (metric)
1 ton (metric)	= 1000 kg
	= 2204.622 lbs
	= 1.1023 ton (short)

FLOWRATE

1 cubic foot per second	= 0.28316 m³/s
	= 448.8 gal (U.S.)/ min (gpm)
1 cubic foot per minute	= 4.72×10^{-4} m³/s
	= 7.4805 gpm
1 gallon (U.S.) per minute	= 6.31×10^{-5} m³/s
1 million gallons per day	= 0.0438 m³/s
1 million acre feet per year	= 39.107 m³/s
1 cubic meter per second	= 35.315 ft³/s (cfs)
	= 2118.9 ft³/min (cfm)
	= 22.83×10^{6} gal/d
	= 70.07 Ac-ft/d

DENSITY

1 pound per cubic foot	= 16.018 kg/m³
1 pound per gallon	= 1.2×10^{5} mg/L
1 kilogram per cubic meter	= 0.062428 lbs/ft³
1 gram per cubic centimeter	= 62.427961 lbs/ft³

CONCENTRATION

1 milligram per liter in water (specific gravity = 1.0)	= 1.0 ppm
	= 1000 ppb
	= 1.0 g/m³
	= 8.34 lbs per million gal

PRESSURE

1 atmosphere	= 76.0 cm Hg
	= 14.696 lbs/in² (psia)
	= 29.921 in. Hg (32°F)
	= 33.8995 ft H_2O (32°F)
	= 101.325 kPa
1 pound per square inch	= 2.307 ft H_2O
	= 2.036 in. Hg
	= 0.06805 atm
1 Pascal (Pa)	= 1 N/m²
	= 1.45×10^{-4} psia
1 inch of mercury (32°F)	= 3386.4 Pa
(60°F)	= 3376.9 Pa

ENERGY

1 British Thermal Unit	= 778 ft-lb
	= 252 cal
	= 1055 J
	= 0.2930 Whr
1 quadrillion Btu	= 10^{15} Btu
	= 1055×10^{15} J
	= 2.93×10^{11} kW-hour
	= 172×10^{6} barrels (42-gal) of oil equivalent
	= 36.0×10^{6} metric tons of coal equivalent
	= 0.93×10^{12} cubic feet of natural gas equivalent

1 Joule	= 1 N-m
	= 9.48×10^{-4} Btu
	= 0.73756 ft-lbs
1 kilowatt-hour	= 3600 kJ
	= 3412 Btu
	= 860 kcal
1 kilocalorie	= 4.185 kJ

POWER

1 kilowatt	= 1000 J/s
	= 3412 Btu/hour
	= 1.340 hp
1 horsepower	= 746 W
	= 550 ft-lbs/s
1 quadrillion Btu per year	= 0.471 million barrels of oil per day
	= 0.03345 TW

Appendix C   Saturation Values of Dissolved Oxygen in Water at Various Temperatures

| | Chloride Concentration in Water (mg/L) | | | Difference per |
	0	5,000	10,000	100-mg chloride
Temperature in (°C)		*Dissolved Oxygen (mg/L)*		
0	14.6	13.8	13.0	0.017
1	14.2	13.4	12.6	0.016
2	13.8	13.1	12.3	0.015
3	13.5	12.7	12.0	0.015
4	13.1	12.4	11.7	0.014
5	12.8	12.1	11.4	0.014
6	12.5	11.8	11.1	0.014
7	12.2	11.5	10.9	0.013
8	11.9	11.2	10.6	0.013
9	11.6	11.0	10.4	0.012
10	11.3	10.7	10.1	0.012
11	11.1	10.5	9.9	0.011
12	10.8	10.3	9.7	0.011
13	10.6	10.1	9.5	0.011
14	10.4	9.9	9.3	0.010
15	10.2	9.7	9.1	0.010
16	10.0	9.5	9.0	0.010
17	9.7	9.3	8.8	0.010
18	9.5	9.1	8.6	0.009
19	9.4	8.9	8.5	0.009
20	9.2	8.7	8.3	0.009
21	9.0	8.6	8.1	0.009
22	8.8	8.4	8.0	0.008
23	8.7	8.3	7.9	0.008
24	8.5	8.1	7.7	0.008
25	8.4	8.0	7.6	0.008
26	8.2	7.8	7.4	0.008
27	8.1	7.7	7.3	0.008
28	7.9	7.5	7.1	0.008
29	7.8	7.4	7.0	0.008
30	7.6	7.3	6.9	0.008

Note:—Saturation at barometric pressures other than 760 mm (29.92 in.), C'_s is related to the corresponding tabulated values, C_s by the equation:

$$C'_s = C_s \frac{P - p}{760 - p}$$

where C'_s is the solubility at barometric pressure P and given temperature (mg/L), C_s is the saturation at given temperature from table (mg/L), P is the barometric pressure (mm), and p is the pressure of saturated water vapor (mm) at temperature of the water selected from table (mm).

Appendix D Soil Moisture Retention for Various Amounts of Potential Evapotranspiration for a Root Zone Water-Holding Capacity of 4 Inches

PET	0.00	0.01	0.02	0.03	0.04	0.05	0.06	0.07	0.08	0.09
0.00	4.00	3.99	3.98	3.97	3.96	3.95	3.94	3.93	3.92	3.91
0.10	3.90	3.89	3.88	3.87	3.86	3.85	3.84	3.83	3.82	3.81
0.20	3.80	3.79	3.78	3.77	3.76	3.75	3.74	3.73	3.72	3.71
0.30	3.70	3.69	3.68	3.67	3.66	3.65	3.64	3.63	3.62	3.62
0.40	3.61	3.60	3.59	3.58	3.57	3.56	3.55	3.54	3.54	3.53
0.50	3.52	3.51	3.50	3.49	3.48	3.47	3.46	3.46	3.45	3.44
0.60	3.43	3.42	3.41	3.40	3.39	3.38	3.38	3.37	3.36	3.35
0.70	3.34	3.33	3.32	3.31	3.30	3.30	3.29	3.28	3.27	3.26
0.90	3.26	3.25	3.24	3.23	3.23	3.22	3.21	3.20	3.19	3.19
0.90	3.18	3.17	3.16	3.16	3.15	3.14	3.13	3.12	3.12	3.11
1.00	3.10	3.09	3.09	3.08	3.07	3.06	3.05	3.05	3.04	3.03
1.10	3.02	3.02	3.01	3.00	2.99	2.98	2.98	2.97	2.96	2.95
1.20	2.94	2.94	2.93	2.92	2.91	2.90	2.90	2.89	2.88	2.87
1.30	2.86	2.86	2.85	2.84	2.83	2.82	2.82	2.81	2.80	2.79
1.40	2.79	2.78	2.77	2.76	2.75	2.75	2.74	2.73	2.73	2.72
1.50	2.72	2.71	2.70	2.70	2.69	2.68	2.68	2.67	2.66	2.66
1.60	2.65	2.64	2.64	2.63	2.62	2.62	2.61	2.60	2.60	2.59
1.70	2.58	2.58	2.57	2.57	2.56	2.55	2.54	2.54	2.53	2.52
1.80	2.51	2.51	2.50	2.49	2.49	2.48	2.48	2.47	2.47	2.46
1.90	2.45	2.45	2.44	2.43	2.43	2.42	2.41	2.40	2.40	2.39
2.00	2.39	2.38	2.38	2.37	2.36	2.36	2.35	2.35	2.34	2.34
2.10	2.33	2.33	2.32	2.32	2.31	2.30	2.29	2.29	2.28	2.28
2.20	2.27	2.27	2.26	2.25	2.25	2.24	2.24	2.23	2.22	2.22
2.30	2.21	2.21	2.20	2.19	2.19	2.18	2.18	2.17	2.16	2.16
2.40	2.15	2.15	2.14	2.14	2.13	2.13	2.12	2.12	2.11	2.11
2.50	2.10	2.10	2.09	2.09	2.08	2.08	2.07	2.07	2.06	2.06
2.60	2.05	2.05	2.04	2.04	2.03	2.03	2.02	2.02	2.01	2.01
2.70	2.00	2.00	1.99	1.99	1.98	1.98	1.97	1.97	1.96	1.96
2.80	1.95	1.95	1.94	1.94	1.93	1.93	1.92	1.89	1.91	1.91
2.90	1.90	1.90	1.89	1.89	1.88	1.88	1.87	1.87	1.86	1.86

PET	0.00	0.01	0.02	0.03	0.04	0.05	0.06	0.07	0.08	0.09
3.00	1.85	1.85	1.84	1.84	1.83	1.83	1.82	1.82	1.81	1.81
3.10	1.80	1.80	1.79	1.79	1.78	1.78	1.78	1.77	1.77	1.76
3.20	1.76	1.75	1.75	1.74	1.73	1.73	1.72	1.72	1.71	1.71
3.30	1.71	1.70	1.70	1.69	1.69	1.69	1.68	1.68	1.67	1.67
3.40	1.67	1.66	1.66	1.65	1.65	1.65	1.64	1.64	1.63	1.63
3.50	1.63	1.62	1.62	1.61	1.61	1.61	1.60	1.60	1.59	1.59
3.60	1.59	1.58	1.58	1.57	1.57	1.57	1.56	1.56	1.55	1.55
3.70	1.55	1.54	1.54	1.53	1.53	1.53	1.52	1.52	1.51	1.51
3.80	1.51	1.50	1.50	1.49	1.49	1.49	1.48	1.48	1.47	1.47
3.90	1.47	1.46	1.46	1.45	1.45	1.45	1.44	1.44	1.43	1.43
4.00	1.43	1.42	1.42	1.41	1.41	1.41	1.40	1.40	1.40	1.39
4.10	1.39	1.39	1.38	1.38	1.38	1.37	1.37	1.37	1.36	1.36
4.20	1.36	1.35	1.35	1.35	1.34	1.34	1.34	1.33	1.33	1.33
4.30	1.32	1.32	1.32	1.31	1.31	1.31	1.30	1.30	1.30	1.29
4.40	1.29	1.29	1.28	1.28	1.28	1.28	1.27	1.27	1.27	1.26
4.50	1.26	1.26	1.25	1.25	1.25	1.25	1.24	1.24	1.24	1.23
4.60	1.23	1.23	1.22	1.22	1.22	1.22	1.21	1.21	1.21	1.20
4.70	1.20	1.20	1.19	1.19	1.19	1.19	1.18	1.18	1.18	1.17
4.80	1.17	1.17	1.16	1.16	1.16	1.16	1.15	1.15	1.15	1.14
4.90	1.14	1.14	1.13	1.13	1.13	1.13	1.12	1.12	1.12	1.11
5.00	1.11	1.11	1.10	1.10	1.10	1.10	1.09	1.09	1.09	1.09
5.10	1.08	1.08	1.08	1.07	1.07	1.07	1.07	1.06	1.06	1.06
5.20	1.05	1.05	1.05	1.04	1.04	1.04	1.04	1.03	1.03	1.03
5.30	1.02	1.02	1.02	1.01	1.01	1.01	1.01	1.00	1.00	1.00
5.40	1.00	1.00	0.99	0.99	0.99	0.99	0.98	0.98	0.98	0.98
5.50	0.98	0.97	0.97	0.97	0.97	0.97	0.96	0.96	0.96	0.96
5.60	0.95	0.95	0.95	0.94	0.94	0.94	0.94	0.93	0.93	0.93
5.70	0.92	0.92	0.92	0.92	0.91	0.91	0.91	0.91	0.90	0.90
5.80	0.90	0.90	0.90	0.89	0.89	0.89	0.89	0.89	0.88	0.88
5.90	0.88	0.88	0.88	0.87	0.87	0.87	0.87	0.87	0.86	0.86
6.00	0.86	0.86	0.86	0.85	0.85	0.85	0.85	0.85	0.84	0.84
6.10	0.84	0.84	0.84	0.83	0.83	0.83	0.83	0.83	0.82	0.82
6.20	0.82	0.82	0.82	0.81	0.81	0.81	0.81	0.80	0.80	0.80
6.30	0.80	0.79	0.79	0.79	0.79	0.79	0.78	0.78	0.78	0.78
6.40	0.77	0.77	0.77	0.77	0.77	0.77	0.77	0.77	0.76	0.76
6.50	0.76	0.76	0.76	0.76	0.75	0.75	0.75	0.75	0.74	0.74
6.60	0.74	0.74	0.74	0.73	0.73	0.73	0.73	0.73	0.72	0.72
6.70	0.72	0.72	0.72	0.72	0.71	0.71	0.71	0.71	0.71	0.70
6.80	0.70	0.70	0.70	0.70	0.70	0.69	0.69	0.69	0.68	0.68
6.90	0.68	0.68	0.68	0.68	0.67	0.67	0.67	0.67	0.67	0.67
7.00	0.66	0.66	0.66	0.66	0.66	0.66	0.66	0.66	0.65	0.65
7.10	0.65	0.65	0.65	0.64	0.64	0.64	0.64	0.64	0.54	0.63
7.20	0.63	0.63	0.63	0.63	0.63	0.62	0.62	0.62	0.62	0.61
7.30	0.61	0.61	0.61	0.61	0.61	0.61	0.60	0.60	0.60	0.60
7.40	0.60	0.60	0.60	0.59	0.59	0.59	0.59	0.59	0.58	0.58
7.50	0.58	0.58	0.58	0.58	0.58	0.58	0.58	0.57	0.57	0.57
7.60	0.57	0.57	0.57	0.57	0.57	0.56	0.56	0.56	0.56	0.56
7.70	0.56	0.56	0.56	0.55	0.55	0.55	0.55	0.55	0.55	0.55

PET	0.00	0.01	0.02	0.03	0.04	0.05	0.06	0.07	0.08	0.09
7.80	0.54	0.54	0.54	0.54	0.54	0.54	0.54	0.54	0.53	0.53
7.90	0.54	0.53	0.53	0.53	0.53	0.52	0.52	0.52	0.52	0.52
PET	0.05	0.05		PET	0.00	0.05		PET	0.00	0.05
8.00	0.52	0.51		9.00	0.40	0.40		10.00	0.31	0.31
8.10	0.50	0.50		9.10	0.39	0.39		10.10	0.30	0.30
8.20	0.49	0.48		9.20	0.38	0.38		10.20	0.30	0.29
8.30	0.48	0.47		9.30	0.37	0.37		10.30	0.29	0.28
8.40	0.47	0.46		9.40	0.36	0.36		10.40	0.28	0.28
8.50	0.45	0.45		9.50	0.35	0.35		10.50	0.27	0.27
8.60	0.44	0.44		9.60	0.34	0.34		10.60	0.27	0.26
8.70	0.43	0.43		9.70	0.34	0.33		10.70	0.26	0.26
8.80	0.42	0.42		9.80	0.33	0.32		10.80	0.25	0.25
8.90	0.41	0.41		9.90	0.32	0.32		10.90	0.25	0.24

Note—A storage ability equal to 4 in. of water is the combination of the ability of a given soil to store water and the thickness of the soil layer that provides the equivalent of 4 in. of water.

Appendix E Glossary

Abiotic The organism's nonliving environment (e.g., a wastewater tank).

Absorption The assimilation and incorporation of substances by solution in a liquid.

Acid A substance that increases hydrogen ion concentrations in a solution. The ions can react to neutralize a base or an alkaline substance.

Acre-Feet A unit equivalent to the volume that would cover 1 acre of land to a depth of 1 ft.

Activated Carbon A very porous, granular material used to adsorb pollutants from air or water.

Activated Sludge Process A biological wastewater treatment in which microorganisms, suspended in wastewater and air, absorb organic pollutants and convert them to stable substances.

Adsorption A process involving the contact and trapping of water or air pollutants on the surface of a solid material such as activated carbon.

Aeration A process in which air is mixed with water or wastewater for purification.

Aerobic A biochemical process or environmental condition occurring in the presence of oxygen.

Aerobic Microorganisms (Aerobes) Microorganisms that require an aerobic environment to live and multiply.

Air Classifier A device to remove light-weight material from the refuse.

Air Pollution The presence of contaminants in the air.

Air Stripping The use of air to remove dissolved volatile chemicals from a liquid by forming a gas.

Algae Single-cell, microscopic plants (e.g., phytoplankton) that are suspended in water.

Alkali A substance that increases hydroxyl ion concentrations in a solution. The ions can react to neutralize an acid substance.

American System of Units Based on feet, pounds, seconds, and degrees-Fahrenheit ($°F$).

Ammonia A compound of nitrogen and hydrogen in a liquid or gaseous form.

Anaerobic A biochemical process or environmental condition occurring in the absence of molecular oxygen.

Anaerobic Microorganisms (Anaerobe) Microorganisms that live and multiply without free oxygen.

Aquatic Organism An organism that lives in water.

Aqueous Solution A mixture with water.

Aquifer An underground layer of relatively impermeable rock or soil that can yield significant amounts of groundwater for public supply.

Area Method A method of placing, compacting, and covering solid waste in a sanitary landfill without excavation.

Aromatic Compounds Generally refers to aromatic hydrocarbons with a benzene-ring structure.

Ash The incombustible material that remains after a fuel or solid waste has been burned.

Atmospheric Pressure The force per unit area exerted by the prevailing atmospheric conditions.

Atom The smallest part of an element with the same chemical characteristics.

Atomic Weight The mass of an atom measured in atomic mass units (amu). One amu is one-twelfth of a carbon atom.

Autotrophic Organisms Self-nourishing organisms. For example, green plants and a few bacteria use the sun's energy to convert inorganic substances to chemical energy by photosynthesis.

Bacteria Microscopic, single-celled plants without chlorophyll.

Baghouse Filter A fabric device used to remove particulate air pollutants from industrial stack emissions.

Baling A mechanical method by which solid waste is compressed under pressure into bales for transport or land disposal.

Bedrock The solid rock underlying the soil and unconsolidated materials or appearing at the surface where soil and these materials are absent.

Benefit/Cost An analysis and/or comparison of the value benefits in contrast to the costs of any particular project or action.

Benefit/Cost Ratio The value of the benefits to be gained from a project divided by the cost of that project. If the ratio is greater than 1, the project is economically justified; if the ratio is less than 1, it is not economically justified.

Benzene (C_6H_6) The parent compound of aromatic hydrocarbons.

Best Available Technology A general term for the best method currently available to treat and/or dispose of a hazardous waste.

Biochemical Oxygen Demand (BOD) The amount of oxygen required by microorganisms to biodegrade organics in water; a measure of organic contamination.

Biodegradable Readily broken or decomposable into stable or simpler substances through microbial action.

Biogeochemical Cycle The cycle of elements through the biotic and abiotic environment.

Biological Treatment Use of organisms to break down organic compounds, thus lowering the biochemical oxygen demand.

Bioremediation Use of microorganisms to convert harmful chemical compounds into less harmful compounds to clean a contaminated area.

Biosphere The area near the earth's surface where all living organisms are found; includes parts of the hydrosphere, atmosphere, and crust.

Biotic The living environment of organisms.

Bond An interest-bearing certificate issued by the government or a business promising to pay the holder a specified sum on a specified date.

Break Horsepower (BHP) Power transmitted to the pump shaft.

Btu (British Thermal Unit) A unit of measure for the amount of energy a given material contains.

Bulky Waste Large wastes such as appliances, furniture, automobile parts, and large tree parts.

Buy-Back Center A facility where individuals can bring back recyclable materials in exchange for payment.

Capillarity The condition by which a fluid is drawn up in small interstices or tubes because of surface tension.

Capital Costs The funds needed to construct a facility or buy equipment.

Capital Financing The funds needed to finance construction of publicly or privately owned systems.

Capital Recovery Factor The value of a series of end-of-period payments or disbursements.

Capitalized Cost The present sum of money that would need to be set aside at some interest rate to yield the funds required to provide a service.

Carbon Cycle The various ways that natural processes continuously transport carbon back and forth between the atmosphere, biospheres, and oceans.

Carbon Dioxide (CO_2) A colorless, odorless, and nonpoisonous gas produced during thermal degradation and microbial decomposition of solid wastes.

Carbon Monoxide (CO) A colorless, poisonous gas produced during thermal degradation and microbial decomposition of solid wastes when the oxygen supply is limited.

Carcinogenic A substance capable of causing cancer.

Catalyst A substance that modifies a chemical reaction without being consumed.

Cation A molecule with a positive charge.

Centrifugal Pump A mechanical device that adds energy to a liquid using a fast-rotating impeller in a casing; generally used in water and wastewater treatment.

Centrifugation   A process that uses a centrifuge to separate hazardous constituents.

Chemical Compound   A combination of two or more molecules of different elements.

Chemical Oxygen Demand (COD)   The equivalent amount of oxygen required to oxidize any organic matter in a water sample by using a strong oxidizing agent.

Chlorinated Hydrocarbons   Any organic compound that contains chlorine (in addition to carbon and hydrogen) in its chemical structures.

Chlorofluorocarbons (CFCs)   Synthetic organic compounds used as aerosol propellants, refrigerants (now prohibited in the United States), and in the manufacture of plastic foams; known to destroy the ozone layer in the upper atmosphere.

Clarifier (Sedimentation Basin)   A tank in which suspended particles in a liquid agglomerate sink to the bottom under quiescent conditions of flow.

Clay   Very small soil particles.

Clean Air Act   The legislation passed by Congress to ensure the air is "safe enough to protect the public's health."

Clean Water Act   The legislation passed by Congress to protect the nation's water resources; requires the U.S. Environmental Protection Agency to establish a system of national effluent standards.

Closure   The procedure that a solid or hazardous waste management facility undergoes to cease operations and ensure protection of human health and environment.

Coagulation   The addition of certain chemicals to water and wastewater to destabilize very small, suspended particles and thus have them collide, stick together, and form settleable flocs.

Code of Federal Regulations (CFR)   A compilation of all final federal regulations. (The U.S. Environmental Protection Agency's regulations are included in 40 CFR.)

Codisposal (of Wastes)   Household and industrial wastes disposed of in the same landfill.

Coefficient of Roughness   A factor that accounts for the friction causing resistance to flow in a channel or pipe and that depends on material and age of the pipe or lined channel.

Collection Routes   Planned routes followed in the collection of waste from sources such as homes, businesses, industrial plants, and others.

Collection Systems   Collectors and equipment used for the collection of commingled and source-separated waste.

Colloids   Very small particles in water or wastewater that cannot be removed by plain sedimentation or filtration without coagulation.

Combustible Waste   The materials in a waste stream that are combustible.

Combustion   The burning of a substance by combining it with oxygen, which produces heat and sometimes light.

Commercial Solid Waste   Waste that originates from commercial establishments and office buildings.

Commingled Waste   A mixture of all waste components in one container.

Compaction   The process of decreasing the volume of waste or soil by using mechanical means.

Compactor   Any mechanical equipment designed to compress and thereby reduce the volume of waste.

Compost   The relatively stable, decomposed organic material resulting from the composting process.

Composting   The controlled biological decomposition of organic solid waste under aerobic conditions.

Compound   A substance composed of two or more elements.

Comprehensive Environmental Response, Compensation, and Liability Act (CERCLA)   The legislation passed by Congress that authorizes the U.S. Environmental Protection Agency (EPA) to clean up uncontrolled or abandoned hazardous waste sites and respond to accidents, spills, and other emergency releases of hazardous substances. CERCLA provides the EPA with the authority to make responsible parties pay the cleanup costs of remediating a contaminated site.

Compressibility   The change or tendency for change that occurs when a substance is subject to pressure.

Concentration   The presence of a substance in another substance; usually expressed in terms of weight percent or milligrams per liter.

Conditionally Exempt Small-Quantity Generators (CESQG)   Facilities that produce less than 100 kg of hazardous waste, or less than 1 kg of acutely hazardous waste, per calendar month.

Cone of Depression   The lowered surface of the groundwater table, in the form of a cone, resulting from pumping of water.

Construction and Demolition Waste   Materials from the construction, remodeling, repair, or demolition of buildings, pavements, and other structures.

Consumer   An organism that obtains its energy from the organic materials of other organisms, living or dead.

Container   A receptacle in which material is stored, transported, treated, or otherwise handled.

Containment   The process of confining movement of a material at the place where it occurs or some desired location to prevent environmental contamination.

Contaminant   Any unwanted substance or matter in water, soil, or air.

Contaminant Plume   A mass of contaminated groundwater.

Conveyor   A device used for short-distance transport of solid waste.

Corrective Action   The U.S. Environmental Protection Agency's program to address the investigation and cleanup of contamination from solid and hazardous waste facilities.

Corrosivity   The characteristic identifying wastes that can readily corrode metal and dissolve materials.

Cost Analysis   A study of price or expense.

Cost Index   A ratio of the cost of something today to its cost sometime in the past.

Cover Material   The soil or other material used to cover compacted solid waste in a sanitary landfill.

Cradle-to-Grave   A provision of the Resource Conservation and Recovery Act that requires tracking a hazardous waste from generation to ultimate disposal.

Cullet   Clean, generally color-sorted, crushed glass used to make new glass products.

Curbside Collection   The collection of mixed and source-separated wastes from the curbside, where they are placed by residents.

Cyclone   A mechanical air pollution–control device to remove particulates.

Darcy's Law   A formula that expresses the velocity of groundwater flow as a function of the slope of the water table and of the soil permeability in the aquifer.

Decomposers   Organisms that use energy from wastes or dead organisms.

Deep Well   A relatively deep and narrow, vertical excavation drilled to penetrate an aquifer.

Deep-Well Injection   A system used for limited disposal of some liquid hazardous wastes.

Degrade   To lower the quality or usefulness of a substance, or to stabilize it.

Delisting   A site-specific petition process whereby a handler can request the U.S. Environmental Protection Agency to certify that a waste stream generated at its facility meeting a listing description does not pose sufficient hazard to warrant regulation under the Resource Conservation and Recovery Act.

Denitrification   The conversion of ammonia and nitrates into free nitrogen by microorganisms.

Density   The mass of a substance divided by a unit volume.

Deoxygenation   The consumption of oxygen by organisms as they oxidize materials in an aquatic environment.

Detention Time   The theoretical amount of time that an average particle of fluid spends in a container.

Detritus   Dead organic matter such as fallen plant wastes and animal wastes.

Dewatering   The removal of water from sludge.

Digestion   The decomposition of organic waste by microorganisms under controlled conditions.

Dimensions   Describe characteristics such as the length, time, mass, and force of an object.

Dioxin   A synthetic, organic chemical of the chlorinated hydrocarbon class; very toxic.

Disease   Any impairment of normal functions in an organism.

Disposal   The removal and handling of waste or the residual matter after waste has been processed and the recovery of products or energy has been accomplished.

Dissolved Oxygen (DO)   Oxygen gas molecules (O_2) dissolved in water.

DNA (Deoxyribonucleic Acid)   The natural organic molecule in virtually all organisms carrying genetic or hereditary information.

Dose Response   The impact of the pollutant on the receptor.

Drainage Layer   Used in a landfill for dissipating excess water.

Drawdown   The vertical distance between the static water level and the pumping water level in a well.

Drop-Off Center   A method of collecting recyclables or compostibles in which the materials are taken by individuals to collection sites and deposited into designated containers.

Dump   An uncontrolled area where wastes have been placed in an environmentally unsound manner.

Ecology   The study of living organisms and their environment.

Economic Analysis   An examination of the income (i.e., benefit) and cost (i.e., expense) of an undertaking.

Ecosystem   An arbitrarily chosen boundary of a system containing plants and animals and the air, water, and minerals necessary for their survival.

Effluent   The fluid exiting a tank, process, or system.

Electrostatic Precipitator   An air pollution--control device that uses electric field to trap small particulates.

Elutriate   The removal of a substance from a mixture by washing and decanting.

Emergency Planning and Community Right-to-Know Act   The legislation passed by Congress to help communities in the event of a chemical emergency and to increase citizens' knowledge of the presence and threat of hazardous chemicals.

Emission   The discharge of a gas into the atmosphere.

Energy Recovery   The conversion of waste to energy through combustion of raw or processed waste.

Engineering News--Record (ENR)   A weekly magazine for the engineering and construction industry.

Environment   The physical surroundings (e.g., air, water, and land).

Environmental Impact Statement (EIS)   The written report for a proposed project that summarizes the findings of a detailed environmental review process.

Environmental Movement   The public awareness and citizen action regarding environmental issues.

Environmental Protection Agency (EPA)   The federal agency responsible for establishing and enforcing pollution control and environmental quality standards.

EPA Identification Number   A unique number assigned by the U.S. Environmental Protection Agency to each hazardous waste generator, transporter, or treatment, storage, and disposal facility.

Epidemiology   The study of the incidence rate of diseases in real populations.

Erosion and Sediment Control   The use of covers, diversion channels, or detention basins to prevent exposed soil from washing away.

Ethics   A framework for decision-making based on moral values.

Eukaryote   An organism in which each cell has a nucleus.

Eutrophication   The aging of a lake as characterized by high nutrient levels, excessive plants, and bottom sediments.

Evaporation   The conversion from liquid phase to vapor phase.

Evapotranspiration   The combined processes of evaporation of water from the ground surface and transpiration of water by vegetation.

Facultative Bacteria   Bacteria that prefer to use molecular oxygen when available but can survive and biodegrade without it.

Fault   A fracture in a rock mass where opposite sides of the rock have moved with respect to each other.

Federal Insecticide, Fungicide, and Rodenticide Act (FIFRA)   The legislation passed by Congress that provides procedures for registration of pesticide products to control their introduction to the market.

Ferrous Metals   Metals composed predominantly of iron.

Field Capacity   The total moisture that can be retained in a waste sample subject to the downward pull of the gravity.

Filtration   The removal of suspended particles from water or air by using a porous material that traps and retains the particles.

Final Cover   The layer of compacted clay used to close a landfill after it becomes full.

Financial Assurance   The requirements under the Resource Conservation and Recovery Act to ensure that owners and operators of waste treatment, storage, and disposal facilities have the financial resources to pay for closure, postclosure, corrective action, and liability costs.

Flexible Membrane Liner (FML)   A plastic sheet, also called a geomembrane, used as an impermeable liner at a landfill.

Floc   A mass of particles suspended in liquid and produced by chemical or biological action.

Flocculation   The formation of settleable flocs by gentle agitation of water or wastewater after the addition of coagulation chemicals.

Floodplain   The land along a river likely to be inundated by water during a 100-year flood.

Flotation   Causing suspended solids to float to the surface of a tank, where they can be removed.

Flow Rate   The flow movement of a fluid volume per unit of time.

Fluidization   The suspension of particles by the upward velocity of a fluid.

Fly Ash   Small, solid particles of ash and soot generated when coal, oil, or solid wastes are burned.

Food Chain   An interrelated series of living organisms that feed on each other.

Fossil Fuel   Substances formed below the ground (e.g., coal, oil, and natural gas).

Fracture   A crack or split across the surface of a rock.

Frequency Analysis   A statistical method to determine the recurrence interval of storms, floods, or droughts.

Friction Head   Energy lost because of friction between a pipe's inside surface and the flowing water.

Functional Element   The various activities associated with management of solid wastes from generation to final disposal.

Garbage   Food wastes in refuse; usually discarded from kitchens and restaurants.

General Obligation Bond   Issued against taxes received by the governmental entity that issued the bond and backed by the full taxing power of the issuer.

Generator   Any person whose action first creates or produces a hazardous waste, used oil, or medical waste or brings such materials into the regulatory area of the Resource Conservation and Recovery Act.

Geology   The study of the materials from which the earth is made.

Geosynthetics   A general term for synthetic fabrics used as landfill liners.

Granular Activated Carbon   A finely powdered charcoal used for treatment of hazardous waste.

Gravity Flow   The open-channel flow in a ditch, pipe, or streambed with the liquid surface at atmospheric pressure.

Greenhouse Gases   Gases in the atmosphere that absorb infrared energy and raise the air temperature; carbon dioxide, water vapor, methane, chlorofluorocarbons, and other hydrocarbons.

Gross National Product (GNP)   The total value of all goods and services exchanged during a year in a country.

Groundwater Monitoring   The sampling and analysis of groundwater to detect release of contamination from a solid or hazardous waste land unit.

Groundwater   The water beneath the surface of the earth that occupies the pore spaces in soil or the fissures in rock.

Groundwater Table   The interface between the aeration and saturation zones in soil.

Halogen   A compound with chlorine or bromine.

Hammermill   A type of crusher or shredder used to break up waste material into smaller pieces.

Haul Distance   The distance a waste-collection vehicle travels after picking up the waste combined with the distance a collection-vehicle travels after unloading to the location of waste pickup.

Haul Time   The elapsed or cumulative time spent transporting solid waste between two specific locations.

Hauled Container System   Collection systems in which the containers used for storage of wastes are hauled to the disposal site, emptied, and then returned to either their original or some other location.

Hazardous Material   Chemicals that can pose a risk to public health or the environment.

Hazardous Waste   A waste material that may harm living forms and the environment.

Hazen-Williams Equation   A formula for computing pressure losses during distribution of water through pipes.

Heavy Metals   Hazardous elements, including cadmium, mercury, and lead, that may be found in a waste stream.

Herbicide   A chemical used to kill or inhibit the growth of undesired plants.

Heterotrophic Organism   An organism that consumes plants or animals for energy.

Hierarchy   A rank order or dominance pattern; for example, in waste management, the priorities in descending order are source reduction, recycling, waste transformation or combustion, and disposal.

High-Density Polyethylene (HDPE)   A type of low-density plastic used for making beverage containers.

Histogram   A plot of observations and frequency as a bar graph.

Home Hazardous Waste (HHW)   The hazardous waste from homes discarding unused or unwanted chemicals such as cleaning solvents, pesticides, insecticides, or herbicides.

Horsepower (hp)   An expression of energy (1 hp = 550 ft-lbs/min.)

Hydraulic Barrier   Provided in a landfill to prevent water infiltration from reaching the waste.

Hydraulic Conductivity   The permeability coefficient.

Hydraulic Grade Line   A graph of the pressure head in a hydraulic system that consists of sloping straight lines showing a drop in pressure in the direction of flow.

Hydraulic Head   The height of the free surface of a body of water above a given reference point.

Hydraulic Ram   A plunger-type device operated by oil pressure to lift heavy objects.

Hydraulics   The study of water at rest and in motion in tanks, reservoirs, pipes, and pumps.

Hydrocarbon   An organic compound that contains only hydrogen and carbon atoms.

Hydrogen Ions   Hydrogen atoms that have lost their electrons.

Hydrogen Sulfide (H_2S)   A toxic gas with an odor like rotten eggs that is produced from the reduction of sulfates and the decomposition of a sulfur-containing organic material.

Hydrologic Cycle   The cycle of water moving through the environment as rainfall, surface and subsurface flow, and vapor.

Hydrology   The study of the occurrence and distribution of water both on and below the ground.

Hydrostatic Pressure   The force per unit area on the side walls of a tank, tank bottom, dam, or pipe because of standing water.

Ignitable Waste   A hazardous waste that can easily catch fire.

***In Situ* Treatment**   The treatment of contaminated soil or groundwater at the site of an old waste dump.

Incineration   The process by which solid, liquid, or gaseous combustible wastes are burned and changed into gases and inert residues (i.e., ashes); also called combustion.

Incinerator   A facility for burning waste material under controlled conditions.

Industrial Wastes   The wastes generally discarded from industrial operations or derived from manufacturing processes.

Infiltration   In hydrology, a term describing the penetration of water from precipitation into the ground.

Influent   The liquid that flows into a treatment plant or during some other process.

Injection Well   A well into which liquid hazardous waste is pumped for disposal deep underground.

Inorganic   A mineral substance, usually without carbon.

Institutional Waste   The waste material originating in school, hospital, prison, and other public building.

Integrated Solid Waste Management   A practice of using several alternative waste management techniques to manage and dispose of specific components in the municipal solid waste stream.

Interest   A specified cost of borrowing money.

Ion   An electrically charged fragment of an atom or molecule.

Ion Exchange   The reversible exchange of ions of the same charge sign between a solution and an insoluble solid in contact with it.

Joint and Several Liability   A legal provision that permits cost recovery by the U.S. Environmental Protection Agency for a cleanup from the parties responsible for the waste.

Kinetic Energy   The energy inherent in motion or movement.

Lagoon   A pond used for settling or biological stabilization of wastewater or for storage of hazardous waste.

Landfill   An engineered method of solid waste disposal by compaction and then placement of a cover made from impermeable material either on or in the land.

Landfill Gas (LFG)   Methane and other gases produced by decomposition of organic portions in a landfill.

Large-Quantity Generators (LQG)   Facilities that generate more than 1000 kg of hazardous waste or more than 1 kg of acutely hazardous waste per calendar month.

Leachate   A highly contaminated liquid that comes out of landfills with high concentrations of organic and inorganic materials.

Lease   A contract by which one party (e.g., a bank or company) gives another party (e.g., an individual, group of people, or company) the use or possession of machinery, equipment, or a building for a specified period of time and for fixed payments.

Less Developed Country (LDC)   Typically, a country with a low gross national product, high population growth, and low industrialization.

Lethal Dose$_{50}$ (LD$_{50}$)   The amount of a chemical that will kill 50 percent of test animals.

Life Cycle   The various stages of life.

Liner   The membranes that are placed under or on the sides of hazardous waste disposal and chemical storage areas to prevent soil and groundwater contamination.

Listed Wastes   A list of various hazardous wastes included in 40 CFR 261.

Litter   The portion of solid waste that is generated and carelessly discarded by consumers.

Magnetic Separation   A device that uses magnets to separate ferrous materials from mixed waste material in municipal solid waste.

Mandatory Recycling   Programs that, by law, require consumers to separate trash so that some or all recyclable materials are not burned or landfilled.

Manifest System   A procedure mandated by the federal government to monitor or track hazardous waste material from "cradle-to-grave."

Manning's Formula   An equation used for analyzing and designing gravity-flow storm or sanitary sewer pipelines.

Manometer   A device to measure pressure by using the height of a column of liquid.

Manual Separation   The separation of recyclable or compostable materials from waste by handsorting.

Mass Burn   The incineration of raw or unprocessed municipal solid waste in a controlled system.

Mass Density   The ratio of the mass to the volume of a material.

Materials Balance   An accounting of the weights of the materials entering and leaving a facility or processing unit.

Materials Recovery Facility (MRF)   A facility used for further separation and processing of wastes that have been separated at the source.

Mean   The sum of observations divided by the number of observations; also called the average.

Mechanical Separation   The separation of waste into various components by using mechanical means.

Median   The value of the middle datum in a series arranged in increasing or decreasing order.

Medical Waste   Culture and stocks of infectious agents, pathological wastes, human and animal blood, and other body wastes.

Medium-Quantity Generator   A generator that produces between 100 to 1000 kg of hazardous waste per month.

Metabolism   The biochemical process by which living organisms produce energy to survive and reproduce.

Methane   A gas produced through anaerobic decomposition of organic material.

Microbe   A microscopic, single-celled, living organism.

Microorganisms   Microscopically small, living organisms that digest decomposable materials through metabolic activity.

Mode   The observation with the highest frequency in a set of data.

Moisture Content   The weight that is loss (expressed as a percentage) when a sample of solid waste is dried to a constant weight at a temperature of 105°C.

Mole   The mass of a substance divided by its molecular weight.

Molecule   The smallest fragment of a compound that still has the same chemical properties.

Monitoring Well   A well used to sample and assess groundwater for pollution, quality, and aquifer characteristics.

Mulch   Ground or mixed yard wastes or any other organic or inorganic material placed around vegetation to prevent evaporation of moisture and freezing of roots.

Municipal Solid Waste (MSW)   All the waste generated from residential households, commercial and business establishments, institutional facilities, construction and demolition activities, and municipal services.

Mutagenic Chemical   A chemical causing harmful health effects in the next generation.

National Environmental Policy Act (NEPA)   The legislation passed by Congress that established a national policy to "maintain conditions under which humans and nature can exist in productive harmony, and fulfill the social, economic, and other requirements of present and future generations of Americans."

National Pollution Discharge Elimination System (NPDES)   Requires a permit before discharges can be made to the nation's waters.

National Priority List (NPL)   A list of U.S. hazardous waste sites for the Superfund cleanup.

National Recovery Technologies (NRT)   A company developing processes to recover various materials from solid waste.

Nitrification   The conversion of ammonia into nitrates by oxidation through bacterial action.

Nitrogen Cycle   The biogeochemical processes that take nitrogen from the atmosphere, place it into various organic chemical forms, and then return it to the atmosphere.

Nomograph   A chart used to solve equations graphically.

Nonaqueous Phase Liquid   Liquids that do not mix with water.

Nonbiodegradable   A substance that cannot be broken down by biological organisms; for example, plastic, metals, and many chemicals.

Nonferrous Metals   Metals that contain no iron; for example, aluminum, brass, copper, and lead.

Nonpoint Source Pollution   The pollution from general runoff of sediments, pesticides, fertilizer, and other contaminants from urban and rural areas.

Nonrenewable Resources   Resources available in a fixed amount (e.g., fossil fuel, minerals) and not replenished by nature.

Normal Distribution   The frequency data plot in the shape of a bell curve.

Nutrients   The minerals that are essential to life.

Occupational Safety and Health Administration (OSHA)   The federal agency that enforces regulations of the Occupational Safety and Health Act (1970), which deals with protection of employees in the workplace.

Off-Route Time   The time spent by solid-waste collectors on nonproductive activities (from the aspect of the overall collection operation).

On-Site Disposal The disposal of waste matter at the location where it was generated.

On-Site Handling, Storage, and Processing The activities associated with handling, storage, and processing of solid waste at the source.

Open Dumps Solid waste disposal facilities that fail to comply with the Resource Conservation and Recovery Act, Subtitle D criteria.

Organic Compounds Compounds that contain carbon and one other element.

Osmosis The transport of a solvent through a semipermeable membrane separating two solutions of different concentration.

Oxidation Reduction Oxidation is the addition of oxygen or removal of electrons; reduction is the removal of oxygen or addition of electrons.

Particulate Matter (PM) Tiny pieces of matter resulting from the combustion process that can contaminate the environment and have harmful effects on public health.

Parts per Million (ppm) The number of units of one substance that are present in a million units of another.

Pathogen An organism capable of causing disease.

Perched Water Table The upper surface of a body of water perched on an aquiclude that lies above the main water table.

Percolation The flow of water through the pore spaces of soil because of gravity.

Permeable Rock or soil having pores or openings that permit fluids to pass.

Pesticide A chemical used to kill pests.

pH A measure of the hydrogen ion concentration, ranging from 0 (most acidic) to 14 (most basic).

Photosynthesis The natural process by which green plants convert carbon dioxide, nutrients, water, and sunlight energy into food substances.

Pollutant Any substance that may degrade the environmental quality.

Pollution The process of contaminating the air, water, or soil with material that will degrade the environmental quality.

Polychlorinated Biphenyls (PCBs) A group of manufactured chemicals made of carbon, hydrogen, and chlorine.

Polyethylene Terephthalate (PET) A high-density plastic.

Porosity The percentage of voids (i.e., pore spaces) in a unit volume of soil or rock.

Postclosure The period during which owners and operators of closed solid or hazardous waste disposal units conduct monitoring and maintenance to preserve the integrity of the disposal facility and thus protect the environment.

Potency Factor The ratio of incremental lifetime cancer risk to chronic daily intake of a chemical.

Potential Energy The ability to do work that is stored in some chemical or physical state.

Power The rate at which work is done.

Predator–Prey Relationship The predator is the animal feeding on the prey; thus, a feeding relationship exists between the two.

Present Worth The current value of something in terms of dollars.

Pressure A force at a point divided by the area on which the force is acting.

Pressure Head The actual or equivalent height of water above a selected point.

Pretreatment On-site treatment usually done to reduce volume or hazardous characteristics before discharge into municipal sewers.

Primary Producers The organisms that use sunlight energy to make their organic constituents from inorganic compounds.

Prokaryote An organism lacking a distinct nucleus.

Prokaryotic Simple organisms (e.g., bacteria).

Protective Equipment The clothing and other protective devices used by workers for their own safety.

Protozoa Microscopic, single-celled animals capable of movement.

Pump-and-Treat Remediation A process of treating contaminated groundwater by pumping it to the surface for removal of contaminants.

Putrescible Waste subject to biological and chemical decay.

Pyrolysis   A process of breaking down burnable waste by combustion in the absence of oxygen.

Qualitative   Refers to measurements involving purity.

Quantitative   Refers to measurements involving numbers.

Radical   An electrically charged group of atoms that act as a unit in chemical reactions.

Radius of Influence   The horizontal distance from a well to the area where the groundwater table elevation is not affected by pumping.

Rainfall Intensity   The rate of rainfall, expressed as inches or millimeters per hour.

Rational Method   A procedure for estimating peak stormwater runoff rates.

Reactive Waste   Waste material that is explosive, flammable, or highly corrosive.

Reclamation   The restoration of a material to a useful or a better state.

Recycling   The process by which materials otherwise destined for disposal are collected, reprocessed or remanufactured, and reused.

Reduction   The loss of oxygen from and/or the addition of electrons to a compound through a chemical or biochemical reaction.

Refuse   A term often used interchangeably with the phrase solid waste.

Refuse-Derived Fuel (RDF)   The waste material remaining after certain recyclable and noncombustible materials have been removed from municipal solid waste and burned for energy.

Remediation   The process of cleaning a contaminated site, which may include water, soil, or rock.

Residue   The materials remaining after processing, incineration composting, or recycling have been completed.

Resource Conservation and Recovery Act (RCRA)   The cornerstone of legislation passed by Congress to control hazardous and solid wastes from their creation to ultimate disposal.

Resource Recovery   A general term used to describe the extraction and use of materials and energy from waste stream.

Respiration   The process by which organic material is oxidized by living organisms to provide the energy needed for growth and reproduction.

Retention Basin   A small reservoir or pond designed to retain runoff and prevent erosion.

Return Period   The average number of years between storms of specific intensities and durations; also called recurrence interval.

Reuse   The use of a product or waste material more than once.

Risk   The probability of harm or loss; for hazardous waste it may be defined as the chance of encountering the potential adverse effect of human or environmental exposure to a hazard.

Rubbish   A general term used to describe the dry, nonbiodegradable portion of solid waste.

Runoff   The water from rain or snowmelt that flows overland into lakes, rivers, and streams.

Sanitary Landfill   An engineered method for disposing of solid wastes on land that protects human health and the environment.

Saturated Zone   A subsurface zone in which all interstices are filled with water under pressure greater than that of the atmosphere.

Scrap   Discarded or rejected industrial material often suitable for recycling.

Screening   An operation that separates mixed materials of different sizes into two or more size fractions.

Scrubber   An antipollution device that uses a liquid or slurry spray to remove acid gases and particulates from flue gases.

Secondary Consumer   An organism that feeds mostly on animals that feed on plants.

Sediment   Solid matter that has settled down from a state of suspension in a liquid.

Sedimentation   The separation of solids from liquids by gravity.

Separation   The process of sorting wastes into groups of similar materials, such as paper products, metals, glass, and so on, and also of further sorting of materials into more specific categories, such as clear glass and dark glass.

Shredding   The mechanical operations used to reduce the size of solid waste.

Silt   A particle finer than fine sand but coarser than clay.

Size Reduction   The mechanical conversion of solid waste components into small pieces; other methods to reduce size include shredding and grinding.

Sludge   A semiliquid residue remaining from the treatment of municipal and industrial water and wastewater.

Sludge Digestion   The biological stabilization of organic sludge to reduce its volume and prepare it for drying.

Slurry Wall   A structure built by injecting a mixture of clay and water below ground level to prevent contaminated water or other liquid from spreading.

Small-Quantity Generator   Individuals or companies generating less than 100 kg of nonacutely hazardous waste or less than 1 kg of acutely hazardous waste per month.

Soil Liner   A landfill liner composed of compacted soil and used to contain leachate.

Soil Survey Map   A map prepared by the U.S. Soil Conservation Service that shows the soil series.

Solar Energy   Energy derived from the sun.

Solid Waste Management   The planning, design, construction, and operation of facilities to collect, transport, process, and dispose of solid waste.

Solid Wastes   A variety of solid materials (as well as certain liquids in containers) that are considered to be of no use by the owner and are discarded.

Source Reduction   Methods and processes such as good operating practices, technological or product changes, and reuse of materials that minimize the quantity or toxicity of waste being generated.

Source Separation   The segregation of specific materials at the point of generation for separate collection.

Special Waste   Refers to items that require special or separate handling (e.g., household hazardous wastes, bulky wastes, tires, used oil).

Species   The organisms forming a population or group with specific characteristics.

Specific Gravity   The ratio of the weight of a given volume of a substance to the weight of an equal volume of water.

Stack Emissions   The air emissions from a combustion facility stack.

Standard Deviation   A measure of scatter or dispersion of data.

Standard International (SI) Units   A system of measurement based on meters, kilograms, seconds, and degrees-Kelvin (°K).

Static Level   The elevation of the water surface in a well that is not being pumped.

Stationery Container System   Collection systems in which containers for the storage of wastes are located at the point of waste generation (except when they are taken to the collection vehicle).

Steady Flow   Movement of liquid that is constant with time.

Stoichiometry   The process of balancing chemical equations.

Storage   Holding waste for a temporary period, after which it is taken and treated, disposed of, or stored elsewhere.

Subtitle C   The hazardous waste section of the Resource Conservation and Recovery Act.

Subtitle D   The solid, nonhazardous waste section of the Resource Conservation and Recovery Act.

Sulfur Dioxide (SO_2)   A major air pollutant, this gas is formed as a result of burning sulfur-rich fuel.

Superfund   The common name for the Comprehensive Environmental Response, Compensation, and Liability Act to clean up abandoned or inactive hazardous waste sites.

Surety Bond   A guarantee that a surety company will cover the financial responsibilities of the owner or operator of a waste treatment, storage, or disposal facility.

Surface Impoundments   A natural topographic depression, manmade excavation, or diked area formed primarily of earthen materials that is used to treat, store, or dispose of hazardous waste.

Surface Water   The water in streams, lakes, rivers, and ponds or depressions.

Tanks   The stationery devices used to store or treat hazardous waste.

TDS Facility   Any facility pertaining to the treatment, disposal, or storage of hazardous waste.

Throughput Capacity The amount of raw material put through processing or finishing operations in a specific period of time.

Tipping Fee A fee, usually stated in dollars per ton, for the unloading or dumping of waste at a landfill, transfer station, or waste-to-energy facility; also called a disposal or service fee.

Tipping Floor The unloading area for waste delivered to a transfer station, materials recovery facility, or waste combustor.

Toxic Substance A chemical or mixture that may present an unreasonable risk to public health and the environment.

Toxic Substances Control Act (TSCA) The legislation passed by Congress that controls the manufacture and sale of certain chemical substances.

Toxic Waste A waste with a potential for killing, injuring, or damaging the environment.

Toxicity Characteristic The characteristic identifying wastes that are likely to leach dangerous concentrations of toxic chemicals into groundwater.

Transfer The process of transferring wastes from the collection vehicle to a larger transport vehicle, railroad car, or barge.

Transfer Station A permanent facility where waste is taken from smaller collection vehicles for transport by truck trailers, railroad cars, or barges.

Transpiration The process by which the water absorbed by plants, usually through the roots, is dissipated into the atmosphere.

Transporter Any person engaged in the off-site transportation of hazardous waste, used oil, universal waste, or medical waste.

Trash Any material considered to be useless, unnecessary, or offensive that is usually thrown away.

Treatment The processes designed to reduce the volume and/or toxicity of contaminants.

Trickling Filter A biological treatment unit where wastewater flows over a bed of rocks.

Trommel Screen A rotary drum–type classifier for sorting solid waste on the basis of size and density.

Trust Fund A mechanism by which a facility can set aside money to cover financial assurance or responsibility requirements.

Uncontrolled Site Generally, an abandoned, hazardous waste site.

Underground Injection A method for disposal of hazardous waste by pumping it through deep wells into confined, porous aquifers.

Underground Storage Tank A tank that contains an accumulation of regulated substances and that has at least 10 percent of its combined volume underground.

Uniform Flow Flow at a constant surface slope and cross-sectional area.

Vacuum Filter A mechanical device used for sludge dewatering.

Vadose Zone The zone between the ground surface and the permanent groundwater.

Vapor Pressure The pressure exerted at any temperature by a vapor in equilibrium with its liquid or its solid phase.

Velocity The distance traveled per unit of time.

Virgin Material Any basic material used for industrial processes that has not previously been used.

Viscosity The property of a substance to offer internal resistance to flow.

Volatile A substance that is easily converted into vapor at lower temperatures.

Volatile Organic Compound (VOC) Any organic compound that readily evaporates into the atmosphere and causes smog.

Volume Reduction The processing of waste to decrease its volume (by compaction).

Waste Exchange The process of matching manufacturers so that one manufacturer's waste may be another manufacturer's useful material.

Waste Generation The act or process of generating solid waste.

Waste Minimization The reduction of waste at the source.

Waste Stream The waste output from a location or facility.

Waste Transformation The transformation of waste materials from one phase to another (e.g., from solid to gas).

Water Table The static water surface in a well.

Watershed   The land area that contributes runoff to a body of surface water; also called the drainage basin.

Watt   A unit of power.

Weigh Station   A station with platform scales of sufficient size to accommodate the largest vehicle expected to use the disposal or recycling facility.

White Goods   Large items in solid waste, such as broken appliances (e.g., stoves, dishwashers, refrigerators, clothes washers and dryers).

Windrow   A long and narrow pile of organic waste in an open field–composting facility.

Yard Waste   Leaves, grass clippings, prunings, stumps, brush, and other discarded natural organic matter.

Appendix F   For Further Information

Worldwatch Institute
1776 Massachusetts Avenue NW
Washington, D.C. 20036
(202) 452-1999

GOVERNMENT/PUBLIC ORGANIZATIONS

Association of State and Territorial Solid Waste Management Officials
444 North Capitol Street, Suite 388
Washington, D.C. 20001
(202) 624-5828

Solid Waste Information Clearinghouse
P.O. Box 7219
8750 Georgia Avenue, Suite 140
Silver Spring, MD 20910
(301) 67-SWICH

U.S. Environmental Protection Agency
Municipal and Industrial Solid Waste Division
401 M Street SW
Washington, D.C. 20460
(202) 382-6261

U.S. EPA, Region 5 (IL, IN, MI, MN, OH, WI)
RCRA Permitting Branch
77 Jackson Boulevard (HRP-8J)
Chicago, IL 60604
(312) 886-7452

MAGAZINES AND PERIODICALS

BioCycle Magazine
The JG Press, Inc.
419 State Avenue
Emmaus, PA 18049
(717) 967-4135

Bottle/Can Recycling Update
Resource Recycling, Inc.
P.O. Box 10540
Portland, OR 97210
(503) 227-1319

Chemical Engineering Progress
345 East 47th Street
New York, NY 10017
(212) 705-7496

Critical Reviews in Environmental Control
CRC Press
2000 Corporate Boulevard NW
Boca Raton, FL 33431
(800) 272-7737

Environmental Science and Technology
American Chemical Society
155 16th Street NW
Washington, D.C. 20036
(800) 333-9511

Journal of Environmental Engineering
American Society of Civil Engineers
345 East 47th Street
New York, NY 10017-2398
(212) 705-7496

Journal of Hazardous Materials
Elsevier Science B.V.
655 Avenue of America
New York, NY 10010
(212) 633-3750

Materials Recycling Markets
P.O. Box 577
Ogdensburg, NY 13669
(800) 267-0707

MSW Management
Forester Communications
216 East Gutierrez
Santa Barbara, CA 93101
(805) 899-3355

Organic Gardening Magazine
The JG Press, Inc.
Emmaus, PA 18098
(610) 967-5171

Recycling Times
Environmental Industry Associates
4301 Connecticut Avenue, Suite 300
Washington, D.C. 20008
(800) 424-2869

Recycling Today
GIE, Inc., Publishers
4012 Bridge Avenue
Cleveland, OH 44113-3320
(216) 961-4130

Resource Recovery and Conservation
Elsevier Science
655 Avenue of Americas
New York, NY 10010
(212) 633-3750

Resource Recovery Report
5313 38th Street NW
Washington, D.C. 20015
(202) 362-6034

Resource Recycling
P.O. Box 10540
Portland, OR 97210-0540
(503) 227-1319

Scrap Processing and Recycling
Institute of Scrap Recycling Industries
1325 G Street NW, Suite 1000
Washington, D.C. 20005-3104
(202) 466-4050

Solid Waste Report
Business Publishers, Inc.
951 Pershing Drive
Silver Spring, MD 20910
(301) 587-6300

Solid Waste Technologies,
HCI Publications
410 Archibald Street
Kansas City, MO 64111-3046
(816) 931-1311

Waste Age
4301 Connecticut Avenue NW, Suite 300
Washington, D.C. 20008
(202) 244-4700

Waste Management
Elsevier Science, Inc.
660 White Plains Road
Tarrytown, NY 10591-5153
(914) 333-2400

Waste Management and Research
Academic Press
24-28 Oval Road
London NW1 7DX
United Kingdom

Water, Air, and Soil Pollution Journal
Kluwer Academic Publishers Group
P.O. Box 358
Accord Station
Hingham, MA 02018-0358
(617) 871-6600

World Wastes
Argus Communication
6255 Barfield Road
Atlanta, GA 30328
(404) 955-2500

Index

A

abiotic, 2
absolute pressure, 59
absolute temperature, 97
absorption, leachate treatment
   process, 337, 353–354
acceptable risk, 34
access roads, 291
accidents, 398–399, 407–412,
   413, 471
acenaphthene, 437
acenaphthylene, 437
acetic acid, 91, 468, 477
acetic anhydride, 468
acetone, 335, 437, 468, 469, 471
acetonitrile, 468
acid gas, 168, 245, 378
acid rain, 120
acids, 95
   basic solutions *versus,* 95
   corrosivity, 396
   emission limit, municipal
      waste combustors, 233
   hazardous waste, 398, 430
   leachate treatment, 340,
      355–356
   organic decomposition, 8
   reaction levels, 94
   recycling, 273
   site remediation, 477, 479
acreage
   composting, 253, 254
   landfills, 291–293
   transfer stations, 192
acrylic acid, 150, 468
acrylonitrile, 335, 468
adhesives, 150
Administrative Procedures Act
   of 1946, 116
adsorption, leachate treatment
   process, 344

aerators, leachate treatment
      process, 350–352
aerial maps, 290
aerobic systems
   decomposition, 6–8
   leachate treatment process,
      336
agriculture
   contaminants, 438
   hazardous waste exclusion,
      394
   waste, 134–135
air
   combustion, 242–243
   site reclamation, 439, 448
air classification, 225
air knife, 226, 229
air pollution, 120, 168
   burning regulations, 141
   control facilities, 110
   emission limits on municipal
      waste combustors, 233
   hazardous waste storage,
      429–430
   incinerators, 244–245
   landfill regulations and, 363
   recycling and, 276
   site reclamation, 442, 451
   solid waste management,
      129–130
   ultimate analysis, 160
airports, 292
air stripping
   hazardous waste treatment,
      424
   leachate treatment, 337, 344,
      350–352
alcohol, 92, 95
   chemistry, 95
   linear, 468
   site remediation, 477

aldehydes
   leachate treatment, 356
   site remediation, 477
aldrin, 437
algae, 103, 106
aliphatic compounds, 95
alkyl halides, saturated, 469
alleys, 173, 178, 179
aluminum, 114, 438
   cans, 259
   chemistry, 90
   leachate treatment, 356
   packaging, 138
   recovery, 227–228, 229–230
   recycling, 267, 268, 270, 275,
      280
   specific weight, 154
American engineering system
   hydraulics, 58–81
   rainfall intensity, 87
   units of measure, 43–47
American Society of Testing
      Materials (ASTM), 232,
      309
ammonia, 3, 5, 128
   landfill gas, 387
   leachate treatment, 337, 342
   site remediation, 473
ammonium hydroxide, 355
anaerobic systems, 6–8, 336
analysis
   benefit/cost, 26–27
   cost indexing, 25–26
   economic, 22–25
   environmental impact state-
      ments, 35–39
   organic compound, landfills,
      335
   risk, 27–35
   technical, 21, 50–57
angle of repose, 110

anilene, 150
animals
    agricultural waste, 134–135
    dead, 133
    organic compounds, 95
anthracene, 437
antifreeze, 150
antimony, 401, 438
appliances, 133, 173
    component of municipal solid
        waste, 146
    disposal, 146, 321
    particulate control devices,
        247
    recycling, 270
    repair shops, 148
approximations, 49–50
apron conveyor. *See* flight con-
    veyor
aquifers, 83. *See also* ground-
    water
    hydrogeological properties,
        295
    landfill siting limitations, 292
    Safe Drinking Water Act of
        1974, 431–432
area, 489
area method, landfilling,
    298–300
aroclor, 437
aromatic compounds, 95, 244,
    438
    site remediation, 477
arsenic, 32, 128, 397, 401, 438
    chemistry, 90
    leachate treatment, 359
    site remediation, 478
arsenic oxide, 150
asbestos
    disposal, 321
    priority toxic pollutants, 401
asbestos cement pipes, pres-
    sured flow through, 72
ash
    air pollution, 244
    contaminated, 439
    disposal, 322
    domestic waste, composition
        of, 137
    energy content of solid waste,
        164–165
    hazardous waste, 400
    hazardous waste exclusion,
        394

hazardous waste treatment,
    426
incineration, 231, 250
moisture content, 156
particle size distribution, 157
as percentage of solid waste,
    135
pollution from waste manage-
    ment, 130
proximate analysis, 160
residential waste, 133
specific weight, 154
ultimate analysis, 161
waste composition, 139, 162,
    163
Asia, glass recycling in, 271
asphalt
    hazardous waste treatment,
        426
    recycling, 271, 272
asthma, 9
Athens, 111
Atlas Storage and Retrieval Sys-
    tem, 211–212
atmospheric pressure, 59, 92
atomic weight, 90, 91
atoms, 89, 91
automobiles, 133
auto-repair shops, 148
averages, 50–52

B

backhoes. *See* equipment
backyard collection, 179, 189
bacteria, 101–108
    biological properties, 167
    leachate treatment, 337, 343
baghouse collector, 249
bales, 192, 194, 220, 282
bank loans, 18
barium, 438
    chemistry, 90
    hazardous waste, 397
bases
    acid solutions *versus,* 95
    leachate treatment, 355–356
    reaction levels, 94
basins
    leachate treatment, 345–347
batch reactors, 100
battelle pyrolysis reactor, 256
batteries, 9, 148, 150, 151, 263,
    268
    corrosivity, 396

hazardous waste exclusion,
    394
particulate control devices,
    247
reusing, 262
site contamination, 436
beaches, 133
belt conveyors, 212–213, 214
belt filters, 461, 462
benefits, employee. *See* labor
    costs
bentonite, 426, 458
benzaldehyde, 468
benzene, 150, 335, 437, 438
    chemistry, 95, 96
    EPA target chemical, 416
    hazardous waste, 397
    oxygen demand, 468
    toxicity data, 32
benzo[a]anthracene, 437
benzo[a]pyrene, 32, 437
benzo[b]fluoranthene, 437
benzo[ghi]perylene, 437
benzo[k]fluoranthene, 437
Berkeley, California, municipal
    solid waste components,
    146
Bernoulli's theorem, 69
beryllium, 401, 438
beverage-container industry, 138
BHC, 437
bicycle paths, 305
biochemical oxygen demand
    (BOD), 332, 333, 468
biodegradation, 168, 321
biogeochemical cycles, 3–5
biological contaminants, 439
biology
    hazardous waste treatment,
        424, 425
    leachate treatment, 338
    municipal solid waste proper-
        ties, 167–169
    solid waste transformation,
        250–257
biomass, 239
biomedical waste, 322
bioreclamation, 467–476
biosorption, 338
biosphere, 2
bis(2-Chloroethox)methane, 437
bis(2-Chloroethyl)ether, 437
bis(2-Ethylhexyl)phthalate, 437
bleach, 396

blood, 439
boilers, 388
bonds, types of, 17–18
books, 139
boron, 90
Boston, Massachusetts, 111, 146
bottles, returnable deposit law, 141, 259
boxes, 139
bricks
    moisture content, 156
    specific weight, 154
brines, 431
bromine, 90
bromochloromethane, 437
bromodichloromethane, 335, 437
bromoform, 437
4-Bromophenyl phenyl ether, 437
bronchitis, 9
Brown and Ferris Industrial and MSW landfills, organic compound analysis, 335
brownfields, 124–125
bubonic plague, 127
bucket conveyor, 213
buffer zones, 291, 301, 323
bulky waste, 133, 137, 173
bulldozers, 449
burning
    contained. *See* combustion; incinerators
    open, 130, 141
butane, 468, 469
butanol, 471
2-butanone, 437
butylene, 468

C

cadmium, 150, 438
    chemistry, 90
    composting, 129
    hazardous waste, 32, 397, 401, 416
    leachate treatment, 337, 359
    recycling, 276
    site contamination, 436
    site remediation, 478
calcium, 2, 438
    chemistry, 90
    site remediation, 474
calcium carbonate, 93
calcium hydroxide, 355

calcium hypochlorite, 356
Calcutta, India, disposal rates, 140
California Air Resources Board, 245
cancer, 9
cans
    garbage, 171
    recovery, 219
    recycling, 259, 280
    tin, 154, 156
capacitors, electrical, 401
capital
    materials recovery facility (MRFs), 277–278
    present worth calculations, 22–25
capping, hazardous waste site reclamation, 457–459
carbohydrates, 167
carbolic acid, 468
carbon, 2
    aerobic cycle, 4
    biological properties, 167
    chemistry, 89, 90
    composting, 252
    cycle, 4
    energy content, 166
    hazardous waste treatment, 427
    leachate, 333
    organic *vs.* inorganic compounds, 89–90, 95
    proximate analysis, 160
    solid waste analysis, 163, 164
    stoichiometric combustion, 231
    ultimate analysis, 160, 161
carbon adsorption
    granular-media, leachate treatment process, 348
    hazardous waste treatment, 424
    leachate treatment process, 354, 359–361
carbonates
    leachate treatment, 338
    site remediation, 478
carbon dioxide, 3
    biological properties, 167
    hazardous waste gas, 452
    landfill gas, 378–380, 387
carbon disulfide, 335, 437
carbonic acid, 477

carbon monoxide, 130, 244
    air pollution, 120
    chemistry, 92
    landfill gas, 387
carbon tetrachloride, 335, 437
    hazardous waste, 32, 335, 397, 416
    oxygen demand, 468
    site contamination, 437
carboxylic acids, 477
carcinogens, 31–33
cardboard
    moisture content, 156
    particle size distribution, 157, 216
    recycling, 275, 280, 282
    solid waste composition, 137, 140, 162, 163
    specific weight, 154
    ultimate analysis, 161
Carson, Rachel, 12, 115
case studies
    Crown Equipment Corporation hazardous waste reduction, 422–423
    Hancock County, Ohio, sanitary landfill, 365–378
    Marion County Recycling Center, 281–286
    pollution fines, 125–126
    Seymour, Indiana, Superfund site, 480–485
    transfer station revival, 200–203
caustic soda, 355
cells, landfill, 301–302, 369
cellulose
    biological properties, 167
    energy content, 164
    pyrolytic reaction, 255
cement, 426
Centers for Disease Control, 410
centrifugal dewatering, 461
centrifugation, 93, 273
ceramics, 132–133, 169
chelation
    site remediation, 477, 478
chemical energy, 5
Chemical Engineering cost indexes, 25
Chemical Manufacturers Association, 414

chemical manufacturing
  hazardous waste, 398, 403
  process waste, 134
  Superfund Act, 123
chemical oxygen demand
    (COD), 468
  leachate, 332, 333
chemicals
  commercial and industrial
      waste, 133
  contaminated, 439
  environmental audit target,
      416
  Federal Insecticide, Fungi-
      cide, and Rodenticide
      Act, 121
  hazardous wastes, 398–399,
      400, 402
  hazardous waste treatment,
      424–425
  incinerators, 239
  leachate, 332
  production, 418
  public health, effect on,
      127–128
  toxicity data, 32
  Toxic Substances Control Act,
      122
chemistry, 89–101
  acid-base reactions, 94
  combustion, 230–250
  elements, basic information
      on, 90
  energy content, 164–167
  gases, 96–98
  ionization, 93
  leachate treatment, 344–361
  organic compounds, 95
  oxidation and reduction,
      94–95
  oxygen demand, 106–108
  partition coefficients, 95–96
  proximate analysis, 159–160
  site remediation, 476–479
  solid waste composition,
      159–167
  solutions, 92
  stoichiometry, 90–92
  suspensions, 92–93
  ultimate analysis, 160–164
China, solid waste composition
    and output, 137, 139,
    140
chlordane, 150, 397, 437

chloride, 492
chlorinated hydrocarbons
    priority toxic pollutants, 401
    reducing toxicity, 263
chlorinated phenols, 150
chlorination, 354
chlorine, 89, 90, 168, 230
chlorine dioxide, 356
chlorine gas, 356
chlorislane, 481
chlorobenzene, 150, 335, 437
    hazardous waste, 397
    oxygen demand, 468
chloroethane, 335, 437
chlorofluorocarbons, 394
chloroform, 335, 437
    hazardous waste, 397, 398,
        416
    oxygen demand, 468
    toxicity data, 32
chloromethane, 437
2-chloronaphthalene, 437
chlorophenol, 485
2-chlorophenol, 437
2-chlorophenyl phenyl ether,
    437
cholera, 127
Christmas trees, 147
chromium, 32, 438
    chemistry, 90
    hazardous waste, 397, 416
    hazardous waste exclusion,
        394
    leachate treatment, 337, 356,
        357, 359
    site remediation, 478
chrysene, 437
ciliata, 104
Cincinnati, Ohio
    municipal solid waste compo-
        nents, 146
    pollution case study, 125
cities
    brownfields, 124–126
    solid waste composition, 146
clarification
    hazardous waste treatment,
        425
    leachate treatment, 347–348
classification, hazardous materi-
        als, 406
clay
    capacities, 314
    characteristics, 10, 11

hazardous waste gas, 452
hydraulic grade line, 85–86
landfill
    cover, 316
    liners, 308
    siting limitations, 292
    soil, 300
    permeability, 84–85,
        294–295, 311
    surface runoff, 87, 312, 313
    water-holding capacity, 319
Clean Air Acts (1963, 1970 and
        1990 amendments), 120,
        429
cleaning products, 128
    corrosivity, 396
    hazardous waste, 150, 403
Clean Water Act (CWA) of
        1972, 116, 120–121,
        363
climate, leachate estimates, 317
Clinton, Bill, 268
closure, landfill, 361–362
clothing
    disposal rates, 146
    domestic waste, composition
        of, 137
    particle size distribution, 157
    reusing, 262
clover, 86
coagulation
    hazardous waste treatment,
        425
    leachate treatment, 344
coal, 135–136, 139
coal tar, 394
cobalt, 438
Code of Federal Regulations
        (CFR), 116
coffee cups, 257–258
coke, 394, 402
collection
    contamination controls,
        459–460
    household hazardous waste,
        151
    landfill sites, 289
    methods, 176–191
    recycling costs, 277
    routes, 187–191, 208
    solid wastes, 113
    systems and equipment,
        174–176
colloids, 93

combustion
  component of municipal solid
    waste, 143
  conditions, 230–231
  defined, 230
  federal regulations, 233
  heat value, 232–237
  incineration, 231–243
commercial waste, 133–134, 134
  collection routes, 187
  hazardous materials, 148
  storage, 173–174
common law, 115
Community Right-to-Know pro-
    visions, 415
compacted waste
  permeability of, 158–159
  specific weight, 153
compactors, 299, 300
  Hancock County, Ohio, sani-
    tary landfill, 370
  landfill operations, 319–321
composting, 114
  containers, 265–266
  microorganisms, 104
  municipal solid waste compo-
    nent, 143
  public health, 129
  rates, 109
  regional approaches, 15
  solid waste transformation,
    251–255
compounds
  organic *vs.* inorganic, 89–90
Comprehensive Environmental
    Response, Compensa-
    tion, and Liabilities Act
    of 1980 (CERCLA),
    123–124
  case study, 480–485
  described, 435–436
  gas collection and recovery
    systems, 451–460
  hazardous waste disposal, 424
  National Contingency Plan,
    440
  politics, 41
  remedial investigation/feasi-
    bility study, 440–451
  remediation treatment tech-
    nologies, 460–480
  site contamination, 436–440
Comprehensive Environmental
    Response, Compensa-

tion, and Liability Infor-
    mation System (CER-
    CLIS), 125
computer industry, 140
computer programs
  collection routes, 189
  leachate generation estimate,
    315, 317
concentration
  calculations, 44–46
  conversion factors, 490
  solutions and suspensions,
    93
concrete
  pipes, pressured flow through,
    72
  recycling, 271
construction, landfill
  costs, 19, 25–26
  technical analysis, 21
construction debris, 134, 142
  disposal, 322
  hazardous waste examples,
    403
  municipal solid waste compo-
    nent, 137
  recycling, 268, 271–272
consumers, 6
contactors, rotating biological,
    342–343
containers
  beverage, 259
  collection, 180, 187
  food, 262
  glass, 270
  grass, 274
  hauled systems, 175–176
  hazardous waste, 394,
    398–399
  hydrostatic pressure, 60–62
  public health, 127
  recycling, 265
  soft-drink, 269
  source waste storage,
    171–177
  stationary systems, 174–175
copper, 438
  chemistry, 90
  leachate treatment process,
    337, 343, 356, 359
  priority toxic pollutants, 401
  recycling, 270
  site remediation, 477
corn syrup, 468

corrosivity
  hazardous waste, 396
  shipping and packing classifi-
    cation, 406
cosmetics, 150, 403
costs
  Atlas Storage and Retrieval
    System, 211
  benefit/cost analysis, 26–27
  capital, 18–19
  capping hazardous waste, 458
  composting, 255
  flail mills, 218
  indexing, 25–26
  landfill, 326–327
  landfill liners, 308
  materials recovery facilities
    (MRFs), 277–278
  operation and maintenance,
    19–20
  particulate control devices,
    247
  recycling, 264
  site remediation, 443–446,
    446
  solid waste collection, 174,
    180–184
  transfer stations, 192–197
cover, landfill, 302–308, 316,
    326, 380
cracking, soil, 11–12, 295
cranes, 237, 449
creosote, 398, 436
cresol, 397
critical habitat areas, 292
Crown Equipment Corporation,
    422–423
curbside collection, 179
curves, bell-shaped, 53
cyanide, 438
  EPA target chemical, 416
  hazardous waste disposal, 430
  leachate treatment process,
    354, 357
  priority toxic pollutants, 401
  reactivity, 396
  site reclamation, 471
cyclonic dust collector, 249

D
Dade County, Florida, RDF fa-
    cility, 238
Dalton's law of partial pres-
    sures, 98

Darcy's law, 84, 158, 311
DDE, 437
DDED, 437
DDT, 32, 437
dead-end streets, 190
death, risk analysis, 28–30
decomposition
    aerobic and anaerobic, 6–8,
        102
    biological properties, 167
    food chain, 3
    liquid released by, 88
    stages, 334
degreasers, 150, 395, 438
demolition wastes, 134
    disposal, 322
    energy content, 165
    in municipal solid waste, 137
    recycling, 271–272
density
    calculations, 44
    conversion factors, 490
    waste component separation,
        224–226
depressions
    landfills, 311
    precipitation and, 82
designs
    deep well injection, 432–433
    groundwater wells, 454–456
    hazardous waste caps,
        458–459
    hazardous waste storage, 429
    incinerators, 243–250
    landfills
        closure, 364
        cover, 304–308
        regulations, 364
    leachate estimates, 317
    leachate treatments, 352–353,
        356, 357–358, 359
    multiple-chamber incinera-
        tors, 239–242
    trommels, 223
detention time, 48
detritus, 6
dewatering, 460–464
dextrose, 469
diapers, disposable *versus* cloth,
    257
diatomaceous earth, 426
dibenzofuran, 437
dibromochloromethane, 335, 437
1,2-dibromo-3-chloropropane,
    437

1,2-dibromoethane, 335, 437
1,1-dichlorethylene, 335
1,2-dichlorethylene, 335
1,2-dichlorobenzene, 335, 437
1,3-dichlorobenzene, 437
1,4-dichlorobenzene, 335, 397
3,3-dichlorobenzidene, 437
1,1-dichloroethane, 437
1,2-dichloroethane, 335, 437
1,4-dichloroethane, 437
1,1-dichloroethylene, 32
dichloromethane, 416
2,4-dichlorophenol, 437, 468
2,4-dichlorophenoxy acetic acid,
    397
1,2-dichloropropane, 335, 437
dieldrin, 32, 437
diesel engines, 245
diethylamine, 469
diethyl phthalate, 437
dikes, 459–460
dimensions, 43–47
dimethyl aniline, 471
1,4-dioxane, 468
dioxin, 130, 168, 244, 438
    control, 245
    hazardous waste, 398
    incinerators, 245
    sources, 245
    Times Beach, Missouri, 391
    toxicity data, 32
dirt, 169
    moisture content, 156
    screening out, 221–223
    solid waste composition, 162
    specific weight, 154
    trash, 132–133
discharge. *See* flow rate
disease. *See* public health
disk screens, 224
dispersion. *See* scatter
disposal facilities, public access,
    128–129, 372
dissolved air flotation, 337
distillation
    motor oil, 273
    suspensions, 92
distribution, partition coeffi-
    cients, 95–96
diversion, contamination con-
    trol, 459–460
drainage
    landfills, 307, 316
    leachate generation estimates,
        315

drop-off areas, 324–327
drums, hazardous waste storage,
    426, 427, 428
dry cleaners, 148, 401, 438
dumps, 111, 130
dumpsters, 173–174, 176
dust, 169
    fugitive, 451–452
    landfill operations, 323
    particle size distribution, 157
    shredders, 221
    trash, 132–133
dyes, 402
dysentery, 127

E
ecology
    aerobic and anaerobic decom-
        position, 6–8
    benefit/cost analysis, 26
    biogeochemical cycles, 3–5
    defined, 2
    energy, food chains, and me-
        tabolism, 2–3
    Hancock County, Ohio, land-
        fill, 373–374
    photosynthesis, respiration,
        and metabolism, 5–6
    stable ecosystems, 8
economics
    brownfields, 125
    changes in solid waste, 137
    collection methods, 178–184
    collection vehicle size,
        184–187
    landfill costs, 326–327
    materials recovery facilities
        (MRFs), 276–281
    metal recovery, 227–228
    recyclable market, 268,
        274–275
    transfer stations, 194–197
    waste minimization, 418–419
ecosystems, stable, 8
electricity, generation, 388–389
electronics, 133
    particulate control devices,
        247
electronics industry, site con-
    tamination, 436
electroplating
    hazardous waste, 398, 400
    leachate treatment process,
        355
    priority toxic pollutants, 400

electrostatic precipitators (ESPs), 247–250
elements, 89, 90
elevation, potential energy, 68
Emergency Planning and Community Right-to-Know Act of 1986, 124
emergency response
  hazardous waste pretransport regulations, 413
  hazardous waste transportation, 407–412
emissions
  limits, 244
  municipal waste combustors, 233
  particulate control devices, 247, 250
emphysema, 9
encephalitis, 127
endosulfan, 437
endosulfan sulfate, 437
endrin, 397, 437
endrin aldehyde, 437
endrin ketone, 437
energy, 2–3
  aluminum production, 270
  conservation of, 68–71, 98–100
  consumption and greenhouse gases, 4
  conversion factors, 490–491
  hydrologic cycle, 82
  municipal solid waste content, 164–167
  pumps, 78
  recycling, 276
  synthetic organic materials, 168
  waste-to-energy facilities, 110
engineering
  combustion, 230–250
  environmental, 103
  landfill costs, 326
  microorganism systems, 103–105
  recycling facility, 277
Engineering-News Record cost indexes, 25
engineers
  calculations, 43–57
  technical analysis, 21
England, solid waste output in, 139
environment. See also life cycle assessment

containers and, 259
environmental impact statements (EIS), 18, 35–39, 119–124, 130–131
  hazardous waste audits, 415–420
  hazardous waste excavation, 447–448
  landfill suitability, 289–291, 292
  laws and regulations, 115–126
  public health and, 127–131
  recycling and, 275–276
  site remediation and, 443–446
enzymatic degradation, 479
enzymes, 103–104
EPA. See U.S. Environmental Protection Agency (EPA)
epidemiology, 34–35
equalization, leachate treatment, 344
equations, balancing, 90–92
equipment
  collection, 172, 174–176
  composting, 255
  costs, 18, 19–20, 23–25, 25, 326
  Hancock County, Ohio, sanitary landfill, 374–375
  hazardous waste remediation, 447, 449
  landfill, 299–300, 326
  mass burn incinerator, 237–238
  materials recovery facility (MRF), 277
  recycling facility, 277
  scales, 206
  soil support, 12, 293
erosion, 5
errors, measurement, 48
estimates. See approximations
ethane, 468
ethers, 401
ethics, environmental, 39–41
ethyl acetate, 468
ethyl alcohol, 394
ethylbenzene, 335, 437, 438, 468
ethylenediamine, 469
ethylenediaminetetracetic acid, 468
ethylene dichloride, 468

ethylene glycol, 150, 468, 469
ethyl ether, 468
eukaryotic organisms, 101
Europe, packaging in, 263
evaporation, 82, 83
  leachate treatment, 86, 337, 344, 358
  solvents, 273
evapotranspiration, 83–84
  leachate, 86, 312, 315
  water balance method, 315–316, 318
excavation
  hazardous waste sites, 442, 444–451
  landfills, 326
explosions
  commercial and industrial waste, 133
  hazardous waste storage, 429
  methane, 130
  shredders, 220–221
explosives
  gas, Subtitle D (landfill) regulations, 363
  hazardous waste, 398, 402
  shipping and packing classification, 406

F
fabric. See textiles
fabric filters, 247–250
fatty acids, 340
fault areas, 292, 450
feasibility reports, 18
federal government. See regulations
Federal Insecticide, Fungicide, and Rodenticide Act (FIFRA) of 1947, 121
Federal Railroad Administration, 405
Federal Register, 116, 117
fees
  bulky waste, 173
  tip fee, 207
  tire disposal, 272
fences, landfill, 323, 371
fermentation, 95
ferrous metals, 114, 136
  market value, 227
  moisture content, 156
  particle size distribution, 216
  recovery, 229
  recycling, 267, 270

ferrous metals, (cont.)
  solid waste composition, 136, 162
  specific weight, 154
ferrous sulfate, 478
fertilizers, 135, 436
F hazardous wastes, 398
fibers, crude, 167
field capacity, solid waste, 157–158
film, camera, 422
filter press, 461–462, 464
filters, 93
  chemical spills, 484
  contaminated site treatment, 461–464
  hazardous waste treatment, 424, 459
  landfill cover, 306–307
  leachate treatment, 337, 339, 341, 344, 348–349, 349–350, 354
  solvents, 273
finances
  capital financing, 16–18
  closed landfills, 365
  costs, 18–20, 25–26
  economic analysis, 22–25
  Hancock County, Ohio, sanitary landfill, 375, 376–377
  interest tables, 487–488
  operating revenues, 16
financing, 291
fines, pollution, 125–126
fire, hazardous waste, 395–396
fire extinguishers, 401
firefighters, 407–412
fireplaces, 245
fires
  hazardous waste storage, 429
  shredders, 220–221
fish, 8
fixed-film hazardous waste treatment, 336
flail mills, 218–219
flea powder, 150
flies, 127, 128, 167
  biodegradables, 168
  garbage, 132
  landfills, 323
flight conveyor, 212
flocculation, 348, 358–359

flood plains
  landfill siting limitation, 292
  site reclamation, 450
flotation
  hazardous waste treatment, 425
  leachate treatment, 348
flow
  continuity of, 67–68
  energy, conservation of, 68–71
  equalization, leachate treatment process, 346
  in gravity-fed pipes, 75–78
  overview, 64–66
  in pressured pipes, 71–75
  rate calculations, 46–47, 66–67, 490
  water in landfills, 84
fluids
  concentrations, 44–46
  detention time, 48
  disposal, 431–433
  flow rate. See flow
  precipitation, 56
fluoranthene, 437
fluorene, 437
fluorides, 359
fluorine, 90
fluorine gas, 353–354
fly ash, 244, 426
foaming agents, 361
food, 8–9, 132, 168
  biological properties, 167
  commercial and industrial waste, 133
  composting, 250–255
  containers, 262
  disposal rates, 146
  domestic waste, composition of, 137
  energy content, 165
  garbage, 132
  moisture content, 156
  particle size distribution, 157
  as percentage of solid waste, 135
  priority toxic pollutants, 402
  recycling, 267
  solid waste composition, 136, 142, 162, 163
  specific weight, 154
  ultimate analysis, 161
  waste composition, 139

food chains, 2–3, 7
food grinders, 146
food plain maps, 290
formaldehyde, 469
fossil fuels, 402
  carbon cycle, 4
  hazardous waste exclusion, 394
foundation, landfill, 308
France, solid waste output, 139
Freeman Field Industrial Park cleanup, 480–485
fuels
  gas, 114
  oil, 436–437
  yard waste, 274
functions, cumulative, 53–54
fungicides, 148
  Federal Insecticide, Fungicide, and Rodenticide Act, 121
  priority toxic pollutants, 402
furans, 130, 245, 438
furnaces. See incinerators
furniture, 133, 173
  disposal, 321
  hazardous waste examples, 403
furniture polish, 148

G
garbage, 8–9, 132
  biological properties, 167
  commercial and industrial waste, 133
  as percentage of solid waste, 135
  solid waste definition, 110
garbage cans, 171
garbage grinders, 138, 139
garden tools, 270
garden waste. See yard waste
gas(es), 96–98
  air pollution controls, 451
  analyzers, 449
  bacterial growth, 102
  collection and recovery systems, 451–460
  Dalton's law of partial pressures, 98
  distillation, 92
  hazardous waste disposal, 431
  heating by, 135–136
  ideal gas law, 97

incineration, 231, 244–245, 387–388
landfill, 295, 326, 327, 335, 378–380, 381–383, 387–389
migration, 380–381, 383–386
migration controls, 444–445
odor, 322
permeability of compacted waste, 158–159
pressure, 63
remedial technology, 442
shipping and packing classification, 406
solid waste management, 129–130
solutions, 92
Subtitle D (landfill) regulations, 363
from synthetic organic materials, 168
gasification, 230
gas oil, 468
gasoline, 150
  oxygen demand, 468
  site reclamation, 471
  volatile organic compounds (VOC), 436–437
gastroenteritis, 127
gauge pressure, 59
Gauss function, 53
generators, 428
geology
  geological maps, 290
  soil and, 9–12
geomembranes, 305, 308, 326
geotechnical investigations, 18
geotextiles, 305, 306–307
Germany
  packaging, 263
  per-capita generation of municipal solid waste, 140
  petroleum product spill, 473
glasphalt, 270
glass, 114, 169
  disposal rates, 146
  domestic waste, composition of, 137
  energy content, 165
  hazardous waste treatment, 427
  moisture content, 156
  packaging, 138, 259

particle size distribution, 157, 216
recovery, 229
recycling, 267, 268, 270–271, 275, 280
screening out, 221–223
site remediation, 480
solid waste composition, 132–133, 136, 142, 162
wet, 226
glycerine, 468
golf courses, 305
governments, national. See regulations
governments, state and local. See also municipalities
Emergency Planning and Community Right-to-Know Act, 124
emergency response, 407–412
site remediation concerns, 446
grants, 120
grass
  containers, 274
  evapotranspiration, 86
  surface runoff potential, 87
grates, incinerator, 241
gravel
  hazardous waste gas, 452
  landfill siting limitations, 292
gravity
  field capacity, 157–158
  separation, hazardous waste treatment, 424
  specific, VOCs, 438
gravity thickening, 460
greenhouse gases, 4
grinders
  garbage, 138, 139
  wood, 284, 285
groundwater, 82–83
  bioremediation, 105
  contamination, controlling, 439, 454–456
  Hancock County, Ohio, landfill, 373–374
  hazardous waste storage, 429–430
  landfills
    costs, 326
    siting limitations, 292

leachate removal system, 329–332
monitoring, 364
pesticides, 121
pollution pathways, 33
remediation, 104, 442, 471, 475, 478

H
halogenated alkanes, 471
halogenated hydrocarbons, 477
halogenated organic compounds, 430
halogens, 332
Hamburg, Germany, disposal rates, 140
hammer mills, 217–220
Hancock County, Ohio, sanitary landfill, 365–378
haulers, 449
Hazardous and Solid Waste Amendments of 1984 (HSWA), 123
Hazardous Materials Transportation Act (HMTA), 405
hazardous waste
  brownfields, 124–126
  carbon adsorption process, 361
  characteristics, 391–404
  cleanup laws, 123–124
  defined, 392
  development of environmental regulations, 115
  disposal, 322, 430–434
  emergency response, 124, 407–412
  environmental audits, 415–420
  fines, 126
  gas removal, 96
  generation, 392–395
  generator responsibility, 393–395
  households, 150, 268
  incinerator pollution, 244–245, 250
  landfill class, 289
  leachate treatment, 332, 337–338, 343
  listed, 398–400
  microorganism treatments, 103
  minimization, 414–415, 418–423

hazardous waste (*cont.*)
  organic compounds, 95
  overview, 1
  priority toxic pollutants,
      400–403
  public health, effect on, 9,
      127–128
  risk analysis, 27–35
  safety in transport, 412–413
  scientific concepts, 89–108
  screening for, 237
  site remediation, 435–486
  solid waste composition, 137,
      148–151
  source reduction, 263–264
  storage, 427–430
  Subtitle D (landfill) regula-
      tions, 363
  testing, 397
  transportation, 405–413
  treatment, 424–427
Hazen-Williams equation,
    72–75, 80
HDPE. *See* high-density poly-
    ethylene
head loss, 71, 74
hearings, 116
heat
  contaminated site treatment,
      464–466
  plastic melting points, 269
  pyrolysis, 255–257
  site remediation, 479
heat value
  incineration, 232–237
  landfill gas, 387–389
  municipal solid waste, 256
heavy metals. *See* metals, heavy
HELP computer model, 315,
    317
hemicellulose, 167
hepatitis, 127
heptachlor, 437
  hazardous waste, 397
  toxicity data, 32
heptachlor epoxide, 437
heptane, 468
herbicides, 128, 148, 396, 402
heuristic routing, 189–191
hexachlorobenzene, 397, 437
hexachlorobutadiene, 437
hexachloro-1,3-butadiene, 397
hexachlorocyclopentadiene,
    437

hexachloroethane, 32, 150, 397,
    437
2-hexanone, 335, 437
1-hexene, 468
high-Btu gas, 389
high-density polyethylene
    (HDPE), 230, 269
  recycling, 275, 280
  separation, 224
highways, 292
hills, collection on, 190
hiring practices, 180
history, waste management,
    12–13
Hoff-Arrhenius, Van't, 104–105
Holland, composting in, 251
horizontal shaft mills, 218
horsepower. *See* power
hospitals, hazardous waste, 134,
    322, 396, 439
household(s), 132–133, 137
  chemical disposal rates, 140
  cleaners, 148, 150
  collection methods, 176–178
  contaminants, 438
  hazardous waste, 150, 266
  hazardous waste exclusion,
      394
  solid waste composition,
      132–133,
  144–146
  storage, 171–273
hydraulic grade line (HGL), 71,
    85–86
hydraulics
  flow, 64–78
  landfill cover, 308
  landfill layer, 308
  leachate estimates, 317
  permeability of compacted
      waste, 158–159
  pressure, 58–64
  pumps, 78–80
hydrazine, 468
hydrocarbons, 95, 150, 244
hydrochloric acid, 94
  hazardous waste, 398
  leachate treatment, 355–356
  site remediation, 477
hydrogen
  acids, 94
  biological properties, 2, 167
  chemistry, 89, 90
  energy content, 166

landfill gas, 387
solid waste analysis, 163, 164
stoichiometric combustion,
    231
ultimate analysis, 160, 161
hydrogen chloride (HCl), 130,
    244, 245
hydrogen cyanide, 479
hydrogen fluoride (HF), 245
hydrogen gas, 481
hydrogen peroxide, 356, 476
hydrogen sulfide, 3, 169
  landfill gas, 380, 387
  site remediation, 479
hydrology
  cycle, 82–84, 310–315
  Hancock County, Ohio, land-
      fill, 373–374
  landfill cycle, 84–88,
      295–297
hydrolysis, 479
hydrophilic, 96
hydrostatic pressure, 60
hydroxide, 478
hydroxyl group, 89

I

ignitable materials. *See* haz-
    ardous waste
impoundments, surface,
    429–430, 431
incineration/incinerators, 111,
    114, 231–243, 290
  air pollution, 160–161,
      244–245
  case study, 200
  coffee cups, 259
  combustion air, 242–243
  contaminated site treatment,
      465–466
  design guidelines and objec-
      tives, 243–250
  fluid bed combustors, 238–239
  hazardous waste treatment,
      424, 427
  landfill gases, 387–388
  mass burn, 237–238
  multiple-chamber, 239–242
  pollution from landfills, 130
  refuse-derived fuel, 238
  storage system, 209
  tires, 272
Incinerator Institute of America,
    239

Industrial Revolution, 111
industrial waste, 133–135
    collection routes, 187
    contaminants, 438
    dioxin production, 245
    environmental audits,
        415–420
    hazardous, 398–400, 403
        disposal, 431
        generators, responsibilities
            of, 393–395
        storage, 428
    industrial revenue bonds
        (IRBs), 18
    landfill class, 289
    priority toxic pollutants,
        400–402
    site contamination, 436
    water excluded from hazards
        list, 394
infiltration
    landfills, 311–312
    leachate generation estimates,
        315
    water, 82, 84–86
inks, 148
    hazardous waste, 398
    recycling, 276
    reuse, 422
inorganic compounds, 89–90,
    162, 438–439
insecticides, 128
    agricultural waste, 135
    Federal Insecticide, Fungi-
        cide, and Rodenticide
        Act, 121
    leachate, 334
    priority toxic pollutants, 401
    Silent Spring, effect of, 12
    toxicity, 396
insect repellent, 150
insects
    landfills, 302, 323
    public health, 127
In Situ treatment
    biodegradation, 471
    bioreclamation, 467–476
    chemical treatment, 476–479
    contaminated sites, 467–480
    physical methods, 479–480
    site remediation, 442, 443,
        444–445
institutional waste, 134
    site remediation, 446

solid waste component, 137
    storage, 173–174
instruments, pressure measure-
    ment, 63–64
interest tables, 487–488
in-vessel composting, 254
iodine, 2, 90
ion exchange, 337, 344,
    354–355
ionization, 93
iron, 2, 438
    chemistry, 90
    hazardous waste, 398
    leachate treatment, 332, 356,
        359
    oxidation, 94
    pipes, pressured flow through,
        72
    site reclamation, 470, 478
irrigation, 394
isobutane, 468
isobutyl alcohol, 468
isophorone, 437
isopropyl acetate, 468
isopropyl alcohol, 468
isubutylene, 468
Italy, solid waste generation,
    140

J
jacketing systems, hazardous
    waste treatment, 426
jams, shredder, 220–221
Japan, solid waste generation,
    140
jet fuel, 471
Journal of the Sanitary Engi-
    neering Division, ASCE,
    361

K
kerosene, 468, 469
ketones, 477
K hazardous wastes, 398

L
labels
    Federal Insecticide, Fungi-
        cide, and Rodenticide
        Act, 121
    hazardous waste, 407,
        408–410, 413, 428
    recycling, 264, 269
laboratories, 439
labor costs, 18, 278–279

labor unions, 189
lagoons. See also ponds, aerated
        stabilization
    air pollution controls, 451
    hazardous waste storage,
        429–430
    leachate treatment, 341–342
lakes, 83
    algae, 103
    landfill siting limitations, 292
lamps, 150
landfills, 114
    bacterial growth, 102
    case study, 365–378
    classification, 288–289
    closure and postclosure,
        361–362
    cover design, 304–308
    facilities, 323–327
    field capacity, 157–158
    gas
        composition, 387
        migration, 380–381
        production, 378–380,
            381–383
        recovery, 383–386
        utilization, 387–389
    hazardous waste disposal,
        414, 430–431
    hydrogeological properties,
        84–88, 295–297
    leachate
        control, 329–335
        generation rates, 315–316
        treatment, 336–361
    liners, 308–310
    microorganisms, 104
    moisture, 310–315
    operations, 316–323
    overview, 109–110
    performance regulations,
        362–365
    siting, 289–293
    soil properties, 293–295
    techniques, 298–304
    tip fee, 207
land prices, 18
landscaping. See yard waste
land surveys, 18
land use plans, 290
latex paint, 150, 268
leachate
    biochemical oxygen demand
        (BOD), 108

leachate (*cont.*)
    characteristics, 332–335
    collection system, 308,
        371–372
    control, 329
    costs, 326, 327
    cover design, 304
    environmental impact state-
        ments, 38–39
    field capacity, 157–158
    generation, 310, 315–316,
        317, 318
    hazardous waste, 400,
        430–431
    hydrogeological properties,
        295–297, 310–315
    landfill sites, 295–297
    odor, 322
    permeability of compacted
        waste, 158–159
    pollution from, 6, 130
    precipitation and, 88
    pumps, 78
    reducing, 264
    remedial technology, 442
    removal system, 329–332
    solutions, 92
    storage tanks, 63
    treatment, 336–361
lead, 128, 150, 438, 439
    air pollution, 120
    chemistry, 90
    composting, 129
    EPA target chemical, 416
    hazardous waste, 397, 398,
        401
    leachate treatment, 332, 337,
        356, 359
    recycling, 270, 276
    site remediation, 478
lead chloride, 244
leasing, 18
leather
    biological properties, 167
    crude fibers, 167
    disposal rates, 146
    energy content, 165
    hazardous waste exclusion, 394
    hazardous waste from manu-
        facturing, 403
    moisture content, 156
    recycling, 267
    solid waste composition, 136,
        162, 163

    ultimate analysis, 161
legal services, 20, 326
length, conversion factors, 489
Leopold, Aldo, 40
lice, 127
life cycle assessment (LCA),
    257–260
lifestyles, waste generation and,
    140
light bulbs, 263
lignin, 167, 168
lime
    hazardous waste treatment,
        426
    leachate treatment, 355
limestone
    landfill siting limitations, 292
    site remediation, 479
lindane, 397, 437
linear actuator, 65
liners, landfill, 308–310
    hazardous waste disposal,
        430–431
    landfill costs, 326
    site reclamation, 447–448,
        458–459
    soil, 330
lipids, 167
liquefied petroleum gas, 468
liquid injection-incinerator sys-
    tem, 466
liquids. *See also* leachates
    contaminated, 439
    density, separation by,
        224–226
    flow, 64–78
    hydrology, 82–88
    landfill cover design, 304
    oxidation and reduction reac-
        tions, 94–95
    partition coefficients, 95–96
    presssure, 58–64
    pumps, 78–80
    shipping and packing classifi-
        cation, 406
    solutions, 92
    suspensions, 92–93
liquors, pulping, 394
litter
    environmental impact state-
        ments, 37
    landfill, 322
live bottom pit system, 209–210
loaders, 300, 449

loam, 319
lobbying. *See* politics
Love Canal, 115, 391, 414, 435
low-density polyethylene
    (LDPE), 269
lumber, plastic, 269

M
macroencapsulation treatment,
    426
magazines, 139, 146
maggots, 8
magnesium, 2, 90, 438
magnetic separation, 226–229
maintenance
    Atlas Storage and Retrieval
        System, 211
    hazardous waste caps, 458
    landfill costs, 326
    materials recovery facilities
        (MRFs), 279–281
malaria, 127
manganese, 438
    chemistry, 90
    site remediation, 474
manganese oxides, 470
manifest, hazardous waste, 407,
    411
Manning's formula, 75–78
manufacturing
    contaminated ash, 439
    environmental audits,
        415–420
    hazardous waste, 398–402,
        403, 436
        generators, responsibilities
            of, 393–395
        reduction, 420–423
        storage, 428
    product changes, 420
maps, 12, 290
*Maryland Conservation Council
    v. Gilchrist,* 119–120
mass, conversion factors, 490
mass balance, 98–100
mass burn incinerators, 237–238
mass per unit volume, 44–45
mastigophora, 104
materials balance. *See* mass bal-
    ance
materials recovery facilities
    (MRFs)
    economic analysis, 276–281
    public access, 129

mattresses, 133
means, 50–51
measurements, 48–49
  conversion factors, 489
  units of, 43–47
media, contaminated, 439
medical waste, 134, 322, 396, 439
medicine, 103
medium-Btu gas, 388
MEK, 437
membranes
  landfill gas treatment, 388
  leachate treatment, 349–350, 354
  synthetic, 309–310
mercaptans, 356
mercury, 150, 244, 263, 438, 439
  chemistry, 90
  composting, 129
  EPA target chemical, 416
  hazardous waste, 397, 401
  leachate treatment, 332, 356, 359
  scrubbers, 245
  site contamination, 436
  site remediation, 478
mercury manometer, 63–64
metabolism
  ecosystems, 2–3
  photosynthesis, respiration and, 5–6
metal-finishing industry, 357
metals, 169
  commercial and industrial waste, 133
  disposal rates, 146
  emissions limits, 233
  energy content, 165
  ferrous, 114, 136
  hazardous waste, 403, 438
  hazardous waste exclusion, 394
  heavy, 161, 244, 438–439
    composting, 129
    incinerators, 244
    reducing toxicity, 263
    site contamination, 436
    site reclamation, 471
  incinerators, 244
  leachate treatment, 332, 337, 338, 343, 354, 355, 356, 358–359

magnetic separation, 226–229
moisture content, 156
nonferrous, 136
particle size distribution, 157, 216
particulate control devices, 247
priority toxic pollutants, 400, 401
recovery, 229
recycling, 267, 268, 270, 271, 273
screening out, 221–223
shredder, 220
site remediation, 478
solid waste composition, 137, 142, 162
specific weight, 154
technological changes, 421
toxicity, 396
trash, 132–133
ultimate analysis, 161
methane
  bacterial growth, 102
  biological properties, 167
  chemistry, 89, 95, 96
  Hancock County, Ohio, sanitary landfill, 372
  hazardous waste gas, 452
  landfill generation, 378–380, 382, 387
  leachate treatment, 343
  oxygen demand, 468
  solid waste management, 129–130
methanol, 96
methoxychlor, 397, 437
methyl alcohol, 468
methyl bromide, 335, 437, 468
methyl chloride, 335, 437, 468
methylene bromide, 335
methylene chloride, 32, 150, 335, 398, 437, 438, 471
methyl ethyl ketone (MEK), 335, 397, 416, 468
methyl iodide, 335
methyl isobutyl ketone, 469
  EPA target chemical, 416
  oxygen demand, 468
methyl methacrylate, 468
4-methyl-2-pentanone (MLK), 335, 437
Miami County solid waste incinerator, 200–203

microbiology, 101–108
  engineered systems, 103–105
  oxygen demand, 106–108
  permeability of compacted waste, 158–159
microfilm, 150
microorganisms, 103–105
  hazardous waste reclamation, 467–476
  hazardous waste treatment, 424
  solid waste transformation, 250–257
  synthetic organic materials, 168
microwave ovens, 137
migration, landfill gas, 380–381, 383–386
mining
  hazardous waste disposal, 432
  hazardous waste exclusion, 394
  phosphorous, 5
  recycling, 276
mixture rule, hazardous waste, 399–400
moisture
  air classification of solid waste, 225–226
  composting, 251
  decomposition, 334
  energy content of solid waste, 164–165
  field capacity, 157–158
  heat value, 232, 234
  landfills, 310–315
  leachate formation, 329
  proximate analysis, 160
  soil retention, 493–495
  soil swelling and cracking, 11–12
  solid waste analysis, 148, 155–156, 163
  specific weight, 154
  ultimate analysis, 161
monitoring
  groundwater, 364
  landfill gas, 372, 386, 453
  municipal waste combustors, 233
monochlorodifluoromethane, 468
monocyclic aromatics, 402
monoethanolamine, 468

mosquitoes, 127, 128, 323
motor oil, 9, 150, 151, 263, 268
  hazardous waste, 399
  hazardous waste exclusion, 394
  recycling, 272–273
Muddy Creek wastewater plant, 125
mulch, 273
multifamily residences, 173–174
municipalities
  benefit/cost analysis, 26–27
  costs, 18–20, 25–26
  decision-making, basis for, 14–16
  economic analysis, 22–25
  environmental ethics, 39–41
  environmental impact statements, 35–39
  financing methods, 16–18
  responsibilities, 1
  risk analysis, 27–35
  solid waste generation, 133–134, 142
  technical analysis, 21
  waste management decisions, 14–42

N

nails, removing, 285
naphthalene, 437, 468
National Academy of Science, 31
National Association of Diaper Services, 257
National Contingency Plan. See Comprehensive Environmental Response, Compensation, and Liabilities Act (CERCLA)
National Environmental Policy Act (NEPA), 13, 35–39, 119–120, 130–131
National Pollutant Discharge Elimination System, 338, 363
National Priorities List (NPL), 436
National Resources Technology Group, 228, 229–230
National Response Center, 410
nervous system illnesses, 9

neutralization
  hazardous waste treatment, 424, 425
  leachate treatment, 337, 344, 355–356
  site remediation, 479
New Jersey recycling laws, 141, 266
newspapers, 139
  newsprint production, 168
  recycling, 275, 280
  solid waste composition, 146
newtons. See SI (Systeme International)
New York, 111, 140
nickel, 438
  chemistry, 90
  EPA target chemical, 416
  hazardous waste, 32, 401
  leachate treatment, 337, 343, 356, 359
  site remediation, 477
nitrate, 3, 5
nitric acid
  leachate treatment, 355–356
  site remediation, 477
nitrification, 336, 338
nitrobenzene, 150, 437
  hazardous waste, 397
  oxygen demand, 468
nitrogen
  aerobic cycle, 4
  algae, 103
  biological properties, 2, 167
  chemistry, 90
  composting, 252
  cycle, 5
  energy content, 166
  landfill gas, 387
  leachate treatment, 333, 336
  site remediation, 473
  solid waste analysis, 163, 164
  ultimate analysis, 161
nitrogen oxides, 130, 244, 245
  emissions limit, 233
nitrosamines, 402
Nixon, Richard M., 13, 116, 119, 120
noise, landfill, 322–323
nomograms
  Hazen-Williams formula, 73
  Manning formula, 77

North American Response Guidebook, 410
nuclear materials. See radioactive materials

O

Occupational Safety and Health Act of 1970, 120
Occupational Safety and Health Administration (OSHA), 129
  Code of Federal Regulations (CFR), 116
  hazardous waste remediation, 448
oceans, 83, 110
octanol, 468
octanol-water partition coefficient, 95–96
1-octene, 468
odors, 127
  biodegradables, 168
  biology of municipal solid waste, 168–169
  garbage, 132
  landfill, 322, 380
  pollution from waste management, 130
office buildings, 291
office paper, 139
oil, 394
  recycling, 272–273
oil companies
  hazardous waste disposal, 431
  Oil Pollution Act, 124
  Superfund Act, 123
oil filters, 394
Oil Pollution Act of 1924, 124
oil-water separation method, 425
open-space areas, 361–362
operations
  hazardous waste reduction, 420–421, 422–423
  landfill, 316–322, 322–323
  materials recovery facilities (MRFs), 279–281
orchards
  agricultural waste, 134–135
  water-holding capacities, 319
organic materials, 89–90, 95.
  See also composting
  combustors' emissions limits, 233

contaminants, 436–438
landfill odor, 322
leachate treatment, 336
separating from nonorganic, 225
in solid waste sample, 162
organisms
    compounds, 90
    environmental ethics, 39–40
OSHA. *See* Occupational Safety and Health Act/Administration
osmosis, reverse, 337, 344, 349, 354
overburden weight, 157
oxalic acid, 468
oxidation, 94–95
    hazardous waste treatment, 424
    leachate treatment process, 337, 344, 356–358
oxidation/reduction method, 479
oxides, 338
oxidizers
    reactivity, 396
    shipping and packing classification, 406
oxygen, 2
    aerobic process, 102
    biochemical and chemical demand, 106–108
    biological demand. *See* biological oxygen demand (BOD)
    biological properties, 167
    chemistry, 90
    composting, 252
    energy content, 166
    landfill gas, 387
    saturation in water, 492
    solid waste analysis, 163, 164
    ultimate analysis, 160, 161
ozone, 473
    air pollution, 120
    leachate treatment process, 356

P
P2, 414
packaging, 137, 138
    commercial and industrial waste, 133
    disposal rates, 140
    environmental effects, 259

hazardous materials, 406, 407, 408–410, 412
    as percentage of solid waste, 135
    recycling laws, 141
    source reduction approaches, 262–263
packer trucks, 174–175
paint, 9, 148, 150, 268
    exchange programs, 151
    hazardous waste, 401
    ignitability, 395
    product changes, 420
paint strippers, 150
paint thinner, 148
pan conveyor. *See* flight conveyor
paper, 114
    commercial and industrial waste, 133
    disposal rates, 146
    energy content, 165
    household waste, composition of, 137
    moisture content, 156
    particle size distribution, 157, 216
    printing uses, 139
    recycling, 266–268, 275, 276, 280, 282
    screening out, 221–223
    solid waste composition, 136, 140, 142–143, 162, 163
    specific weight, 154
    trash, 132–133
    ultimate analysis, 161
    wet, 226
paper industry, 245, 403
parks, 142
    former landfills, 305, 361–362
    landfill sitings and, 292
    refuse, 133
particle size
    distribution, 156–158, 214–216
    incinerators, 244
    landfill filters, 306–307
    waste component separatio 221–224
particulates
    air pollution, 120
    incinerators, 247–250

partition coefficients, 95–96
parts per million, 45
pasture, surface runoff, 313
pathogens, 127, 129
PCBs. *See* polychlorinated biphenyls
PCE. *See* perchloroethylene
pen composting method, 252
pentachlorophenol, 250, 397, 437
1-Pentene, 468
percentage concentration, 45–46
perchloroethylene (PCE), 469, 477
percolation, 83
    leachate generation, 315
    water, 82
periodicals, 139
permanent wilting percentage (PWP), 314–315
permeability, 84
    compacted waste, 158–159
    defined, 10
    soil, 294–295, 311
permits
    costs, 18, 326
    environmental regulations, 117–118
    Hancock County, Ohio, landfill, 374
    Resource Conservation and Recovery act, 122–123
peroxide, 406
pesticides, 128, 148, 149, 245, 437
    agricultural waste, 135
    Federal Insecticide, Fungicide, and Rodenticide Act, 121
    hazardous waste, 398, 401, 438
    leachate treatment, 334, 356, 361
PETE. *See* polyethylene terephthalate
petroleum products
    hazardous waste, 398, 400, 431
    plastics, 269
    site reclamation, 471, 473
    volatile organic compounds (VOC), 436–437

Pewamo Series soil type, 12
pH, 94
    bacterial growth, 102
    composting, 252
    leachate treatment, 337, 354
    site remediation, 477
P hazardous wastes, 398–399
phenanthrene, 437
phenols, 437, 438, 469
    leachate treatment process,
        356, 361
    oxygen demand, 468
    priority toxic pollutants, 402
    site contamination, 436
    site remediation, 477
Philadelphia, 111
phosphates, 3
    leachate treatment, 359
    site remediation, 470, 473,
        474, 478
phosphoric acid, 477
phosphorous, 2
    aerobic cycle, 4
    algae, 103
    chemistry, 90
    cycle, 5
photochemical smog, 245
photoprocessing companies,
    150, 421–422
photosynthesis, 3–4, 5–6
phthalate esters, 401
phthalates, 438
phthalic anhydride, 468
physical properties
    contaminated site remedia-
        tion, 479–480
    solid waste, 153–159
physics, 89–90
    mass balance, 98–100
    reactors, 100–101
piezometer, 63–64
pigments
    hazardous waste, 398
    priority toxic pollutants, 402
pipes
    closed system, continuity of
        flow, 67–68
    conservation of energy, 68–71
    flow rate, 46–47, 66–67
    gravity flow, 75–78
    hazardous waste disposal,
        430
    landfill gases, 381, 386
        costs, 326

leachate collection, 308, 330,
    331
pressurized, flow in (hy-
    draulic grade line),
    71–75
pistons, 64
pit and crane system, 209
pits, hazardous waste storage,
    429–430
placards, hazardous waste trans-
    portation, 450
plants. See vegetation
plastics, 114
    biological properties, 167
    commercial and industrial
        waste, 133
    contaminants, 438
    density separation, 224
    disposal rates, 146
    domestic waste composition,
        137
    dust suppressants, 452
    energy content, 165
    moisture content, 156
    organic compounds, 95
    particle size distribution, 157,
        216
    pipes, pressured flow through,
        72
    priority toxic pollutants, 400,
        401
    recycling, 267, 268–270, 275
    screening out, 221–223
    site reclamation, 426
    soft drink containers, 259
    solid waste composition, 136,
        140, 142, 162, 163
    specific weight, 154
    synthetic organic materials,
        168
    ultimate analysis, 161
plug flow reactors, 100
pneumatic conveyor, 213–214,
    215
poisons. See hazardous waste
Poland, waste composition in,
    137, 139, 140
police, 407–412
politics
    bonds, selling, 17
    environmental ethics, 39–41
    environmental impact state-
        ments, 39
    landfill siting, 288, 291

local decision-making, 14–15
pollution, 2. See also environ-
    ment, regulations, haz-
    ardous materials, site
    reclamation
abating older sources,
    123–124
air, 120, 168
    hazardous waste site recla-
        mation, 451
    incinerators, 244–245
    remedial technology, 442
aquifers, 83
benefit/cost analysis, 26
case study, 125–126
development of environmental
    regulations, 115–119
hazardous waste in municipal
    solid waste, 148–151
hazardous waste storage,
    429–430
leachate, 332
pollution control revenue
    bonds (PCRBs), 18
prevention, 415, 417–418
recycling, 276
risk analysis, 27–35
solid waste management,
    129–130
water, 8, 120–121
Pollution Prevention Act of
    1990, 9, 124
polychlorinated biphenyls (PCBs),
    122, 244, 437, 439
    hazardous waste, 32, 401
    hazardous waste disposal, 430
    leachate treatment, 356, 361
polycyclic aromatic hydrocar-
    bons, 402
polyethylene terepthalate
    (PETE), 224, 230, 269
    recycling, 275, 280
polymerization process, 478
polymers process, 348
polynuclear aromatic hydrocar-
    bons, 356
polyprophylene spheres, 451
polypropylene (PP), 269
polystyrene (PS), 269
polyvinyl chloride (PVC), 168,
    269, 401
ponds
    hazardous waste storage,
        429–430

landfill, 311
landfill siting limitations, 292
leachate treatment, 336
porosity
defined, 10, 87
leachate estimates, 317
potassium, 2, 438
chemistry, 90
potassium cyanide, 468
potassium permanganate, 356
power
calculations, 44
conversion factors, 491
flail mills, 218–219
hydraulic pressure, 64
landfill gas generation,
388–389
pneumatic conveyors,
213–214
pozzolan-Portland cement, 426
precipitation, 82
chemical, 337
hazardous waste treatment,
424, 425
landfills, 84, 310
leachate formation, 130, 317,
329
leachate treatment, 337, 338,
344, 355, 358–359
site reclamation, 470, 474,
478
suspensions, 92–93
preprocessing, 205–214
composting, 252
receiving and storage areas,
207–212
refuse conveying, 212–214
source reduction, 263–264
weigh stations, 205–207
pressure
calculations, 43
computation, 60–62
conversion factors, 490
Dalton's Law of Partial
Pressures, 98
head, 62–64
hydrostatic, 58–60
ideal gas law, 97
mechanical advantage, application for, 64
potential energy, 68
prices, recycled materials,
275
printers, 422

printing industry, 148
hazardous waste examples,
403
waste composition, 139
problems
air classification, 225–226
shredding machines, 220–221
processing, solid waste
components separation,
221–230
particle size distribution,
214–216
public health and safety,
128–129
shredding and size reduction,
217–221
process waste, 134
Procter & Gamble, 257
products
exchange programs, 151
hazardous waste reduction,
420
life cycle stages, 258
reusing, 262
prokaryotic organisms, 101
propane, 468
propellants, 401
property values, 192
propionic acid, 468
propylene, 468
propylene oxide, 468
proteins, 167
protozoa, 103, 104
proximate analysis, 234
public drop-off area, 324–327
public health
biomedical waste, 322
environmental effects,
127–131
hazardous waste gas, 452
issues, 8–9
site remediation, 443–446,
446
Subtitle D (landfill) regulations, 363
transporting hazardous waste,
412–413
pulping liquors, 394
pumps
centrifugal, 78
flow equalization, leachate
treatment, 347
groundwater control, 454
hydraulics, 78–80

positive-displacement, 78
site reclamation, 447, 449
PVC. See polyvinyl chloride
pyrene, 437
pyridine, 397
pyrolysis, 230, 255–257, 424

Q
quality circles, 421

R
radicals, 89, 93
radioactive materials
commercial and industrial
waste, 133
disposal, 432
exclusion from hazardous
waste list, 394
shipping and packing classification, 406
site reclamation, 449
radioactive wastes
treatment, 427
railroads, hazardous waste transportation, 194, 407
rainfall intensity, 87
rats, 128, 167
garbage, 132
landfills, 302
RCRA. See Resource Conservation and Recovery Act
reaction rate, temperature and,
104–105
reactivity, hazardous waste, 396
reactors, 100–101
completely stirred (CSTR),
343–344
Reagan, Ronald, 116
reagents, 356
receiving areas, solid waste,
207–212
recharge areas, 83
reclamation, 421–422
recreation areas, 142
recurrence interval, 56–57
recycling
disposal rates, 146
facilities, 290
Hancock County, Ohio, sanitary landfill, 372
hazardous waste, 150–151,
418
exclusion, 394
reduction, 421–422

recycling (*cont.*)
  Marion County Recycling
      Center, 281–286
  overview, 264–276
  particle size distribution,
      156–158
  public access, 129
  public drop-off areas,
      324–327
  rates, 109
  regional approaches, 15
  solid waste composition, 135,
      141, 143
  transfer station revival,
      200–203
reduction, 94–95
  hazardous waste treatment,
      424
  leachate treatment process,
      356–358
refineries, 134
refrigerants, 401
refuse
  conveying, 212–214
  defined, 110
  fuel incinerators, 238
regions, solid waste in
  composition, 148
  moisture content, 155–156
  plans, 15
regulations
  air pollution, 244–245
  asbestos disposal, 321
  changes in composition of
      solid waste, 141
  composting, 251
  emission limits, 244
  environmental, 35–39,
      115–126
  hazardous waste transporta-
      tion, 405–407
  heavy metal disposal, 161
  history of, 110–112
  incinerators, 233
  landfill siting limitations, 292
  leachate treatment, 330, 338
  ocean dumping, 110
  recycling, 141, 266
  safety of workers, 129
  Subtitle D (landfill), 362–365
  waste diversion, 171
Remedial Investigation/Feasibil-
      ity Study (RI/FS)
      process, 440–451

remediation, groundwater, 104,
      105
repairs, 140
reproduction, bacterial, 101–102
residence time. *See* detention
      time
residential waste. *See* house-
      holds
Resource Conservation and Re-
      covery Act (RCRA), 15,
      116, 288, 392
  asbestos disposal, 321
  development, 122–123
  hazardous waste disposal, 424
  siting limitations, 292
  solid waste definition, 110
  Subtitle C (transportation),
      405–407
  Subtitle D (landfill), 362–365
Resource Recovery Act of 1970,
      122
resource recovery systems
  public health and safety,
      128–129
  solid wastes, 113–114
respiration, 5–6, 6
retention time. *See* detention
      time
return period, 56–57
reuse
  hazardous waste management,
      150–151
  hazardous waste reduction,
      422
revenue. *See* finances
reverse osmosis, 337, 344, 349,
      354
Reynolds number, 225
risks analysis, 27–35
rivers, 83, 292
Rivers and Harbors Act of 1899,
      118–119
roads. *See also* transportation
  hazards, 128
  landfill access, 291
  landfill costs, 326, 327
Robosort, 230
rocks, 9, 221–223
rodenticides, 121
rodents, 128, 167
  diseases, 8–9
  landfill operations, 323
  public health, 127
roll-out containers, 171, 172

Rome
  history of waste disposal, 111
  rate of waste disposal, 140
rotary kiln incinerator, 465
rotifers, 103
routes, collection, 187–191, 208
rubber
  biological properties, 167
  disposal rates, 146
  energy content, 165
  moisture content, 156
  organic compounds, 95
  priority toxic pollutants, 400
  recycling, 267
  solid waste composition, 136,
      162, 163
  ultimate analysis, 161
rubbish
  defined, 132–133
  specific weight, 154
Ruetgers-Nease Corp., 125
runoff, surface, 82
  environmental impact state-
      ments, 37
  hazardous waste, 459–460
  landfills, 86–87, 312–313
  leachate generation estimates,
      315
  pollution from landfills, 130
rust removers, 396
rusty objects, 128

S
Safe Drinking Water Act
      (SDWA) of 1974, 121,
      431–432
salt mining, 432
salts, 92, 355
sand
  capacities, 314
  characteristics, 10, 11
  hazardous waste gas, 452
  landfill cover, 316
  landfill siting limitations, 292
  permeability, 84, 311
  site remediation, 478
  surface runoff, 87, 312, 313
  water-holding capacities, 319
Santa Barbara oil spill, 115
Sao Paolo, Brazil, 148
sarcodina, 104
saturated zone. *See* groundwater
saturation, 92, 492
scales, 206

scatter analysis, 52–56
schools, waste from, 134
Schweitzer, Dr. Albert, 40
scrapers, 299, 300, 449
screens
    landfills, 301
    separation of waste compo-
        nents, 221–224
scrubbers, 245–246
SD. *See* scatter
sea level, atmospheric pressure
    at, 59
seams, synthetic membranes, 310
seasons, 147
Secretariat of Communications
    and Transportation of
    Mexico, 410
Secretary of Transportation, 405
sedimentation, 93, 337, 344,
    358–359
seismic impact areas, 292, 450
selenium, 90, 438
    hazardous waste, 397
    priority toxic pollutants, 401
    site remediation, 478
semitrailers, 20–21
semivolatile compounds, 437
separation, waste, 112, 221–230
separators, 157
settlement, landfill, 307
sewage sludge, 142
    biochemical oxygen demand
        (BOD), 108
    co-composting, 252
    hazardous waste exclusion,
        394
    incinerators, 239
sewage treatment plants, cost in-
    dex, 25
sewers, gravity flow in, 75–78
Seymour, Indiana, Superfund
    site, 480–485
shear shredder, 219–220
shipping, hazardous materials
    classification, 406
shoe polish, 150
shop displays, 138–139
shopping bags, 262
shoulder barrel, 171
shredders/shredding, 157
    municipal solid waste, 256
    permeability of waste, 159
    solid waste, physical process-
        ing of, 217–221

signs
    hazardous waste transporta-
        tion, 450
    landfill, 323, 367
Silent Spring, 12, 115
silicate, 426
silicon, 90
silt
    capacities, 314
    characteristics, 10, 11
    permeability, 311
    site remediation, 478
    surface runoff potential, 87
silver, 150, 438
    chemistry, 90
    hazardous waste, 397, 401
    leachate treatment, 356
    reclamation, 421–422
    recycling, 270
SI (Systeme International)
    conversion factors, 489–491
    hydraulics, 58–81
    newtons, 58
    rainfall intensity, 87
    units of measure, 43–47
sites
    contamination, 436–440
    landfills, 289–293
    preparation costs, 326
    reclamation, 447, 449, 471
    selection costs, 18
    transfer stations, 197–200
size reduction, 217–221
slope, landfill area method,
    298–299
sludge, 133, 142
    activated, 339
    activated, leachate treatment
        process, 336
    co-composting, 252
    dewatering, 460
    disposal, 322
    hazardous waste, 398, 400
    incinerators, 239
    landfill class, 289
    leachate treatment, 344,
        355–356, 359
    solid waste composition, 137
    solid waste definition, 110
slurries, 460
socioeconomic status, waste
    generation and, 146
soda ash, 355
sodium, 89, 90, 438

sodium acrylate, 469
sodium alkylbenzenesulfonates,
    468
sodium alkyl sulfates, 468
sodium cyanide, 468
sodium dihydrogen phosphate,
    477
sodium hypochlorite, 356
sodium sulfate, 478
sodium sulfite, 356
soft-drink containers, 259, 269
    density separation, 224
    particle size distribution,
        156–157
soil
    capacities, 314
    contaminated, 439–440
    geology of, 9–12
    landfill cover material,
        302–304
    landfill sites, 293–295
    leachate estimates, 317
    moisture retention, 493–495
    moisture storage, 315
    permeability, 84, 311
    pollution pathways, 33
    recycling, 286
    site reclamation, 471
    surface runoff, 87, 313
    vapor extraction system, 480
    water-holding capabilities,
        319
    water partition coefficient,
        95–96
Soil Conservation Service
    (USDA)
    classifications, 10, 11
    survey maps, 12, 290
solar energy, 5
solidification method,
    425–427
solids
    gas removal, 96
    solutions, 92
    suspensions, 92–93
solid waste
    biological transformation,
        250–257
    collection, 174–191
    combustion, 230–250
    composition, 132–152
    contaminated, 439
    defined, 110
    landfill disposal, 288–328

solid waste (*cont.*)
  life cycle assessment (source assessment), 257–260
  management. *See* waste management
  physical processing, 214–230
  preprocessing, 205–214
  properties, 153–170
  source reduction, 262–287
  storage, 171–174
  transfer stations and transportation, 192–200
  water infiltration, 87–88
Solid Waste Disposal Act of 1965, 112, 121–122
solutions (solutes), 92, 93
  leachate treatment process, 337, 353–354
  oxidation and reduction reactions, 94–95
  removing iron, 94–95
solvents, 92, 128, 148, 149
  hazardous materials reduction, 421
  hazardous waste, 398
  ignitability, 395
  priority toxic pollutants, 401, 402
  recycling, 273
  site remediation, 427, 481
  volatile organic compounds (VOC), 436–437
sorption, 425–426
sorting, 282, 283
source reduction, 262
  hazardous wastes, 420–421, 422–423
sources
  discarded materials, 112
  life cycle assessment (LCA), 257–260
  solid waste, 132–135
  storing solid waste, 171–174
Southside River-Rail Terminal, Inc., 125
soybean oil, 468
special interest groups. *See* politics
specific weight, 153–155
Spiegel, S. Arthur, 125
spills
  emergency response, 407–412
  hazardous waste, 398–399
spot removers, 150
springs, 83

stabilization, 425–427
stain removers, 148
stains, 150
starch, 167
states. *See* governments, state and local
static pile composting, 253–254
steam stripping
  hazardous waste treatment, 424
  leachate treatment process, 337, 352–353
steel, 92
  hazardous waste, 398
  packaging, 138
  pipes, pressured flow through, 72
  recycling, 270, 275, 280
stoichiometry, 90–92
storing
  hazardous waste, 427–430
  hazardous waste before transportation, 413
  leachate, 63, 315
  solid waste at source, 112–113, 171–174
  solid waste before processing, 207–212
  transfer stations. *See* transfer stations
stormwater. *See* sewers
street, waste disposal in, 111
street refuse, 133
street sweepings, 137, 142
stripping, solvents, 273
Sturgeon, Missouri, chemical spill site, 482–484
styrene, 335, 437, 468
suctoria, 104
sulfates, 3, 169, 359
sulfide, 478
sulfur, 2
  aerobic cycle, 4
  biological properties, 167
  energy content, 166
  solid waste analysis, 163, 164
  stoichiometric combustion, 231
  ultimate analysis, 160, 161
sulfur dioxide, 130, 244, 245, 356
sulfuric acid, 150
  hazardous waste, 398
  leachate treatment process, 355–356
  site remediation, 477

sulfur oxides, 120
sulphur, 90
Superfund Act. *See* Comprehensive Environmental Response, Compensation, and Liabilities Act of 1980 (CERCLA)
Superfund Amendments and Reauthorization Act (SARA), 415
supplies, cost, 19–20
surface impoundments, 429–430, 431
suspensions
  defined, 92–93
  leachate treatment, 337, 348, 361
swamps, 83
swine, waste disposal to, 111
symbols, elements, 89, 90
synthetic membranes, 308–309
synthetic organic materials, 95, 168

T
tanks
  hazardous waste storage, 429
  leachate treatment process, 345–347
  reactors, 100–101
tare weight, 206
taxes, financing waste management through, 16
Taylor, Paul, 39–40
1,2,7,9-TCDD. *See* dioxin
TCE
  site remediation, 477
technology
  changes in composition of solid waste, 135–140
  contaminated site treatment, 460–480
  hazardous waste reduction, 421
  site remediation, 424–425, 425–427, 440–443, 444–446
temperature
  acid gas combustion, 245
  air stripping, 352
  bacterial growth, 102
  biochemical oxygen demand (BOD), 106
  composting, 251
  decomposition, 334

microorganisms, 104–105
odors, 168–169
particulate control devices, 250
pyrolysis, 255–257
scrubbers, 245–246
volatility of heavy metals,
438–439
terichloroethane, 437
testing
hazardous waste, 397
synthetic membranes, 309
1,1,2,2-tetrachloroethane, 335,
437
tetrachloroethylene, 335
hazardous waste, 32, 397, 416
oxygen demand, 468
tetrahydonaphthalene, 468
textiles
crude fibers, 167
disposal rates, 146
domestic waste, composition
of, 137
energy content, 165
moisture content, 156
particle size distribution, 157
priority toxic pollutants, 400
recycling, 267
solid waste composition, 136,
162, 163
ultimate analysis, 161
thallium, 401, 438
thermal destruction method,
464–466
thermoplastic microencapsulation method, 426
Three (3) M Company, 418–419
tiles, leachate removal, 330
time
collection methods, 178–184
money value, 22–25
Times Beach, Missouri, 115,
391, 435
tin cans
moisture content, 156
specific weight, 154
tipping fee, 207, 375
tipping floor, 192, 210–211, 237
tires, 133, 262
compaction, 319
disposal, 321
public health, 127
recycling, 268, 271–272
toluene, 335, 437, 438
hazardous waste, 400, 416
oxygen demand, 468

Topcoat, 371
topographic maps, 290
topsoil, 286, 306, 316
towers, packed column air stripping, 352
toxaphene, 397, 437
toxic dumps, 391
toxicity, hazardous waste, 396,
397
toxic pollutants, priority list,
400–403
Toxic Release Inventory, 124
Toxic Substances Control Act
(TSCA), 122, 401
toys, 133
tractors, 299
transfer stations, 128–129, 181
case study, 200–203
costs, 192–197
operational control, 197–200
public access, 129
solid waste, 114
vehicles and equipment, 194
transition phase, landfill gas
generation, 378
transmission fluid, 150
transpiration, 82, 83–84, 86
transportation
costs, 192–197
hazardous waste, 405–413,
450
leachate, 336
pollution pathways, 33
refuse conveying, 212–214
solid waste, 114
transfer vehicles, 194
transportation maps, 290
Transport Canada, 410
trash, 132–133
treatments
contaminated sites, 460–480
hazardous waste, 424–427
leachate, 336–361
aerobic systems, 339–344
biological, 338
physical and chemical,
344–361
trees, 142
tree stumps
disposal, 321
trench method, 300
1,2,4-trichlorobenzene, 437
1,1,1-trichloroethane, 32, 335,
437
1,1,2-trichloroethane, 335, 437

trichloroethane (TCA), 422
trichloroethylene (TCE), 32,
150, 335, 397, 469
trichlorofluoromethane, 335
2,4,5-trichlorophenol, 437
2,4,6-trichlorophenol, 437
2,4,5-trichlorophenoxypropionic
acid, 397
1,2,3-trichloropropane, 335
1,1,1-trichloroethane, 416
triethanolamine, 468
triochloroethylene, 416
trommel, 221–223
trucks, hazardous materials labels, 407, 408–410
typhoid fever, 127
typhus, 127

U
ultimate analysis, 234
ultrafiltration process, 337
underground storage tanks, 429
United Kingdom
domestic waste composition,
137
solid waste generation, 140
United States
Congress, 13
chemical production, 418
domestic waste composition,
137
materials generation and recovery, 267
measurement conversion factors, 489–491
packaging materials, 263
solid waste generation, 139,
140
universities, waste from, 134
urea, 468
U.S. Army Corps of Engineers,
26, 374
U.S. Coast Guard, 405
U.S. Department of Transportation (DOT), 405, 406,
410
hazardous waste transportation, 450
U.S. Environmental Protection
Agency (EPA), 15, 116
asbestos disposal, 321
ash classification, 250
brownfields, 125
carcinogens, potency factor
(PF), 31–33

U.S. Environmental Protection
    Agency (EPA), (*cont.*)
Clean Water Act, 121
cost indexes, 25
creation of, 116–117
environmental audits,
    415–420
Hazardous and Solid Waste
    Amendments of 1984,
    123
hazardous waste
    characteristics, 395–397
    "derived from" rule, 400
    generators responsibility
        limits, 393–395
    injection wells, 431–434
    pretransport regulations,
        412–413
    transportation safety,
        412–413
    treatments endorsed, 427
heavy metal disposal, 161
heuristic routing program,
    189–191
incinerator design, 243
leachate collection, 330
leachate generation estimates,
    315
pesticides, banned, 149
recycling center costs,
    278–279
solid waste estimates, 109,
    141
solid waste reduction guide-
    lines, 262
Subtitle D (landfill) regula-
    tions, 362–365
Superfund program, 435–436
target compound/analyte list,
    437–438
toxic pollutants, priority,
    400–403
waste minimization, 414, 415
U.S. Federal Aviation Adminis-
    tration (FAA), 405
U.S. Federal Highway Adminis-
    tration (FHA), 405
U.S. Pollution Prevention Act of
    1990, 415
U.S. Public Health Service-Ten-
    nessee Valley Authority
    project, 253
U.S. Soil Conservation Service,
    297

U.S. Weather Bureau, 82
user fees, 16
utilities
    costs, 20
    landfills, 291
    transfer stations, 192

V
vacuum filters, 461, 462, 463
vacuum, partial, 59
vadose zone, 82–83
valences
    chemistry, 89
    elements, 90
vanadium, 438
vapors, 95–96
variation, coefficient of, 53
varnish, 150
vegetation
    biogeochemical cycles, 3–5
    evapotranspiration, 86, 312,
        313
    landfill cover, 305–306
    leachate estimates, 317
    organic compounds, 95
    precipitation and, 82
    soil support, 12, 293, 295
    surface runoff, 313
    water-holding capabilities,
        319
vehicle maintenance shops, 403
vehicles. *See* equipment
    collection, 113
        cost effectiveness, 180–184
        cost-effectiveness, 22–25
        hauled container systems,
            175–176
        route layouts, 187–191
        weighing, 153–154
    composting, 255
    cost analysis, 23–25
    economic analysis, 184–187
    leachate transportation, 336
    receiving and storage areas,
        207–212
    size needs, 20–21
    solid waste transportation,
        114
    stationery container systems,
        174–175
    transfer, 194
velocity
    conversion factors, 489
    potential energy, 68

vent systems, 454
    landfill gas, 380–381,
        383–386
vertical shift mills, 217–218
veterinary pharmaceuticals, 398
video industry, 140
vineyards, 134–135
vinyl acetate, 335, 468
vinyl chloride, 32, 335, 397,
    437
Vinylcycler, 230
viruses, 40
vitrification, 427, 479–480
volatile compounds, 437
    proximate analysis, 160
volatile organics compounds
    (VOC), 337, 436–438
volatility, leachate, 332
volume
    conversion factors, 489
    ideal gas law, 97
    specific weight, 154

W
wages. *See* labor
Wagner Seed Company, 126
warfarin, 150
waste management, 288–328
    defined, 110
    development of, 110–112
    ecology, 2–8
    environmental audits,
        415–420
    facilities, health and safety at,
        128–129
    functional elements, 112–114
    geology and soil, 9–12
    hazardous waste
        characteristics, 391–404
        disposal, 430–434
        minimization, 414–415,
            418–423
        storage, 427–430
        transportation, 405–413
        treatment, 424–427
    history, 12–13
    landfill closures, 361–362
    landfills, 362–365
    leachates, 329–335, 336–361
    municipal decision-making,
        14–42
    partition coefficients, 95–96